Amateur Astronomy

Amateur Astronomy

a comprehensive and practical survey

consultant editor Colin Ronan
revised and updated by Storm Dunlop

NEWNES BOOKS

First published 1979 as *Encyclopedia of Astronomy*. This revised and updated edition, entitled *Amateur Astronomy*, first published by Newnes Books, a division of The Hamlyn Publishing Group Limited, 84–88 The Centre, Feltham, Middlesex, TW13 4BH and distributed by The Hamlyn Publishing Group Limited, Rushden, Northants, England.

Copyright original edition © The Hamlyn Publishing Group Limited 1979
Copyright new material © Newnes Books 1984.

All rights reserved. No part of this publication may be reproduced, stored in a retrieval system, or transmitted in any form or by any means, electronic, mechanical, photocopying, recording or otherwise, without the permission of the Publisher and the copyright owners.

ISBN 0 600 35667 1

Phototypeset by Input Typesetting Ltd, London
Printed in Italy by AMILCARE PIZZI ARTI GRAFICHE S.p.A. CINISELLO B. (MILANO - ITALIA) - 1984

Contents

Contributors

Consultant Editor:
Colin Ronan, MSc, FRAS, author of several books on astronomy and the history of science. Contributor and advisor to the *Encyclopedia Britannica*; editor of the *Journal of the British Astronomical Association*; past Council member of the Royal Astronomical Society; member of the International Astronomical Union and the Science Committee of the National Film Archive.

Editor of this edition:
Storm Dunlop, FRAS, FRMetS, scientific writer and translator. Secretary of the British Astronomical Association; member of American Association of Variable Star Observers. Principle interests: planetary geology and variable stars.

Contributors to the original edition:
Heather Couper, FRAS, planetarium lecturer, National Maritime Museum, Greenwich. Council member of the British Astronomical Association and Junior Astronomical Society; presenter of television series on astronomy. Principle interests: galactic and extragalactic astronomy.

Peter Gill, Editor with scientific publishing company. Chairman of Solar Commission of International Union of Amateur Astronomers; member of Royal Astronomical Society of Canada. Principle interests: the Sun, Moon and variable stars.

Dr Ian Robson, FRAS, lecturer in astronomy, Preston Polytechnic. Author of many research papers on astrophysics and observational cosmology. Currently engaged in far infra-red astronomy.

Dr William Somerville, FRAS, lecturer on galactic and extragalactic astronomy, Department of Physics and Astronomy, University College, London. Author of many research papers and review articles on physics and astronomy. Currently engaged in research into the interstellar medium.

How to use this book

In astronomy, perhaps more than in most other disciplines, practice cannot for long remain divorced from theory – at least, not in the minds of those whose interest is more than casual. In its previous edition, however, this book was uncompromisingly theoretical throughout. This put it beyond the reach of beginners in practical astronomy, unless they were already conversant with recent developments in general physics.

To overcome these difficulties, the most relevant parts of the previous edition have been retained in a comprehensively revised and updated form, and an entirely new approach has been adopted to the presentation of practical material. This is intended particularly to assist the practising amateur astronomer, whether new to the subject, or advanced.

This practical material can be easily identified on leafing through the book: it consists of self-contained sections, the pages of which have a broad tinted border.

Terms which are included in the Glossary (page 249) are shown in small capital letters the first time they occur in the text (e.g. ABSOLUTE ZERO).

Technical terms which are explained in the text are indicated by the use of bold type (e.g. **parallax**). Such references may be located by use of the index.

All photographs and diagrams are orientated with north at the top, except where any confusion might arise due to reversed relief, or for other considerations. In such cases the orientation is given.

The constants (and derived data) used throughout this book are those adopted at the 16th Congress of the International Astronomical Union at Grenoble in 1976.

All quantities are expressed in Systeme Internationale (SI) units.

Preface

This encyclopedia is designed to give an up-to-date picture of the universe for amateur astronomers and others who are interested in astronomy in this latter part of the twentieth century. Every aspect of the subject is discussed and illustrated in a way which can be understood by a reader with no scientific training. What is more, this second edition of what appeared originally as the *Encyclopedia of Astronomy* also contains some additional material to assist the amateur observer – hence its change in title.

Many people have helped on the original project, and special thanks are due to Mr Iain Nicolson for preliminary discussions, Dr Russell Cannon, Dr Garry Hunt, Professor Malcolm Longair, Mr David Malin, Dr Simon Mitton and Miss Elizabeth Sim for the photographs, Mijnheer Willebrordus Tirion for the star charts, Mr Gerald Hodgkinson for his original table of interstellar substances, Mr Gordon Taylor for information on minor planets, Mrs Enid Lake for locating and checking sources, and Mr Storm Dunlop for his editorial assistance.

The revision of the whole work has been the responsibility of Mr Storm Dunlop and I am most grateful to him for all the effort he has put into this.

COLIN A. RONAN

CHAPTER ONE

Introduction

Astronomy is usually said to be the oldest of the sciences, and so it is. Even in the earliest times, when man was just embarking on his pilgrimage towards civilization, his curiosity was aroused by the Sun, Moon and stars. As soon as primitive science was born, the heavens were naturally its first subject. But if astronomy is the oldest science, it has lost nothing of its fascination over the thousands of years since man first began to rationalize the heavens. Today it is still the most dynamic science, forging ahead to the limits of our understanding as it grapples with the problems raised by strange new celestial objects like quasars and black holes. Its scope is vast, covering nearer worlds like the Moon and extending outwards to the very edge of space and time.

Modern astronomy may seem a far cry from the primitive star-lore of our earliest ancestors, yet it rests on the same basic evidence – observation of the night sky. Certainly we no longer seriously believe that the celestial bodies are gods or spirits, and that their appearances spell out death and disaster, or even peace and plenty. We see now that there is no warrant for believing that the heavens can be used for divining the future. But still we observe the skies from Earth or from out in space, to gather our raw material so that we can discover how the physical universe works.

A mere glance at the night sky will show the vastness of the task facing the astronomer. The heavens seem to be crowded with stars – at least this is the appearance they give on a clear night away from city lights. There are simply myriads of them, arranged apparently at random. Yet as familiarity with the night sky grows, certain patterns become evident. First of all, besides the separate stars, we see that a hazy band of pearly coloured light stretches right across the sky, and this is observable wherever we may be. In the northern hemisphere it appears as an uneven band with some black patches in it here and there; in the southern hemisphere it is somewhat the same, although there are more black patches and, in addition, two small pieces of it look as though they have broken off and gone adrift into space. The Greeks named it *galaxias* or Milky Way, which has given us the modern term 'galaxy'.

The second thing we notice is that the stars are not scattered evenly over the sky; they are arranged in groups or patterns, known as constellations. In early civilizations these patterns were recognized and referred to by imagining that they represented familiar objects, animals and characters from myth and legend. For instance, from Sumer we have the bull and the lion, and from ancient China a tortoise, five chariots and the Purple Palace. As each civilization came into being, it took some of the constellations of its predecessors, and added some of its own. Today we use a total of eighty-eight and specify their boundaries by international agreement and with scientific precision.

With the recognition of constellations, it was also noticed that the stars themselves differed in brightness. In a desire to classify them, the brightest stars were given names, but for referring to others a rather cumbersome system was used at first. Then in 1603, when Johann Bayer published his star atlas *Uranometria*, astronomers were at last given a simple method. Bright stars were all referred to by the letters of the Greek alphabet; the brightest star in each constellation was designated alpha (α), the next brightest beta (β), and so on. About 1300 stars were covered in this way; the rest, when there was a need to refer to them, were later designated by numbers, Thus a star may have a name, usually of Greek or Arabic origin – Canopus, Procyon, Algol or Aldebaran, for example – and it will always have a designation. Thus, Sirius, the brightest star in the constellation Canis Major (the Greater Dog), is alpha (α) Canis Majoris, the second brightest star in Canis Major is beta (β) Canis Majoris, and so on. One of the nearer stars to us, however, has no name and is not even bright enough to rate a Greek letter; it is known as 61 Cygni.

The difference in brightness of the stars is due both to their varying distances from us and to their own very real differences in intensity. However, to the astronomers of earlier civilizations, to whom the stars appeared to be fixed either to the inside of the dome of the sky or to the inside of a sphere of the heavens, there was no question of the stars being at different distances; the brightness differences they noticed had, therefore, to be taken as real differences in intensity. The brightest stars were those to which names were given and they clearly seemed to be the most significant. Thus, in the second century BC, when the Greek astronomer Hipparchus compiled a catalogue of stars, he referred to the brightest stars as those of the first magnitude – the first importance – to those less bright as being of the second magnitude, and so on, down to the dimmest he could observe, which he termed the sixth magnitude.

To divide the stars visible with the unaided eye into six brightnesses was no personal idiosyncrasy of Hipparchus. It was due, although he did not realize it, to an inbuilt human physical and psychological mechanism which operates on any stimulus from outside. This – the Weber-Fechner law – shows that we recognize changes on a LOGARITHMIC scale. Thus, the brightness differences between one magnitude and the rest are not simply one, two, three, four and five times a given amount; they differ in a more complex way. In his six magnitudes Hipparchus responded to this law: his magnitude 1 was 100 times brighter than his magnitude 6, with the result that each magnitude is 2·5119 – or just over two-and-a-half times – brighter than the one below it, since 2·5119 multiplied by itself four times is 100. (Mathematically $\sqrt[5]{100} = 2{\cdot}5119$.)

Hipparchus numbered stellar magnitudes backwards – or so it seems – since stars of magnitude 2 are 2·5119 times dimmer, not brighter, than those of magnitude 1. However, this is purely a consequence of saying that stars of the first importance are stars of the first magnitude, those of secondary importance of second magnitude, and so on. Astronomers have found it convenient to continue this approach, only now, after Norman Pogson's establishment in 1856 of precise numerical relationships between one magnitude and the next, it is found that some stars are in fact brighter than those which Hipparchus designated as first magnitude. The magnitude scale therefore goes to 0 for stars 2·5119 times brighter than magnitude 1, and −1 for stars 2·5119 times brighter than magnitude 0. Thus while the star Spica is magnitude 1, Vega is magnitude 0, and Sirius is −1·4. On the same scale, the full Moon has a magnitude of −12·5 and the Sun −26·7.

The ancient idea that the stars are fixed to the inside of a sphere persisted for a very long time, at least from the time of Homer until the 1570s, a span of some 2 300 years. However, in the late sixteenth century this belief was replaced by the concept of an infinite universe, and the face of astronomy changed. When this was coupled with other new ideas about the heavens, the problem of determining the distances of the stars became very pressing: it was a challenge which had to be met. The basic principle of determining distances in space was not in doubt, but the only problem was how to make observations with sufficient precision to detect the very small angles involved. In the event it was not until more than two centuries after the invention of the telescope in about 1608 that the first stellar distance was successfully measured, although some inspired guesses were made before this. In 1839 Friedrich Bessel found the distance of 61 Cygni, but the angle he had to measure to do this was only 0·35 arc seconds (arc sec., modern value 0·29 arc sec.) One arc sec. is $\frac{1}{60}$ of one arc minute, which itself is $\frac{1}{60}$ of a degree, so Bessel's angle was only one ten thousandth part of a degree. Yet this was only a beginning: today determinations are 100 times more precise, involving angles equivalent to measuring the thickness of a human hair at a distance of 300 metres (m).

The successful principle was based on the surveyor's method of triangulation, originally devised for determining distances on Earth to inaccessible points. The astronomical adaptation is shown in Fig. 1·1. Observations of a star are made at six-monthly intervals – that is, from opposite sides of the Earth's orbit round the Sun, giving a base-line of almost 300 million kilometres (km). From each point, measurements are taken of the observed position of the star against the background of more distant stars. Since two different positions are used, the star appears to shift its position with reference to the background stars. (You can obtain a similar effect by holding up a finger at arm's length, and looking at it first through one eye, and then through the other. Your finger will appear to shift in relation to background objects.) This 'parallactic shift' can be measured and the angle SXE_1 (or SXE_2) is known as the **parallax** of the star. Once known, this, together with the base-line distance E_1E_2, allows the star's distance to be calculated.

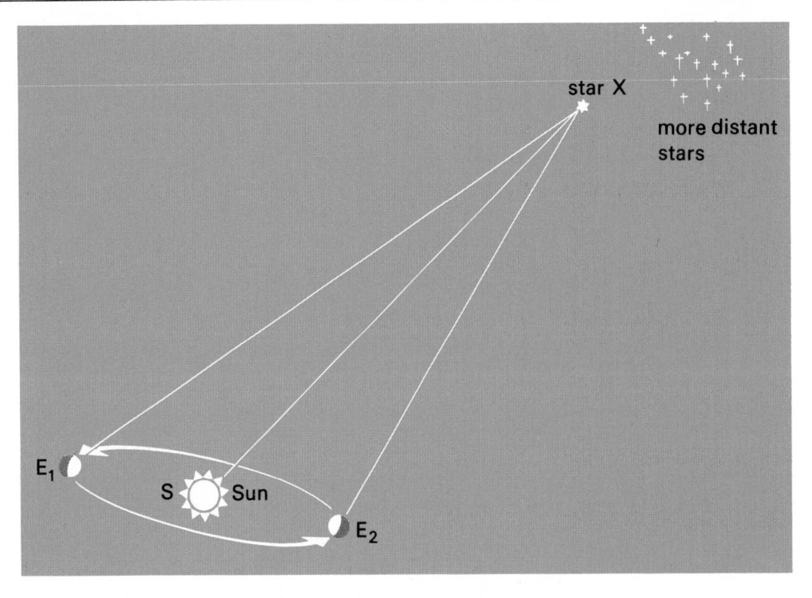

Fig. 1·1 Determining the distance of a nearby star using the ends of the Earth's orbit as a base-line.

The distances of stars are so great compared with distances on Earth that miles or kilometres are too small to be convenient. For instance, if we express the distance to the nearest star, α Centauri C, or Proxima Centauri, in kilometres, we find we are dealing with the number 40 570 700 000 000, which is cumbersome. Even if we write it in the index notation where 100 is expressed as 10^2, 1 000 as 10^3, 1 000 000 as 10^6 and so on, we still have $4{\cdot}057 \times 10^{13}$ km which is hard if not impossible to imagine. We need to have some way of scaling the number down, and one of the most convenient ways is to replace kilometres by **light-years**. A light-year is a distance, not a period of time: it is the distance light travels in one year, and to all intents and purposes is $9{\cdot}5 \times 10^{12}$ km. On this scale, the distance of Proxima Centauri is 4·3 light-years.

The professional astronomer tends to favour a different, and slightly larger unit, the **parsec** (pc). The parsec is that distance at which a star would have a parallax of one second of arc, and it is equal to 3·26 light-years. For the more distant stars the **kiloparsec** (kpc), a unit of one thousand parsecs, is used; and for the most distant realms of space, there is the **megaparsec** (Mpc), one million parsecs.

The method of determining parallax using six-monthly sightings from Earth – trigonometrical parallax – is only effective for the nearer stars.

Part of a chart of the northern hemisphere constellation drawn up by the great observational astronomer Johannes Hevelius in 1645 and published by him in his Tabula Selenographia.

Beyond about 30 pc the angles become too small to be measurable with precision, and beyond 300 pc too small to detect; more indirect methods have then to be used. These involve analysing either the star's light, in some stars their periodic variations, and in even rarer cases the brightness of exploding stars. Something may also be done by measuring space velocities, but details must wait until we discuss the stars themselves (Chapter 3). However, the distances of the Sun and Moon are, of course, small compared with the stars and trigonometrical parallax methods may be used. Indeed the Moon is so close that its distance can be determined using observations made from two different points on the Earth's surface. Nevertheless, in the interests of obtaining their distances with as great an accuracy as possible, radar methods are now used. Pulses of radio waves are shot out into space and the time taken for them to bounce back to Earth is measured. Since radio pulses travel at the same speed as light, the distance of the Sun and Moon can be obtained with great precision.

One of the earliest problems that faced astronomers was posed by the wandering stars or **planets**. As soon as the stars had been grouped into constellations, it was noticed that a few of the brighter stars seemed attached to no particular constellation, but weaved their way across the sky quite independently. Their motions were complex; sometimes they moved forwards, sometimes backwards, and on certain occasions they stood still. It fell to the Greeks to try to systematize these motions, and for a variety of reasons they settled on explanations involving a series of circular orbits centred on the Earth. Systematized and described in the second century AD by the Greek astronomer Ptolemy, this Ptolemaic system satisfied the astronomical world until the sixteenth century. Then new explanations of old ideas were in fashion and Nicolaus Copernicus proposed a new model in which the planets orbited around the Sun. Copernicus had no proof, but his proposal was favoured by mathematicians and led to a vast amount of observational and mathematical research. Within the next 150 years, the old Earth-centred (geocentric) universe gave place to a Sun-centred (heliocentric) one, and due to the work of Tycho Brahe, Johannes Kepler, Galileo Galilei and Isaac Newton, the orbits of the planets were found to be ELLIPSES and their motions to be governed by universal gravitation. It

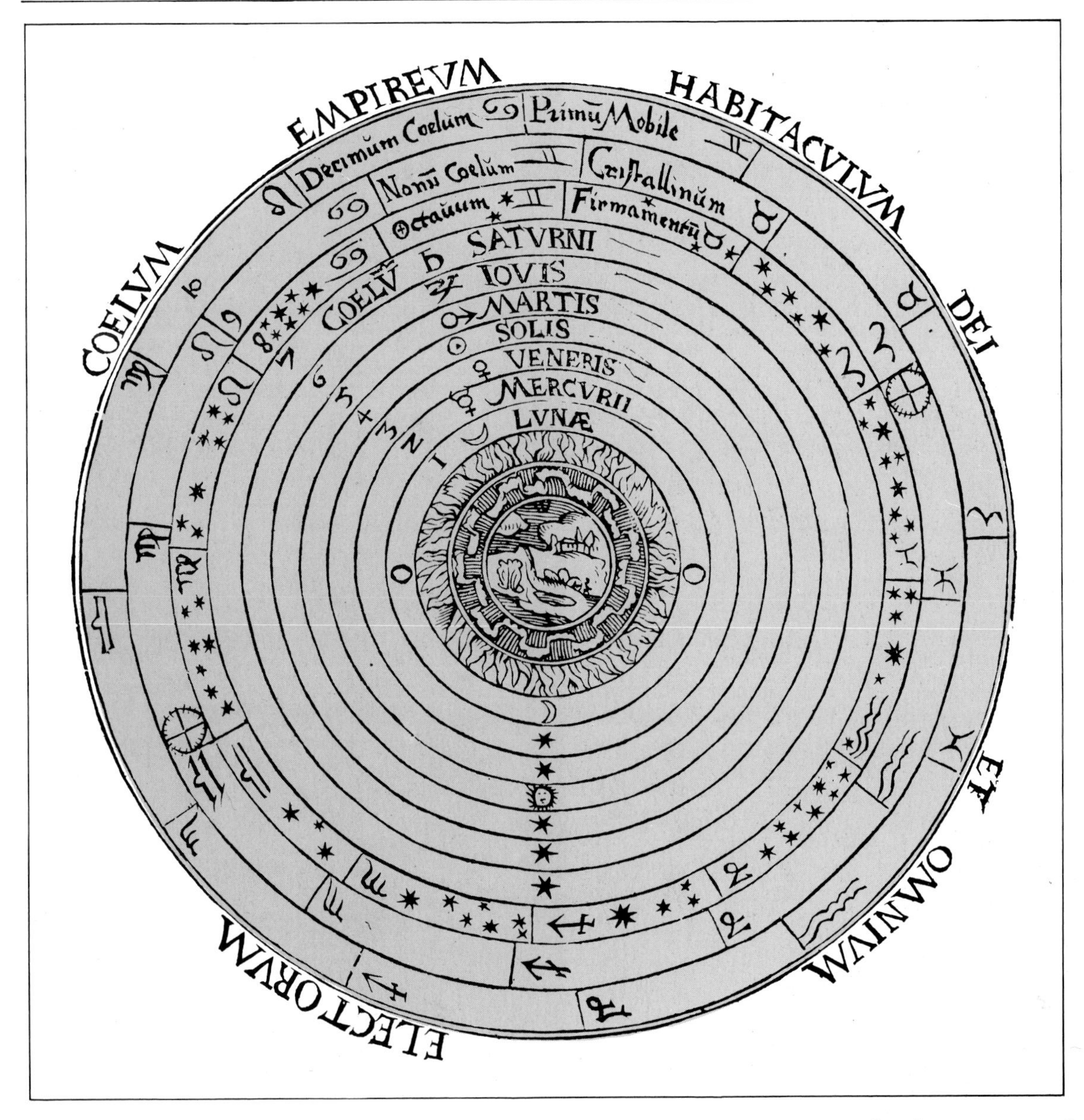

A plan of the sphere of the heavens. The stars lie on the outermost sphere, beyond which lies Heaven. In the centre is the Earth with the 'elements' earth, water, air and fire. Between are the transparent spheres which were supposed to exist and on which the Moon and the planets were thought to be attached. From Peter Apian, Cosmographia, *published in Antwerp in 1539.*

was a vast stride forward that also brought in its train the realization that the Sun and Moon were ordinary physical bodies, not special celestial entities as had once been believed.

It was during this exciting period of research that the telescope was invented and first used in astronomy. The telescope does two things: firstly, it grasps more light than the unaided eye can do, and so reveals visible objects too dim to see without it; secondly, it magnifies distant objects so that their images are spread out and more detail can be detected – in other words it increases ANGULAR RESOLUTION. Its effect, as Galileo was forthright in pointing out, was that man began to realize that the universe was far more astonishing than he had previously imagined. First of all it was discovered that the planets were visible because they reflected sunlight; only the stars emitted light on their own account. There were, then, at least two classes of bodies in the universe. Moreover, some planets had satellites orbiting round them, in essence similar to the way the Moon orbits the Earth. Again, the Moon was seen to possess mountains and craters, valleys and plains, and questions arose about the nature of the other planets – had they surface features too, and were they perhaps inhabited?

But undoubtedly the most significant advance brought about by the telescope was its ability to show a whole range of new objects in the universe. Not only were there planets and stars, but the number of stars was far greater than anyone had supposed. Subsequent research has shown that the universe is vast indeed, and populated with other material besides just stars and planets. Out in space, among the stars, are vast gas clouds or **nebulae**, often extending for as much as 30 pc across space. Some of these glow brightly because of hot stars embedded in them; others are dark and can be noticed only because they blot out the stars behind them, looking like great holes in space. These are the cause of the dark patches observed in the Milky Way. The use of radio telescopes has made it evident that not only is there material in between the stars, but also there are other clouds of dark gas, never detected before.

The stars and nebulae have also been found to be organized in a grand scheme. The stars themselves are either separate or in associations and clusters, although these groups have nothing whatever to do with the constellations. In fact, we now know that the stars of a constellation are not physically connected, constellation patterns being purely fortuitous arrangements observed from Earth; observed from far out in space they would look entirely different. We recognize **open clusters**, like the Pleiades or the one in Scorpius, which may contain as few as 15 or 20 members, or as many as 2 000, and **globular clusters**. Visible only with a telescope, the globulars are spherically shaped concentrated collections of hundreds of thousands or even millions of stars. They contain no dust or gas, both of which may be present in an open cluster.

All these occupants of space – planets, stars, open clusters, globular clusters, the wisps of dust and gas between the stars and the vast nebulae – are all part of a giant star island, known as the Galaxy. Shaped like a pair of dishes placed rim to rim with a bulge at the centre (Fig. 1·2), it has a diameter of some 30 000 pc (30 kpc, about 98 000 light-years), and at its centre its thickness is around 41 kpc (13 000 light-years).

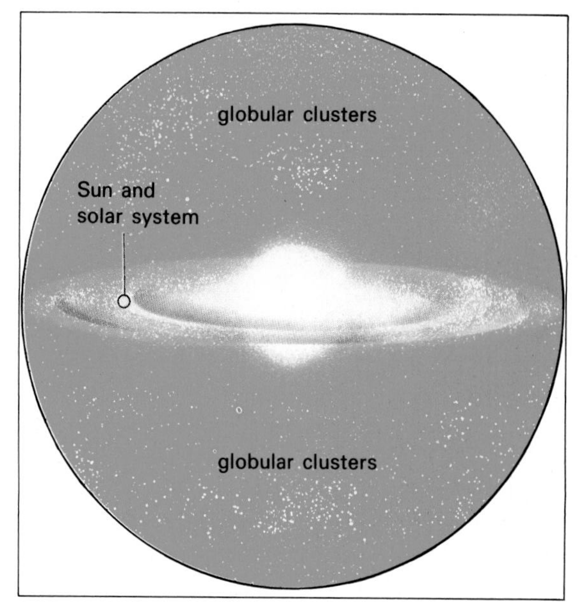

Fig. 1·2
An artist's impression of our Galaxy surrounded by a halo of globular clusters, and showing the position of the Sun and the solar system.

The Sun, with its planetary system, is situated rather more than 10 kpc from the centre of the Galaxy, a little nearer to the edge than we are to the centre. When we look towards the rim or the centre we are seeing the denser regions of the central plane, the regions we call the Milky Way. The star associations, open clusters and nebulae lie mainly in the plane of the Galaxy, but not so the globular clusters. These lie around the Galaxy, forming a kind of spherical halo, also about 30 kpc in diameter. The main part of the Galaxy is not only disc-shaped: observed from outside it would look like a giant catherine wheel or pin-wheel, with the Sun lying in one of its spiral arms.

In spite of the fact that the Galaxy contains some 100 000 stars, to say nothing of the gas and dust in it, it is nowhere near the whole of the universe; it is only a minute unit among literally millions of other galaxies. Some of these galaxies are spirals like our own, others have no spiral arms but seem to be merely giant conglomerations of stars having little or no dust and gas. There are still others which are irregular, like the two Magellanic Clouds which can be seen from the southern hemisphere, and also a number that are small but intensely active, emitting atomic particles, radio waves and other radiation. All are moving outwards into space, so that it seems as if the entire universe is expanding, probably having once begun as a tiny compact mass of material which exploded out into space ten or twenty thousand million years ago.

To obtain the kind of picture of the universe we now have has meant centuries of painstaking observation and measurement, and the precise expression of the results of this work in numbers. This obsession of astronomy with numbers is vital if the science is not to be one of pure speculation, as a simple question such as whether or not the stars are fixed in space will show. In 1718 Edmond Halley (of Halley's comet fame) compared his observations of three stars with observations made in Greek times. He found discrepancies too gross to be put down to observing errors, and was forced to conclude that these stars had actually moved in space. What was true of his three stars has been found to be true of all stars.

Today we refer to the motion of an individual star as its **proper motion**, which is defined as its motion across the sky. It is measured in arc seconds because the movement is very small; Barnard's star, which has the greatest proper motion of any star, moves only 10·3 arc sec. per year. Proper motion, however, does not specify the entire motion of a star; there is still the question of a star's movement towards or away from us. Such **radial motion** cannot be observed directly because stars are too far away to show any noticeable change in size when they approach or recede. The spectroscope, a device unknown in Halley's time, has to be used to determine this.

To reach his conclusions Halley had to measure star positions carefully; he also had to express them in numerical form, just as the Greeks had done. The way they did so was very similar to that used today, and is the method we shall adopt in the text of this

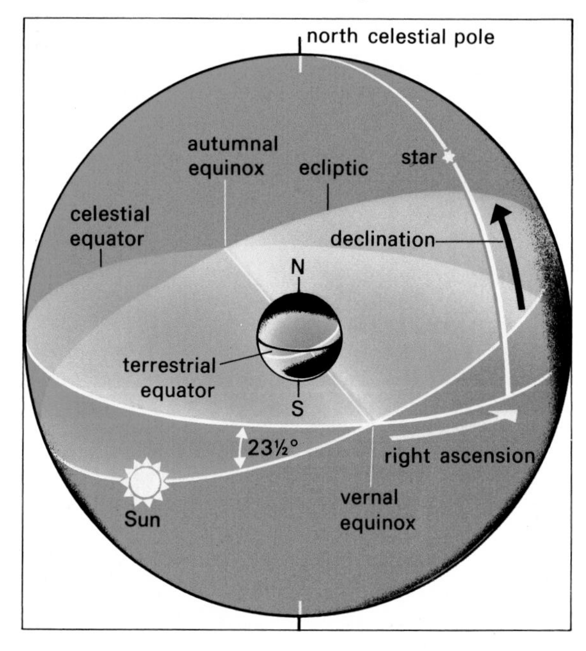

Fig. 1·3, far right: Celestial co-ordinates from which the positions of celestial objects can be specified precisely.

book. It is a method using coordinates, a celestial equivalent of terrestrial longitude and latitude, except that in astronomy these coordinates are known as **right ascension** and **declination**. For such measurements of position, it is convenient to think of the stars as being fixed to the inside of a sphere with the Earth in the centre, even though we know otherwise. Figure 1·3 illustrates the system. The **celestial equator** is a projection on this sphere of the terrestrial equator. North and south of the terrestrial equator we have latitude, north and south of the celestial equator we have declination. East and west along the terrestrial equator longitude is measured; eastwards along the celestial equator we measure right ascension. Terrestrial longitudes begin on the Greenwich meridian, but on the celestial sphere the starting point is different: it depends not only on the celestial equator but also on another circle, the **ecliptic**.

The ecliptic is the Sun's apparent path in the sky. In the northern-hemisphere summer the Sun is north of the celestial equator, and in the southern-hemisphere summer it is south of the celestial equator. The points where the Sun and the ecliptic cross the celestial equator are the times of **equinox**, those two days in the year when day and night are equal. Right ascension is measured eastwards from the Spring or **vernal equinox**, when the Sun moves north of the celestial equator. It is measured, as previously noted, along the celestial equator. Measurements along the ecliptic are measurements of **celestial longitude**, and north and south of it, **celestial latitude**. Halley and the Greeks used those coordinates, but now right ascension and declination are found to be more convenient.

The ecliptic, however, is used for defining the Sun's path and enters into calculations of **eclipses**, those times when the Sun and Moon are aligned in the sky. There are two kinds of eclipse: those of the Moon and those of the Sun. Figure 1·4 shows how the Sun, Earth and Moon are aligned for a lunar eclipse, when the Moon passes either fully or partially into the shadow cast by the Earth. Since the Moon orbits the Earth once a month, one might expect a lunar eclipse every month, but this does not occur because the Moon's orbit is tilted with respect to the ecliptic. A solar eclipse (Fig. 1·5) takes place when the Earth comes into the Moon's shadow, which happens when Sun, Moon and Earth are in alignment. Eclipses of the Sun can be of two kinds – **annular** and **total**. If the Moon, in its elliptical orbit round the Earth, is at greater than average distance, its main shadow – the **umbra** – will not quite reach the Earth. Then the observer on Earth will not see the whole of the Sun's disc eclipsed, but only the central area of it: a ring or annular section of the Sun is still visible. At a total eclipse, when the umbra covers an area of the Earth, observers within that area will see the Sun's disc completely obscured; those outside the area but in the secondary shadow (the **penumbra**) will see the partially eclipsed Sun, where the disc appears with only a section blotted out.

At a total solar eclipse the Sun's light is reduced nearly 800 times compared with that of full Moon, bright stars are visible, birds go to roost, animals prepare for rest, the air temperature begins to drop, and surroundings take on an eerie cardboard look. The Sun's tenuous atmosphere, the **corona**, appears as a pearly coloured light, and bright pink flame-like

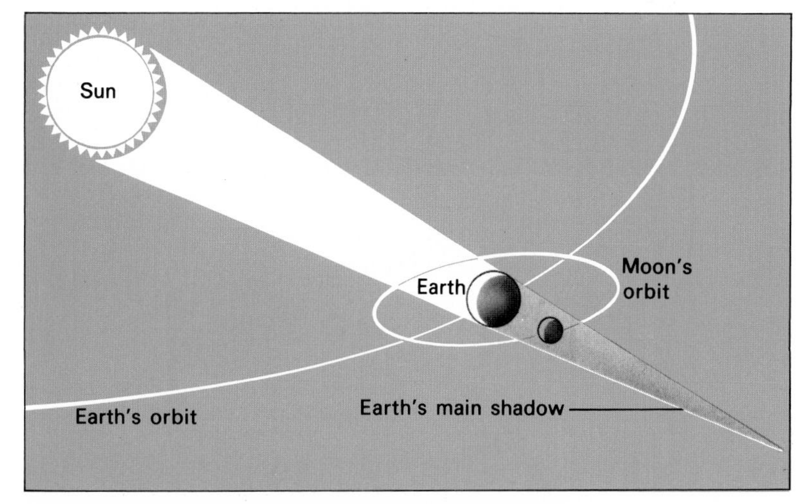

Fig. 1·4, above: *A diagram of an eclipse of the Moon, which happens when the Moon passes into the Earth's shadow.*

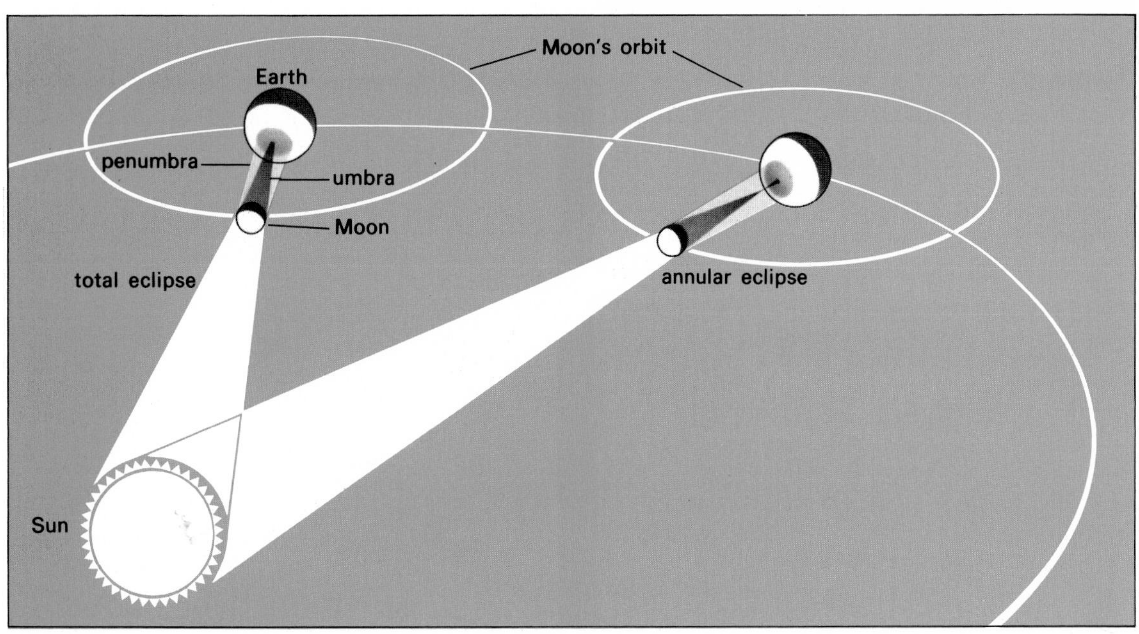

Fig. 1·5
A total eclipse of the Sun (left) and an annular eclipse (right) when a small ring of the Sun's disc is still seen when the Moon blots out the central portion.

Northern hemisphere solar eclipse tracks for the period 1977–2006.

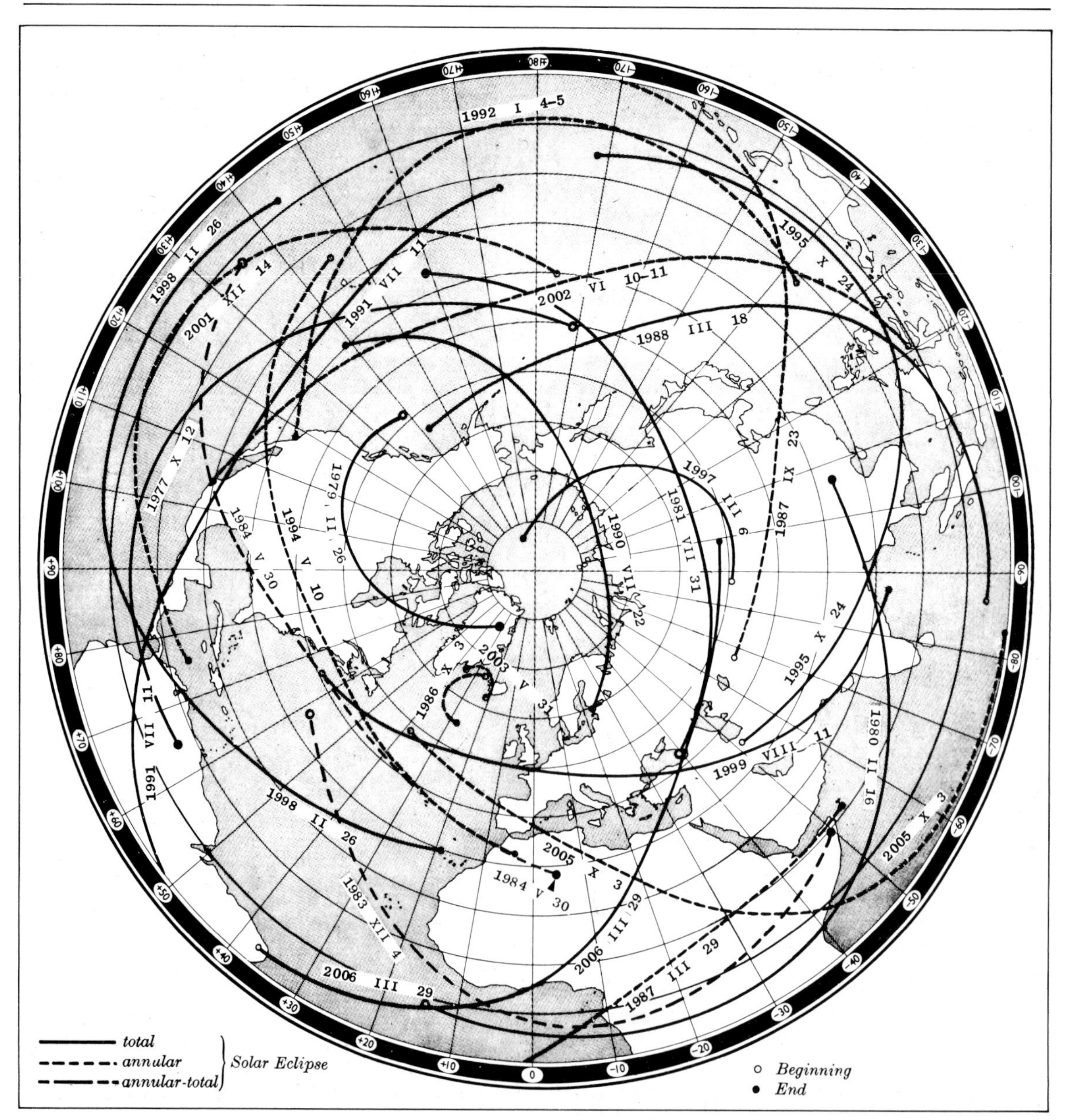

Below:
Kitt Peak National Observatory, in Arizona, like most modern observatories, is sited on a mountain to obtain the most favourable atmospheric conditions.

The 'diamond ring' effect, photographed at the 1973 June 30 eclipse from a ship off the coast of West Africa. Despite the spurious reflections caused within the camera lens, the striking effect shows clearly.

prominences may sometimes be observed protruding over the edge of the Moon's disc. Since the Earth and Moon are both moving in space, the Moon's shadow is continually moving and a total solar eclipse never lasts long for any one observer (unless he is in an aircraft moving with the shadow). Totality usually occupies no more than a few minutes, and a little over 7 minutes is the maximum. As the Moon moves across the Sun's disc, just before and just after totality, the mountainous nature of the Moon's surface gives it a serrated edge which allows patches of bright sunlight still to reach Earth-based observers, and the appearance of the Sun is like a curved string of bright jewels – the so-called **Baily's Beads** effect. Sometimes one last gap allows a final burst of sunlight, giving rise to the **diamond ring** effect. A word of warning here: for observing a solar eclipse always look through a *very* dark filter – welder's goggles, for example, or four or five completely dark photographic negatives – except during totality. As a general rule *never* look directly at the Sun, even with the unaided eye. With binoculars or a telescope permanent blindness will result, and projection methods simply *must* be used (see p. 82). The Moon, on the other hand, has no light of its own and is visible only because it reflects the light of the Sun, which is sufficiently dimmed in the process to make direct observation through a telescope quite safe. The fact that the Moon only reflects light is the reason why it displays phases, each phase depending only on the relative positions of the Sun and Moon with respect to an observer on Earth (Fig. 1·6).

The very short duration of a total solar eclipse underlines the fact that, even on Earth, we are always observing from a moving platform. This can be of use when measuring trigonometrical parallax, but it does make for some complications when we are measuring star positions. This is because right ascension (or celestial longitude, for that matter) is measured from the vernal equinox, the point where the ecliptic and celestial equator cross one another, and this crossing point is continually moving. The movement appears as a westward motion of the vernal equinox, and the effect of this 'backwards' motion is called **precession**, or more explicitly, 'precession of the equinoxes'. It is caused mainly by

Fig. 1·6
The phases of the Moon as seen from out in space and by an observer on Earth.

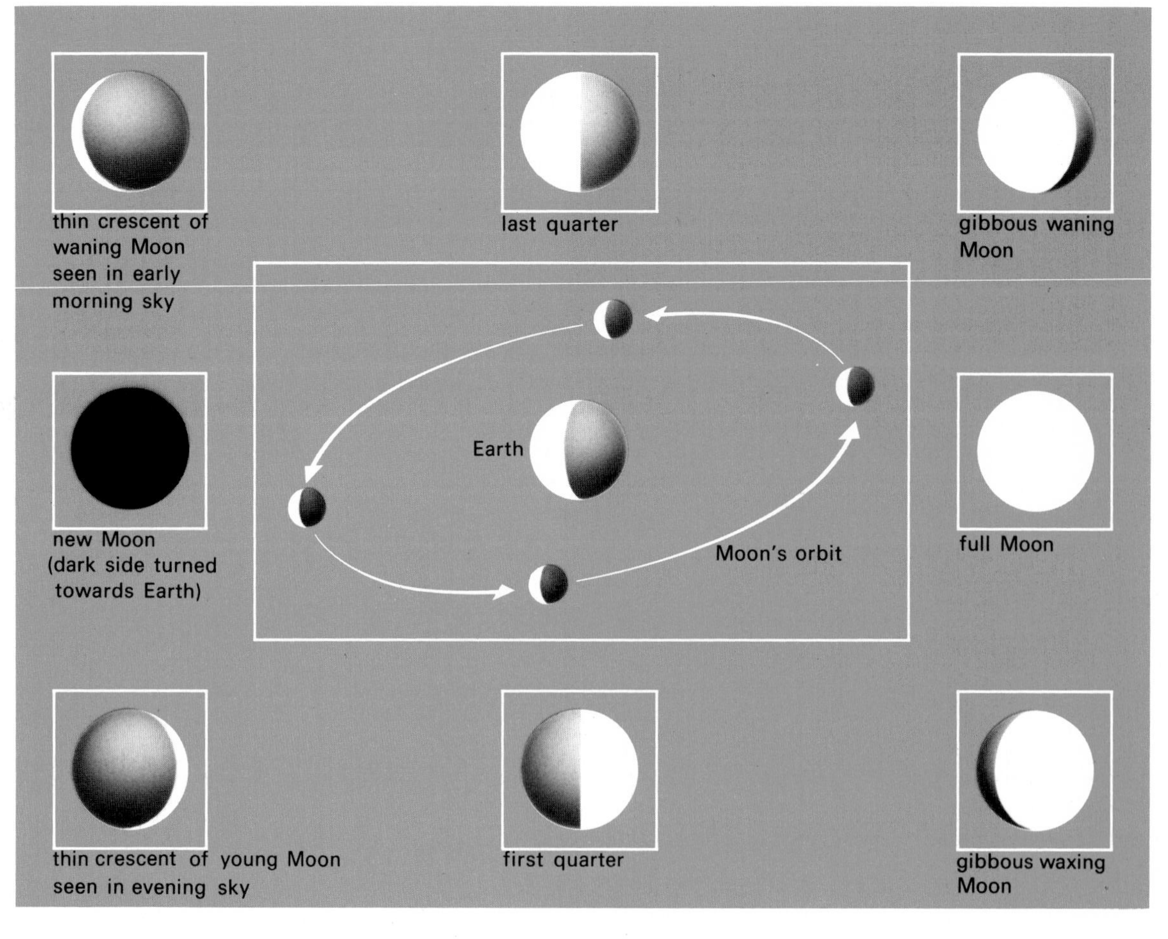

the gravitational effects of the Sun and Moon on the Earth, but there is another component, due to the gravitational pull of the planets, which reduces the luni-solar effect. Combined, they move the vernal equinox 50·2619 arc sec. per year. This is because the gravitational pulls cause the Earth's axis to precess in space, just like a gyroscope does, and the result of this is that the direction in which the axis points also changes. It sweeps out a circle in the sky, taking 25 800 years to complete one rotation (Figs. 1·7 and 1·8). Thus, although the north pole now points towards the star Polaris, in 7 600 AD it will point close to Alderamin (α Cephei) and in 14 800 AD close to Vega (α Lyrae). The Earth's axis also 'wobbles', due to a lunar pull. This nodding or **nutation** has a period of 19 years, and although very small – it amounts to no more than a movement of the celestial pole of 9·23 arc sec. – it is important in parallax determinations.

Fig. 1·7
Diagram of the way the Earth's axis moves with respect to the Earth's orbit, causing precession of the equinoxes.

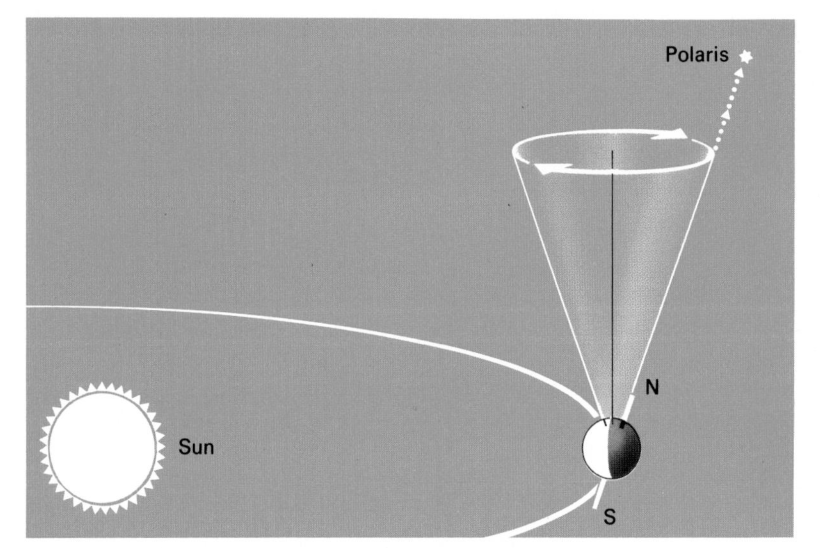

Another effect of observing from a moving platform is that we see different constellations in the night sky at different seasons of the year, and also see all celestial bodies rise and set once every day due to the earth's axial rotation. For these reasons the determination of times has, until recently, always been the province of the astronomer. The Earth orbits the sun once a year, the **tropical year** being that time in which the Earth has completed one orbital circuit; observed from Earth it is the time between the Sun's apparent complete circuit of the stars. Its length is 365·2422 mean solar days. A **mean solar day** is the average time the Earth takes to rotate once and is so called because the motion of the Sun is used to determine the civil measurement of the day. A 'mean Sun' has to be used because the actual Sun does not move regularly across the sky for the simple reason that the Earth's orbit is an ellipse, making the Earth itself move at varying speeds as it orbits. The difference (apparent solar time – mean solar time) is known as the **equation of time**, apparent solar time being the time measured, for example, by a sundial.

The rotation of the earth can also be measured with respect to the stars and this gives **sidereal time**. Since the stars appear to make 366 revolutions in a year – 365 due to the rotation of the Earth on its axis, and one extra rotation because the Earth has orbited once round the Sun – the sidereal day is shorter than the solar day. Its length is 23 hours (h) 56 minutes (m)

4·1 seconds (s) of mean solar time. This is why the stars appear to rise about 4 minutes earlier each night, and it is the time reckoning frequently used by the astronomer.

In this century the Earth's rotation has been found to be somewhat irregular. Time determination using the Earth's axial rotation is, therefore, not entirely satisfactory. In consequence **ephemeris time** is now used, so called because it is based on the orbital motion of the Sun, Moon and planets whose positions are given in an ephemeris or set of tables. Since there are 31 556 925·9747 mean solar seconds in a tropical year of 365·2422 mean solar days, the duration of one second of ephemeris time is 365·2422 divided by 31 556 925·9747.

In the past, observatory clocks and the Earth's rotation made time determination solely the astronomer's responsibility, but now the physicist is involved because mechanical clocks have been replaced by atomic clocks. These have helped to give the precision which has led to irregularities in the Earth's rotation being not only recognized but also measured. Atomic clocks give UNIVERSAL TIME, which is the same as **Greenwich mean time**: it differs slightly from ephemeris time by an amount which, in 1979, amounted to 50s.

An additional consequence of being on the moving space-platform Earth, is that we are surrounded by a blanket of atmosphere. While this is very necessary for supporting life – and so supplying astronomers! – it does mean that our observations are limited. One reason is that the air is always in motion, thus distorting the images of celestial bodies. The twinkling of the stars is one result that can readily be seen, but the telescopic observer finds other effects too, and is never able to use his telescope to full effect. The second main reason for the limitation of observations is that the atmosphere only lets through some of the radiation from space, not all. As human beings we are accustomed to studying the heavens by observing the light emitted by celestial bodies or, more recently, by observing the radio waves emitted too. Yet there is other radiation which the atmosphere absorbs. That other radiation exists we already know: the Sun's heat radiation (infrared) is a common experience, while its ultraviolet light, which cannot be seen, is detectable by the discoloration of our skin we call sun-tan. Yet there are other radiations that do not reach us.

Light, it has been found, may best be described as a wave, or rather as a series of waves, light of different colours having different wavelengths. The shortest waves give us the sensation of violet light, those a little longer, blue light, and so on through the ELECTROMAGNETIC SPECTRUM – green, yellow, orange – to the longest wavelength light which we see as red. According to the wave theory of light, these are electromagnetic waves, all of which travel at the speed of light, but it turns out that light waves are but a tiny fraction of the entire range of **electromagnetic radiation**. Thus, examining the universe with light alone means that we are confining ourselves to a minute fraction of the whole range that should be available. Our atmosphere does let some non-visual radiation through, as we have already seen, but for complete coverage we must go out into space.

Now, at last, in the latter decades of the twentieth century, we may receive and analyse the entire radiation available from space. This gives us a richer and more fantastic universe than man has ever known or imagined, a universe which covers an astounding range of sizes, from the minute electron with a diameter of only 5·6 thousand million millionths (10^{-15}) of a metre, up through man ($1\frac{3}{4}$ metres), the Earth ($12\frac{3}{4}$ million metres), the Sun (14 thousand million metres) to the Galaxy (9 thousand million million (10^{15}) metres) and to the most distant observable depths of space (17 hundred million million million million (10^{26}) metres): a total range of 1 to 10^{41} (or 100,000,000,000,000,000,000,000,000,000,000,000,000,000 or one hundred thousand million million million million million million) – a staggering number however one writes it. This is the universe we shall now examine in some detail.

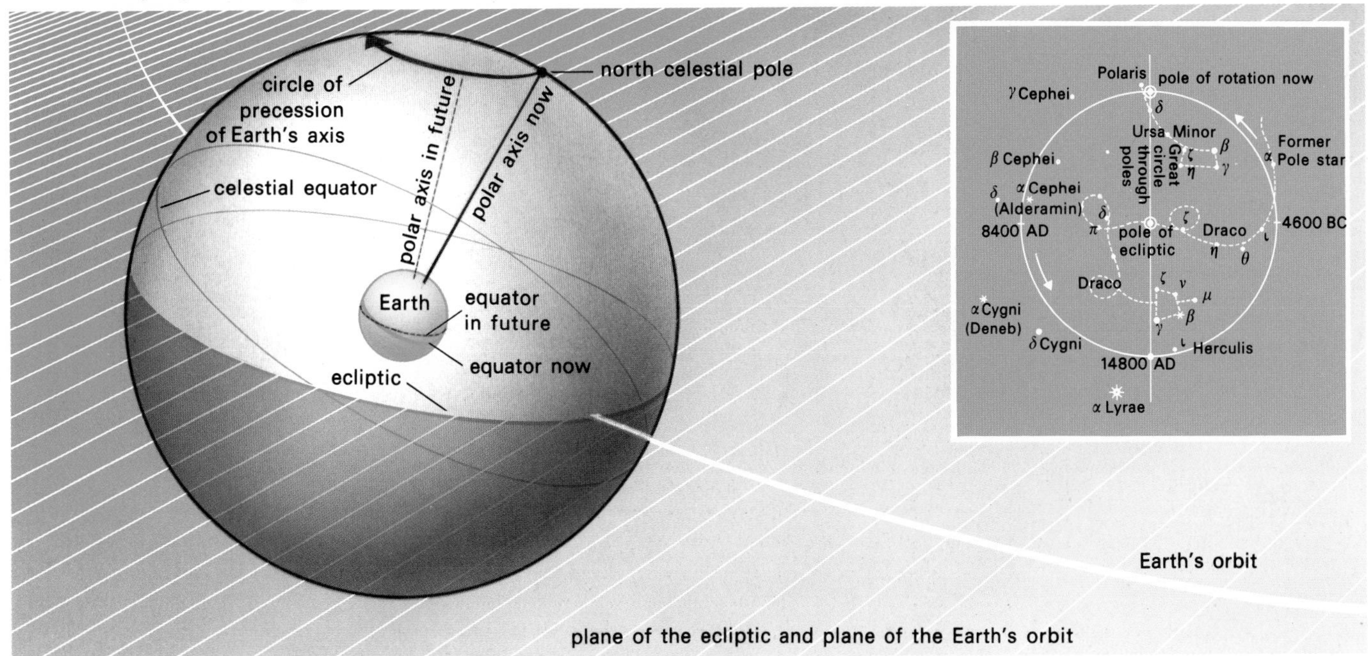

Fig. 1·8
Diagram of the way the Earth's axis moves with respect to the celestial sphere. The inset shows how this causes different stars to do duty as a pole star.

Telescopes for the Amateur Astronomer

Although most amateurs do own some form of optical equipment, it is quite possible to enjoy observing with the naked eye alone, and a considerable amount of truly scientific work can be carried out in this way, particularly in the fields of meteors (page 162–3) and variable stars (page 56–7). Nevertheless there are obviously limitations, and anyone interested in observational astronomy will want to evaluate the different types of instruments with a view to purchasing the one most suitable for his or her requirements.

Binoculars

Binoculars are more versatile than a telescope, and as they are already owned by many beginners in astronomy they are an ideal first instrument. If being chosen primarily for astronomy, then a low magnification combined with a moderate-sized aperture is advisable to give maximum light-grasp and a fairly wide field. A pair of 7 × 50s (magnification 7x, aperture 50mm) enables stars down to at least eighth magnitude to be seen, and covers a field about 6½° in diameter. Such a pair may be used very satisifactorily for observations of variable stars (of which many hundreds would become visible) and for the tracking of artificial satellites (page 96–7).

Quite apart from these uses, binoculars are ideal for general sweeping of the sky, and there is a lot to be said for undertaking a systematic survey of the constellations to familiarize oneself with the sky before concentrating on more specialized work. They are also useful for identifying the brighter double stars, clusters and nebulae. Finally, they are very useful as an auxiliary instrument for finding and identifying the fields to be examined by a larger telescope.

Any binoculars will give even better performance if they are mounted in some manner, and for the largest sizes this is quite essential. Wide-aperture binoculars are used by many workers who concentrate upon searching for comets (page 150–1) and novae (page 56–7). The very largest apertures range up to 125–150mm, but such enormous sizes are only available as ex-military equipment and are very rarely encountered. In any case, very few observers are proficient enough to be able to make proper use of them.

Telescopes

Some form of telescope is essential for most areas of observation and the general principles upon which the diverse types operate are described on pages 229–232. In amateur work the refractor and Newtonian reflector remain predominant, although the compact and relatively portable Schmidt-Cassegrain and Maksutov types are becoming increasingly popular. The minimum useful aperture is about 75mm for a refractor and 150mm for a reflector; smaller telescopes abound but they should be discarded in favour of a good pair of

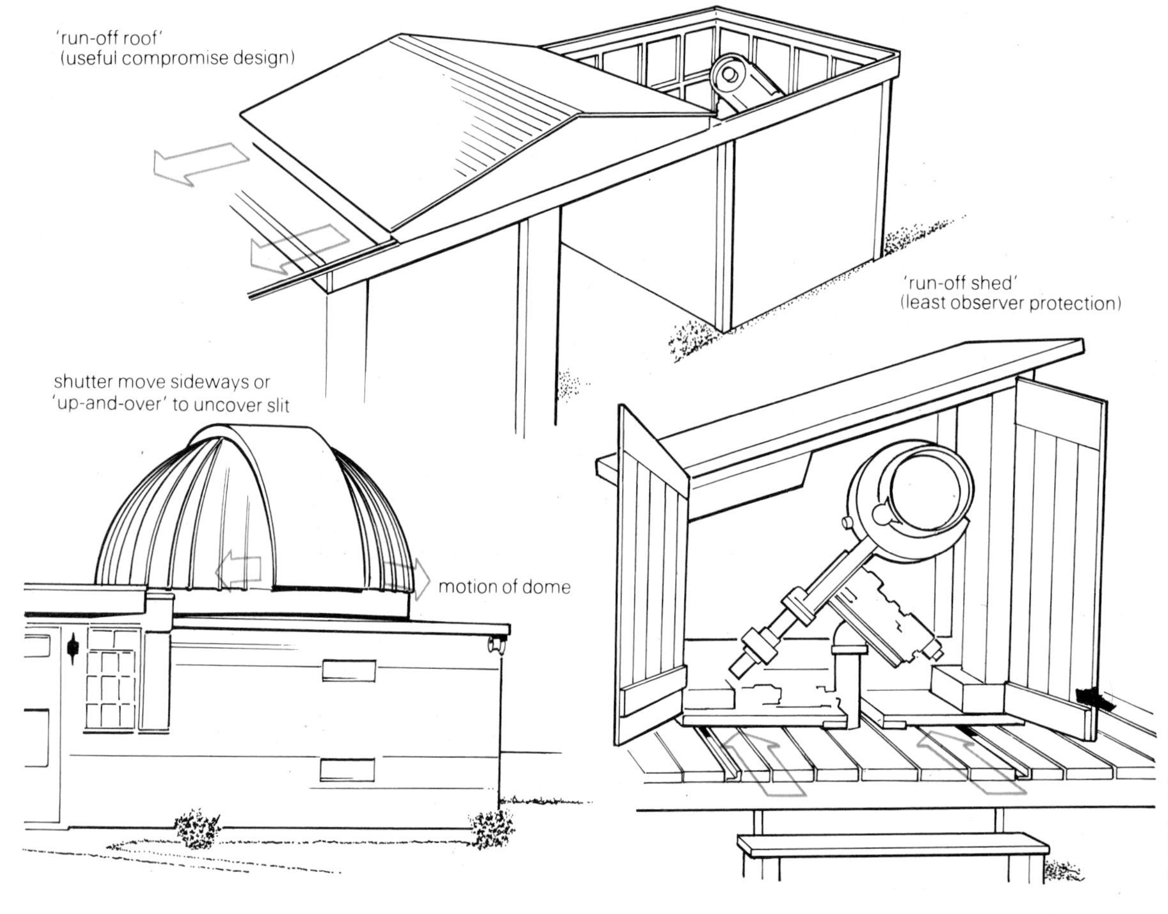

Amateur observatories may range from reasonably simple ones such as those where either the roof (top) *or the whole super-structure* (below right) *moves out of the way when in use, to proper domes* (below left), *which give full protection to both telescope and observer at all times.*

Right:
An altazimuth reflector of the type known as a Dobsonian is particularly easy to construct from ordinary materials.

Far right:
The long focal ratios of most refractors make them very useful for solar work and, in the larger sizes, for double star measurements and planetary observation.

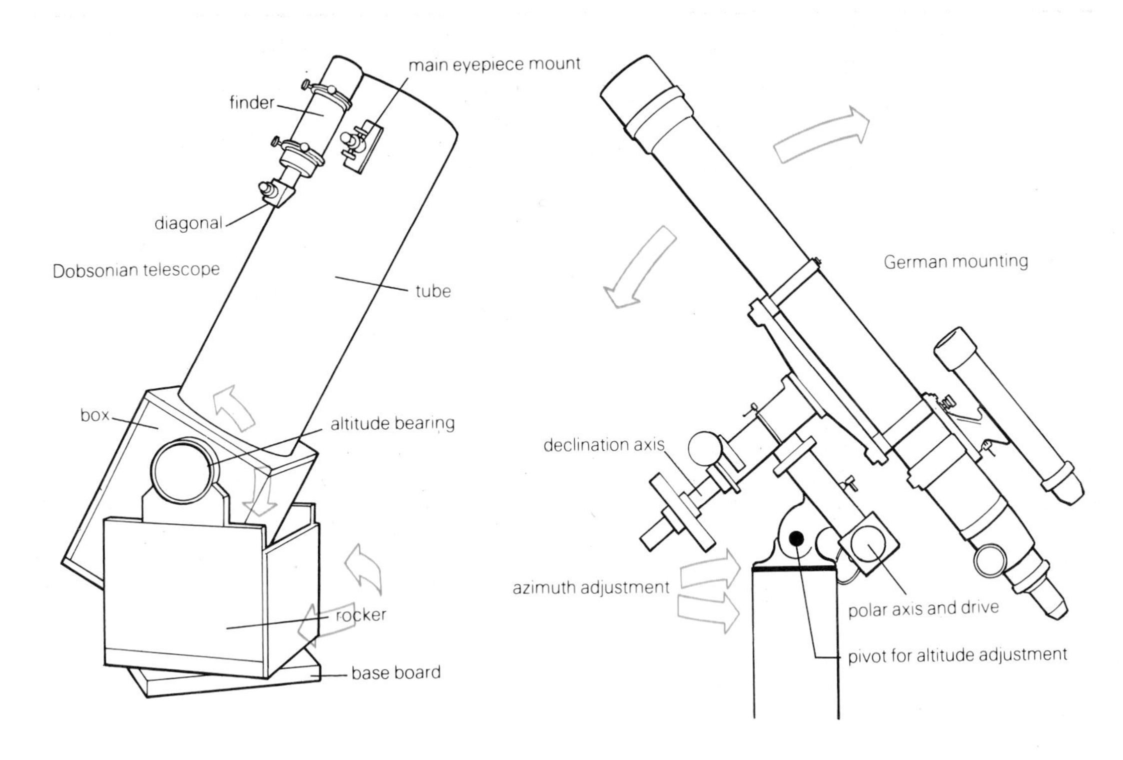

binoculars for all serious observing purposes. Aperture for aperture, 'ordinary' Newtonian reflectors are usually much cheaper than refractors or the compound types, and this is why they tend to predominate at the larger sizes.

A 75mm refractor is widely regarded as the ideal instrument for the beginner. It is sufficient to show considerable detail on the Moon, the belts of Jupiter and the rings of Saturn, countless double stars, clusters and nebulae, and can be used to advantage for observation of the Sun by one of the safe methods described on pages 82–83, and certain variable stars. However, if the observer is certain that his interests lie in areas where light-grasp is likely to be the most important consideration – particularly variable star work – a reflector might well be the first choice, especially as one of about 150mm aperture would be comparable in cost to the 75mm refractor. In the larger sizes reflectors are usually the first choice, on the grounds of cost if nothing else. They may also be made of shorter focal lengths, and thus have 'faster' f-ratios than equivalent refractors; for this reason they are favoured for photographic work on, for example, nebulae and galaxies.

Cassegrain reflectors and the Dall-Kirkham form are easier for amateurs to construct, and are particularly suitable for planetary work, where high magnifications and good light-grasp may be required to discern detail on small planetary disks. The Maksutov and Schmidt-Cassegrain types usually have fairly long focal ratios (f/12 or more) and are thus best suited to examination of restricted fields such as planets. The Maksutovs are particularly suitable for solar work when a proper full-aperture filter is fitted. Both of these types have undoubtedly become so popular because of their comparative portability, which means that their owners can often dispense with a proper observatory.

Somewhere in between binoculars and the usual form of telescopes come the types sometimes described as 'comet-seekers' and 'rich-field telescopes'. Strictly speaking there is only one true 'richest-field telescope' – one which shows the greatest number of stars at a time – but the term has been extended to wide-aperture, wide-field telescopes. Such equipment is frequently used, along with the large binoculars, for comet and nova searching, and may consist of short-focus reflectors as well as specially computed and manufactured achromatic refractors with fast focal ratios.

Telescope-making itself is a very popular branch of amateur astronomy, with every type of optical system, including the most complex ones, having been made by enthusiasts, as well as instruments of very large aperture – 600mm and more. Even for those with little mechanical aptitude it is well worth considering the purchase of the finished optical parts for, say, a 150mm Newtonian reflector, and the construction of a simple, wooden tube and mounting for them. Such an instrument offers a moderate aperture capable of good results for a reasonable cost. Objectives for refractors may also be purchased, but the mounting of these requires a greater degree of engineering skill than is needed for the mirrors in a reflector. Anyone taking up telescope making should perhaps be warned that it has a tendency to become all-absorbing and leave little time for actual observing!

Mountings, observatories and accessories

There has been a move recently for reflectors to be mounted in simple, altazimuth mountings – particularly in the so-called 'Dobsonian' mount – and this design is well worth consideration by any beginner for its simplicity and cheapness. It can even share some advantages with other altazimuth designs for certain types of work where the field orientation is not critical, or when the somewhat faster movement from one portion of the sky to another might be an advantage. However, most serious observers will require a proper equatorial mount which can be driven easily to follow the sidereal motion, as well as allowing the use of setting circles which can be essential for locating faint objects in the absence of a finder chart. The portable Maksutov and Schmidt-Cassegrain instruments

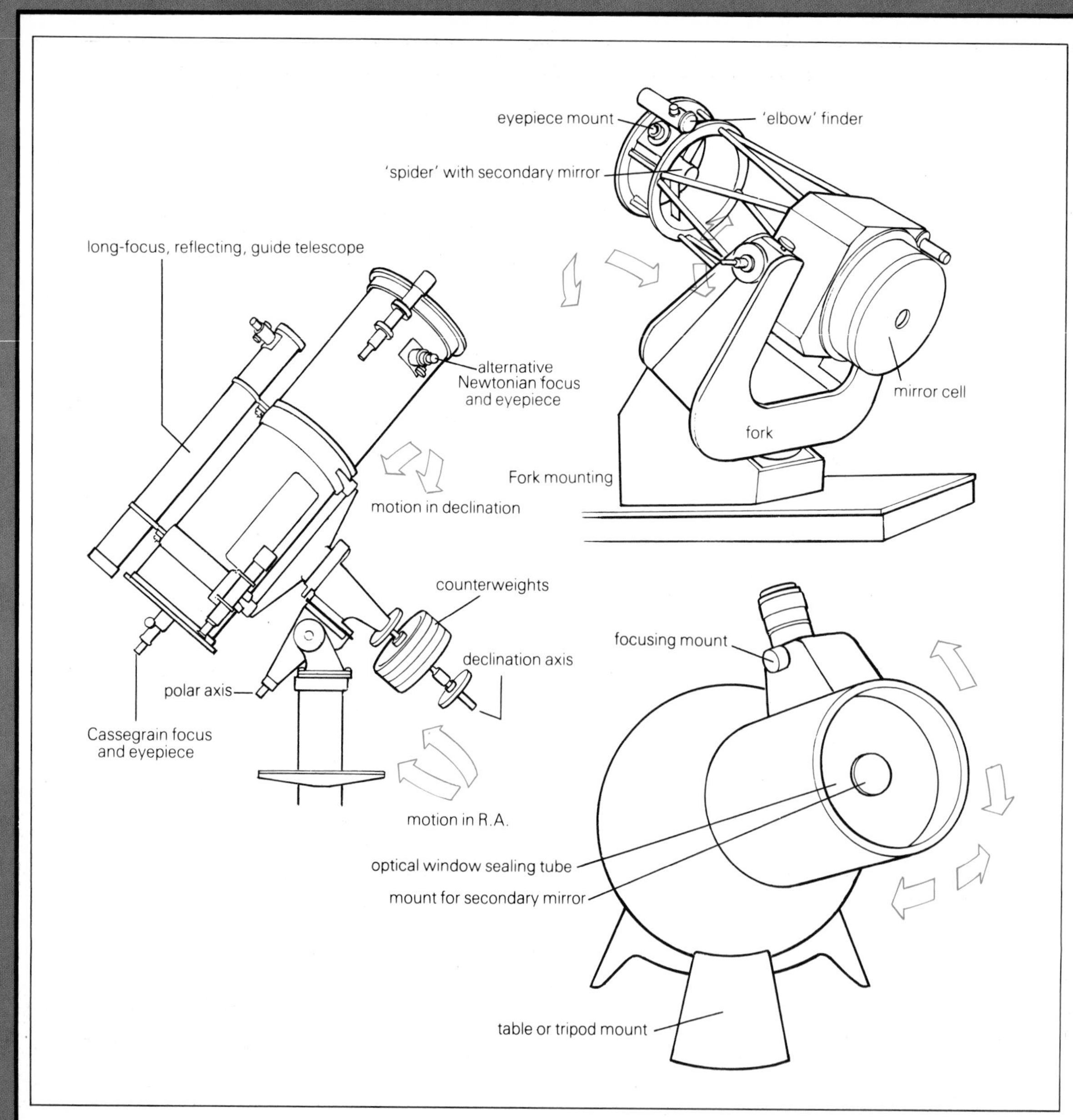

Top:
The generally greater light grasp of reflectors is an advantage in studying faint nebulae and galaxies, as well as for many photographic purposes.

Far left:
This is a typical commercially available Cassegrain telescope which, with its long focal ratio and considerable light-grasp, would be very suitable for planetary observation.

Left:
Some telescopes are specifically designed to be fully portable, as in the case of this small (105mm f/4.2) reflector on a ball and socket mount.

usually offer these facilities, but with refractors and reflectors a fixed permanent mount will be required.

There are many different forms of mounting which may used, but that known as the German mount is one of the best for refractors, while the Fork mounting is particularly suitable for reflectors and is the type most frequently chosen for Maksutovs and Schmidt-Cassegrains.

A permanently-mounted telescope obviously needs to be housed in an observatory. However, even the portable forms require a firm base, and, more important, the observer will benefit from the protection – especially from cold winds – afforded by a proper observatory, and will be able to observe for longer periods and make less errors. There are, of course, many other advantages to having an observatory, such as having everything to hand when observing, but space can be a major problem.

Amateur observatories come in all shapes and sizes, from sheds which are literally lifted up and over the telescope, to fully rotating hemispherical domes. Perhaps one type which deserves wider use is that with a 'run-off' roof: this is basically four walls covered by a roof which is rolled back out of the way when the equipment is to be used.

It is now quite common for equatorially-mounted telescopes to be provided with an electric variable-speed drive for motion in right ascension, together with a similar drive – usually slightly less sophisticated – to position the telescope in declination. Some amateurs have even linked microcomputers to their telescopes to give complete control of the drives on both axes, thus providing for any required rate, or combination of rates, of motion. (Driving may be required on both axes simultaneously to follow comets and minor planets, and there are also sidereal, solar and lunar rates which may be required.) Such sophistication can also provide for push-button setting of the telescope on to the required object, in addition to the digital readout of position which some telescopes already possess. However, it should never be forgotten that all this electronic gadgetry is of little value unless it, and the telescope, is properly and fully used. Many of the most accomplished amateur observers may be found using binoculars or simple telescopes without drives or setting circles. Furthermore, complicated electronic drives may still be inadequate to compensate for a fast-moving comet or minor planet, and the observer may have to resort to using hand-operated slow motion controls. As always, it is the experience of the observer, not the equipment, which is all-important.

It is quite common practice for various pieces of auxiliary equipment to be mounted on the main telescope. The most common items are wide-field

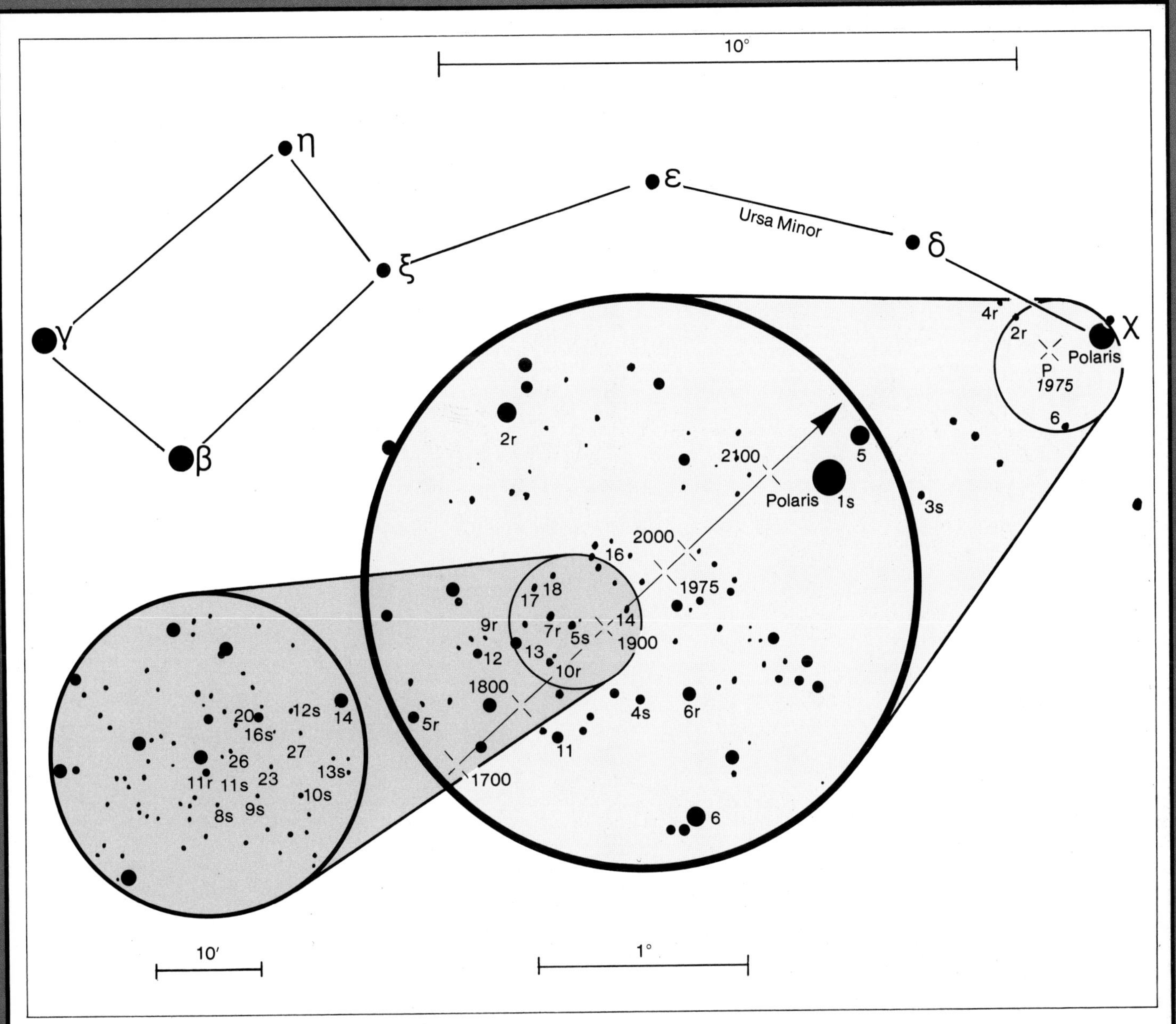

Top:
The North Polar Sequence of stars of known magnitude, part of which is shown here, is a useful means of determining a telescope's performance, and runs from Polaris (1s) at mag. 2.1, to star 16s at mag. 15.5.

and narrow-field finder telescopes, and a long-focus instrument of some form for use as a guide telescope when the primary instrument is being used for photography. Various wide-field cameras may be attached to the mount, thus benefiting from the accurate drive, while occasionally a specialized solar instrument may also be placed on the same mounting; all other telescopes and finders are covered for safety when it is in use.

The simplest and most essential accessories for any telescope are, of course, the eyepieces, and here there are many different types which can be used. One of the simpler and more common types is the achromatic Ramsden eyepiece (or the variant known as the Kellner). More highly-corrected, more complex and more expensive eyepieces of the Orthoscopic and Plösl types are also popular for high magnifications, with Erfle and König forms finding favour for wide fields. With any eyepieces, but most especially with the complex types containing many glass/air surfaces, it is important that they should be provided with anti-reflection coatings, as otherwise the light losses can be quite considerable. For the beginner, three eyepieces are normally sufficient: a low power for general work, a medium power for observation of the Moon and planets, and a high power for use on the rare occasions when the turbulence of the atmosphere permits. A Barlow lens, which increases the magnification that can be obtained with a given set of eyepieces, is also very useful – especially for lunar and planetary work.

Photographic equipment intended for use on the telescope is usually in the form of a standard SLR camera body mounted at the prime focus, but some observers go to the trouble of manufacturing special cameras, or holders for plates or – more usually nowadays – sheet film. These may incorporate special optical arrangements for guiding by means of a star at the edge of the telescope field – a method which is more reliable than using a separate guide telescope, which is subject to unavoidable flexure and consequent misalignment. A somewhat similar method of offset guiding is usually employed in conjunction with photoelectric photometers, where the star under observation must be kept on the limiting diaphragm during the measurements.

There are many items of specialized and complex equipment which have not been mentioned here, some of which are described in the relevant sections of this book.

However, it is perhaps worth emphasising again that astronomy is one of the few sciences where not only can a great deal of enjoyment be gained, but also a serious contribution to knowledge may be made with only the very simplest of equipment.

Practical Observing

Celestial objects chosen for observation at any time will depend to some extent upon one's position on the Earth. An observer's latitude has a considerable effect upon the part of the sky which is visible, quite obviously, and also upon the length of the night – especially upon the length of the time when the Sun is 15° or more below the horizon, which is when astronomical twilight is not present. (At latitudes above 51½° astronomical twilight persists all night at some time during the year.) From the equator – at least in theory – all the stars in the sky are visible, and – again in theory – all stars appear to rise and set. (In practice atmospheric absorption and refraction complicate the issue.) Further north and south certain stars will be circumpolar and remain above the horizon all the time, circling the celestial poles. In the north the relatively bright star α Ursae Minoris is close to the true pole, and thus provides a useful marker, but unfortunately in the south there is no corresponding conspicuous 'pole star' to use as a similar guide.

The observer's position on the Earth will also have an effect upon the visibility of various objects, because of their relationship to the ecliptic or to the Sun. Observers in the northern hemisphere will best be able to see planets, for example, when they are north of the ecliptic. However the most favourable oppositions of Mars take place at a time of year when Mars is well south of the ecliptic, so that southern observers have the best conditions for observing.

Naturally, similar considerations also apply to objects other than planets, particularly when they are close to the Sun in the sky, which can markedly affect their visibility at times close to sunrise and sunset. This can mean that at times observers at the highest latitudes stand the best chance of confirming discoveries of some objects such as novae and comets.

Other factors being equal, general planetary observation is best undertaken from low latitudes, where the high altitude which the bodies attain minimizes atmospheric effects, particularly absorption and scintillation. When high magnifications are desirable, as is certainly the case with planetary work, this is an obvious advantage.

Proximity to the Sun is a factor in the observation of objects such as variable stars which need to be followed continuously. They may be so situated that there are periods of the year when they are completely invisible, and such 'seasonal gaps' are quite unavoidable. However, observations shortly after sunset and just before dawn can help to shorten these breaks in coverage (those made by the considerably fewer dedicated observers who work in the early morning being particularly valuable).

The Moon has a somewhat similar effect to that of the Sun and can cause serious interference to some observations, particularly those of nebulae and galaxies, comets, and faint variable stars, especially around the time of Full Moon. Observers of such objects become resigned to these interruptions, but are known to welcome the occurrence of lunar eclipses which enable them to snatch a few valuable observations.

Facing page: *The south polar region. Sigma Octantis is the nearest moderately bright star, but it cannot be used as easily as Polaris in the north for the purposes of alignment of a telescope.*

Circumpolar star trails. 2-hour exposure on Kodak High Speed Ektachrome (up-rated to ASA 400), taken using a 50 mm focal length, f/1·4 Zeiss Planar lens, from the dark sky of Arizona. Note the bright star α Ursae Minoris, the Pole Star, which is situated approximately 1° from the true north celestial pole.

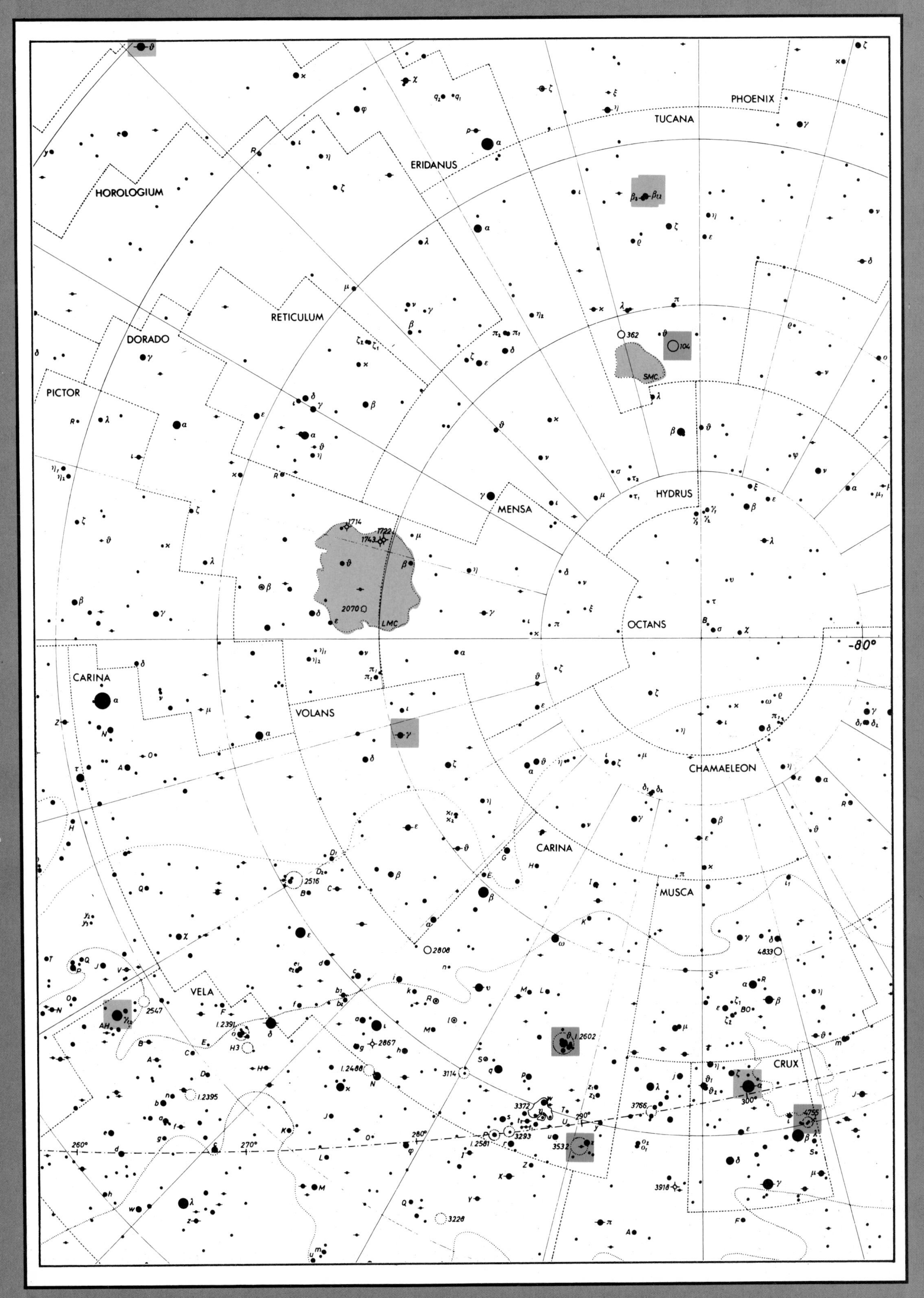
HOROLOGIUM
ERIDANUS
TUCANA
PHOENIX
RETICULUM
DORADO
PICTOR
HYDRUS
MENSA
OCTANS
-80°
CARINA
VOLANS
CHAMAELEON
CARINA
MUSCA
VELA
CRUX
SMC
LMC
260°
270°
280°
290°
300°

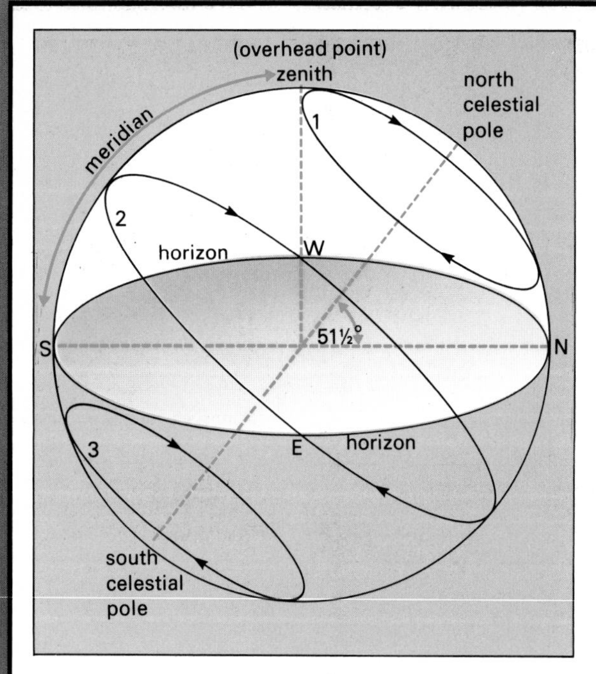

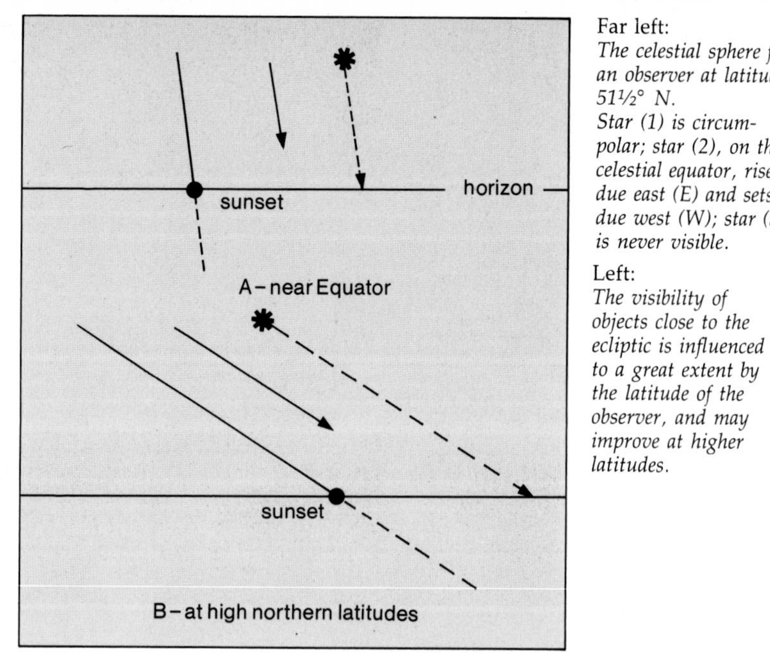

Far left:
The celestial sphere for an observer at latitude 51½° N. Star (1) is circumpolar; star (2), on the celestial equator, rises due east (E) and sets due west (W); star (3) is never visible.

Left:
The visibility of objects close to the ecliptic is influenced to a great extent by the latitude of the observer, and may improve at higher latitudes.

Moonlight is a nuisance because it lessens the contrast between the objects being examined and the sky background by causing the latter to be increased in brightness. It can also contribute to an observer failing to become fully dark-adapted. Dark-adaption is a most important factor in efficient observing, and consists of the formation over a period of about 20–30 minutes of a pigment in the retina of the eye which gives greater visual sensitivity under low-light conditions. It should not be confused with the nearly instantaneous adjustment of the size of the pupils of the eyes to changes in brightness. True dark-adaptation requires low light-levels for it to come into force, so that ample time must be allowed for this to take place before attempting the most demanding observations. Dark-adapted vision is most sensitive to wavelengths centred in the green region, but is comparatively insensitive to long wavelengths. For this reason, red light is used by astronomers to illuminate charts and notebooks and find their way around. Bright lights will destroy the sensitivity at any time.

Even the most casual observer will find that it is well worth while keeping a proper notebook for observations, and for any work of a serious or scientific nature it is absolutely essential. Even for beginners it is helpful to be able to look back to earlier observations – if only to learn by the mistakes. Quite apart from the actual drawing or other observational details which may be required, it is usually necessary to give full information about the equipment being used and the sky conditions. Naturally each type of observation will require certain specific details, and it is important to ensure that none of these are forgotten. Errors inevitably occur, even with the most experienced observers, and the more information there is, the easier it is to detect them. Most important of all, the date and time must be noted in full (being only too easy to forget). The time must be expressed in Universal Time rather than any local time, which may or may not be subject to the additional complication of some daylight-saving adjustment. Universal Time is just that, being used and understood by all astronomers, everywhere.

Observations should be made whenever possible, not just for the sake of coverage of the objects being studied, but also because as with all fields of endeavour, practice does make perfect; and even in so simple a thing as finding one's way around the sky, increasing confidence will bring greater enjoyment. For it should never be forgotten that although amateur astronomers may be making a very valuable contribution to knowledge, with professional astronomers welcoming their observations, it *is* a hobby. The satisfaction that one derives comes principally from the making of observations rather than from an appreciation of their value; whether or not they are scientifically useful is a secondary consideration.

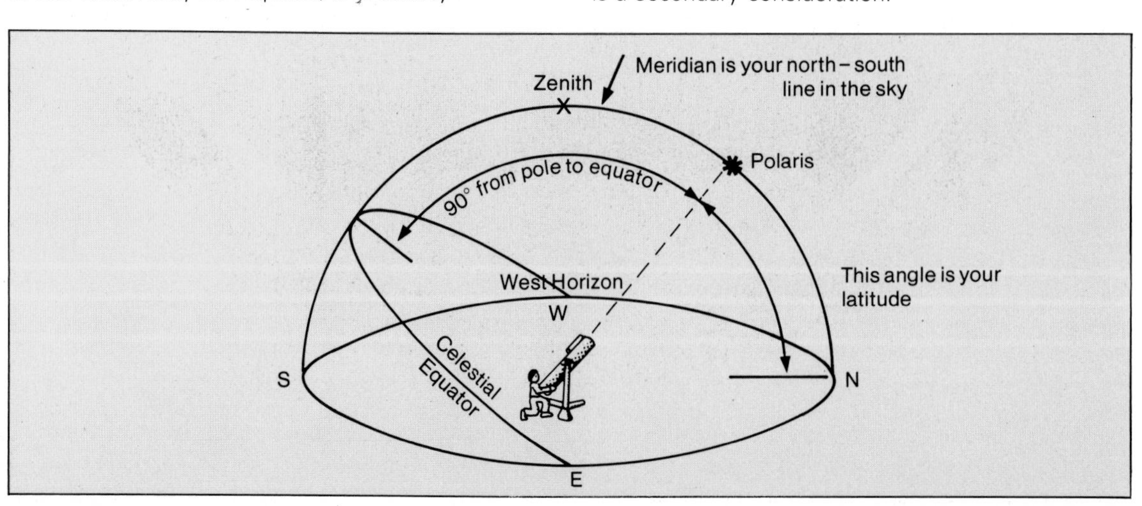

The correct alignment of the polar axis of an equatorial telescope is of great importance, especially for critical work such as long-exposure photography.

This composite of 16 photographs of Saturn taken by Stephen M. Larson with the 1.55 m telescope at Catalina Observatory in Arizona is probably the best image obtained from Earth. However, it shows little more than can be seen by experienced visual observers.

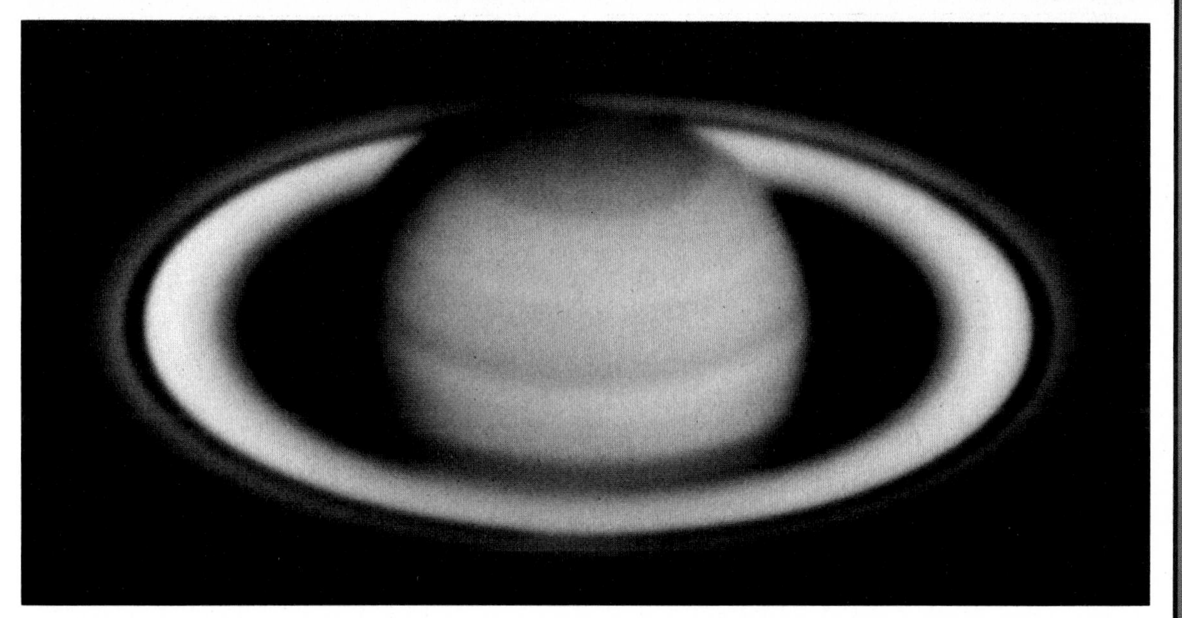

A Planisphere is a simple device which shows the exact region of the sky above the horizon at any hour of the night (or of the day).

CHAPTER TWO

Motions of celestial bodies

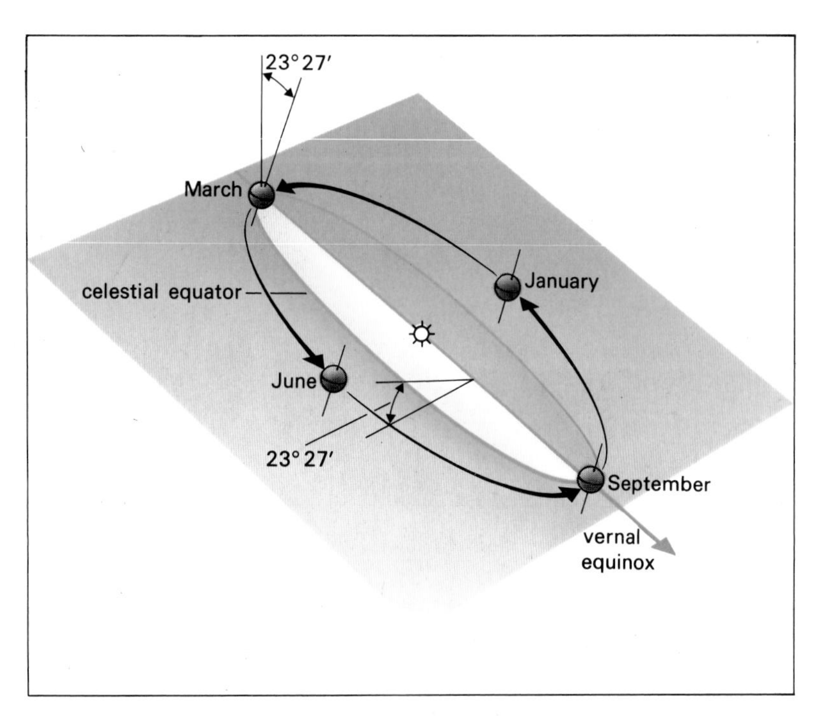

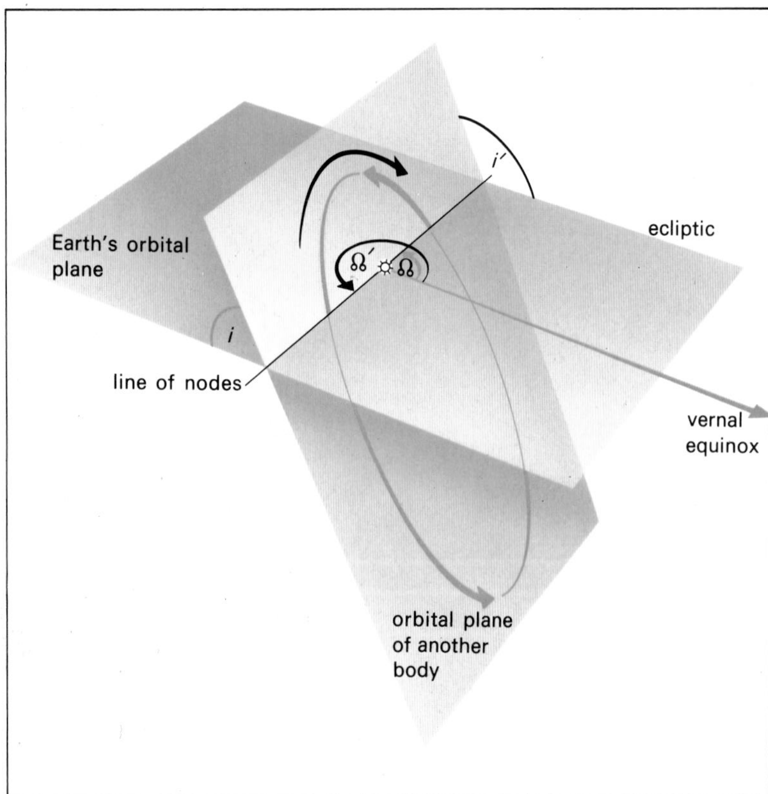

All the objects comprising the Solar System are bound by the Sun's gravitational field and move round it in some kind of orbital path. However, because of the mass of the planets, especially of Jupiter, the centre of mass of the whole system lies just outside the surface of the Sun.

In order to describe these orbits and their position in space, several quantities or elements must be given, and Figs. 2·1 to 2·3 show how these are defined. The plane of the Earth's orbit is taken as the basis for the orbits of the other planets but for planetary satellites the basic reference is to the planet's equator, even though this is usually inclined to its orbit.

The points at which a planet's or satellite's orbit crosses these reference planes are known as **nodes**. The closest and most distant approaches of an orbiting body to the primary body around which it moves are denoted by the prefixes 'per-' and 'ap-'. Thus **perihelion** is the closest point and **aphelion** the most distant of a body orbiting the Sun, **perigee** and **apogee** similar points for a body orbiting the Earth. Recently, though, the more general terms **periapsis** and **apoapsis** have become common.

Looking down on the Earth's north pole, its axial rotation is anticlockwise, while its orbital motion and the motion of all the planets around the Sun is in the same direction. This is termed **direct** rotation and direct orbital motion, while the word **retrograde** is applied to movement in the opposite direction. Objects which orbit with retrograde motion have inclinations greater than 90° (Fig. 2·2).

The calculation of the position of any body in its orbit was not satisfactorily carried out until the early 1600s when Kepler developed his three laws of planetary motion (Figs. 2·4–2·6). Strictly speaking, these 'laws' are relationships derived from observation; the physical reasons for such planetary motion had to await Newton's theory of universal attraction, which took both distance and the masses of the bodies into account. Newtonian theory is sufficiently accurate to predict the motion of all the planets, except that of

Fig. 2·1, top: *The Earth's orbit provides a reference plane in space, while a fixed direction is given by the vernal equinox where, as seen from the Earth, the Sun appears to cross the celestial equator from south to north.*

Fig. 2·2, left: *The orbital plane of any body intersects the ecliptic at a line of nodes. The angle, Ω, of the ascending node (south to north) is measured from the vernal equinox. Objects with retrograde motion have inclinations i′ greater than 90°.*

Mercury, which requires the further refinement offered by Einstein's general theory of relativity.

Once its orbit is known, the position of any body at any time may be established if the **longitude**, which is measured from the ascending node, at any given instant is available. Conversely, from several observations of the position of a body at known times, the orbit and period may be derived. Under certain circumstances, when one of the bodies has little mass, Keplerian methods may suffice, but usually the masses and consequent disturbing effects (PERTURBATIONS) of the planets must be taken into account. By means of electronic computers it is possible to include all the planetary perturbations, and positions of the planets have been calculated for periods of 500 000 years before and after the present, with such studies showing very long-term variations in the orbital eccentricities.

The **axial rotation periods** given in the various tables are **sidereal periods**, rotations measured with respect to the background of the 'fixed' stars. These differ from the **synodic periods** which bring the body into the same axial position relative to the Sun (Fig. 2·7). In the case of the Earth, the mean sidereal axial period is $23_h56_m4{\cdot}1_s$, and the mean solar day is $24_h3_m56{\cdot}6_s$. A similar relationship exists between a satellite's sidereal and synodic orbital periods, and for the Moon, these are approximately 27·322 days and 29·531 days respectively. In the case of a planet, one synodic period brings the Sun, Earth and planet back into the same relative positions (Fig. 2·8). Due to the differing orbital motions of the planets, the Earth may be overtaken by, or overtake, another planet, and at such times the planet appears to retrograde, or move from east to west against the background of stars, contrary to its usual motion. Terms for planetary positions relative to the Earth are illustrated in Fig. 2·10.

Details of the orbits of the planets are given in Table 2·1, while those of satellites, minor planets and comets are covered in their respective sections. The basic unit of distance within the Solar System is the **astronomical unit** (au) which is the average distance of the Earth from the Sun. From highly accurate radar measurements of various objects, such as the planet Venus and minor planets which closely approach the Earth, the latest value for this essential unit is

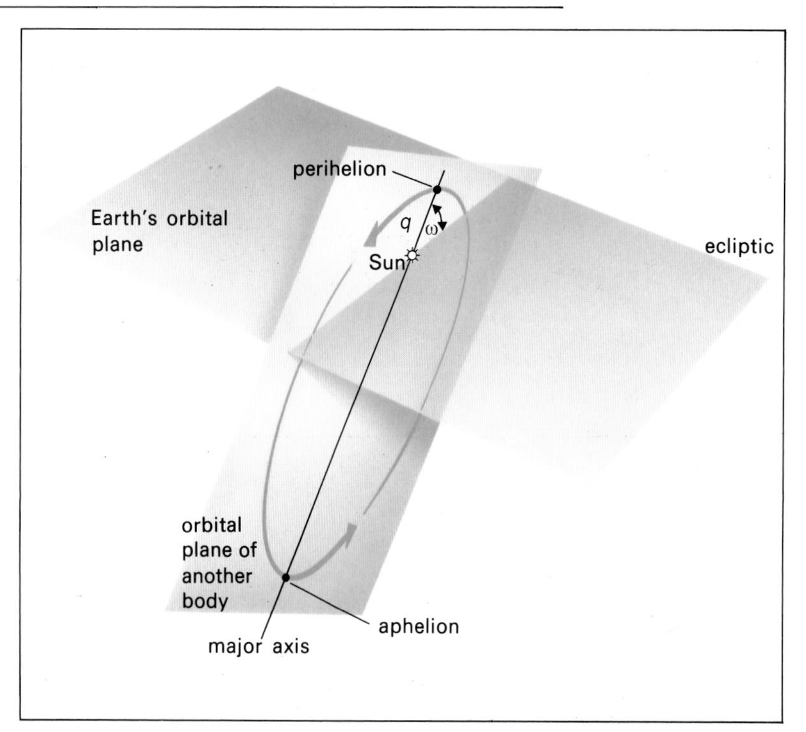

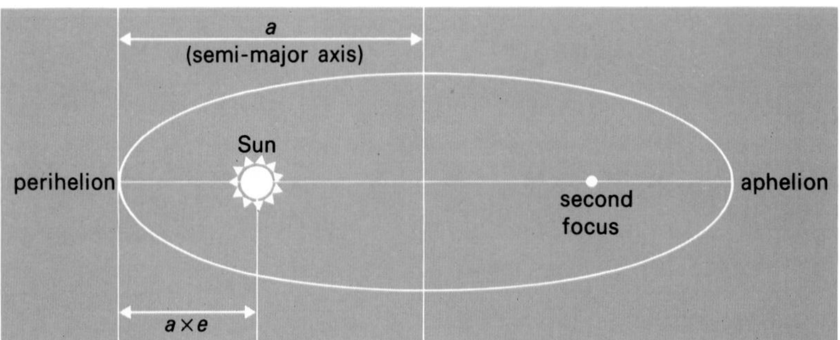

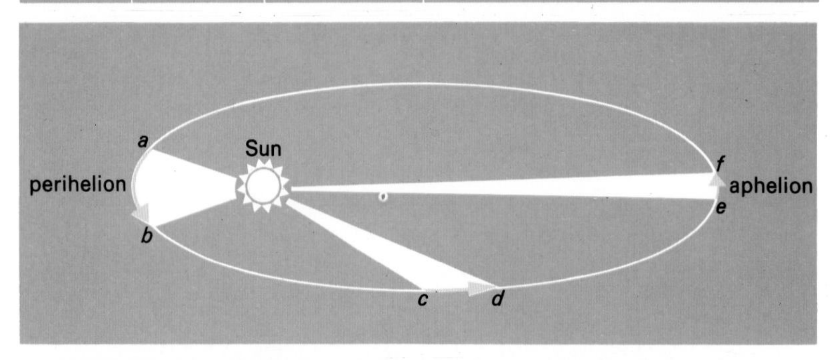

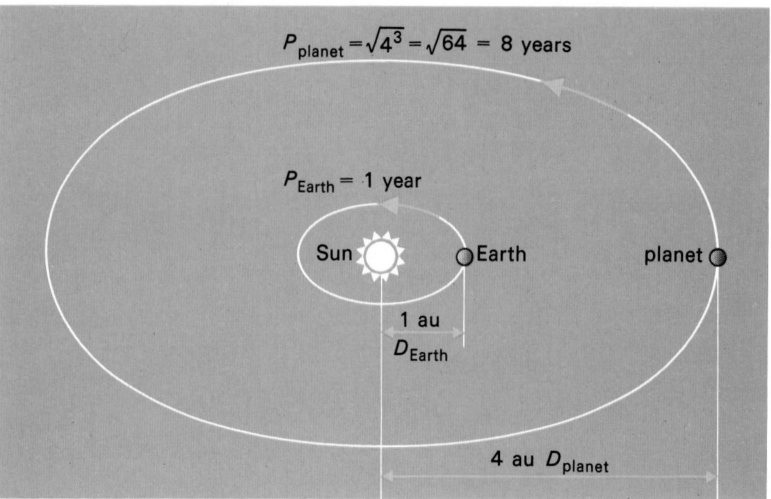

Fig. 2·3, top right: The orientation of an object on its orbital plane is given by the angle of perihelion, ω, measured from the ascending node. The shape of the ellipse is given by the perihelion distance q, and the eccentricity (see Fig. 2·4).

Fig. 2·4, centre top: Kepler's First Law. The planets (and other bodies) move in ellipses, with the Sun at one focus. Mathematically, the perihelion distance q = a (semi-major axis) × e (eccentricity).

Fig. 2·5, centre below: Kepler's Second Law. The radius vector (line joining Sun and planet) sweeps out equal areas in equal times. When close to the Sun the body moves faster, so that distance a–b is greater than c–d, itself greater than e–f.

Fig. 2·6, right: Kepler's Third Law. The square of the periods is proportional to the cube of the distances. (Mathematically P^2/D^3 is the same for every orbit.) Counting the Earth's distance as 1, then for a planet at 4 times the Earth's distance from the Sun, $P^2 = D^3 = 4^3 = 64$, so the period $P = \sqrt{64}$ or 8 years. Certain minor planets actually have periods close to this.

Fig. 2·7 right: *Due to the motion of a planet in its orbit (from A to B and B to C), the sidereal rotation period (A to B) with respect to the stars, may be shorter than the synodic period (A to C) relative to the Sun.*

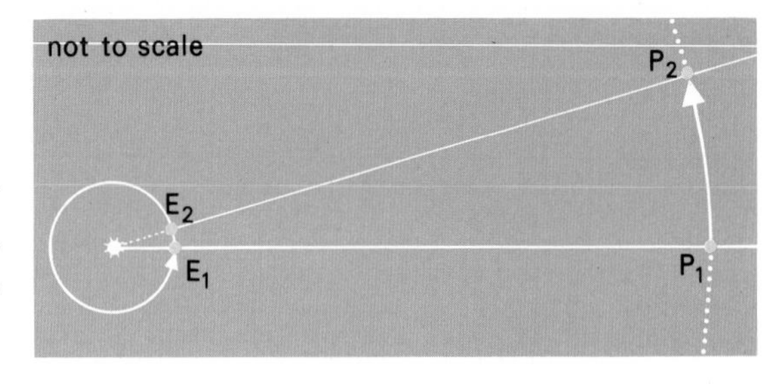

Fig. 2·8 right: *The synodic period of a planet. In the time a distant planet takes to orbit from P_1 to P_2, the Earth completes one orbit from E_1 to E_1 and must travel on to E_2 to give the same position relative to the Sun.*

Fig. 2·9 top right: *As seen from the Earth (E_1 . . . E_9) a planet moves against the background stars, appearing to retrograde from P_4 to P_6. Although the path is shown open for clarity, depending upon the orbital inclinations and positions, it may form a loop.*

149 597 870 km (Fig. 2·11). Once one planetary distance has been established all of the others follow from the application of Kepler's theory.

The Titius-Bode 'law'

An attempt to explain the distances of the planets was made by Johann Titius in 1772 and this was publicized by Johann Bode, with whose name alone the idea was linked for a considerable time. It is now generally known as the Titius-Bode 'law' although it was only obtained by fairly arbitrary numerical manipulation. The figures obtained are given in Table 2·2 together with the actual planetary distances. It will be seen that, with the exception of the value of 2·8 au, the agreement is good for the planets which were known at the time – that is, out as far as Saturn. The discovery of Uranus in 1781 seemingly confirmed the 'law' and encouraged the search for the 'missing' planet at 2·8 au. Ceres, which was discovered accidentally, and other minor planets apparently filled this gap. When the search began for the planet which was perturbing Uranus, the Titius-Bode relation was used to indicate the distance at which it would be found. In the event, however, Neptune's distance of 30·06 au does not agree well with the predicted 38.8 au, and the discrepancy in the case of Pluto is too great (approximately 39·5 au against 77·8) for the 'law' to be any longer accepted. Various attempts have been made to suggest alternative relations, but as yet no satisfactory theory has been devised to account for the formation of the planets at specific distances. However, the Titius-Bode relationship served a useful purpose, as it encouraged the search for other objects.

Table 2·2 **The Titius-Bode 'law'**

planet	**distance (au) from Sun** 'predicted'	**actual**
Mercury	0·4	0·39
Venus	0·7	0·72
Earth	1·0	1·00
Mars	1·6	1·52
—	2·8	—
Jupiter	5·2	5·20
Saturn	10·0	9·54
Uranus	19·6	19·18
Neptune	38·8	30·06
Pluto	77·2	39·4

Table 2·1 **Planetary orbits**

planet	**mean distance** (millions of km)	**sidereal period** (d)	**inclination to ecliptic**	**eccentricity**
Mercury	57·91	87·969	7°00′15·6″	0·2056302
Venus	108·21	224·701	3°23′39·9″	0·0067835
Earth	149·60	365·256	—	0·0167184
Mars	227·94	686·980	1°50′59·3″	0·0933847
Jupiter	778·34	4 332·59	1°18′15·6″	0·0484648
Saturn	1 427·01	10 759·20	2°29′21·2″	0·0556194
Uranus	2 869·60	30 684·9	0°46′23·4″	0·0472585
Neptune	4 496·67	60 190·3	1°46′19·5″	0·0085888
Pluto	5 900·22	90 470	17·2°	0·25

(The values given for Pluto are approximate: inclinations and eccentricities are exact for 1978)

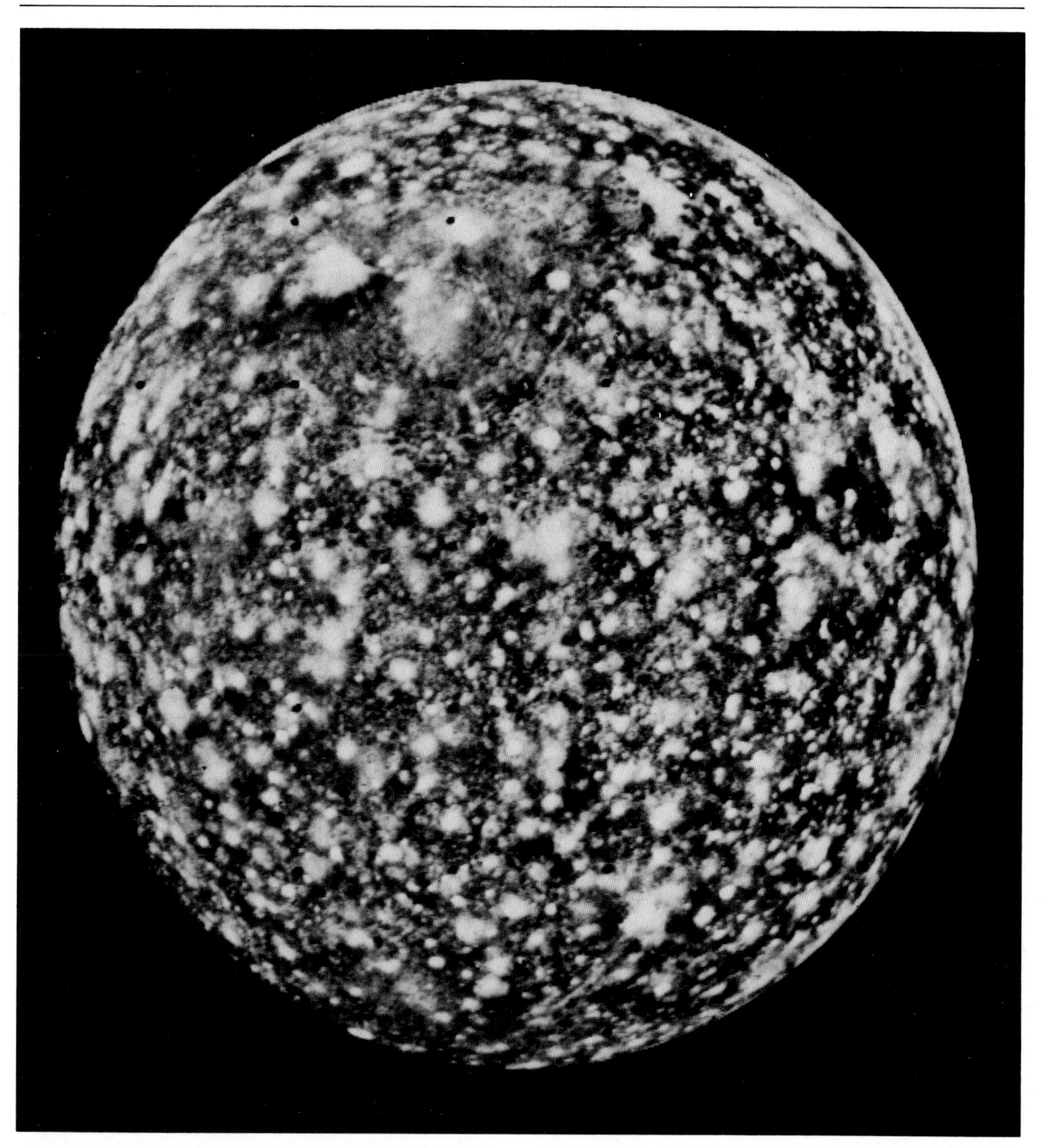

By the regularity of their motions, the four major moons of Jupiter provided important evidence for universal laws governing the motions of celestial bodies (this is the icy, crater-covered Callisto).

Fig. 2·10
Terms describing the relative positions of the Earth and other planets. For this purpose the Earth may be considered as stationary.

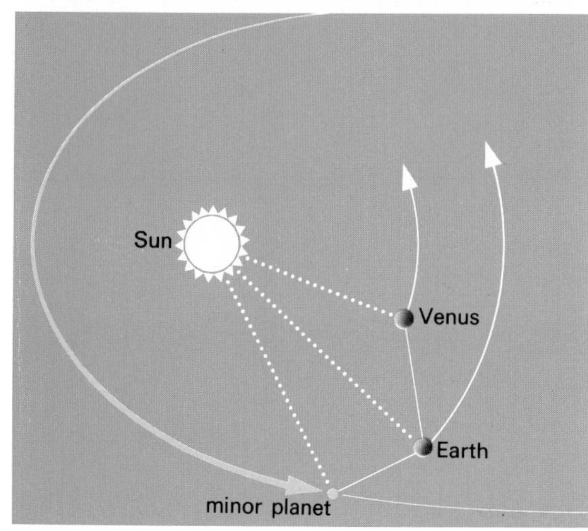

Fig. 2·11 above: *Determination of the astronomical unit. Determination of the Earth-Venus or Earth-planetoid distance by radar, allows the Earth-Sun distance (which cannot be directly measured) to be calculated by simple trigonometry, as all the angles in the triangle are already known.*

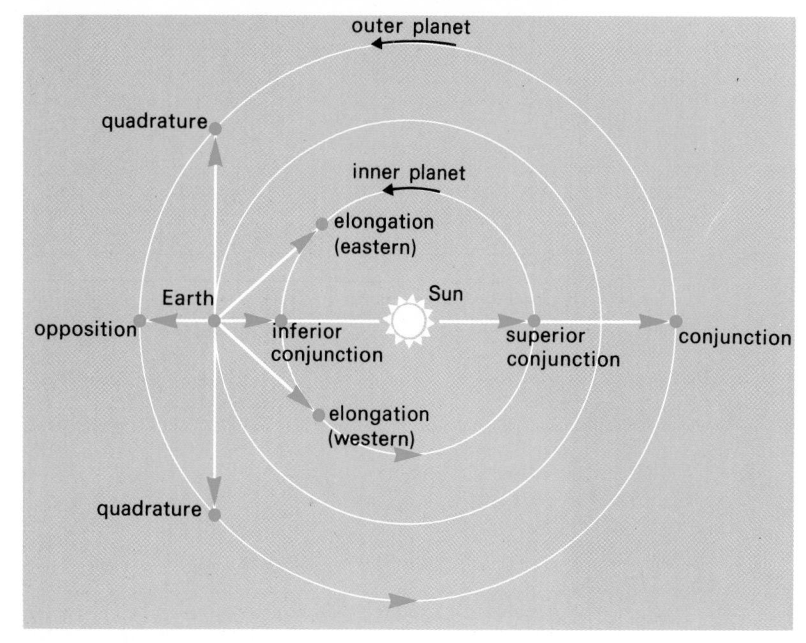

Astronomical Photography

Photography is used in nearly every aspect of astronomy and in its simpler forms need not involve the use of very complicated or expensive equipment. Any type of camera which allows a time exposure may be used on a rigid tripod or other support to obtain photographs of star fields. An exposure of 15–20 seconds on film of 200 ASA, with a standard 50mm lens, at about f/1.8, will show stars down to about the naked-eye limit. Such photographs are interesting in their own right, but even though the star images will be trailed on all but the shortest exposures, due to the Earth's rotation, they may still be used for some purposes such as variable star work.

Long trailed exposures are also quite commonly used in work on meteors, or in fireball patrols. Here the main consideration is the recording of the fast-moving objects, and as long as the star patterns are recognizable, so that the meteoroid's path may be determined, trailing is of little consequence. The amount of trailing will, of course, depend upon three factors: the length of the exposure, the focal length of the lens being used, and the declination of the area being photographed, motion being least at the celestial poles and most at the equator.

The next degree of sophistication is the simple driven mount. There are numerous forms which this can take, some very simply constructed. The actual drive may be of many kinds, ranging from those where the observer turns a handle at a suitable rate – not as difficult as it may first seem – to those with a full electrical drive. All driven mounts must be properly aligned with the celestial pole, and this is all the more important for long photographic exposure with telescopes. With the comparatively short focal lengths of lenses used on cameras alone it is not quite so critical. Some very spectacular photographs of the Milky Way have been obtained with driven cameras, but a lot of specific work may also be undertaken. The lenses which some amateurs already have for everyday photography with single-lens reflex cameras can be especially suitable for particular studies. The 135mm lens, for example, the commonest telephoto lens, conveniently covers a nova search area (page 56–7) and is therefore the standard equipment for this work.

The use of aerial cameras and lenses (usually from ex-government sources) and similar equipment may be regarded as the next stage of complexity. These cover wide fields at fairly wide apertures – up to as much as 150mm diameter. Some of the military designs may require modification or filters for astronomical work, but this sort of equipment is generally excellent for wide-field cometary and minor planet work (pages 148–151). This type of

This photograph of the centre of the Milky Way in the direction of Sagittarius was taken by R. McNaught using high-speed Ektachrome film (160 ASA) and a 30-minute exposure, with a 35mm camera mounted on a driven and guided telescope.

camera can be used alone on a suitable driven mount, but it is more commonly mounted on a full-sized telescope which may then be used for guiding; or the camera may itself be provided with a guide telescope.

Photography through a telescope is rather different, and also takes a number of forms. It is quite possible to use any camera, set to infinity and held to the telescope eyepiece, for short exposures. However, the only suitable subject is the Moon, which with its easy availability, abundance of light – allowing very short exposures with no image motion – and large size can be photographed with acceptable results.

For most subjects it is necessary for the camera body, without its lens, to be properly and rigidly coupled to the telescope. The exact way in which the combination is used depends upon the object being photographed and the image scale required. At prime focus the focal length of the mirror determines the scale at the film plane and this is the principle usually adopted for photography of comets, star fields and galactic and extragalactic objects. For lunar and planetary work some form of magnification is normal unless the focal length of the telescope is very long, with a large image scale, as in some Cassegrain and Maksutov forms. Projection of an enlarged image on to the film plane through a high-quality eyepiece is a method frequently adopted, and a Barlow lens is also very suitable if properly employed. In most cases some experimentation is required to determine the best method for the particular object being photographed, and usually some form of compromise is necessary. A large image scale will minimize grain problems (allowing for substantial print enlargement), but will require a longer exposure, with all the consequent problems of atmospheric conditions and the necessity for accurate drives and guiding.

The most important factor in astronomical photography is probably the maximum aperture available. This will govern the faintest stars that can be photographed in a given time, and in most applications is more important than the 'speed' of the equipment. This applies specifically to point sources such as stars, which remain as points on the film whatever the focal ratio. Extended objects such as nebulae are theoretically better being photographed with faster telescopes or lenses, ideally like the Schmidt types, but here again there is usually an important gain from the greater light-gathering power and increased resolution of a larger instrument.

Astronomical photography may become highly specialized. Many advanced amateurs have built special cameras – sometimes to take plates or sheet film – for mounting on to their telescopes. These can provide better coverage of the field provided by their telescope than the standard 24 × 36mm rectangle available on 35mm film. Others use specialized techniques of film treatment and exposure, such as those discussed elsewhere, particularly hypersensitization and the cooled-emulsion method (pages 176–177). Even the most advanced astronomical photographers occasionally find themselves making use of conventional cameras and lenses, although they do mount them on highly sophisticated telescopes and driven mounts.

The great Orion nebula (M42), NGC 1976. A 22-minute exposure on Kodak Ektachrome EL (ASA 400), taken on 1978 November 26, by Ron Arbour, using a 215mm f/6.3 reflector.

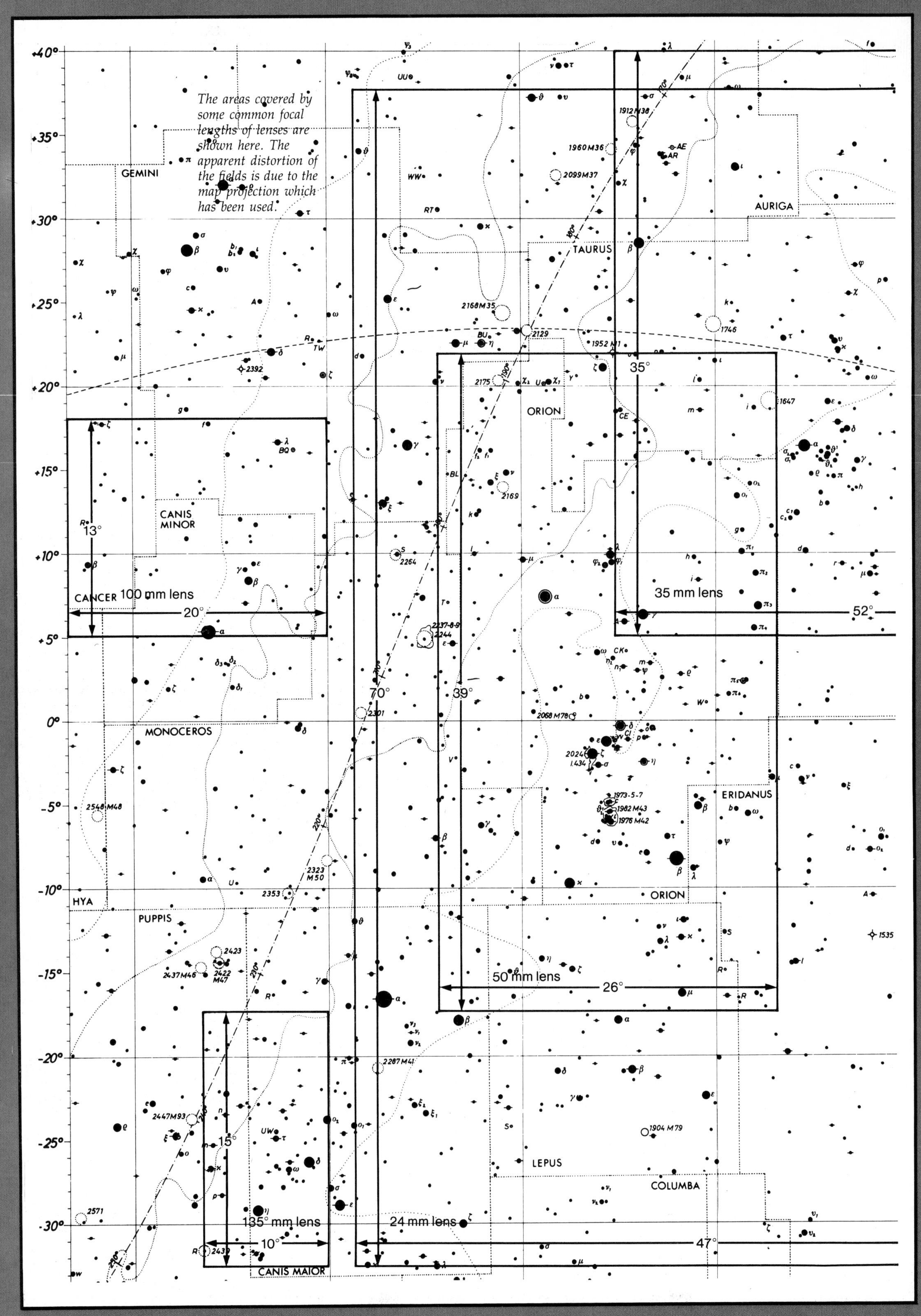

The areas covered by some common focal lengths of lenses are shown here. The apparent distortion of the fields is due to the map projection which has been used.

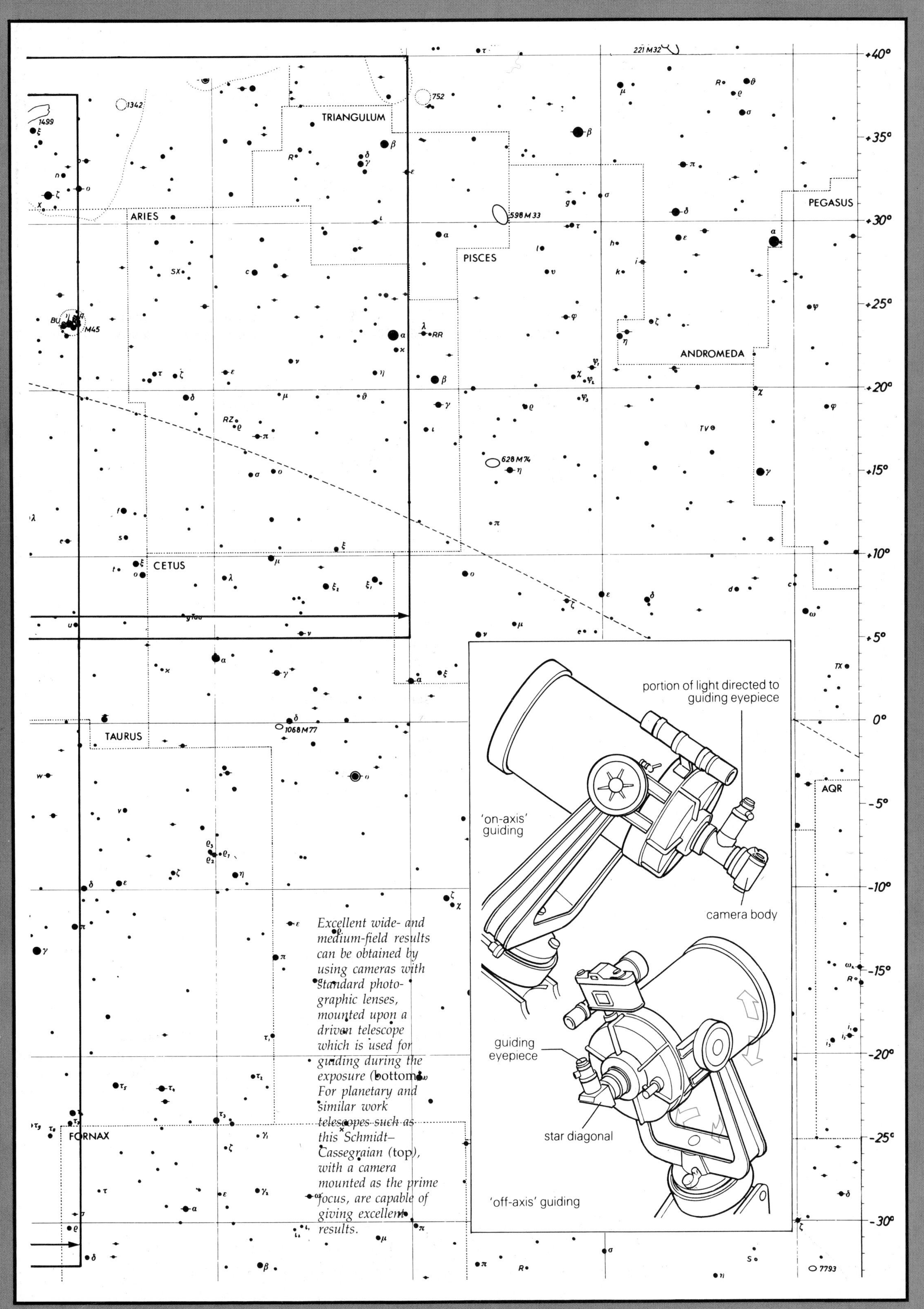

Excellent wide- and medium-field results can be obtained by using cameras with standard photographic lenses, mounted upon a driven telescope which is used for guiding during the exposure (bottom). For planetary and similar work telescopes such as this Schmidt–Cassegrain (top), with a camera mounted as the prime focus, are capable of giving excellent results.

Star charts

The star charts show the stars visible on a clear night without a telescope, as well as some other objects such as galaxies visible only with binoculars or, in some cases, a small telescope. The different sizes of dot refer to stars of different magnitude. In addition special symbols are used for variable stars and double stars, while open clusters and globular clusters have their own symbols, as do planetary nebulae, diffuse nebulae, and galaxies. The boundaries of the Milky Way are shown by dotted lines, each dotted line being an isophote, i.e. showing those parts which are equally bright. The galactic equator (the central line of the Milky Way) is indicated by a line of dots and dashes, the numbers along it giving galactic longitude.

The ecliptic (a line of dashes) indicates the Sun's apparent path in the sky, and it is within the Zodiac – a band some 7° wide centred on the ecliptic – that the Moon and planets appear to move.

The Sun

The Sun moves through all the constellations of the Zodiac in the course of a year, and its approximate positions are as follows:

January	Sagittarius
February	Capricornus
March	Aquarius
April	Pisces
May	Aries
June	Taurus
July	Gemini
August	Cancer
September	Leo
October	Virgo
November	Libra
December	Scorpius

These positions are not exact, since towards the close of a month the Sun may be approaching or entering the next zodiacal sign or constellation. Moreover they do not coincide with the positions used by astrologers for the simple reason that precession has caused the

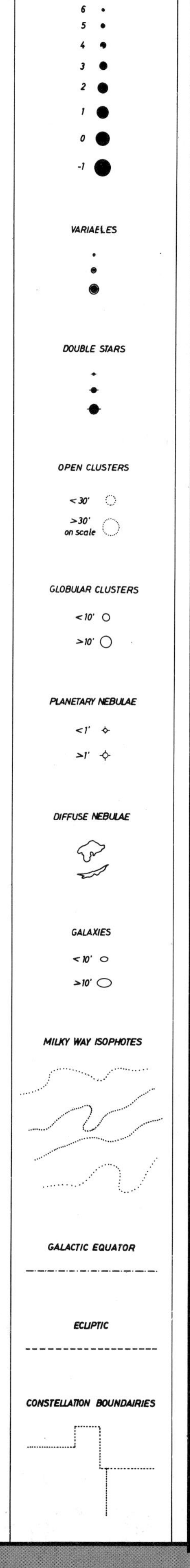

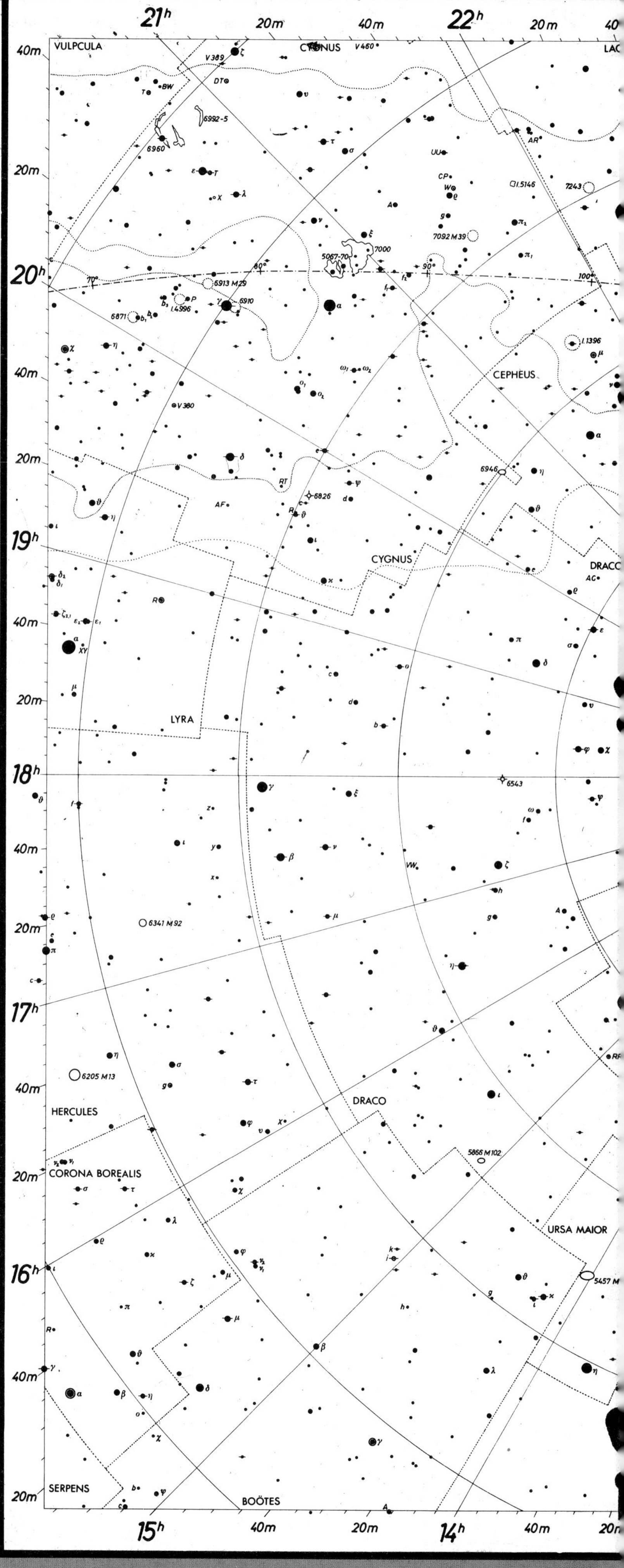

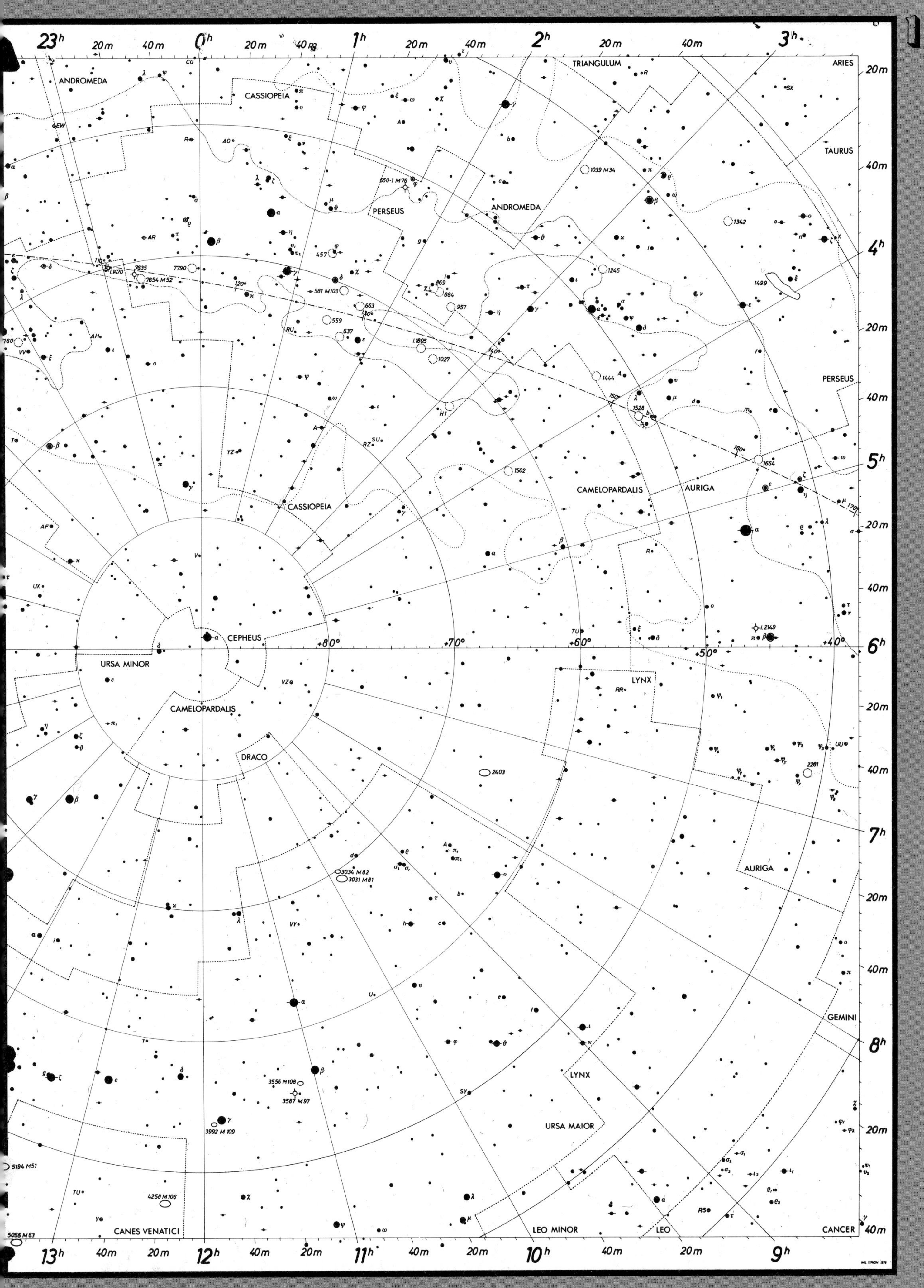
1
23h
0h
1h
2h
3h
ANDROMEDA
CASSIOPEIA
TRIANGULUM
ARIES
TAURUS
PERSEUS
ANDROMEDA
4h
5h
CAMELOPARDALIS
AURIGA
CASSIOPEIA
CEPHEUS
URSA MINOR
CAMELOPARDALIS
DRACO
LYNX
6h
7h
AURIGA
GEMINI
8h
LYNX
URSA MAIOR
CANES VENATICI
LEO MINOR
LEO
CANCER
13h
12h
11h
10h
9h
+80°
+70°
+60°
+50°
+40°
650-1 M76
1039 M34
7654 M52
581 M103
3034 M82
3031 M81
3556 M108
3587 M97
3992 M109
5194 M51
4258 M106
5055 M63

Earth's axis to shift, and has thus changed the zodiacal sign in which the Sun is found at any particular time compared with two thousand and more years ago, although that sign is still used in astrology.

The Moon and planets

It is not possible to include the Moon or the planets on these charts. The Moon moves round the Zodiac once every four weeks (27·231 days), but the motions of the planets are less easy to describe since each has a different period and, combined with the Earth's orbital motion, performs movements which sometimes show apparent standstills and backward (retrograde) motions.

However, Mercury and Venus, being closer to the Sun than the Earth, are always near to the Sun in the sky, appearing as Evening or Morning Stars. On rare occasions they transit across the face of the Sun and some future transit dates are:

Mercury

1986 November 13
1993 November 6
1999 November 15

Venus

2004 June 7
2012 June 5

The other planets may be observed in the night sky when they are not on the sunward side of the heavens, although not all can be seen without optical aid. Mars, Jupiter and Saturn are clearly visible when above the horizon at night, while Uranus may just be picked out on a clear moonless night if one knows precisely where to look, although it will only appear as a very dim star of almost sixth magnitude. Neptune can only be observed with optical aid (magnitude 7·7) and Pluto requires a telescope of at least 300 mm aperture.

To find the positions of the planets for any date, the reader should consult the *Astronomical Almanac* or other annual astronomical data publications such as the *Handbook of the British*

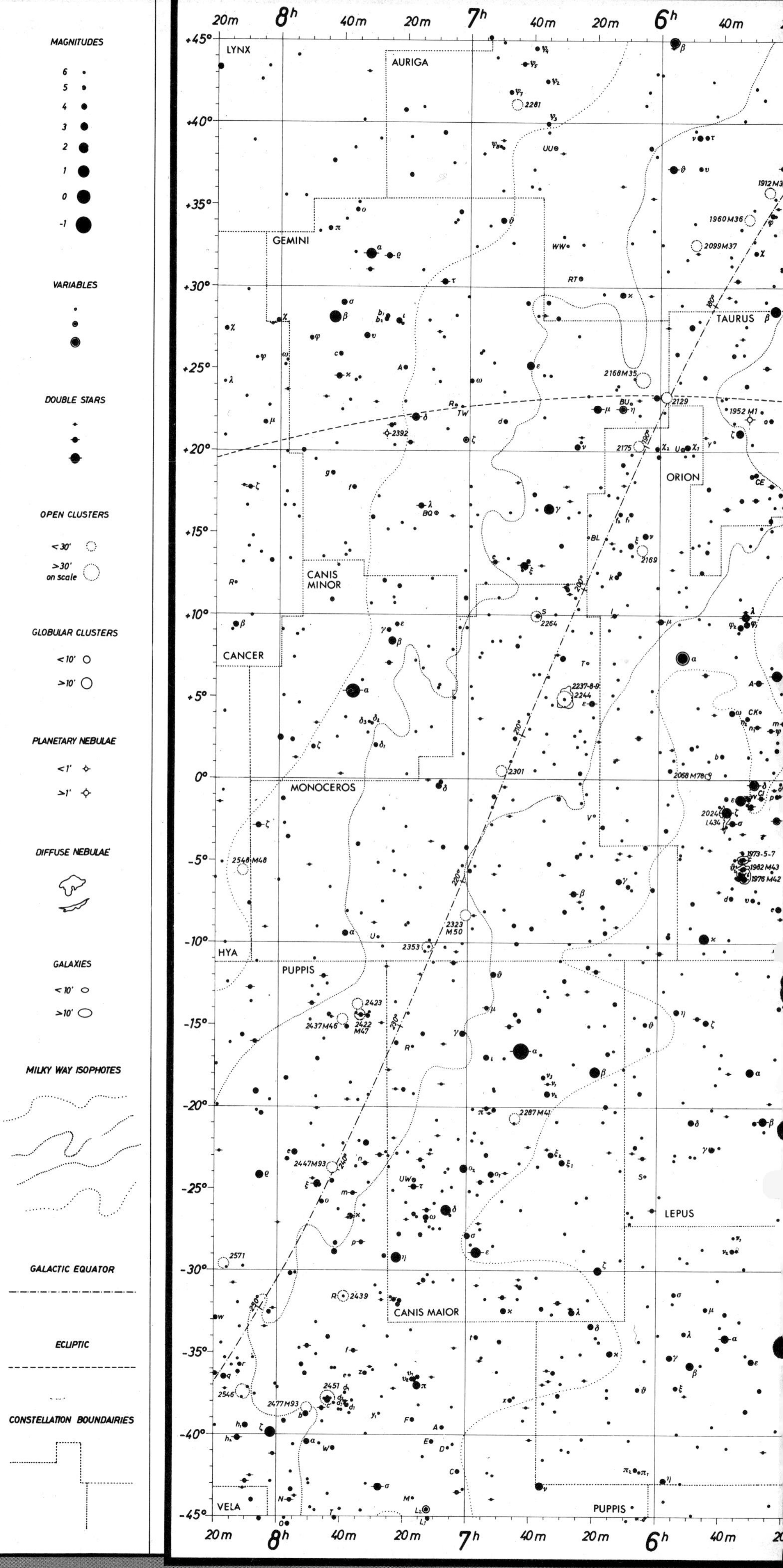

2

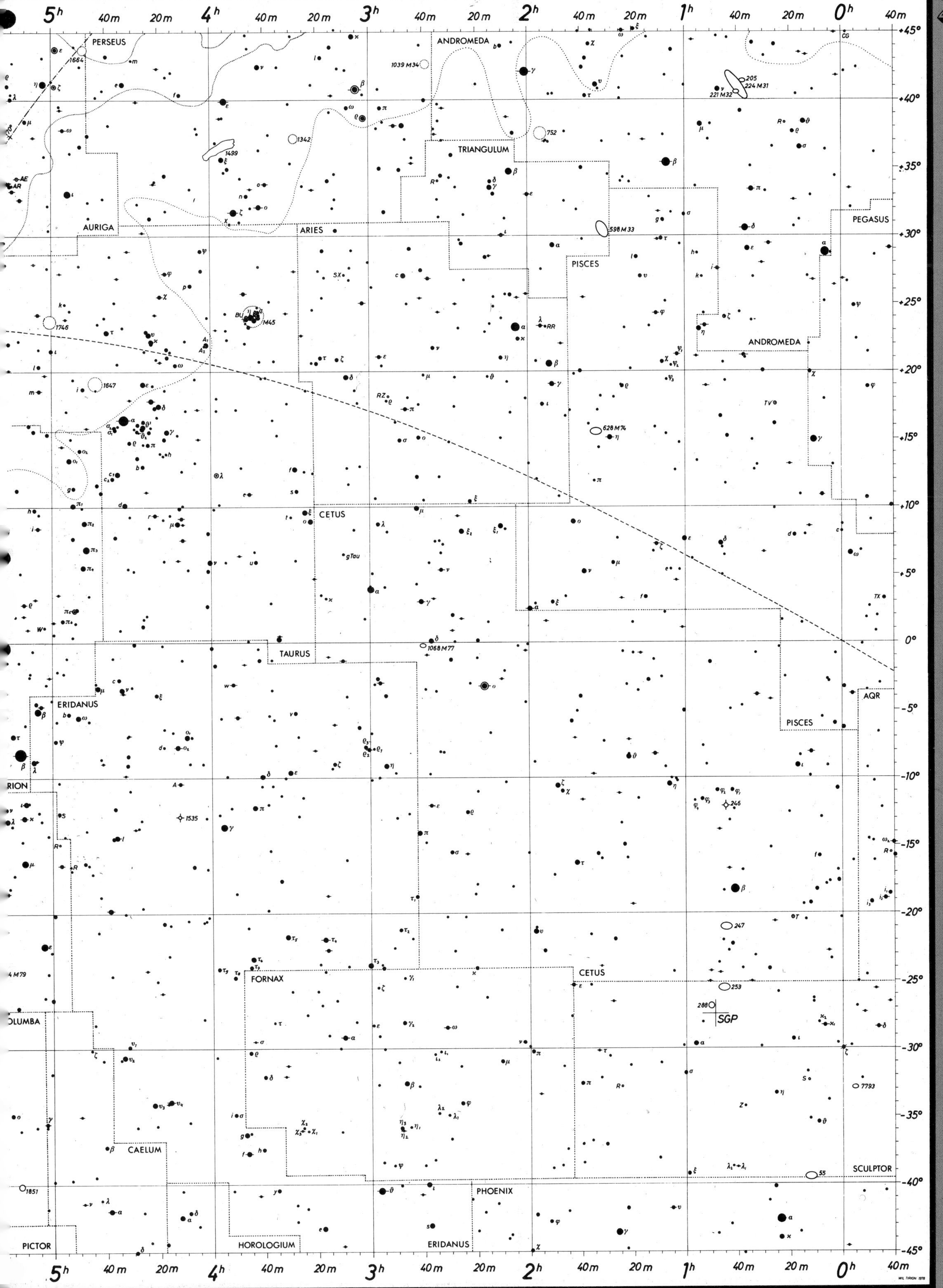

5h 40m 20m 4h 40m 20m 3h 40m 20m 2h 40m 20m 1h 40m 20m 0h 40m
+45° +40° +35° +30° +25° +20° +15° +10° +5° 0° -5° -10° -15° -20° -25° -30° -35° -40° -45°
PERSEUS
ANDROMEDA
1039 M34
205
224 M31
221 M32
1664
1342
1499
752
TRIANGULUM
AURIGA
ARIES
598 M33
PEGASUS
PISCES
1746
M45
ANDROMEDA
1647
628 M74
CETUS
gTau
TAURUS
1068 M77
ERIDANUS
PISCES
AQR
ORION
1535
246
247
M79
FORNAX
CETUS
253
288
SGP
COLUMBA
7793
CAELUM
SCULPTOR
1851
55
PHOENIX
PICTOR
HOROLOGIUM
ERIDANUS
WIL TIRION 1978

Astronomical Association. A few newspapers such as *The Times* of London give a monthly star chart which also details such planets as are visible to the unaided eye.

The constellations

The constellations visible in the night sky vary throughout the year. As a rough guide, the constellations to be seen due south (from the northern hemisphere) or due north (from the southern hemisphere) will be those on the opposite side of the celestial sphere to the Sun. Their right ascension will be the right ascension of the Sun plus 12 hours. The table below lists these right ascensions:

date		*right ascension*
January	5	7h
	18	8h
February	2	9h
	17	10h
March	5	11h
	21	12h
April	6	13h
	23	14h
May	8	15h
	23	16h
June	7	17h
	22	18h
July	6	19h
	21	20h
August	5	21h
	21	22h
September	7	23h
	24	00h
October	10	1h
	29	2h
November	10	3h
	25	4h
December	8	5h
	22	6h

These figures give the right ascensions of constellations which are due south at midnight. Earlier in the evening, constellations westwards (i.e. with smaller values of right ascension) of these will appear due south; after midnight, those to the east (i.e. with larger values of right ascension) will be on the meridian.

The constellation boundaries on the charts are those drawn up by the International Astronomical Union but, of course, the main constellation patterns

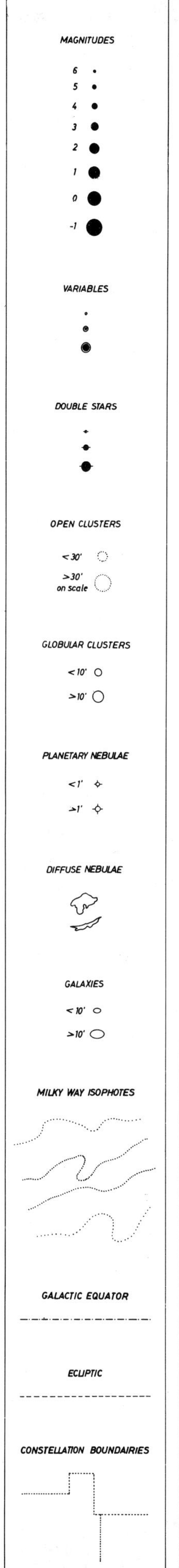

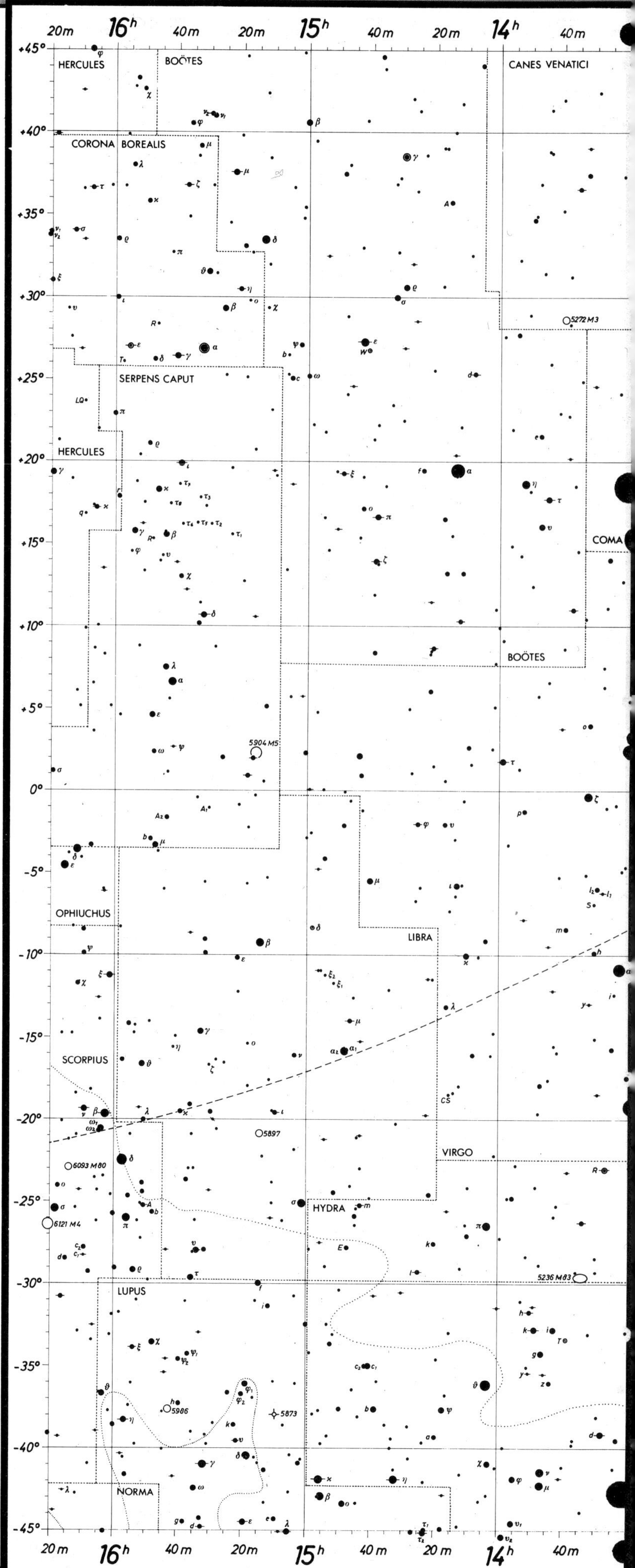

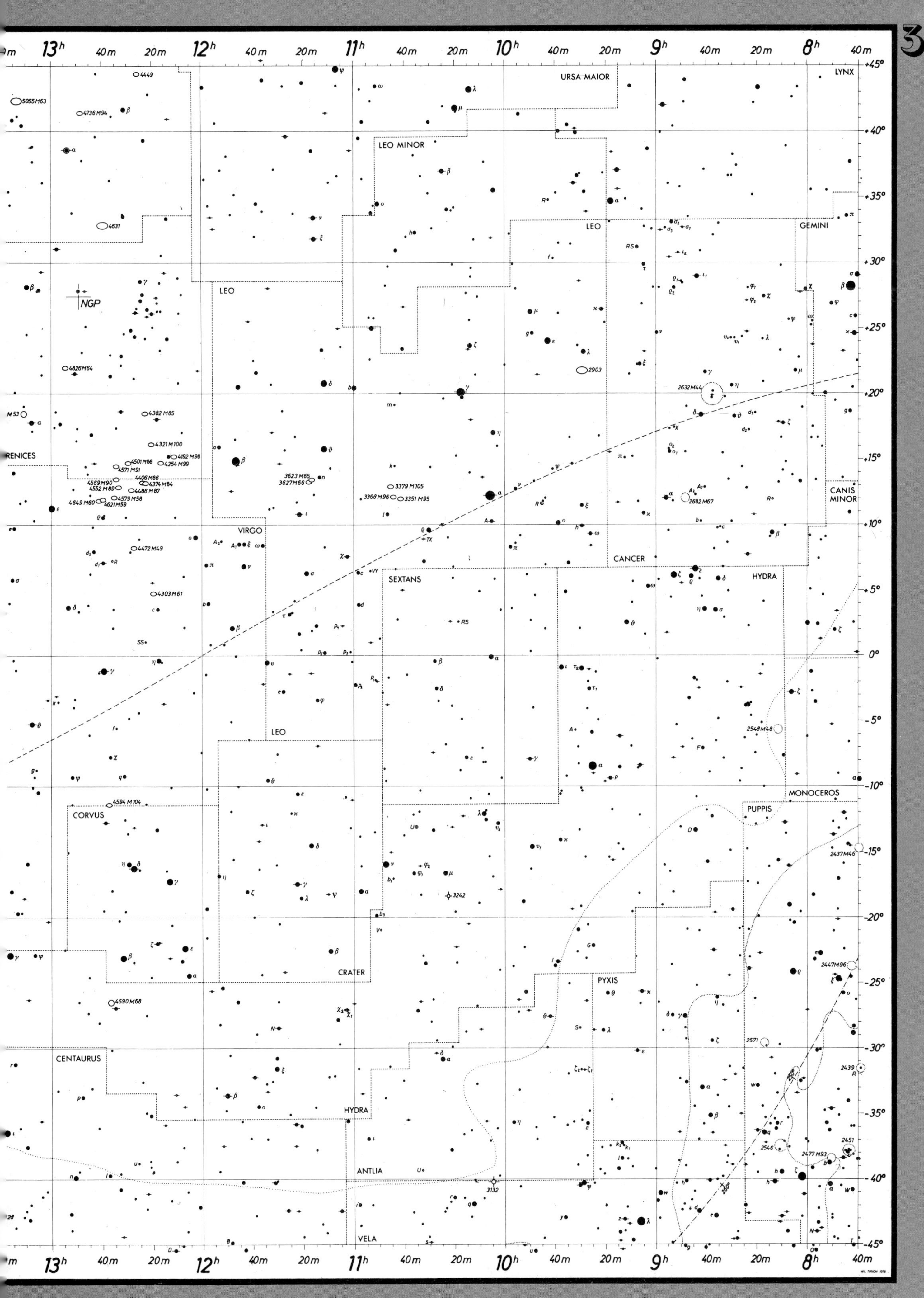
3
URSA MAIOR
LYNX
LEO MINOR
LEO
GEMINI
NGP
CANCER
CANIS MINOR
VIRGO
SEXTANS
HYDRA
MONOCEROS
PUPPIS
CORVUS
CRATER
PYXIS
CENTAURUS
ANTLIA
VELA
13h
12h
11h
10h
9h
8h
+45°
+40°
+35°
+30°
+25°
+20°
+15°
+10°
+5°
0°
-5°
-10°
-15°
-20°
-25°
-30°
-35°
-40°
-45°

we recognise come from considering only the brighter stars. For those unfamiliar with the night sky the best thing is to use an easily recognisable constellation as a starting point, and then use the charts to find others. For the northern hemisphere, the constellations of the Plough (the seven brightest stars of Ursa Major), the 'W' of Cassiopeia, and Orion are the most convenient starting points. For readers in the southern hemisphere, who should turn the charts upside down, convenient guiding constellations are Crux and Centaurus, while the Larger Magellanic Cloud is also a useful indicator.

Using charts and atlases

Once one has become accustomed to them, star charts such as these, where the stars appear black against a white background, are just as easy to use as those which more nearly represent the night sky. Similarly most atlases, even those prepared from photographic plates, show black stars against a white background. Observers are able to mark the objects in which they are interested, or to plot the positions of comets, minor planets and other objects. Indeed the charts reproduced here are available in a larger-sized, modified version which is used by many amateurs for such purposes, and some examples are shown elsewhere in this book. Sometimes, and especially because they are less likely to cause loss of dark-adapted vision when illuminated by a red light at night, charts with black backgrounds are preferred for use 'at the telescope'. Several atlases are available in both versions so that the most suitable ones may be used.

Epochs

Because of the effects of precession (pages 15–16) the positions of all celestial objects as measured by right ascension and declination (page 13) are continually changing, since the coordinates themselves are fixed relative to the Earth. For many accurate scientific observations it is therefore necessary for the

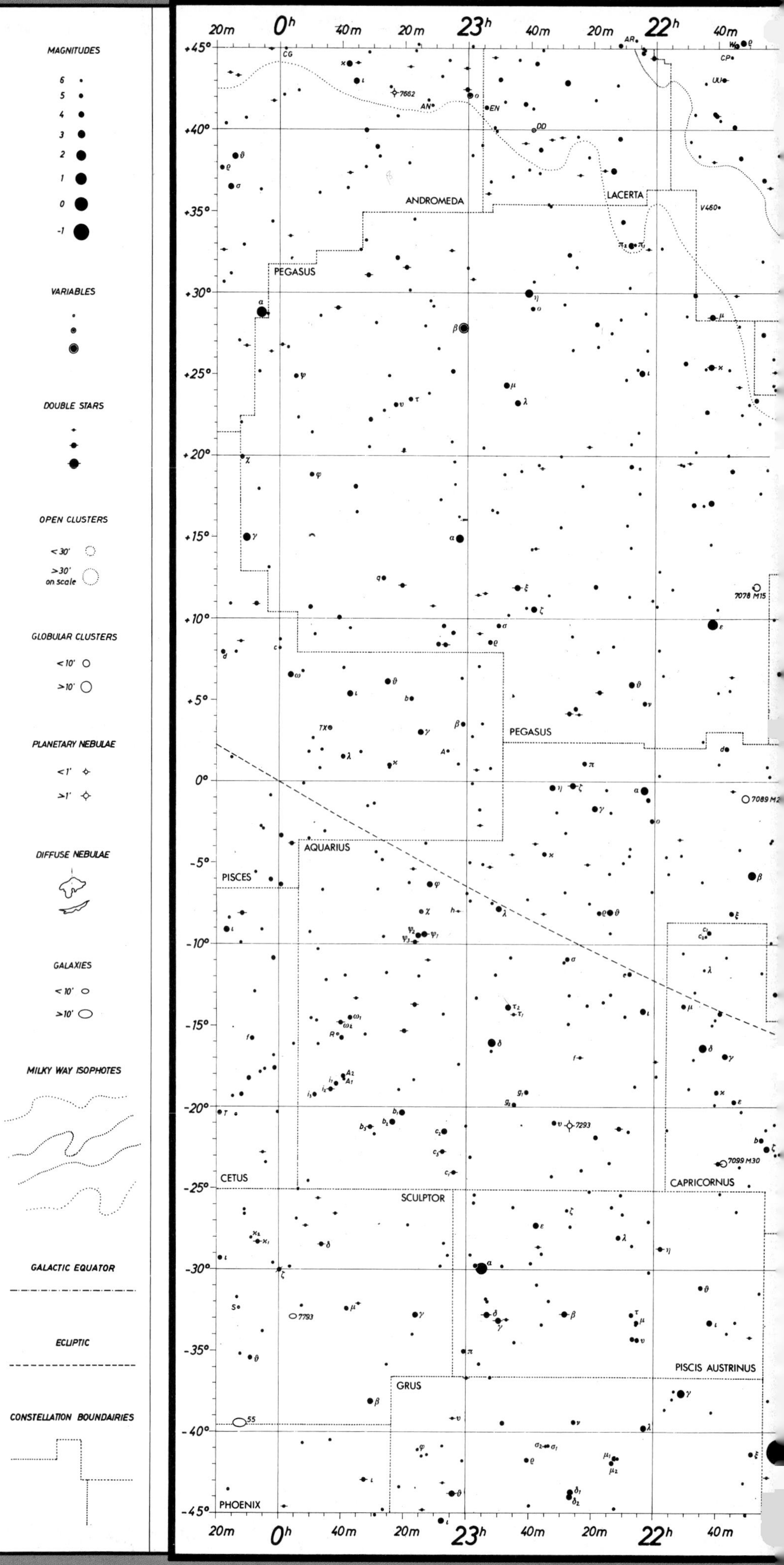

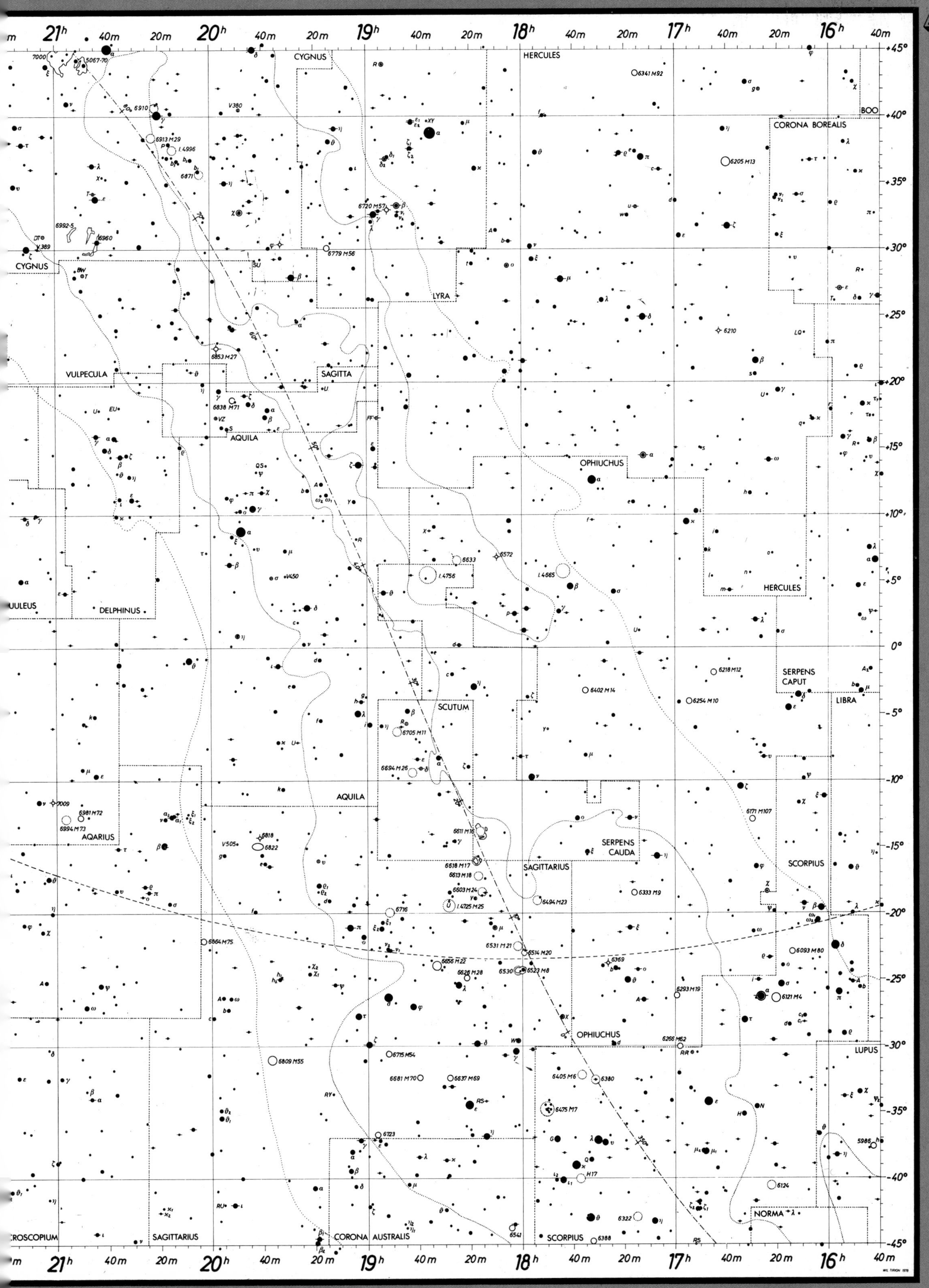
CYGNUS
HERCULES
CORONA BOREALIS
BOO
LYRA
VULPECULA
SAGITTA
AQUILA
OPHIUCHUS
DELPHINUS
HERCULES
SERPENS CAPUT
LIBRA
SCUTUM
AQUILA
AQUARIUS
SERPENS CAUDA
SAGITTARIUS
SCORPIUS
OPHIUCHUS
LUPUS
NORMA
CORONA AUSTRALIS
SCORPIUS
SAGITTARIUS
21h
20h
19h
18h
17h
16h
40m
20m
+45°
+40°
+35°
+30°
+25°
+20°
+15°
+10°
+5°
0°
-5°
-10°
-15°
-20°
-25°
-30°
-35°
-40°
-45°
6913 M29
6853 M27
6838 M71
6720 M57
6779 M56
6341 M92
6205 M13
6218 M12
6254 M10
6402 M14
6705 M11
6694 M26
6611 M16
6618 M17
6613 M18
6603 M24
6531 M21
6514 M20
6523 M8
6656 M22
6626 M28
6981 M72
6994 M73
6864 M75
6809 M55
6715 M54
6681 M70
6637 M69
6333 M9
6494 M23
6171 M107
6093 M80
6121 M4
6293 M19
6266 M62
6405 M6
6475 M7
WIL TIRION 1978

precise position to be calculated for the actual date in question, and many of the large professional catalogues give the appropriate precessional corrections, which are applied to the listed positions for a given epoch. This is a particular instant of time which is used as a reference point. In the case of star charts and atlases where constant redrafting is impossible, and where the changes usually take some considerable time to become apparent, the epochs are usually 50 years apart. These charts are therefore drawn for the epoch of 1950·0, that is 00·00 hours on January 1, 1950, and are similar to those used by most amateurs. It has been decided that from 1984 all calculations and preditions will be based upon the new epoch of 2000·0, and star charts drawn for this epoch are becoming available. As the constellation boundaries are defined, by international agreement, in terms of Right Ascension and Declination, the designated areas of the particular constellations are slowly drifting with respect to the stars. This may mean that at the borders some individual stars with names referring to a particular constellation may now be found in a neighbouring constellation. In practice, however, this does not lead to any real confusion. It should perhaps be emphasized, as it is sometimes misunderstood, that precession does not affect the positions of the stars relative to one another – these only alter due to proper motion (page 12).

Intermediate epochs are frequently used for other calculations, and this sometimes causes confusion. They do not have any special significance, however, being merely convenient reference points in time for calculation purposes. In cometary orbital computations, for example, the quoted epoch is not necessarily related to a particularly important point on the orbit, unlike the time of perihelion.

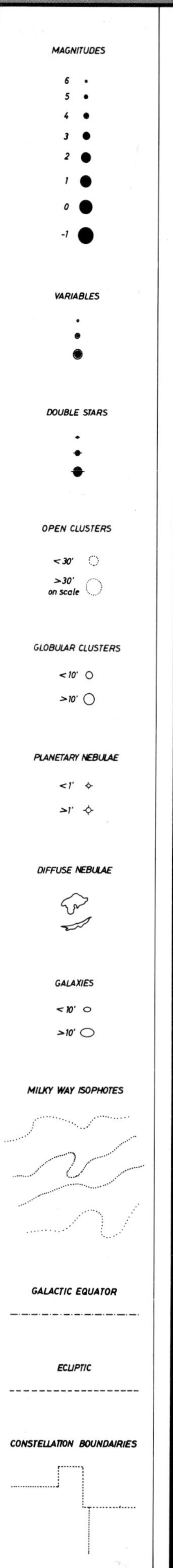

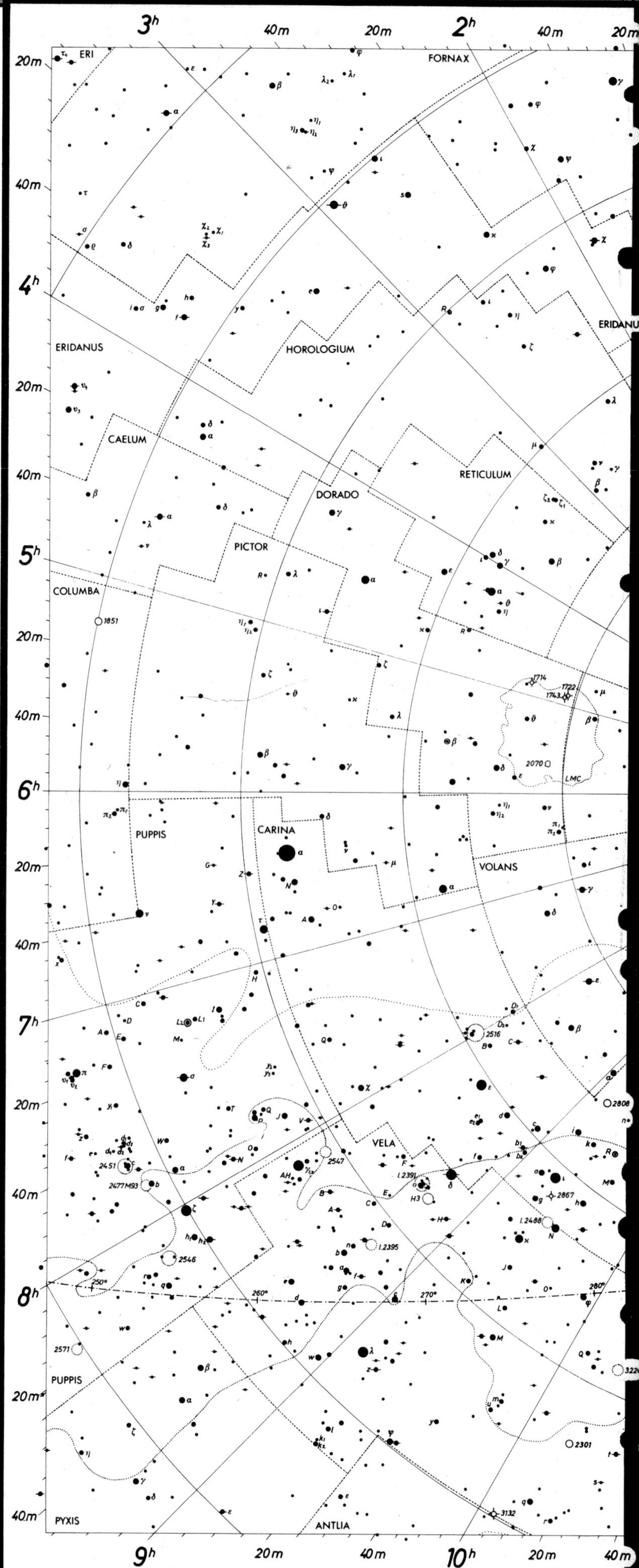

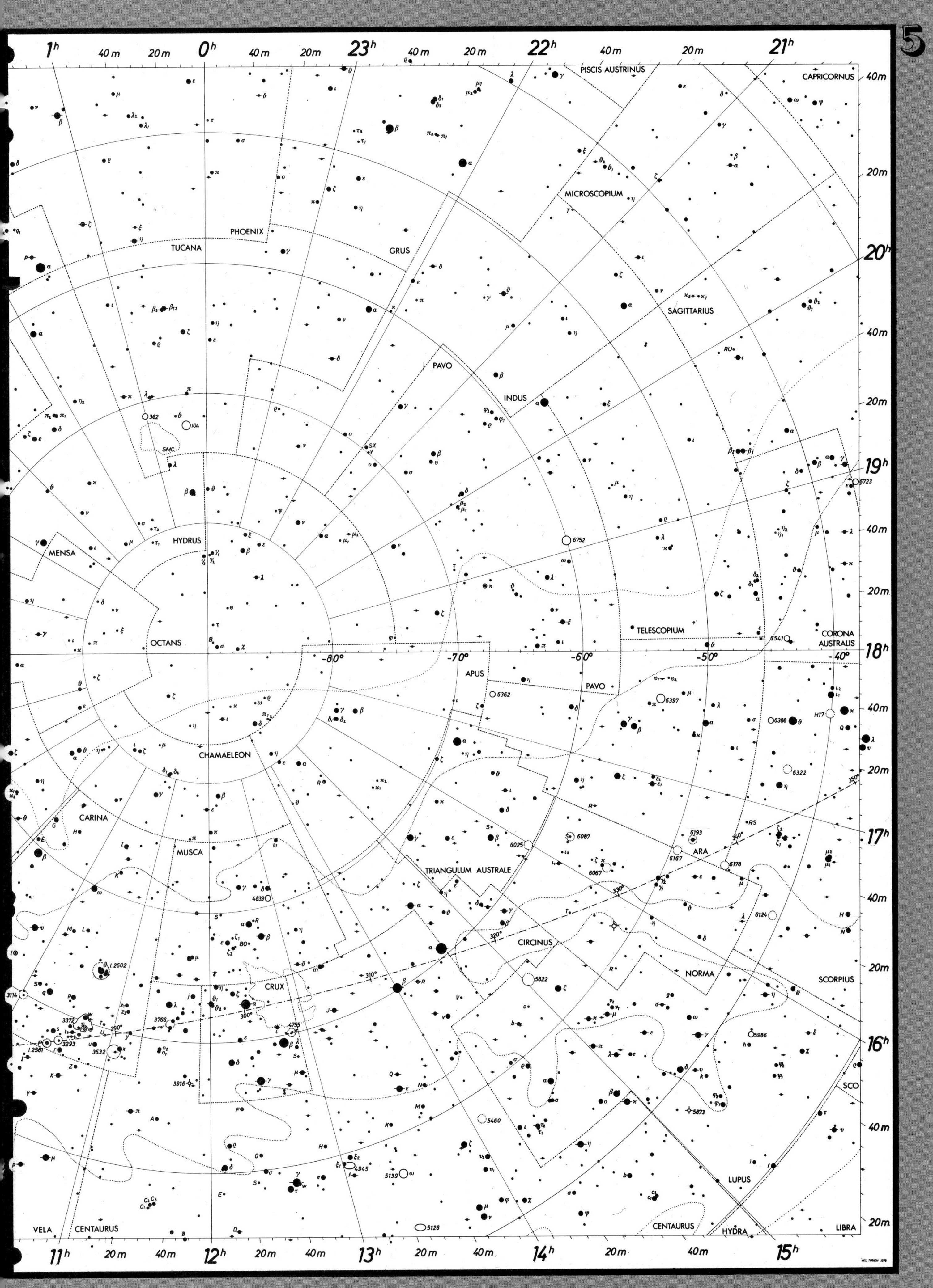

5
1h
40 m
20 m
0h
23h
22h
21h
20h
19h
18h
17h
16h
11h
12h
13h
14h
15h
PISCIS AUSTRINUS
CAPRICORNUS
MICROSCOPIUM
PHOENIX
TUCANA
GRUS
SAGITTARIUS
PAVO
INDUS
SMC
HYDRUS
MENSA
OCTANS
TELESCOPIUM
CORONA AUSTRALIS
APUS
CHAMAELEON
CARINA
MUSCA
ARA
TRIANGULUM AUSTRALE
CIRCINUS
NORMA
SCORPIUS
CRUX
LUPUS
SCO
VELA
CENTAURUS
HYDRA
LIBRA
-80°
-70°
-60°
-50°
-40°

CHAPTER THREE

The Stars

Armed with a knowledge of the climatic record of the Earth for the past 4000 million years and the current power output of the Sun, the mechanisms of its internal energy generation processes have been deduced. These nuclear furnaces have provided the light and heat which have supported life on Earth for this considerable period of time. A comforting thought, but what of the future? How long will the Sun radiate its life-preserving energy, will ice ages reappear or, even worse, will the Sun gradually cool until the entire Earth is cocooned in a mantle of ice; will all life expire?

To answer these chilling questions astronomers must have more information. They need to study not only the Sun but many stars and from their observations hope to detect patterns which will indicate trends in the life cycles of stars. Knowing these, astronomers can compose pictures of how stars are born, flourish in youth, then grow old and die. We will learn that not all stars expire by a slow fading decay. A few are truly spectacular; their lives are consumed in a brief fiery span, and their brilliant hues of blue and white are shining beacons in the Galaxy. When their fuel is exhausted they explode with cataclysmic violence, the light from which is brighter than that from 1 000 million Suns.

How may astronomers study other stars? They are so distant that even the largest optical telescopes can not reveal their shapes. They still appear as twinkling points of light, but even the naked eye can see that these twinkling points are not identical; they differ in brightness. This is a combination of different intrinsic brightnesses and the varying distances of stars from the Earth. Another difference apparent to the naked eye is the colours of stars. The stars Aldebaran (α Tauri) and Betelgeuse (α Orionis) have orange/red hues and these may be contrasted with the blue and white of Rigel (β Orionis) and Sirius (α Canis Majoris). Intermediate are the yellow stars such as Capella (α Aurigae) and our Sun. Experience tells us that when objects are heated, they first glow dull red, then bright red and then white, so perhaps the colours signify stars of differing temperature. Further investigation leads us into the realm of **astrophysics**, whereby astronomical observations together with knowledge of the laws of physics enable the secrets of stars to be understood.

The manner by which astronomers classify the brightness of stars as judged by the naked eye was discussed in Chapter 1. Let us now expand this discussion in a quantitative form. Firstly, we wish to classify the apparent brightness of stars, known as their **apparent magnitudes**, m. It is convenient and also necessary for rigorous classification and study to relate magnitude determinations to a standard. This removes differences between the observers' eyes and equipment. Also, because the human eye, photographic plates and photoelectric devices have responses which vary differently with wavelength, magnitudes determined by those techniques are referred to as visual, photographic and photoelectric and you will often find them expressed as m_{vis}, m_{pg}, m_{pe}. To obtain more information concerning the spectral distribution of the light from astronomical sources, certain specific wavelengths of study have been selected by the use of colour filters. The most common of these are called U, B and V, which respectively transmit only ultraviolet, blue and visible light. Apparent magnitudes measured with these filters are designated by m_U, m_B, m_V or more usually just U, B and V. The UBV system of the 1950s, still in use today, has filters with the following characteristics:

apparent magnitude	**wavelength of peak transmission**
name (colour)	**(nm)**
U (ultraviolet)	360
B (blue)	420
V (visible)	540

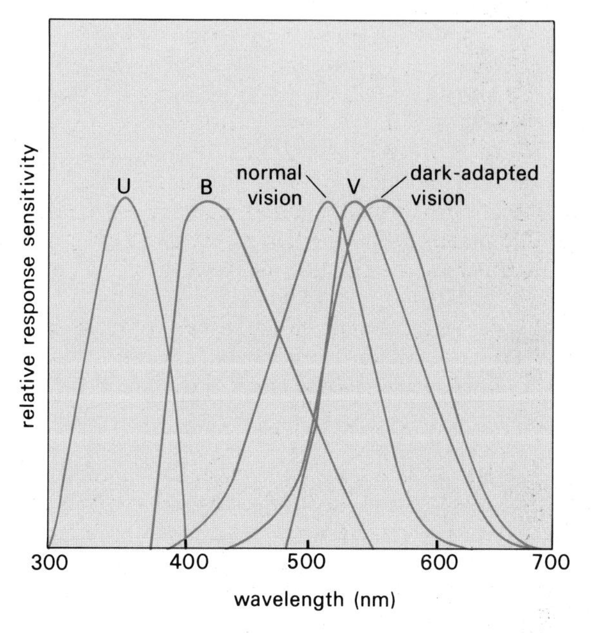

Fig. 3·1
Response curves for the filters used in the UBV system of astronomical photometry. These curves show how the transmission of the filters varies with wavelength. The spectral response of the normal and dark-adapted human eye are also shown for comparison.

Figure 3·1 shows the relative transmission of the UBV system.

Further standard photometry magnitudes are R (red), I (infrared) and the more recent infrared magnitudes of J, K, L, M and N. The latter five are selected to have peak transmission in the narrow windows where infrared radiation is least absorbed by the atmosphere.

By observing in many colours, a useful parameter known as the **colour index** may be obtained. This is defined as the difference between magnitudes measured at two different colours. With the UBV system, two very widely used colour indices are (U-B) and (B-V). These are closely related to a star's luminosity and temperature, both of which are intrinsic parameters astronomers wish to know to learn more about stellar evolution.

Nuclear reactions deep in the interior of a star provide energy which is transported by a very slow process to the outer layers. This energy is eventually radiated into space and the rate at which it is emitted is known as the **luminosity**, L, of the star. This depends mainly on the size and surface temperature of the star, increasing in proportion to the radius, R, of the star squared and to the fourth power of the temperature, T. We can write this mathematically as

$$L \propto R^2T^4.$$

Luminosity is the intrinsic property which we would like to determine, but how bright the star appears to us obviously depends on its distance. This is usually unknown and so we begin by classifying the stars according to apparent brightness. The apparent brightness is the extent to which the power output of the star has been diluted by the distance the light has travelled on its journey to the Earth.

To compare the luminosities of stars we need to be able to determine their distances. Usually, this is impossible to do by a direct means, as will be discussed later. However, accepting this difficulty, it has been possible to obtain distances and thus luminosities. Another term which is often used for luminosity is **absolute magnitude**. This is defined to be equal to the apparent magnitude if the star in question is viewed from a distance of 10 pc. This then classifies the intrinsic brightness of stars on the same scale. (Ten pc is chosen because it fits in conveniently with the way stellar distances are measured.) Absolute magnitudes are written with a capital M, any subscript having the same meaning as before – for example, M_v is the absolute magnitude through a V filter. Absolute magnitudes range from −9 for the most luminous down to +15 for dark dwarf stars. As with apparent magnitudes, an increasing negative number indicates a more luminous star.

A most important relationship may be obtained from the apparent and absolute magnitudes. This is called the **distance modulus**. It is given by

$$m - M = 5 \log d - 5;$$

(where 'log' means 'the logarithm of'). Obviously, if the distance d is known, then this equation enables the absolute magnitude, M, to be deduced from a measurement of the apparent magnitude, m. It was from the use of the direct distance-determining technique of trigonometric parallax that the absolute magnitudes of nearby stars were obtained, setting the basis for the most important diagram in astronomy, the Hertzsprung-Russell (H-R) diagram holding the innermost secrets of stellar evolution.

For the millions of other stars whose distances can not be determined by direct studies because they are too great, this equation is the key. It will be seen later that the temperature of a star may be deduced from its spectrum; the absolute magnitude can also be estimated and by application of the equation the distance may be found from the measured value of the apparent magnitude. The distance modulus determination is a crucial means of finding the distances to other galaxies.

A word of caution; in the above determination of apparent magnitude no account has been taken of the presence of absorbing matter in the space between the stars. Astronomers now know that interstellar dust abounds in space and this dims the light from distant stars, making the star appear fainter than it really is. By measuring the interstellar reddening of the stars' spectra a correction factor may be applied. The effects of dust in the Galaxy and reddening are described on page 184. We shall now investigate the stars in the sky and learn of their differing and often exotic features as they pour forth energy to brighten our night sky.

Stars as radiating bodies

By studying the radiation from stars, astrophysicists have been able to deduce such properties as stellar temperatures, sizes and compositions. They do this by a process called SPECTROSCOPY, the detailed wavelength analysis of radiation; in our case the light emanating from the star. This light energy was originally supplied by nuclear reactions deep in its core. It very gradually filters outwards through the gas of which the star is composed. Eventually, the light energy escapes from the surface region with a wavelength pattern or **spectrum** (Fig. 3·2) which is governed by the temperature, density and chemical composition of the surface layer.

Typical stellar spectra are called absorption-line spectra, that is, they consist of a continuous background of light crossed by dark absorption lines in particular wavelengths. In 1802, William Wollaston, an English chemist, used a prism to split up the light from the Sun into its respective colours. He found that the bright continuous spectrum was crossed by a few dark lines. Joseph Fraunhofer, in 1814, made more detailed measurements and found that the solar spectrum was literally full of dark lines, totalling over 600. It was left to Robert Bunsen and Gustav Kirchhoff to make a brilliant analysis using the spectroscope – a device invented for detailed analysis of a spectrum by using a prism to split up the light, the resulting colours being viewed through a telescope. By a series of elegant experiments, they found that chemical elements when burned did not give a continuous bright spectrum like the Sun, but a series of separate bright lines. They also showed that different elements emitted different patterns of bright

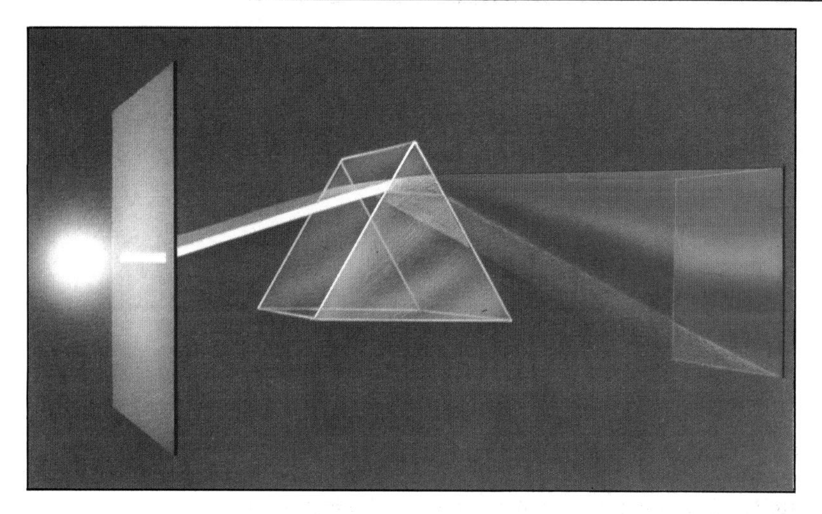

Fig. 3·2
When white light is passed through a prism it is split into many colours known as a spectrum. This rainbow-like effect was first studied in 1665 and 1666 by Sir Isaac Newton. Light of different wavelengths is deviated by varying amounts as it passes through the glass prism, red light being deviated least. This simple experiment shows how white light is a mixture of all colours between red and violet.

lines. These patterns were the fingerprints of the chemical elements, each set uniquely defining a certain chemical element. One such example was sodium, whose prominent lines lie in the yellow part of the spectrum giving a sodium lamp its characteristic yellow glow. Bunsen and Kirchhoff realized that bright lines represented the emission of energy. Therefore, if an emitting system is losing energy and yet remains bright, there must be a continuous supply of energy, in their case the hot Bunsen flame. It is now known that this emission of light is caused by atoms of the gas being raised to what is referred to as an 'excited state' by being supplied with energy. The atoms return to their ground state and emit light in the process. The amount of energy input to the gas and so the degree of excitation of the atoms determines the number of bright lines emitted in a pattern. Hence, a bright line emission spectrum reveals not only the chemical elements present in the gas but also the degree of excitation of the atoms. This must have some bearing on the local energy environment of the atoms, that is, the temperature.

Kirchhoff, puzzled by the difference between the solar spectrum and the spectra of hot gases, performed further experiments. He found that if he heated a solid, it gave off a bright continuous spectrum. If he then placed a sodium flame between the incandescent solid and his spectroscope he saw a continuous spectrum crossed with dark lines, whose positions in the spectrum were identical to where the bright lines of a sodium flame would appear. This crucial experiment marked the beginning of astrophysics. Astronomers could now scrutinize stellar spectra and thereby determine the chemical composition of stars. The science of astronomical spectroscopy was born. Kirchhoff realized that the dark lines are absorption lines caused by the light from a continuous spectrum passing through a gas at a lower (though perhaps still very high) temperature. The light is now absorbed by the atoms of the gas, the patterns of the lines and number of lines in the pattern revealing the chemical composition and degree of excitation (temperature) of the atoms present. In 1864, using the new technique of stellar spectroscopy, nine elements were identified in the star Aldebaran (α Tauri), and progress was swift.

Spectroscopy also brought a new dimension to distance measurement. Due to the work of Christian Doppler and Hippolyte Fizeau it was found that the

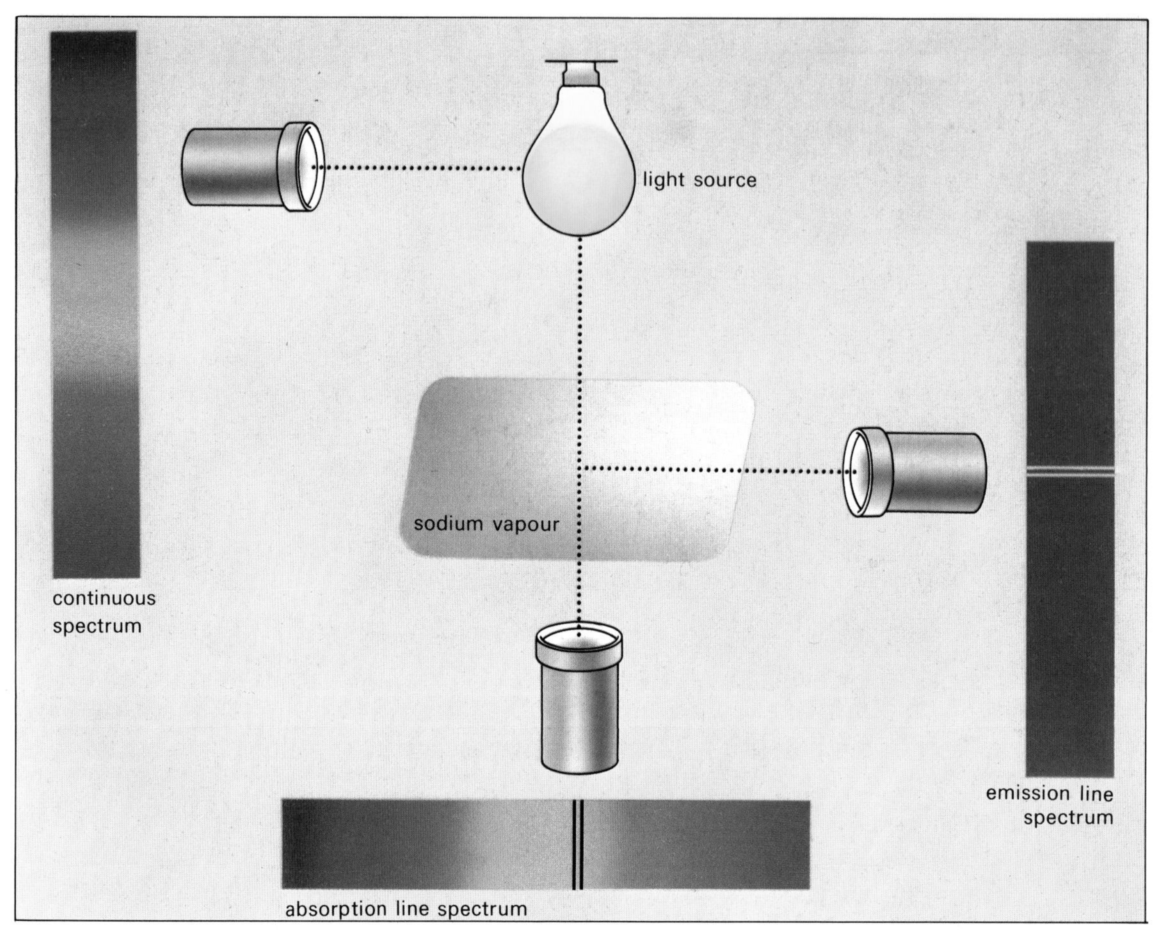

A demonstration of Kirchhoff's Laws of spectral analysis, whereby the continuous spectrum from the hot source has superimposed dark absorption lines of the cooler sodium vapour, which itself emits a bright line emission spectrum characteristic of the elements of the vapour. The absorption lines and emission lines occur at identical wavelengths.

position or wavelength of lines could alter. If a star or other radiating source was moving towards the observer he saw a shift of lines towards the blue end of the spectrum, and if it were moving away, a shift to the red or *redshift*. The amount of this shift was a measure of the **radial velocity** or speed of approach or recession. Moreover, as stars never show any change of size due to such radial motion, even in a telescope, this *Doppler shift* has proved to be the only way to detect such motion (Mathematically, if λ_0 is the normal (rest) wavelength, and $\delta\lambda$ a small change due to radial motion, then if c is the velocity of light and v the velocity of the source, $v = c\delta\lambda/\lambda_0$.)

Temperature

As science advanced, especially in the realms of atomic physics, more was understood about the interpretation of stellar spectra. It was known from studies of the Sun that the continuous part of the spectrum came from an opaque region of the stellar atmosphere known as the surface layer. This layer is not solid but gaseous and as the height in the atmosphere increases the temperature falls. The absorption lines come from those cooler regions and so one can obtain a surface temperature or an atmospheric temperature. In order to compare different stars a common term having a precise definition must be employed. This is the **effective temperature** (T_{eff}) and is defined to be the temperature of a black body whose total luminosity is the same as that of the star in question. A black body, an object which is both a perfect emitter and absorber of radiation, is a familiar concept to physicists. The effective temperature is therefore a physically meaningful parameter if it can be accurately measured. This is particularly important for the hottest stars which emit the bulk of their power at ultraviolet wavelengths, which cannot penetrate the Earth's atmosphere. However, for the majority of stars T_{eff} can be well determined and lies on the KELVIN TEMPERATURE SCALE between 3 000 K and 40 000 K.

At these high temperatures many atoms will be stripped of one or more of their ELECTRONS (**ionized**), the degree of ionization depending on the temperature and the density in the stellar atmosphere. At higher temperatures greater ionization will occur but as the density increases it becomes easier for an ionized atom to recapture an electron. Therefore, as stellar atmospheres become more dense ionization is inhibited. The degree of ionization in a stellar atmosphere is a key parameter in the appearance of a spectrum.

Because there is a large range of effective temperatures for stars and as this determines the degree of ionization of the elements and thus the appearance of the spectrum, we should expect to find a wide variety of stellar spectra. This is, in fact, the case and the range is even larger when one considers stars whose chemical composition is very different from the majority of stars known as normal stars. The Sun is an example of a normal star.

To bring order to the multitude of observed spectra, a classification scheme was devised, principally by Harvard College Observatory during the early years of this century. This culminated in the

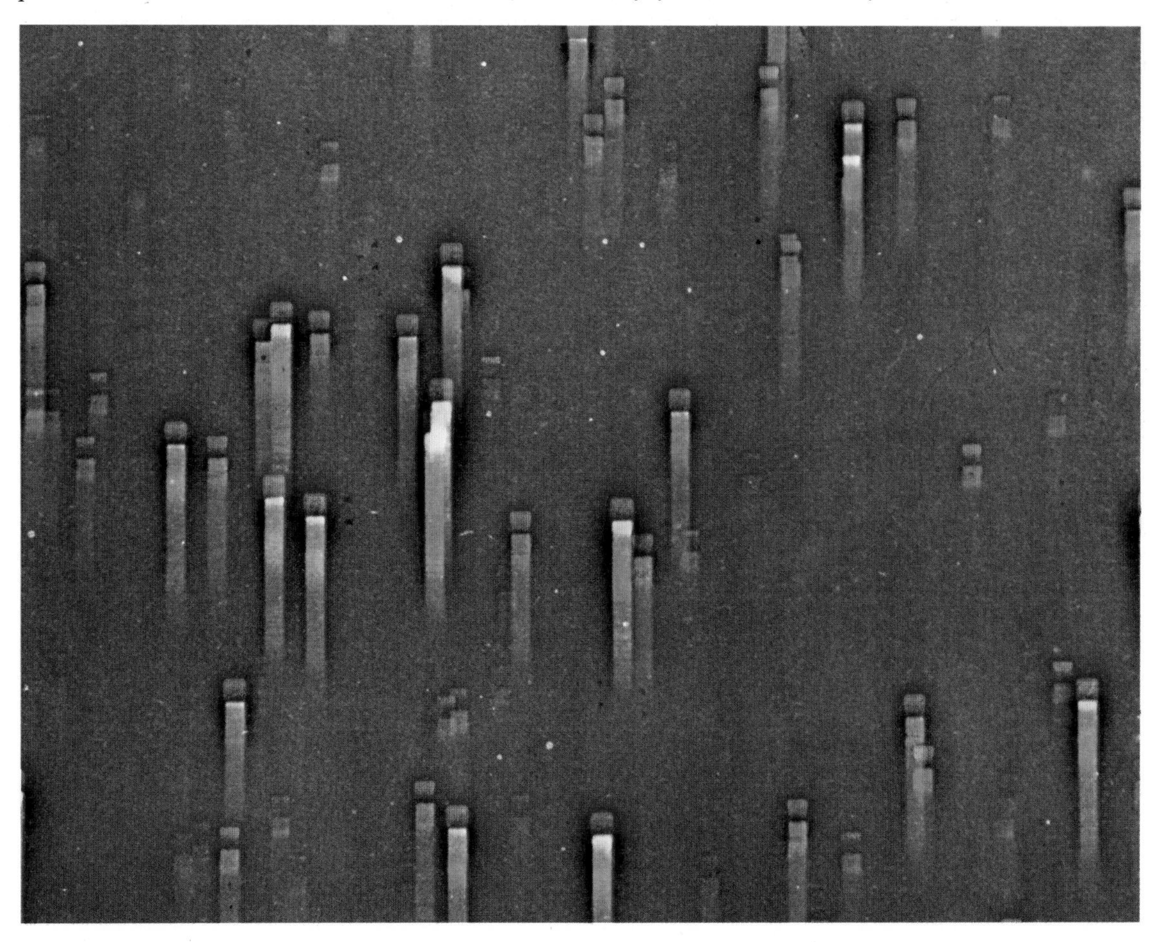

A low dispersion spectrum of the Hyades star cluster obtained with a device known as an objective prism: a large prism placed at the front aperture of the telescope thus enabling the spectrum of each star to be obtained simultaneously.

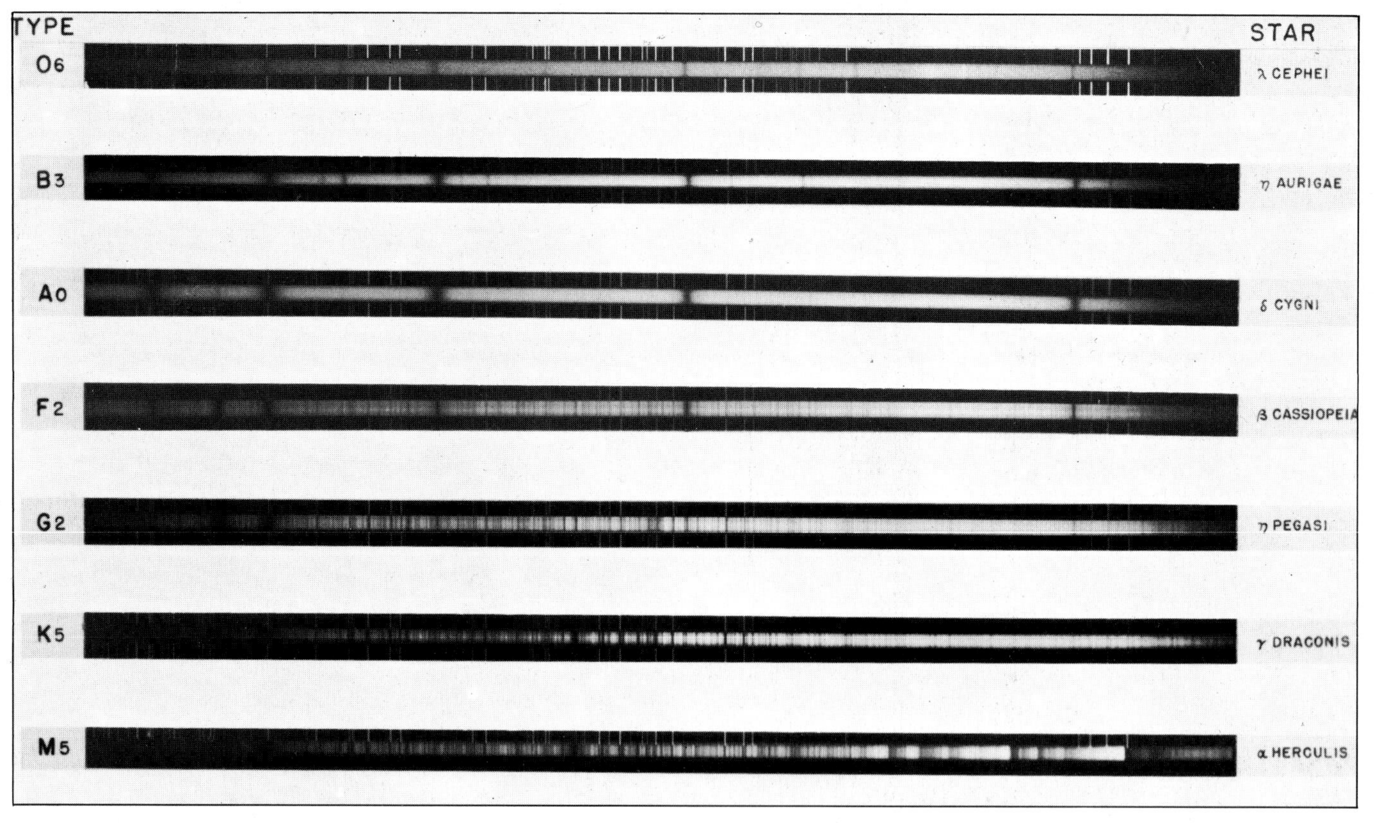

Examples of stellar spectra which illustrate the main features which typify the spectral classification from class O to M.

production of the Henry Draper catalogue containing nearly a quarter of a million stellar spectra. The stars were originally classified on the basis of the strength of the hydrogen absorption lines, beginning at class A and continuing alphabetically. With the discovery of some stars having emission lines class O was added to precede A and finally, as our knowledge of stellar spectroscopy improved, the entire sequence was re-classified. There are now seven main classes. They are from hottest to coolest O, B, A, F, G, K, M. A well-known mnemonic ensures that this sequence, once learned, is never forgotten. This is 'O Be A Fine Girl, Kiss Me'. Each class is further divided into ten sub-classes ranging from zero to 9 in decreasing temperature, except O which goes from 5 to 9·5. Additional cool classes of R, N (carbon stars) and S (zirconium oxide stars) have now been added, named because of the presence of these elements in their stellar atmospheres.

How are the actual spectra classified? This depends on the absence or presence of certain spectral lines along with the ratios of intensities of particular lines. It should be remembered at this point that the absence of lines from a particular element does not necessarily mean it is not present in the stellar atmosphere. It could be that conditions are not suitable for its emission at wavelengths which may be observed from the surface of the Earth. The classification scheme listed below was devised by William Morgan and Philip Keenan.

Astronomers like to have their little differences from other scientists, especially in the labelling of ionized atoms. Physicists label ionized atoms by indices of n+ where n denotes the number of electrons lost by the atoms, for example, C^{2+} indicates a carbon atom which has lost two electrons and is thus left with a positive charge of 2 units. Astronomers use roman numerals to describe the state of the atom with the important difference that the neutral atom, one that has neither lost nor gained electrons, is designated I – for example, C I is neutral carbon (C or C^0 to the physicist). For the above example C^{2+} becomes to the astronomer CIII. An example of the spectra of each class is shown above.

O 'Ionized helium stars'; lines of ionized helium (He II) appear in the spectra. They are very rare in the Galaxy and possess $T_{eff} \sim 40\ 000$ K. An example is Alnitak (ζ Orionis) at O9·5.

B In general, neutral helium stars. He II disappears after B5. Singly ionized oxygen, nitrogen, and so on, replace the more highly ionized forms; the intensity of these decreases rapidly from B5 as the lines of neutral hydrogen (HI) strengthen. $T_{eff} \sim 16\ 000$ K and examples are Bellatrix (γ Orionis) at B2 and Rigel (β Orionis) at B8.

A Hydrogen dominated spectra; neutral helium is replaced by very strong lines of HI which attain their maximum strength about A2. Ionized metal lines of calcium (Ca II), iron (Fe II), chromium (Cr II) and titanium (Ti II). Neutral metal lines increase in strength through the class. $T_{eff} \sim 8\ 500$ K and examples are Vega (α Lyrae) at A0, Sirius (α Canis Majoris) at A1 and Altair (α Aquilae) at A7.

F 'Ionized calcium stars' because these lines become very intense as the class progresses. Neutral hydrogen lines continue to fade but are still strong. Fine and very numerous lines of neutral and singly ionized metals proliferate in the spectra. $T_{eff} \sim 6\ 500$ K and examples are Canopus (α Carinae) at F0 and Procyon (α Canis Minoris) at F5.

G 'Solar type stars'; spectrum dominated by lines of Ca II and neutral metals. Iron is very abundant. H I continues to fade and Ca I appears as class progresses when the molecular bands of CN and

CH begin to appear. $T_{eff} \sim$ 5 500 K and examples are Rigel Kent (α Centauri) at G2, the Sun at G2 and Capella (α Aurigae) at G8.

K Mettallic lines of neutral metals (especially iron) continue to dominate and increase in intensity. Appearance of molecular bands of TiO and K5. $T_{eff} \sim$ 4 000 K and examples are Arcturus (α Boötis) at K2 and Aldebaran (α Tauri) at K5.

M 'Titanium oxide stars'. Numerous bands of TiO and other molecular species and neutral metals are prominent, especially Ca and Fe. $T_{eff} \sim$ 3 000 K and examples are Antares (α Scorpii) at MI, Betelgeuse (α Orionis) at M2 and Barnard's Star at M5.

The continuous radiation from stars has a wavelength distribution very similar to that of a black body emitting at the effective temperature of the star. As we have seen for spherical black bodies the power output is given by

$$L \propto R^2 T^4.$$

Thus, hotter bodies always emit more radiation than do cooler bodies of the same size. (A black body twice as hot emits sixteen times more power.)

This shows that as the black body temperature increases, the maximum of the radiated power appears at shorter wavelengths. A black body of temperature 300 K (a dark room) has a spectrum which has maximum intensity at a wavelength of 10 μm (infrared). A star of temperature 3 000 K has peak emission at 1 μm = 1 000nm (far red) and for a very hot star of 30 000 K this peaks well into the ultraviolet at 100nm.

We can now see that this range of temperature can explain the colours of the stars in the sky: very hot stars appear bluish; hot stars emit nearly equal amounts of radiation at the wavelengths at which the human eye responds and we see this mixture as colourless or white; cool stars, on the other hand, have a very red appearance. The image of the Hyades taken with an objective prism spectrograph is shown on page 47. The spectrum of each star may be examined and its spectral type determined. This technique is very useful as a single picture simultaneously records the spectra of many stars. Also, by making colour photometric measurements of stars it is found that the colour indices (B-V) and (U-B) are well correlated with temperature and spectral class for normal (dwarf type) stars.

Stellar luminosity varies with temperature but it is also proportional to the surface area of the star (that is, to the radius squared). For a fixed temperature, if the star is doubled in radius the luminosity increases by a factor of four. So we are still unable to deduce the size of stars. However, for a few hundred stars which are sufficiently close for their distances to be well determined we can, using the distance modulus (and correcting for any interstellar absorption effects), calculate the absolute magnitude, M, to an accuracy of at least 0·5 magnitudes. We can, therefore, construct graphs of absolute magnitude versus temperature. Absolute magnitude is synonymous with luminosity and so the radii of these stars may be determined. A plot of this type was first published by Ejnar Hertzsprung and Henry Norris Russell in 1913 and the diagram now bears their name, abbreviated to **H–R diagram**.

An H-R diagram is shown in Fig 3·3. It is immediately apparent that stars are not scattered at random across the diagram but fall into well defined zones. The most prominent of these in called the **main sequence**. Here the increase in luminosity is a steadily increasing value or **function** of both temperature and size. From our knowledge of luminosity it is obvious that stars which are intrinsically faint but have high temperatures must be very small in size. These are referred to as **white dwarfs**. On the other hand, stars which are cool but very luminous must therefore be very large and are known as **red giants**.

It was suggested in an earlier section that if the absolute magnitude of a star could be deduced then its distance could be determined. We now see such a method exists because if we can accurately determine the spectral class, then the H-R diagram will reveal the absolute magnitude. The distance is then found. It sounds simple but there is a snag. If one refers to Fig. 3·3 for, say, a KO star, what value of the absolute magnitude do we choose, +6, 0 or −6? It is not sufficient just to determine the spectral class – we need to know the type of star. It is big or small? Spectroscopy once again comes to the rescue because the densities in the atmospheres of giant and dwarf stars vary enormously. This density difference causes subtle but noticeable changes in the spectra. An experienced observer can not only tell the spectral class of a star but also if it is a giant or dwarf. The inclusion of the sizes of stars into a classification scheme fell to the Yerkes observers Morgan and Keenan. They developed what is referred to as the

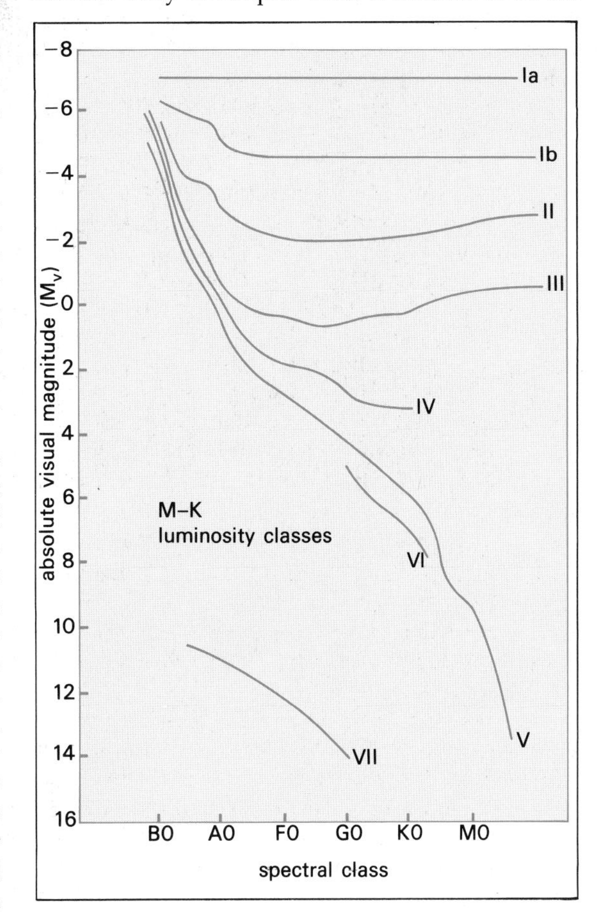

Fig. 3·4
The Morgan-Keenan (Yerkes) scheme whereby stars are classified according to their luminosities and temperatures.

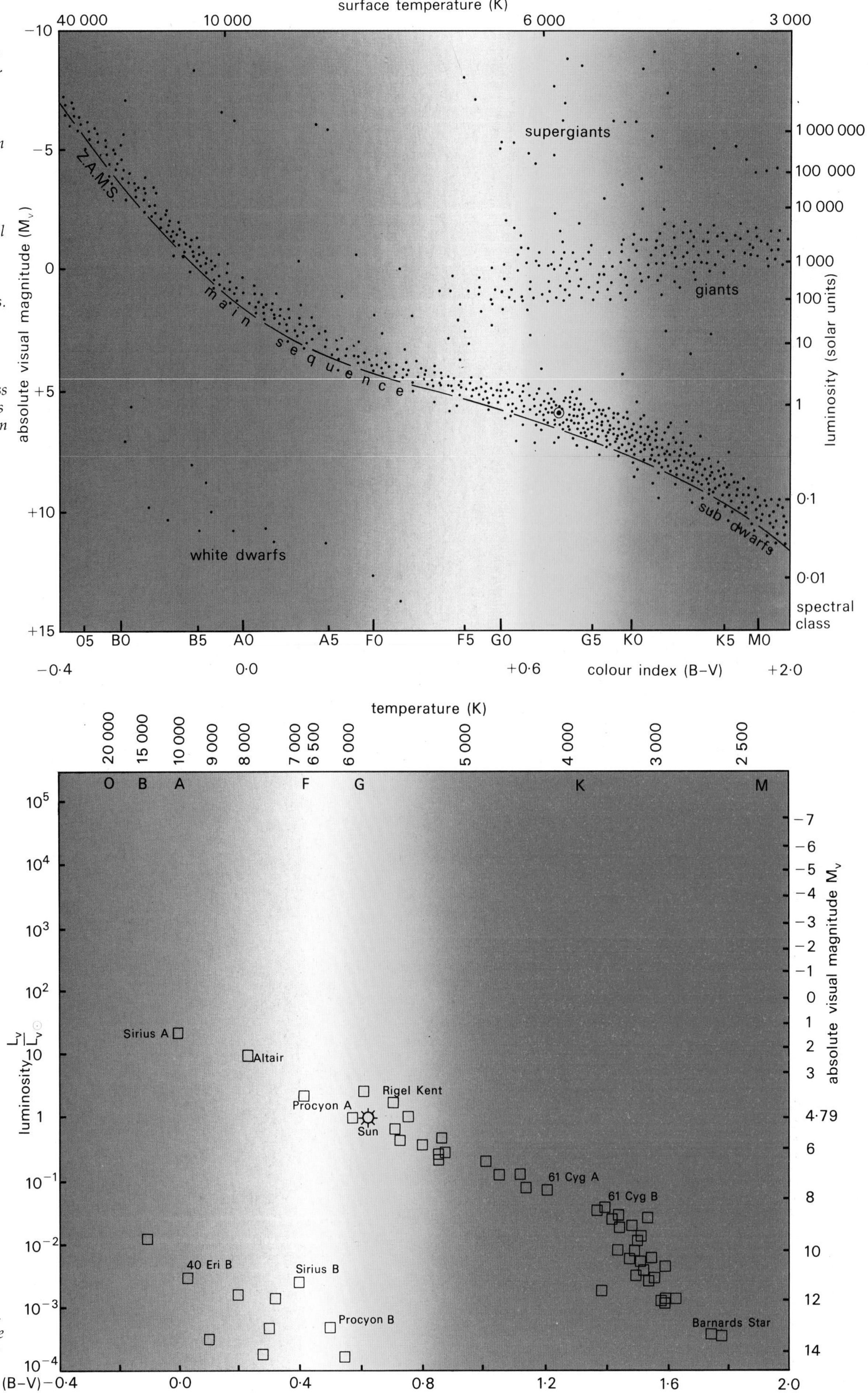

Fig. 3·3
A schematic representation of an H–R diagram showing the names of the major population zones. The zero-age-main-sequence (ZAMS) is the line on the diagram where newly formed stars of differing mass attain an equilibrium, whereby their self-collapsing gravitational energy is balanced by the energy liberated from hydrogen fusion reactions in their cores. All labelled regions represent particular phases of stellar evolution. Whether a particular star will pass through various phases depends primarily upon its mass.

Fig. 3·5
H–R diagram for the nearest stars showing that in the solar neighbourhood of 6·5 parsecs, virtually all the stars except Altair, Sirius and Procyon are less luminous than the Sun. They are also smaller and cooler.

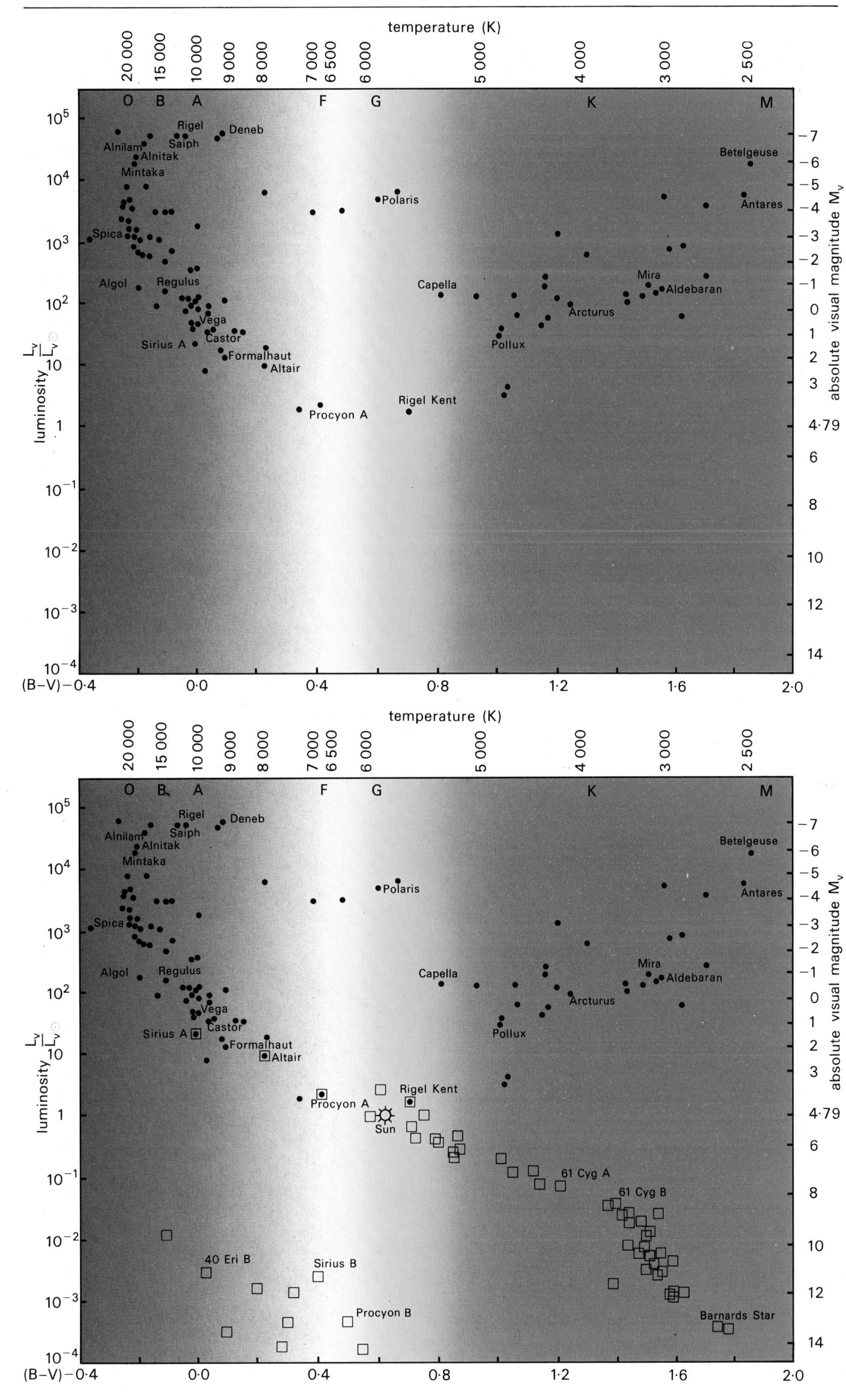

Fig. 3·6
H–R diagram for the brightest stars in the sky. These are all more luminous than the Sun and are at great distances from us. They still appear bright because of their very high intrinsic brightness, classified in astronomical terms either by absolute magnitude or luminosity.

Fig. 3·7.
Hertzsprung-Russell (H–R) diagram showing the positions of the nearest and the brightest stars in the sky based on measurements of their temperatures and luminosities. Notice how stars follow well defined zones of the diagram.

Table 3·1 **The mean values of temperature, luminosity, mass and radius for particular types of star**

spectral class	temperature (K)	luminosity ($L_\odot$)	mass ($M_\odot$)	radius ($R_\odot$)
main sequence, luminosity class V				
O5	40 000	5×10^5	40	18
B0	28 000	2×10^4	18	7
A0	9 900	80	3	2·5
G2	5 770	1	1	1
M0	3 480	0·06	0·5	0·6
supergiant stars, luminosity class I				
B0	30 000	3×10^5	50	20
A0	12 000	2×10^4	16	39
G0	5 700	6×10^3	10	106
M0	3 000	3×10^4	16	500
condensed stars – white dwarfs, approximate values around A	10 000	0·01	0·7	0·01

MK luminosity classification (Fig. 3·3). The current form of this scheme is that stars are grouped into seven classes, designated by a roman numeral after the spectral classification. These are as follows:

I (a), I (b)	supergiant
II	bright giant
III	normal giant
IV	sub-giant
V	main sequence dwarf
VI	sub-dwarf
VII	white dwarf

It should be noted, though, that not all stars may be classified into these distinct classes.

The nearby stars whose distances have been determined by direct means form a basis for the above scheme. It is interesting to compare the H-R diagrams

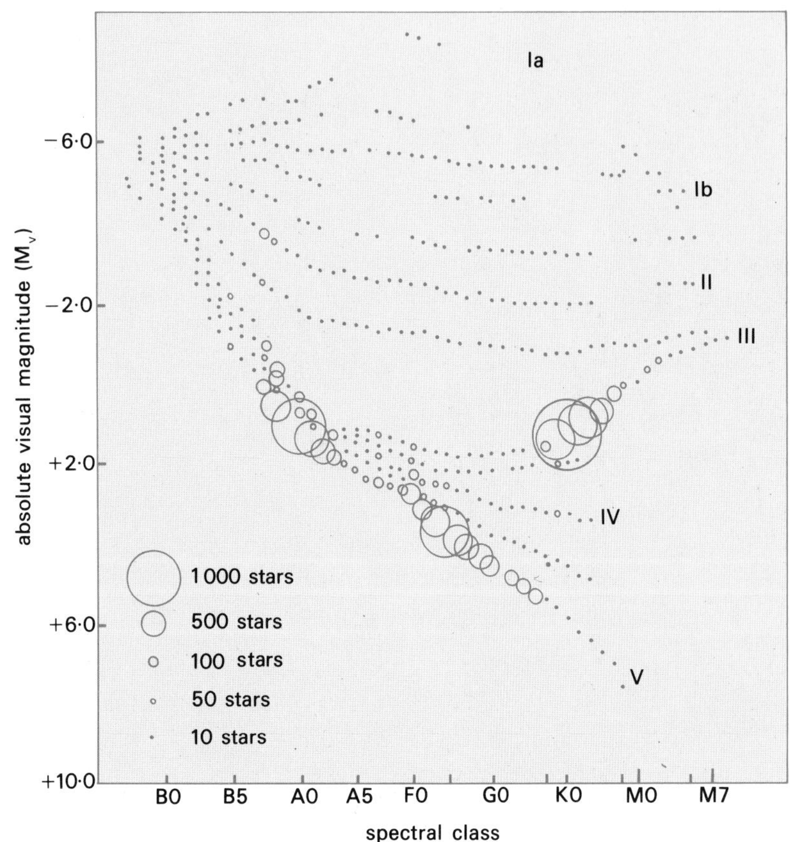

Fig. 3·8 H–R diagram revealing the number density of stars and their respective population zones. The size of each symbol is proportional to the number of stars included in each sample. Luminosity classes are also indicated on the diagram. It is very apparent how stars crowd along the main sequence and giant branch.

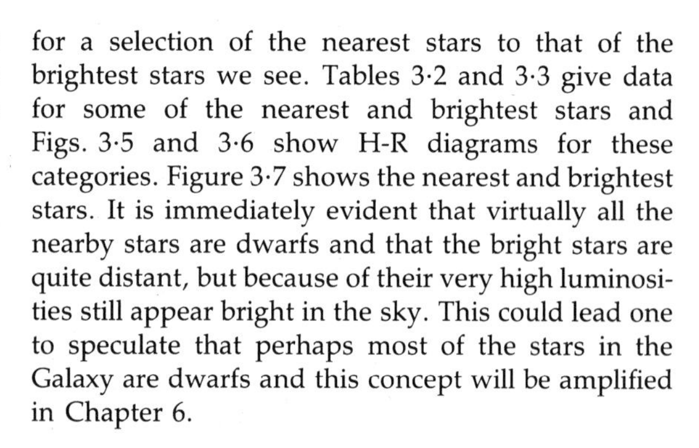

for a selection of the nearest stars to that of the brightest stars we see. Tables 3·2 and 3·3 give data for some of the nearest and brightest stars and Figs. 3·5 and 3·6 show H-R diagrams for these categories. Figure 3·7 shows the nearest and brightest stars. It is immediately evident that virtually all the nearby stars are dwarfs and that the bright stars are quite distant, but because of their very high luminosities still appear bright in the sky. This could lead one to speculate that perhaps most of the stars in the Galaxy are dwarfs and this concept will be amplified in Chapter 6.

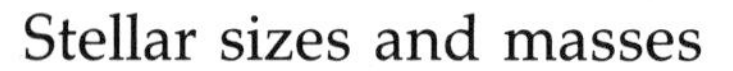

Stellar sizes and masses

We are now in a position to investigate the ranges of stellar sizes. This is usually expressed in terms of the solar radius ($R_\odot$) to give some feeling for their enormity. (Remember the radius of the Sun is 6·96 × 10^5km.) The range is truly phenomenal from white dwarfs of radii 0·01$R_\odot$ to red giants of radii up to 1 000 $R_\odot$ Antares (α Scorpii), an MI supergiant, has a radius 500 $R_\odot$ If we substituted Antares for the Sun in the Solar System, the atmosphere of the giant would extend to engulf all the inner planets, including Mars.

Although a star disc can not be resolved by any ground-based telescope, interferometric studies by the Michelson and speckle techniques (see Chapter 9) allow the sizes of one or two giant stars to be directly determined. The findings are in complete agreement with other methods. A more precise technique is to analyse the light curves of eclipsing binaries. If it is possible to calculate the lengths of their orbits, then the durations of the eclipses yield the diameters of the two stars. The only direct method of determining the masses of stars is through the study of the motions of individual stars in a binary or multiple system, where the stars are orbiting around each other. These, of course, must be observed under favourable conditions and, in general, the determination of stellar masses is a difficult process. For a binary star system, the two stars orbit around their common centre of mass. The orbital parameters are directly connected with the masses of the two stars by equations derived from the law of gravitation. If

Table 3·2 **The twenty nearest stars**

star	apparent visible magnitude m_v	absolute visible magnitude M_v	spectral class	luminosity class	proper motion (arc sec per year)	distance (pc)	mass ($M_\odot$)	radius ($R_\odot$)
Proxima Centauri C	11·05	15·45	M5		3·85	1·31	0·1	
Alpha Centauri A	−0·01	4·3	G2	V	3·68	1·34	1·1	1·23
Alpha Centauri B	1·33	5·69	K5	V	3·68	1·34	0·89	0·87
Barnard's Star	9·54	13·25	M5	V	10·31	1·81		
Wolf 359	13·53	16·68	M8		4·71	2·33		
HD 95735	7·50	10·49	M2	V	4·78	2·49	0·35	
Sirius A	−1·45	1·41	A1	V	1·33	2·65	2·31	1·8
Sirius B	8.68	11·56	WD*	VII	1·33	2·65	0·98	0·022
UV Ceti A	12·45	15·27	M5		3·36	2·72	0·044	
UV Ceti B	12·95	15·8	M6		3·36	2·72	0·035	
Ross 154	10·6	13·3	M4		0·72	2·90		
Ross 248	12·29	14·8	M6		1·59	3·15		
ε Eridani	3·73	6·13	K2	V	0·98	3·30		0·98
L789–6	12·18	14·60	M7		3·26	3·30		
Ross 128	11·10	13·50	M5		1·37	3·32		
61 Cygni A	5·22	7·58	K5	V	5·21	3·40	0·63	
61 Cgyni B	6·03	8·39	K7	V	5·21	3·40	0·60	
ε Indi	4·68	7·00	K5	V	4·69	3·44		
Procyon A	0·35	2·65	F5	IV	1·25	3·50	1·77	1·7
Procyon B	10·7	13·0	WD*	VII	1·25	3·50	0·63	0·01

*white dwarf

sufficient orbital parameters can be determined, the two star masses may be deduced.

There are two distinct classes of binary star appearing to us as observers. In **visual binaries,** the stars are sufficiently close and bright so that their individual orbits can be observed directly. These orbits will have the appearance of ellipses traced out on the sky relative to the 'fixed' star background. The ellipses may be nearly circular or highly elongated depending upon the individual star masses and the inclination of the plane of the orbit of the pair to our line of sight. In practice, this data has been gathered for less than a hundred binary systems.

If the stars are too distant to be resolved by telescopes, the periodic Doppler shift of the lines in the spectrum reveals their binary status. When the plane of their orbits is nearly in our line of sight, we observe the stars eclipsing each other. In such an **eclipsing binary** the masses may be deduced to a reasonable accuracy; otherwise the masses cannot be obtained.

A third method of mass determination has been applied to some condensed stars. We shall learn that white dwarfs have a very high density and this results in the light emitted by the star suffering an observable **gravitational redshift**. The wavelength shift (usually written $\delta\lambda$) of a spectral line from its natural place (written λ_0 since λ denotes the wavelength) in a low density situation (for example, laboratory studies or the spectra of normal stars) to the wavelength at which it is observed in the spectrum of the white dwarf is directly related to the mass M and the radius R of the white dwarf:

$$\delta\lambda/\lambda_0 = GM/Rc^2.$$

Therefore, if the radius of the white dwarf can be obtained from its luminosity, then the mass M may

Table 3·3 **The twenty brightest stars**

star		apparent visible magnitude m_v	absolute visible magnitude M_v	spectral class	luminosity class	distance (pc)
Sirius	α CMa	−1.45	+1·41	A1	V	2·7
Canopus	α Car	−0·73	+0·16	F0	Ib	60
Rigel Kent	α Cen	−0·10	+4·3	G2	V	1·33
Arcturus	α Boo	−0·06	−0·2	K2 p	III	11
Vega	β Lyr	0·04	+0·5	AO	V	8·1
Capella	α Aur	0·08	−0·6	G8		14
Rigel	β Ori	0·11	−7·0	B8	Ia	250
Procyon	α CMi	0·35	+2·65	F5	IV	3·5
Achernar	α Eri	0·48	−2·2	B5	IV	39
Hadar	β Cen	0·60	−5·0	B1	II	120
Altair	α Aql	0·77	+2·3	A7	V	5·0
Betelgeuse	α Ori	0·80	−6·0	M2	I	200
Aldebaran	α Tau	0·85	−0·7	K5	III	21
Acrux	α Cru	0·9	−3·5	B2	IV	80
Spica	α Vir	0·96	−3·4	B1	V	80
Antares	α Sco	1·0	−4·7	M1	Ib	130
Pollux	β Gem	1·15	+0·95	K0	III	11
Fomalhaut	α PsA	1·16	+0·08	A3	V	7·0
Deneb	α Cyg	1·25	−7·3	A2	Ia	500
Mimosa	β Cru	1·26	−4·7	B0	III	150

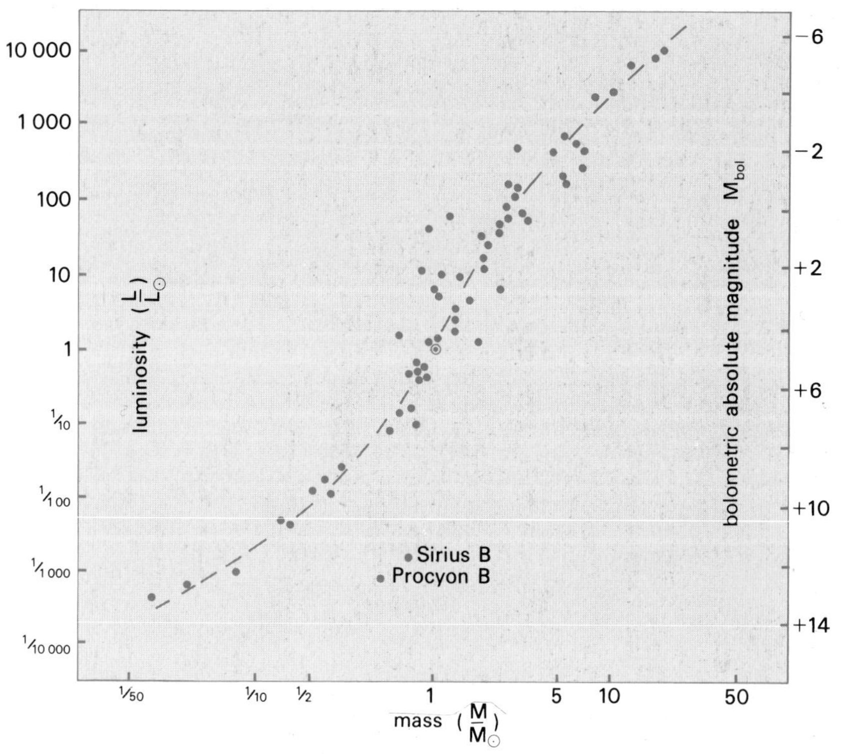

Fig. 3·9
The observed relationship between luminosity, L, and mass, M, expressed in solar units, obtained from observations of stars in binary systems. (For main sequence stars the approximation log L = 4 log_{10} M may be applied).

be found. Only a few white dwarfs have been studied by this means. Fortuitously, a few others happen to be members of nearby binary systems (for example, Sirius B and Procyon B) and so masses have also been determined by the binary orbit method.

When the range of masses is studied a surprising fact emerges. Although only about 200 stellar masses have been obtained using the above techniques, these range from white dwarfs to giants and so cover the entire range of stellar types. However, we find the masses exist in a narrow range, roughly between 0·05 to 50 times the mass of the Sun ($M_\odot$). When dealing with masses for stellar or galactic bodies it is usual to refer to the mass of the Sun as a standard comparison.

Figure 3·9 shows a plot of luminosity against mass for masses deduced from the binary star technique. This reveals the all-important **mass-luminosity relationship** so vital to the understanding of stellar evolution. It is apparent that the luminosity, L, has a definite dependence on the mass, M, of a star. Roughly speaking, L depends on the fourth power of the mass, (that is $L \propto M^4$), for stars whose mass lies in the range 0·3–20$M_\odot$. This rule is found to be applicable to the vast majority of class V stars but not white dwarfs. These appear too faint for their masses, thus betraying their intrinsically different physical forms.

Densities

We have now pieced together information which has allowed us to obtain temperatures, luminosities, masses and sizes of stars. From the last two we can calculate their mean densities. This calculation will be rather crude because it is known that stars do not have anything like a constant density from their centres to the extremes of their outer envelopes. We shall see later that the density rapidly decreases as we move outwards from the centre. However, to get a feel for overall densities look at Table 3·4.

Table 3·4 **Typical properties of certain stellar types**

star type	mass (M)	radius (R)	density (ρ) kg per m³
Sun (G2 V)	2×10^{30} kg	7×10^5 km	$1{\cdot}4 \times 10^3$
red giant (M0 I)	16 $M_\odot$	500$R_\odot$	2×10^{-4}
white dwarf (A2 VII)	1 $M_\odot$	0·015$R_\odot$	5×10^8

To demonstrate the extreme density of a white dwarf, a matchbox full of white dwarf material would weigh about 10 000 kg (equivalent to a double decker bus). Such condensed bodies have amazing properties, as we shall see later.

Chemical composition

From the absorption line spectra astronomers are also able to deduce the relative proportions (**abundances**), of the differing chemical elements observed in stellar atmospheres. These abundances reveal the chemical composition of only a strictly limited portion of the stellar atmosphere, that in which the absorption lines were formed, and provide no information about the chemical composition of the star's interior. Physical conditions in the stellar atmosphere can also be such that although certain elements may indeed be present we cannot observe their lines with Earthbound telescopes. The determination of abundances from stellar spectra is a complicated process and one must first calculate the physical conditions of temperature and pressure in the atmospheric region where the lines are being formed. Only then can the strength of the lines be used to indicate the abundance of each chemical element. The composition of the atmospheres of most normal stars turns out to be more or less the same as that of the Sun's. A compilation is shown in Table 3·5.

Table 3·5 Abundances of the most numerous elements in the universe

element	atomic number (Z)	mass number (A)	abundance by number	per cent by mass
hydrogen	1	1	92·06	73·4
helium	2	4	7·83	25·0
carbon	6	12		
nitrogen	7	14	0·1	1·13
oxygen	8	16		
neon	10	20		
magnesium	12	24	0·01	0·28
silicon	14	28		
sulphur	16	32		
iron	26	56	0·004	0·16
The remainder to Z = 103, A = 256			—	0·004

To a first approximation, it is true to say that all stars are vast globes of a mixture of hydrogen and helium. Yet subtle differences do occur and it is these which enable astronomers to pursue the questions relating to stellar evolution. In Chapter 6 we shall see that stars may be broadly grouped into what are

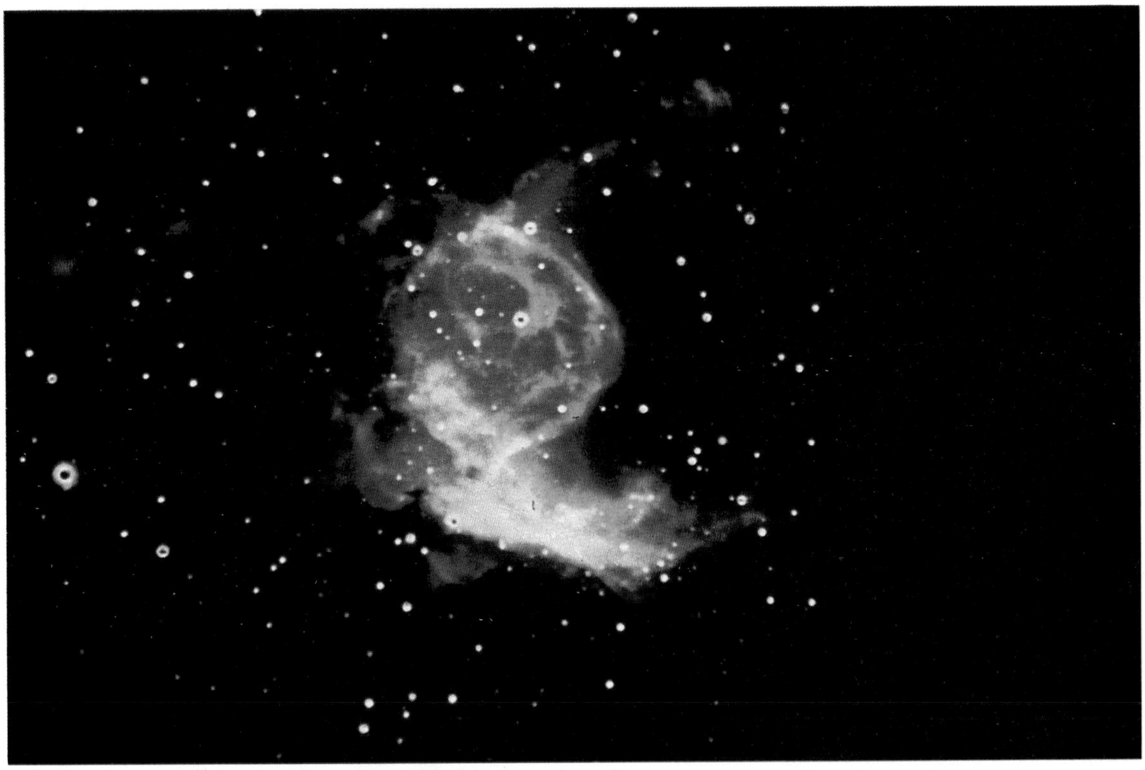

This nebulosity, NGC 2359, surrounds a very hot Wolf-Rayet star to which it owes its luminosity.

known as population classes. Class I is for stars which are metal rich whilst class II is for those that are metal deficient. Metal-rich stars are believed to be very young on an astrophysical timescale, while metal-deficient stars are the very oldest in space.

Peculiar stars

As with any large assemblage of data, peculiarities are bound to occur and this is certainly the case with stars. Although we have up to now been referring to absorption line spectra, some stars reveal emission lines. These emission lines may be superimposed on what is otherwise a normal absorption line spectrum such as the **Shell stars** and **P Cygni** classes. It is believed that these hot stars have very extended atmospheres, sometimes showing expansion. Alternatively, the emission lines, usually of ionized helium, carbon, nitrogen and oxygen are broad and very few absorption lines are seen. These are known as **Wolf-Rayet stars** and with average effective temperatures estimated at between 40 000 K and 50 000 K are the hottest stars known. Other strange features observed in stellar spectra have been found to be caused by strong and often variable magnetic fields. Such stars are those known as **Peculiar A stars**.

Rotation

One further piece of information may be gleaned from very careful study of the lines in a stellar spectrum. This is the rate of rotation of the star, which can be measured using the Doppler effect. A spectral line appears broadened as one side of the star approaches the observer and radiation is shifted to a lower wavelength, while the opposite side of the star recedes and radiation is seen at a longer wavelength. Because we see radiation from the whole disc of the star, the line appears equally red and blue shifted compared with radiation emitted by the centre of the stellar disc. A sharp line may then be grossly broadened under conditions of rapid rotation and although the measurements are tricky and difficult to analyse it is believed that some stars rotate with extreme rapidity. There appears to be a general trend for hotter stars to be the most rapidly rotating. The Sun, however, rotates at a mere 2 km per s, while some O and B stars have rotation velocities as high as 250 km per s. It is thought that the extended clouds of gas surrounding the Shell stars resulted from being thrown off due to extreme rates of rotation, perhaps up to 500 km per s.

Lifetimes and energy sources

How do the observable stellar factors of temperature, luminosity, mass and radius relate to the internal structure of a star? Can we answer the fundamental questions of when stars were born, how long they will live and how they will die? We can, but in doing so we must explore the province of physics applied to astronomical situations.

Stars continuously radiate energy and the mass-luminosity law tells us that for main sequence stars the rate at which they lose energy (their luminosity) increases as the mass increases. Does this imply that stars of different masses have varying lifetimes? From our inspection of the H-R diagram, we know that stars are not scattered at random but populate well-defined zones. We thus believe that certain relations exist between luminosity and surface temperature of stars. Could the H-R diagram represent a form of evolution scheme for stars? How can we begin to determine the ages of stars?

Let us consider the Sun. Inspection of fossil remains on the Earth reveal that it has radiated

Variable Stars

One of the most worthwhile fields of observation is perhaps that of variable stars, and here amateur observations are not only important, but frequently essential. So many variable stars are now known that with limited time and restricted access to telescopes, most professional work must be confined to specific aspects of variability, and it is often only amateurs who can follow stars' changes over a long period of time. Even so, the vast majority of individual variables go unstudied.

Some of the brightest stars are variable and can be followed by the naked eye, although most of them have fairly small amplitudes which can be difficult to determine accurately. With even a small pair of binoculars, however, many thousands of stars are available for study, and the numbers rise so greatly with increased apertures that no observer can hope to cover more than a small fraction of the known objects.

There are so many different classes and subclasses of stellar variability that they cannot all be described here. However, the most popular types are probably the long-period variables (periods over 100 days and reasonably regular light-curves) the semiregulars (more erratic behaviour but still with some periodicity), and the eruptive stars, particularly the dwarf novae (sudden outbursts with 'periods' of between a few tens of days to years, depending upon the individual stars) as well as other related types. True novae and supernovae are, of course, observed whenever they happen to be visible. Of the other classes the most important are probably the eclipsing binaries, where the orbital plane of the two individuals is so aligned in space that, as seen from Earth, each star periodically eclipses the other. The analysis of observations of any of these objects not only provides information about the particular star involved, but also adds to what is known about the nature and causes of variability in general, of the various phases of stellar evolution (page 61), and of the changes occurring within certain types of binary systems.

Although for historical reasons the system of nomenclature used for variable stars is a little complicated, the majority are known by single or double letters, e.g. R Scuti or WW Aurigae, or by a number preceded by the letter V, e.g. V1500 Cygni, which is the official name for Nova Cygni 1975 (page 65). The positions and characteristics of many of the brighter variable stars are listed in celestial handbooks and shown on many charts and atlases. Detailed charts of individual fields can be obtained from the organisations that specialize in variable star observation, together with magnitude sequences of comparison stars. As the magnitude of a variable is usually obtained by comparing its brightness with two comparison stars, one slightly brighter and the other slightly fainter than the variable, it is important that these 'official' magnitude sequences are used, as magnitudes derived from other (or mixed) sources may not be consistent with one another. The number of observations which are made of a particular object will depend upon the actual class of object, or at least should do so if bias is to be avoided. Slow variables such as the long period stars should only be observed about once every ten days, whereas eruptive objects can be estimated perhaps every hour when they are caught on the rise to maximum. Although a form of light-curve may be constructed from observations by an individual observer, there are inevitable personal errors. For this reason it is usual for observations to be reported to one of the amateur groups – just as with many other types of observation – which then prepare mean light-curves using many other estimates, and also carry out further analyses. They may then pass the observations to professional workers for further detailed examination.

The professionals, in their turn, may request specific coverage of particular objects, perhaps to coincide with periods of observation by other specialized telescopes or satellites operating in any of the many regions of the electromagnetic spectrum.

Many amateurs are now acquiring photoelectric equipment, and this can have many uses in the study of variable stars, not least in the determination of accurate sequences of magnitudes for objects to be studied by other means. In addition such equipment may be used to detect both low-amplitude and short-term variations which are difficult, if not impossible, to follow by other methods. All sorts of classes come within this category, including RR Lyrae stars (pages 67 and 172), so that in the end there are very few types of variables that are not studied by amateurs in some way or other. In the field of eclipsing binaries photoelectric work, especially when combined with professional spectroscopic determinations, can provide an amazing amount of information about the stars, their absolute sizes, masses, orbits, limb darkening and so on.

Quite apart from the derivation of light-curves, the actual discovery of certain types of variable is very important, and many amateurs carry out visual or photographic patrols aimed at the detection of novae and supernovae. Supernova work will be discussed elsewhere (pages 212–3); but for the discovery of novae, similar visual observation methods are used as for the detection of comets (page 150) – i.e. the use of large binoculars or rich-field telescopes and the memorizing of star patterns

Fig. 8·2
Opposite: *A detailed chart and magnitude sequence for the variable star R Scuti, which may conveniently be observed during the evening in late summer to early autumn. (See star chart, p. 41, for the general location of the area.) Magnitude estimates made at intervals of a week or so should show the star's variation.*

Right: *Novae generally occur close to the galactic equator, and in an organized nova patrol the search areas are arranged accordingly.*

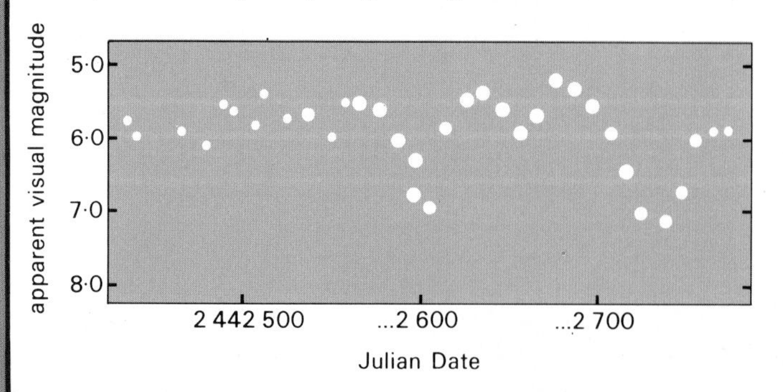

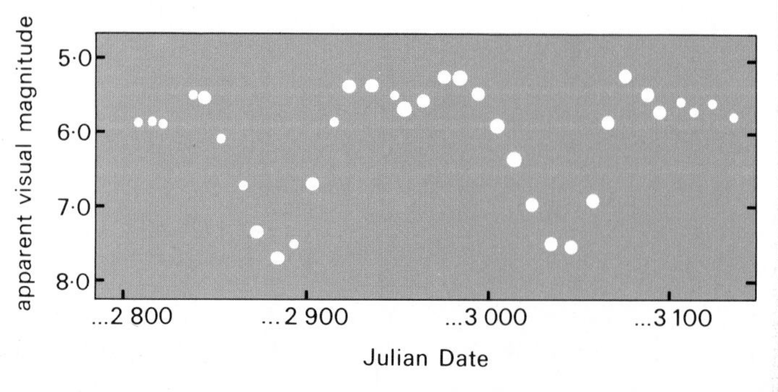

A light curve of R Scuti, based on BAA observations for 1975 and 1976.

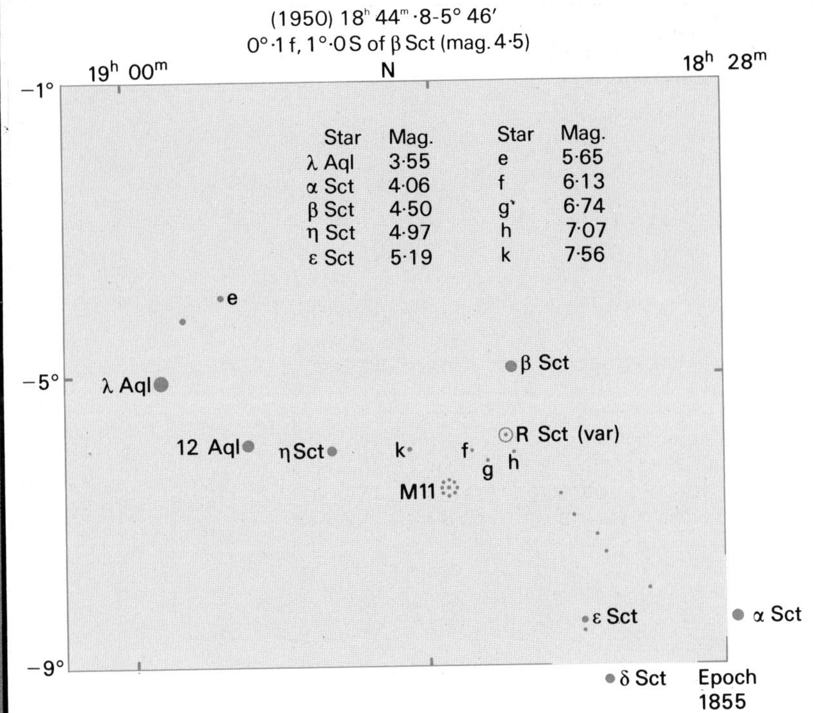

over wide areas of the sky. A few observers become so proficient that they are able to detect 'intruders' almost anywhere in the heavens. Others, particularly those participating in the organized patrols, may only learn one or two specific areas of the sky.

In a similar manner the photographic patrols divide up the sky into areas, these then being allocated to individual observers for coverage. The photographs are taken by conventional cameras and lenses (typically a single-lens reflex with a lens of 135mm focal length) and are always taken in pairs to enable blemishes etc. to be detected. They are then examined and compared with 'master' images. As speed is of the essence many ingenious methods of doing this have been devised, and any suspect object is reported immediately so that confirmation may be sought from another observer before a worldwide alert is issued. Needless to say, novae are of great interest to all professional astronomers and they are frequently prepared to interrupt their particular project to devote some of their hard-won telescope time to confirmed objects, obtaining spectra or photometric measurements.

Although many variables are discovered by photography, the method is not widely used to follow brightness changes, partly because of the time (and expense) involved, and also because of the problems of the different response of films to that of the human eye – although allowance for this may be made without too much difficulty. However, photography does have the great advantage that several variables may be recorded on one exposure, and frequently nova patrol photographs are used to derive magnitudes of objects in their fields. These can form a useful supplement to visual observations of the same objects.

Variable stars

(This list gives a number of bright, or famous, variables shown on the Star Charts. Comparison star charts are required for proper study, but their variation may still be noted.)

Name		Remarks
R	And	Long-period variable (about 400 days), mags 6 – 15
η	Aur	Eclipsing binary (period 27 years), mags 3·5 – 4·5
ζ	Aur	Eclipsing binary (period 32 months), mags 5·0 – 5·5
γ	Cas	Irregular brightenings when shell of material is shed
ρ	Cas	Irregular fades, mags 4·1 – 6·2
δ	Cep	Prototype of Cepheid variables (period 5·36 days), mags 3·9 – 5·0
μ	Cep	Semiregular, deep red star, mags 3·6 – 5·1
o	Cet	(Mira) Famous long-period variable (period 330 days, mags 2·0 (on occasions) – 10
R	CrB	Normally mag. 6·3, unpredictable fades down to 14 – 15
W	Cyg	Semiregular, mags 6·5 – 8·5, red
χ	Cyg	Long-period variable (period 406 days), extreme range 3·3 – 14·2
ß	Lyr	Prototype subclass of eclipsing variable (period 12·91 days), mags 3·3 – 4·2
ß	Per	Prototype subclass of eclipsing variables (period 2·87 days), mags 2·1 – 3·4
L2	Pup	Semiregular, mags 2·6 – 6·0
R	Sct	Semiregular, deep and shallow minima frequently alternate, mags 5·3 – 7·9

energy at the sáme rate as now for the past $4{\cdot}5 \times 10^9$ years. We now also believe that many other stars are about 10^{10} years old and these enormous ages immediately focus our attention on the source of the stellar energy. In the case of the Sun, we can multiply the solar luminosity $L_\odot = 3{\cdot}8 \times 10^{26}$ WATTS (W) by its minimum radiating lifetime ($4{\cdot}5 \times 10^9$ years) to arrive at the total energy emitted at least during the lifetime of the Earth. This turns out to be a staggering $5{\cdot}5 \times 10^{43}$ JOULES (J).

We can then ask what energy sources could satisfy the above requirement. The simplest case is to consider the thermal energy of the Sun; the heat output from the slow cooling of hot, central regions. As these regions cool then for a sphere of gas (the Sun) to maintain gas pressure to balance the weight of the outer layers, the central region must contract somewhat. The entire star then also contracts. Lord Kelvin performed a calculation of the above process, now frequently referred to as Kelvin-Helmholtz contraction. He found that the total amount of heat the Sun could have accumulated in contracting to its present radius is just the present GRAVITATIONAL POTENTIAL ENERGY of the Sun. If this is the total available energy, we may calculate the solar lifetime (using thermal energy) by dividing this by the solar luminosity. This turns out to be only 30×10^6 years, nearly 150 times less than the known age.

With our current preoccupation with the 'energy crisis' on our globe, alternative sources immedately spring to mind. Two major categories are fossil or chemical energy and nuclear energy. Coal is one form of fossil fuel and it is interesting to calculate the lifetime of a coal-burning Sun. As all chemical forms of energy yield roughly the same output for each unit consumed, this calculation will also serve for oil- and gas-burning Suns. Let us allow the Sun to be composed of 90 per cent coal! We know the coal-burning power output and so we can figure out how long the Sun can last burning at the required rate necessary to produce the present luminosity. The result is a disaster as it turns out to be a mere few thousand years, short by a factor of a million. We must, therefore, turn to nuclear energy.

The gradual understanding of the structure of the atom and radioactivity combined with Einstein's principle of equivalence of mass and energy, showed the enormous energy release possible from nuclear reactions. From this famous Einstein relation of $E = mc^2$, where E is the energy, *m* the mass and *c* the velocity of light, it is seen that if only a small quantity of mass is converted into energy, the output is tremendous. For instance, the solar luminosity can be supplied by the complete conversion of 4×10^9 kg of matter into energy every second. Although this is a sizeable amount of material it is utterly negligible compared to the total mass of the Sun.

For most of a star's lifetime its energy comes from the conversion of hydrogen into helium. Four hydrogen nuclei (protons) are fused to form one helium nucleus, with the emission of energy in the form of neutrinos and gamma-rays. The direct process, the **proton-proton chain**, occurs above 10^6K, and another using carbon as a CATALYST, the carbon-nitrogen-oxygen or **CNO cycle**, above about 2×10^7K. In the later stages of evolution 3 helium nuclei may be fused to give one carbon nucleus in the **triple-alpha process**, at temperatures above about 2×10^8K. (Alpha-particles are heluim nuclei.) At even higher temperatures oxygen-, magnesium-, and silicon-burning may occur. The process cannot continue beyond iron, however, and all heavier elements are believed to have been formed in the brief instants when massive stars explode as supernovae (page 63).

The life histories of stars

We have discussed the nuclear power stations of stars and how they shine, so let us now explore their life histories, that is stellar evolution. Three time scales of stellar evolution may be distinguished – the dynamical, thermal and nuclear. In our present context, the latter is defined to be the time taken for nuclear processes inside the star to change the chemical composition significantly. For a star on the main sequence it is the time required to convert all the hydrogen in its centre into helium. The time scale is proportional to the mass, M, divided by the luminosity of the star. We know from the mass-luminosity law that the luminosity of a normal star is proportional to about M^4. Therefore, we see that although a massive star begins with more fuel, it consumes it much faster, and as the **nuclear time scale** varies as $1/M^3$, the more massive (more luminous) stars have much shorter lifetimes on the main sequence. It emerges that this is the longest portion of any star's life and for stars of $20\ M_\odot$ the nuclear (main sequence) lifetime is a mere million years compared with 10^{10} years for stars of solar mass.

This is a vital clue in accounting for varieties of H–R diagrams which we get for different star clusters. Figures 3·10 and 3·11 show H–R diagrams for two types of star cluster in our Galaxy. The first depicts the Hyades, an open (galactic) cluster, and the second is M3, a globular cluster (see page 169). Astronomers believe that all the stars in a cluster

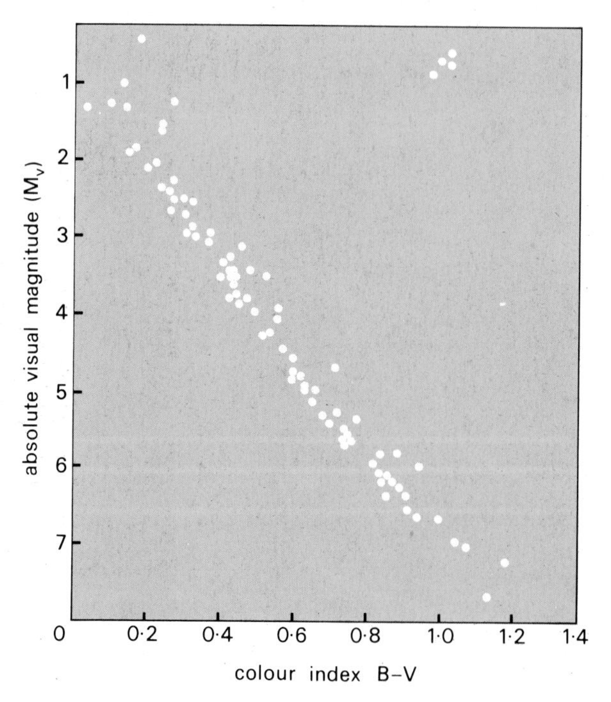

Fig. 3·10 far right: H–R diagram for the Hyades star cluster. The colour index (B-V) is the difference in apparent magnitudes in the blue and visible regions and is a measure of stellar temperature. The main sequence for the Hyades is seen to be well defined. This diagram for this particular cluster is most important for the determination of cosmic distances.

The central region of the Great Nebula in Orion, M42. This is an enormous cloud of gas and dust, about 500 parsecs distant, out of which stars are being born. Arrows show the location of T-Tauri stars. The circle reveals the position of an optically invisible object which, however, emits very strongly in the infrared portion of the spectrum. It is believed that this is a protostar about to embark on its collapse to a T-Tauri stage.

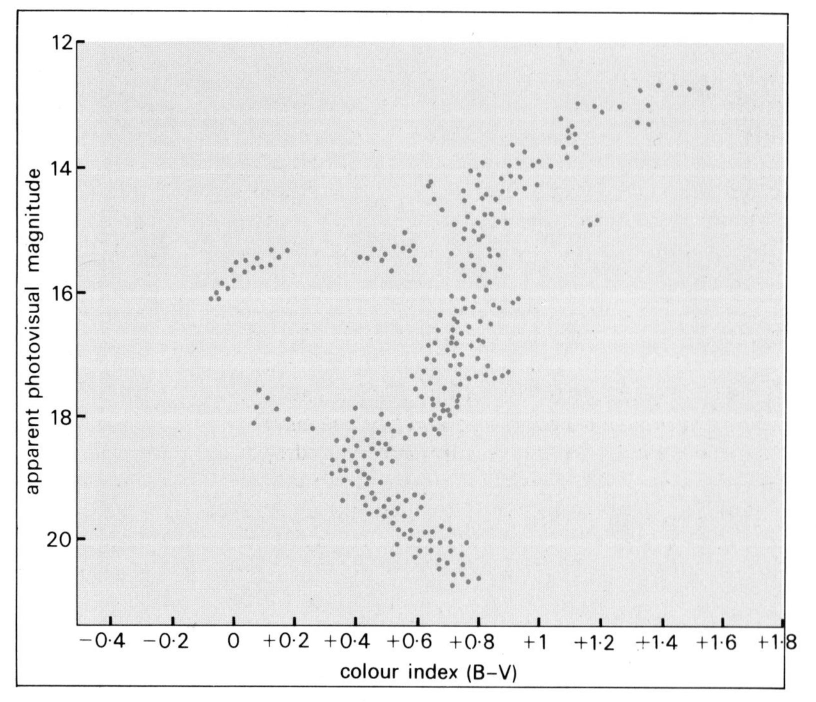

Fig. 3·11
H–R diagram for the globular cluster M3. This shows a protracted main sequence and well developed and heavily populated giant branch. Horizontal branch stars are found typically in globular clusters, but are absent from galactic (open) clusters. This is primarily because of the greater age of globular cluster stars.

formed at roughly the same time out of the same interstellar material. However, even a fleeting scrutiny of the two H–R diagrams exposes their conspicuous difference. Why is this? The predominant reason is the age of the clusters. Consider a very old cluster, approximately 10^{10} years old. Assuming a wide range of stellar masses were originally formed in the clusters, the more massive stars will have long ago converted all their hydrogen into helium. We would, therefore, not expect to see any highly luminous O, B or A stars. The main sequence would merely peter out as illustrated. On the other hand, in a young cluster, massive stars may not have had time to evolve from the main sequence. The cluster would have a long and unbroken main sequence. The length of the main sequence of a cluster of stars thus serves as a good indication of the age of the cluster. A schematic H–R diagram for several clusters is shown in Fig. 3·12 and from this we can say that all globular clusters are extremely old, while galactic clusters have varying ages but some are mere youngsters on an astronomical time scale; h and χ Persei may have an age of only 10^7 years.

The **thermal** (or Kelvin-Helmholtz) **time scale** is the time taken for energy to diffuse from the centre to the surface of a star. From our discussion of opacity and energy transport inside a star, we know that in the case of the Sun this time is about 10^7 years.

The **dynamical time scale** is much shorter, being the time required for the whole star to be aware of the absence of a pressure support. As an example, suppose the nuclear furnaces in the centre of the Sun were extinguished. With the sudden removal of the pressure support, the surface will be aware of this calamity in the time taken for a pressure wave to travel from the centre to the surface. For the Sun, this is about 30 minutes! Due to the great differences in these time scales, various processes playing alternate roles in the course of stellar evolution may be separated and studied in isolation.

The birth of stars

Star formation is not well understood but the basic picture assumes that they are born in dense clouds of interstellar material. Such clouds collapse under their gravitational fields and condensations or **protostars** form. As these protostars continue slowly to contract to a star they use the gravitational potential energy liberated for internal heating and radiation. The protostar heats up and eventually resembles a star. It is very possible that T–Tauri type stars are in this pre-main-sequence phase of evolution; here the surface resembles an ordinary star, although the interior has not yet settled down to steady nuclear burning. Steady burning occurs only when the internal temperature eventually becomes sufficient for hydrogen fusion reactions to commence. The star then stops contracting: it has arrived on the main sequence (Fig. 3·13). The thermal contraction period before arriving on the main sequence is brief, perhaps lasting only 10 000 years for a 20 M⊙ star. As a result we would not expect to find many stars in this phase on an H–R diagram, and we do not.

This demonstrates a crucial point in the use of an H–R diagram. If a region is well populated on the diagram, we can infer that stars spend a substantial fraction of their lives in these zones. Such regions are the main sequence and giant branches. If a region is sparsely populated or vacant, it may then be assumed either that conditions are not suitable for stars to exist there or that stars pass through such areas in a time which is short in an astronomical context, that is, less than a million years.

Fig. 3·12 below: *Composite H–R diagram for ten open clusters and one globular cluster, M3. As the main sequence is followed from top left to lower right, turn-off points for various clusters are observed. Because very hot, massive stars have short lifetimes on the main sequence, this progression of the turn-off from the main sequence identifies clusters of greater age. M67 is one of the oldest known open clusters with an age of about four thousand million years; this is to be contrasted with h and χ Persei of only ten million years. The globular cluster M3 is believed to have an age of twelve thousand million years. By studying such diagrams for many clusters, astronomers find clues to help them with the puzzle of stellar evolution.*

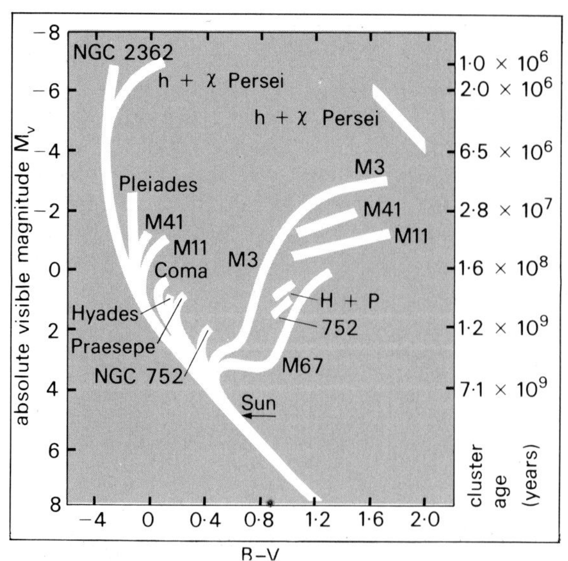

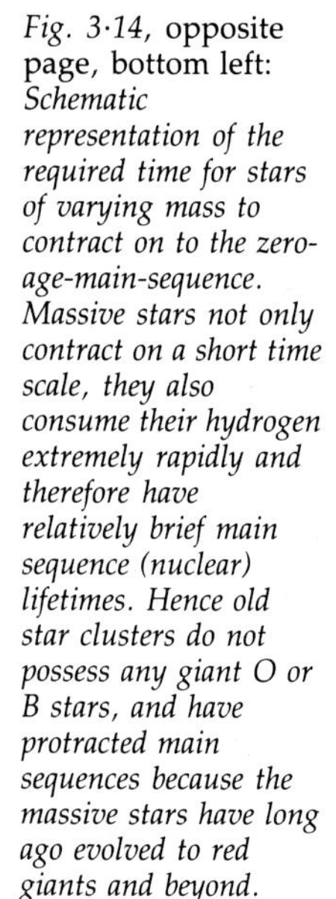
Fig. 3·14, opposite page, bottom left: *Schematic representation of the required time for stars of varying mass to contract on to the zero-age-main-sequence. Massive stars not only contract on a short time scale, they also consume their hydrogen extremely rapidly and therefore have relatively brief main sequence (nuclear) lifetimes. Hence old star clusters do not possess any giant O or B stars, and have protracted main sequences because the massive stars have long ago evolved to red giants and beyond.*

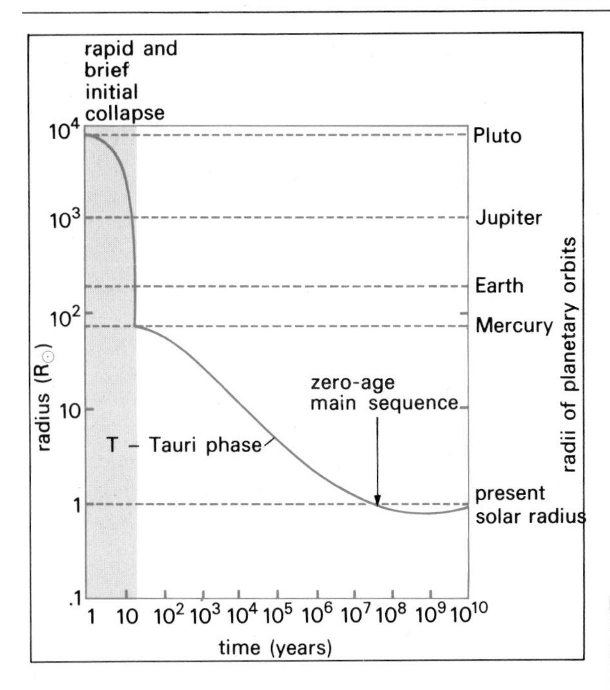

Fig. 3·13 far left: The radius of the Sun during various stages of its history. Notice the very brief initial collapse, lasting a mere twenty years, followed by the slow contraction of T-Tauri stage until arrival on the zero-age-main-sequence with the onset of equilibrium hydrogen core burning.

Evolution

When the protostar arrives on the main sequence and hydrogen burning is supplying the radiating energy of the star, it is then said to be on the zero-age main sequence, ZAMS (Fig. 3·14). A star will stay on the main sequence band of the H–R diagram as long as it is converting hydrogen to helium in its core. This represents a long nuclear time scale. As evolution progresses, hydrogen is converted into the heavier element, helium, which sinks to the core. Eventually the core becomes depleted in hydrogen fuel but enriched with helium 'ash'. For massive stars convection brings in fresh supplies of fuel to the core and burning continues, but in moderate and small mass stars hydrogen is burnt in a thin shell surrounding the helium-rich core. In due course all stars, massive, moderate and small, suffer from fuel depletion and, as a consequence, the energy supply is reduced, pressure falls and the core contracts. This causes a rise in temperature, a subsequent rise in the reaction burning rate and, thus, an increase in pressure: the core contraction is halted. During this readjustment phase, the entire star contracts and is said to be evolving off the main sequence. Such contraction releases gravitational potential energy, half of which is available for internal heating while the other half is radiated away.

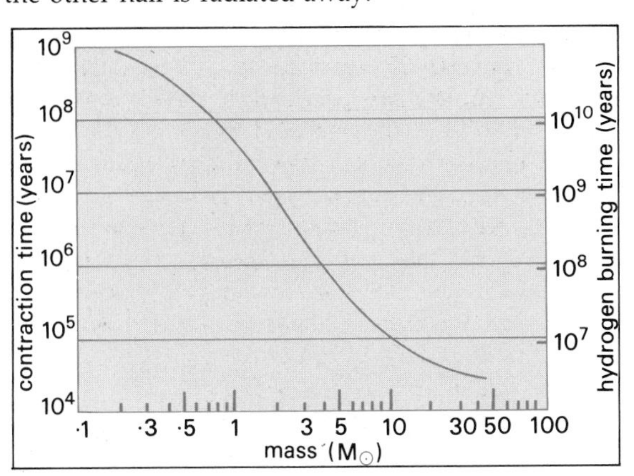

At this stage the contraction of the envelope is halted by the new energy supply and it now expands, while the core continues to slowly contract, becoming hotter in the process. As hydrogen is burnt in successive shells moving outwards from the centre of the star, the helium core continues to increase until a stage is reached when it is so large that it collapses due to the pressure of the overlying layers. This instability occurs on a thermal time scale (page 60) and is accompanied by a rapid expansion of the stellar envelope. The star then moves quickly to the right of the H–R diagram (and as expected few stars are seen in this zone of avoidance known as the Hertzsprung Gap), eventually to become a red giant.

The envelope has now cooled and become convective, the radiative energy is supplied by hydrogen shell burning and the helium core is still contracting and heating. What happens next depends on the mass of the star (Fig. 3·15). For large and moderate

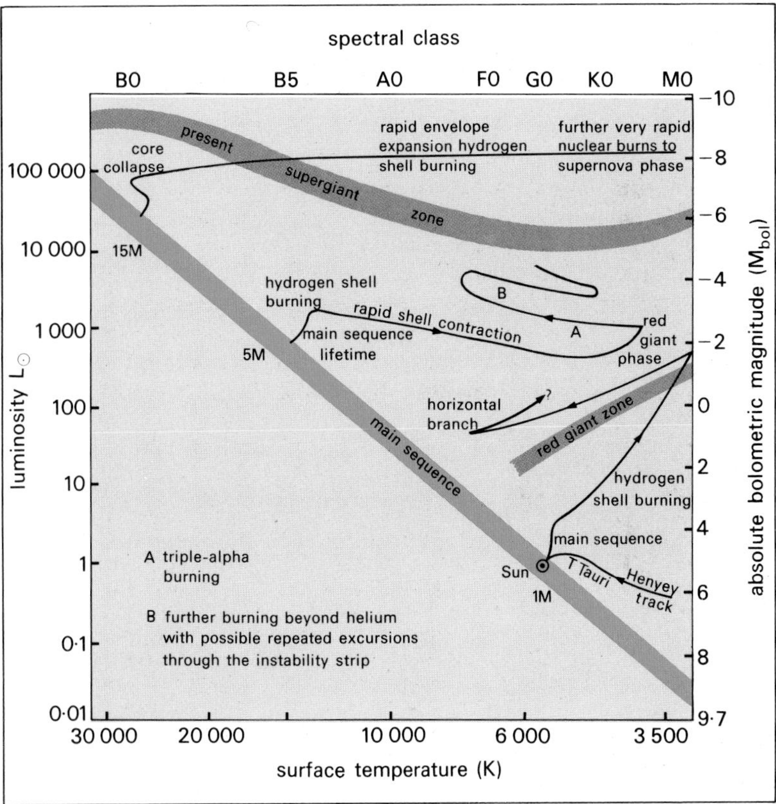

mass stars the core temperature increases until it is about 10^7 K when helium can fuse to carbon by the triple-alpha process. This supplies energy which halts the contraction and the star settles down to another stable period, although lasting much shorter than the main sequence lifetime. Eventually helium in the core runs out as the core becomes carbon rich and begins to contract. Further evolution proceeds and may entail higher orders of nuclear burning, involving oxygen and silicon, for example, but the duration of these phases is extremely short. Sector diagrams of a main sequence star and of the envelope and core of a red giant are shown in Figs. 3·16–3·18.

Fig. 3·15 above: H–R diagram showing possible evolutionary tracks of stars of varying mass. Many theoretical models are constructed to attempt to describe the ways in which stars evolve, all are sensitive to the mass and initial chemical composition of the stars, and for late stages of evolution they become very complex in nature. The models attempt to reproduce observations of the H–R diagram of stellar clusters.

Death

It is during these burnings that the very heavy elements in the universe are synthesized, probably

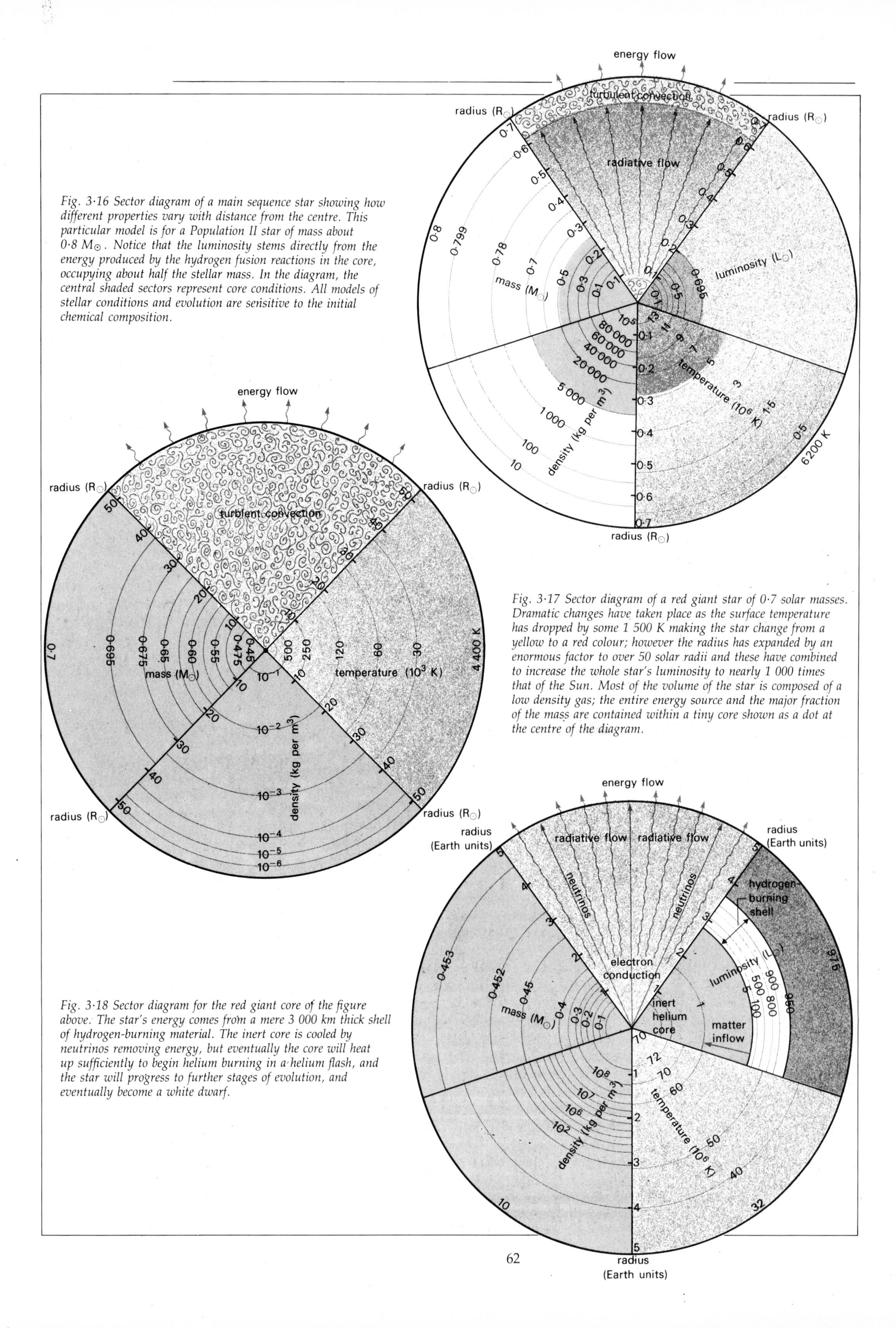

Fig. 3·16 Sector diagram of a main sequence star showing how different properties vary with distance from the centre. This particular model is for a Population II star of mass about 0·8 $M_\odot$. Notice that the luminosity stems directly from the energy produced by the hydrogen fusion reactions in the core, occupying about half the stellar mass. In the diagram, the central shaded sectors represent core conditions. All models of stellar conditions and evolution are sensitive to the initial chemical composition.

Fig. 3·17 Sector diagram of a red giant star of 0·7 solar masses. Dramatic changes have taken place as the surface temperature has dropped by some 1 500 K making the star change from a yellow to a red colour; however the radius has expanded by an enormous factor to over 50 solar radii and these have combined to increase the whole star's luminosity to nearly 1 000 times that of the Sun. Most of the volume of the star is composed of a low density gas; the entire energy source and the major fraction of the mass are contained within a tiny core shown as a dot at the centre of the diagram.

Fig. 3·18 Sector diagram for the red giant core of the figure above. The star's energy comes from a mere 3 000 km thick shell of hydrogen-burning material. The inert core is cooled by neutrinos removing energy, but eventually the core will heat up sufficiently to begin helium burning in a helium flash, and the star will progress to further stages of evolution, and eventually become a white dwarf.

in a supernova explosion when the core burnings occur explosively. This process is by no means fully understood but the following gives the general picture. Neutrinos remove significant fractions of the generated energy so that the pressure support is drastically weakened, then the core rapidly contracts, the temperature rises, reaction rates increase dramatically, and neutrinos are even more copiously emitted. The reactions may next become endothermic by iron burning, so refrigerating the core. Eventually, and very rapidly, the cooled core implodes, crashing in on itself, and the outer layers of the star are ejected outwards with the cataclysmic violence of a **supernova** explosion.

End products of stellar evolution

A supernova is an awe-inspiring phenomenon because the energy output is so incredibly vast: for about a week the light yield from its violence often outshines the entire light of its parent galaxy. When one remembers that the galaxy probably contains some 10^{11} stars, we realize the staggering enormity of a supernova explosion.

The outer layers flung off by the imploding core (which may or may not leave a supercondensed body behind), expand into the surrounding interstellar medium to become a supernova remnant. This remnant is very hot, has intense magnetic fields and may emit radio, optical and X-ray radiation due to **synchrotron emission** (a process by which very high-speed electrons, spiralling round in a magnetic field,

Above:
A supernova explosion observed in 1959 in the galaxy NGC 7331. The first photograph reveals no trace whatsoever of the pre-supernova event which is clearly visible in the second plate. The enormous power of the supernova process is well demonstrated as the eruption was that of a single star, yet it shines like a beacon in the spiral arms of the 5 Mpc distant galaxy.

The planetary nebula NGC 7293, the Helix. A spectacular sight in a telescope, this glowing shell of hydrogen gas is 0·5 parsecs in diameter. It is the entire expelled envelope of a star's late phase of evolution in which it is believed that red giants lose mass in violent outbursts, suffering a final collapsing metamorphosis into a white dwarf star in the process.

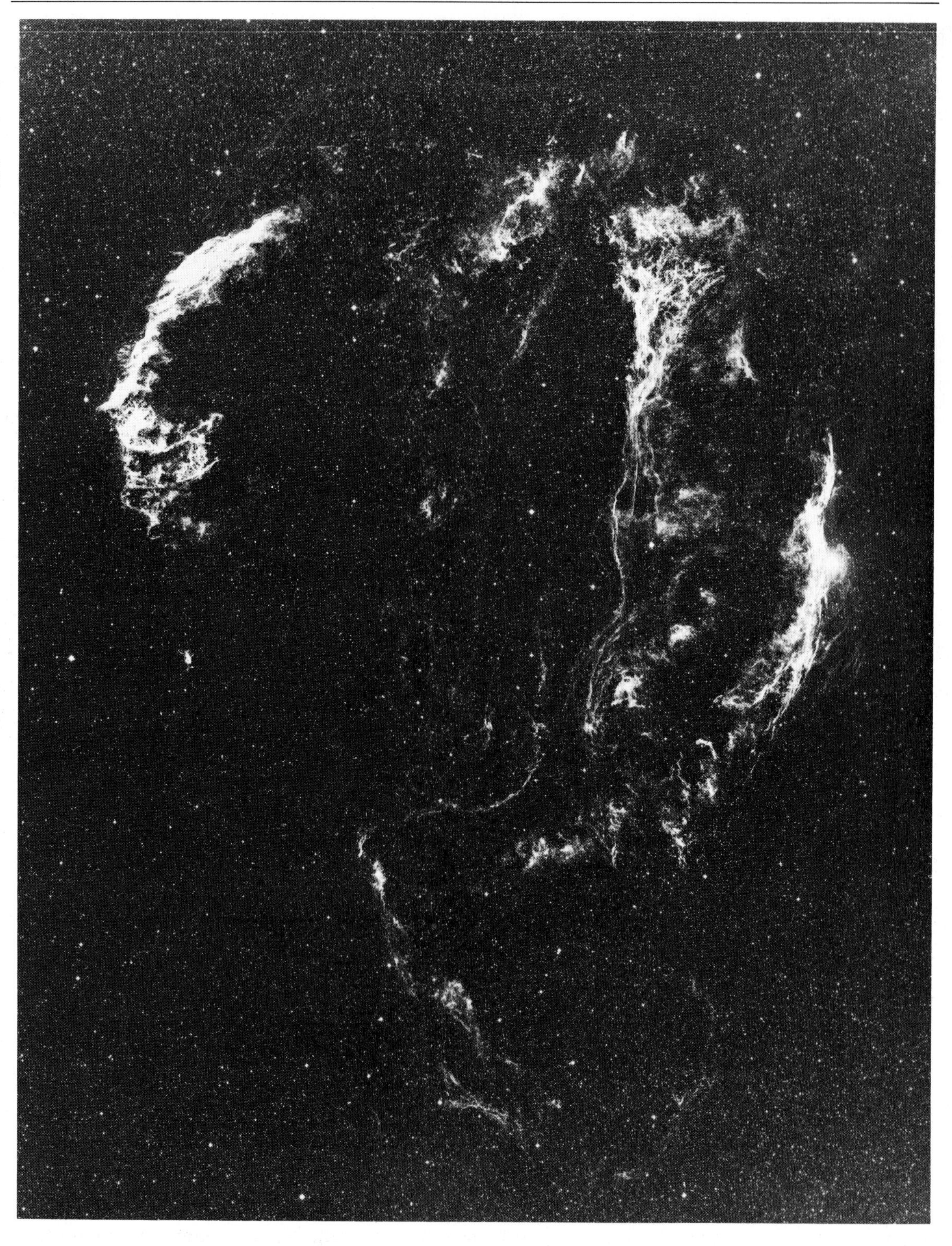

The Cygnus Loop supernova remnant. This shell spans about three degrees on the sky and is believed to be nearly 600 parsecs distant. It is the glowing diffuse remains of a star which exploded about sixty thousand years ago.

emit radiation, and so-called because it was first observed in an artificial nuclear particle accelerator known as a synchrotron). As the remnant ages it cools, the expansion slows and it becomes optically faint; after about a million years only a diffuse patch of radio emission marks the death of a star in the galaxy. The Cygnus Loop, the remnant of a supernova which exploded in our Galaxy about 10 000 years ago, is shown above.

We know supernovae are rare, only occurring about once every 100 years in our Galaxy, so we must conclude that most stars do not die with such violence, but by some other more peaceful process.

Condensed remnants

It is believed that many low mass stars end up as white dwarfs. These have cores of degenerate material, which behaves quite unlike normal material. In such a core, when the temperature becomes sufficiently high for helium core burning to begin, the subsequent rise in temperature does not cause an increase in pressure and a halt to the contraction. Instead, the temperature continues to go up and the reactions burn faster, but still the pressure does not increase. The core continues to contract and progressively heat up. The thermostatic control has obviously malfunctioned and is only reinstated when the temperature is sufficiently high for the gas to revert from a degenerate to a normal state. Core contraction then ceases. This extremely rapid runaway nuclear burn is termed the **helium-flash**, and takes only seconds to occur. The red giant star then readjusts on a thermal time scale of a few thousand years and the star rapidly crosses the H–R diagram to the horizontal branch. Precise evolutionary tracks beyond this stage are extremely complex and can not be plotted accurately.

In 1939, Subrahmanyan Chandrasekhar calculated that the maximum mass attainable by a white dwarf was 1·4 $M_{\odot}$, yet the masses of the stars we are discussing exceed this by factors of two to four. How can these become white dwarfs? From observations we know that stars lose mass during their lives, either by a continuous solar wind type of process (like P-Cygni stars), or by spasmodic mass loss following pulsations (like Mira-type variables). We also observe phenomena termed **planetary nebulae**, which are very hot stars (temperature about 10^5 K) surrounded by a cool expanding shell of gas. We believe these result from a major instability in which a shell containing a large amount of mass is ejected, thereby exposing the deeper, hotter regions of a star. These processes of mass loss can, it is thought, reduce the star below the critical mass limit, and when the nuclear fuels have been exhausted, the star slowly cools, radiating away its stored internal energy to become a white dwarf (see page 70).

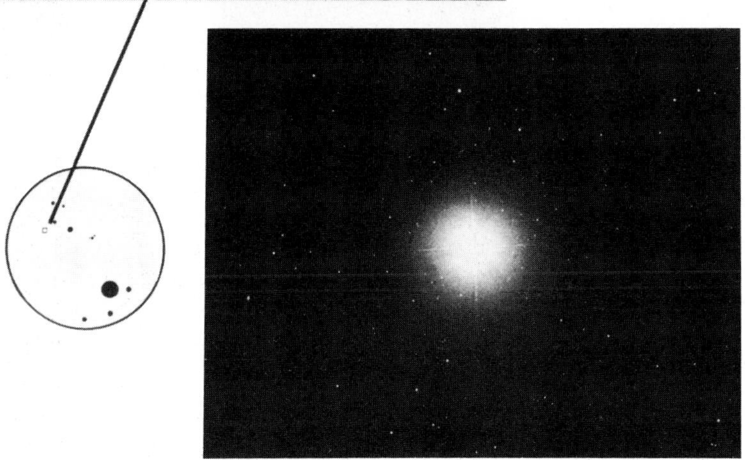

Two views of the same area of sky before and at maximum light of Nova Cygni 1975. The pre-nova star is invisible even on Palomar Sky Survey plates with faintest magnitude equal to 21. At maximum brightness the nova attained second magnitude, therefore it had brightened by a phenomenal nineteen magnitudes.

Variable stars

The depopulated zones of the H–R diagram between the main sequence and giant branch are the domains of the intrinsic variables. These are stars whose luminosity varies due to particular internal changes. Their study is important for the clues they reveal for stellar evolution. Variables come in several categories, but two distinct groups are the explosive variables and the regular variables. Thousands of such stars have now been catalogued and some, such as Polaris (α UMi), are visible to the naked eye. The modern detection technique is to photograph a star field and then compare this, usually electronically, with a photograph of the same star field taken at a different time. Stars which have changed in brightness may then readily be picked out.

We have already met two examples of explosive variables in supernovae and planetary nebulae. **Novae** are another. They are believed to be close binary stars in which mass is being transferred from one component to a small secondary star orbiting so close that the pair are nearly in contact. The secondary is thought to be a condensed star, probably a white dwarf or neutron star (see page 70), and the inflowing mass forms a disc around it. This disc sometimes becomes unstable and large flows of material are then pulled by gravitation on to the superdense body. When this happens the material is dramatically heated and may become explosive, resulting in an outburst or flaring which we refer to as a nova (Fig. 3·19). Spectroscopic study of the sudden increase in luminosity reveals the abundances and temperatures of the chemical elements involved, and Doppler shifts of the spectral lines give the velocity of the exploding material. It is now thought that nova outbursts may recur in the same systems after an interval of some thousands of years. Less violent but more frequent flares occur in the **recurrent novae** with intervals of a few decades. The class of stars known as the U Geminorum stars show even more frequent outbursts of still less intensity, their time-scales being of the order of a few tens to hundreds of days.

Novae are not well understood, but the general idea of close-contact binaries is intriguing because of the strong tidal forces they exert on one another, and, if initially formed with quite distinct masses, they will evolve on differing time scales. Study of the evolution of such objects has become very topical within the past few years with the discovery of many

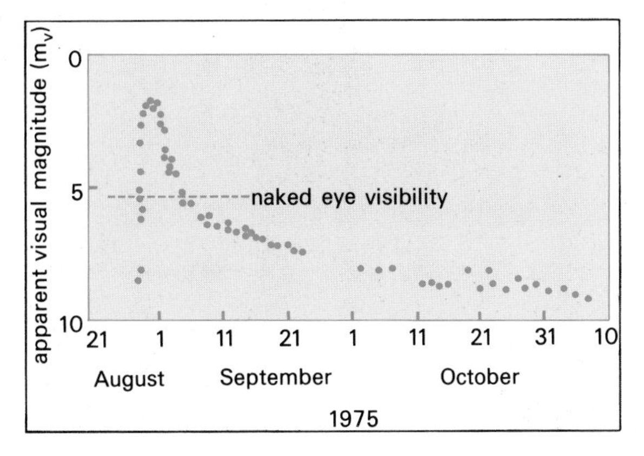

Fig. 3·19
A record of the light curve of Nova Cygni 1975 showing the very steep rise to maximum and subsequent fall before a gradual fading proceeded. Nova Cygni was observable with the naked eye for about a week.

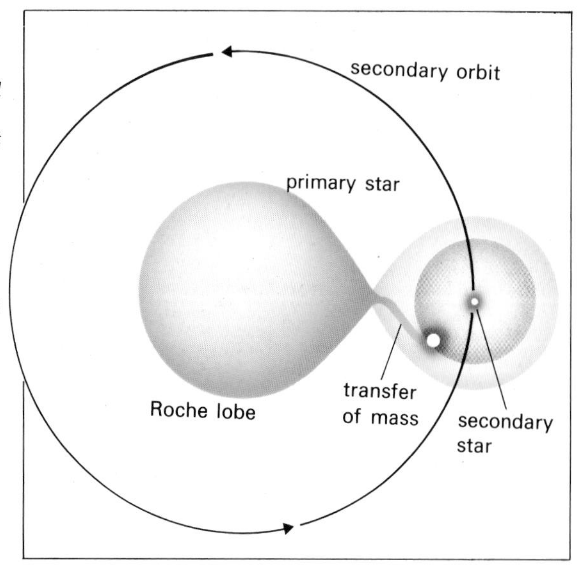

Fig. 3·20 The overall gravitational field in a binary star system has a particular geometrical shape and regions in space may be traced out over which the gravitational field has the same value; these are known as **equipotential surfaces**. *One special case is indicated on the diagram by the tinted area, the two enclosed regions are known as* **Roche lobes**. *An important consideration for close binary stars is that if one fills its Roche lobe, matter may then flow through the lobe onto its companion star, this is then referred to as a semi-detached (β-Lyrae type) binary.*

X-ray variable stars. In these objects, the X-ray variability is found with periods from fractions of a second to days. The former times are typical of pulsation periods of neutron stars, while periods of days represent a binary eclipse period. Some X-ray stars also appear as X-ray novae or flare stars and it is thought that just as in novae, matter pulled gravitationally from one star on to a superdense companion star causes this energetic emission.

As far as regular variables are concerned, it is now widely believed that their variation may be accounted for by pulsations of the stellar surface. Their appearance along what is referred to as the instability strip of the H–R diagram leads astronomers to suspect that these variables are passing through a relatively brief phase of their lives, during which the interior of the star may be in equilibrium but a sub-surface layer exists which is not. It is this layer which produces the oscillations; energy is being stored as the surface contracts and is released when the surface expands. This surface pulsation can be directly observed by measuring the Doppler shift of the spectral lines, demonstrating that the stellar atmosphere really is moving inwards and outwards. This change in surface area of the star is manifest by a change in luminosity which we observe as a regular brightening and fading.

Cepheid variables

The best-known and most-studied group of pulsating stars is the Cepheids, named after the first star of this type observed. This was δ Cephei and its **light curve** (variation of brightness with time) is shown in Fig. 3·21. All Cepheids have light curves similar to this with periods ranging from just over a day to about fifty days. From their position on the H–R diagram, it is apparent that Cepheids are yellow supergiant stars and therefore exceptionally luminous. It is this property which enables them to be detected at large distances and they have been instrumental in determining not only the size and structure of our Galaxy but also the distances of nearby galaxies. This is due to the famous period-luminosity relation for Cepheids, which means that from a measurement of their period the luminosity may be deduced, and, thus, their absolute magnitude. A measurement of the apparent magnitude of the Cepheid then gives its distance (from the distance modulus equation).

The period-luminosity law for Cepheids was discovered in 1908 by Miss Henrietta Leavitt who was studying the Magellanic Clouds. At that time the distance of these clouds was unknown (although we now know they are neighbouring galaxies 50 kpc distant), but they were clearly sufficiently remote for all their stars to be considered to be at the same distance. She noticed that the brighter Cepheids had correspondingly longer periods and because of their equal distance, this showed the existence of a direct link between period and luminosity. However, the distance of at least one Cepheid was required to determine the absolute relation for the group. None are sufficiently close to show trigonometrical parallaxes but studies of star clusters in the early 1960s

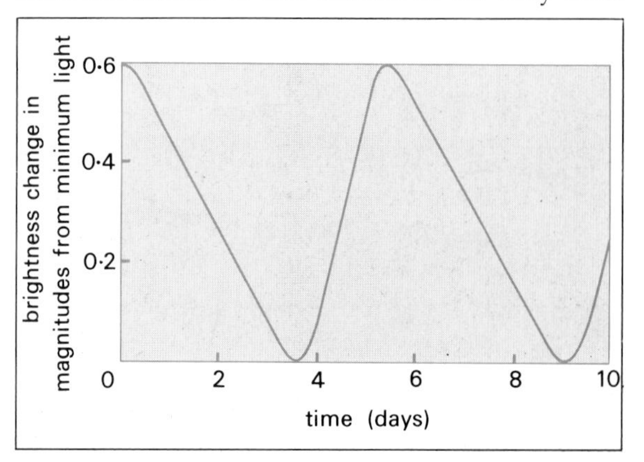

Fig. 3.21 far right, above:
The 5·31 day regular periodic variations in the optical brightness of the star δ Cephei. This was the first star studied in a class of variable, pulsating stars now referred to as Cepheid variables. The pulsation period and the optical period is directly proportional to their luminosity.

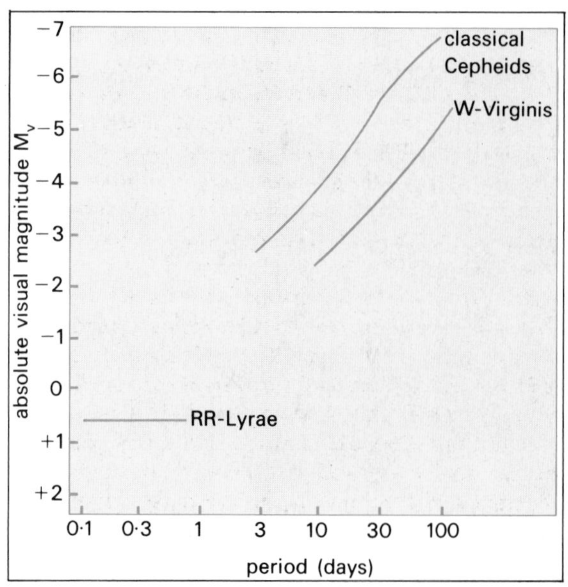

Fig. 3·22 far right, below:
The relationship between the absolute magnitude and optical period for classical Cepheids, W-Virginis and RR-Lyrae variable stars. A measurement of the period for a distant Cepheid enables its absolute magnitude to be derived and a measurement of its apparent magnitude then yields the distance of the Cepheid by application of the distance modulus equation. Cepheids are important tools in the measurement of the distances to galaxies out to a range of about 6 Mpc.

established the relation reproduced in Fig. 3·22. Because of their high luminosity and tell-tale signature, Cepheids are valuable tools for distance determinations of nearby galaxies and for establishing the basis of observational cosmology.

The understanding of Cepheids in general became much clearer in 1944 when Walter Baade made the discovery of stellar populations (page 166) and found that Cepheids could also be included in this scheme.

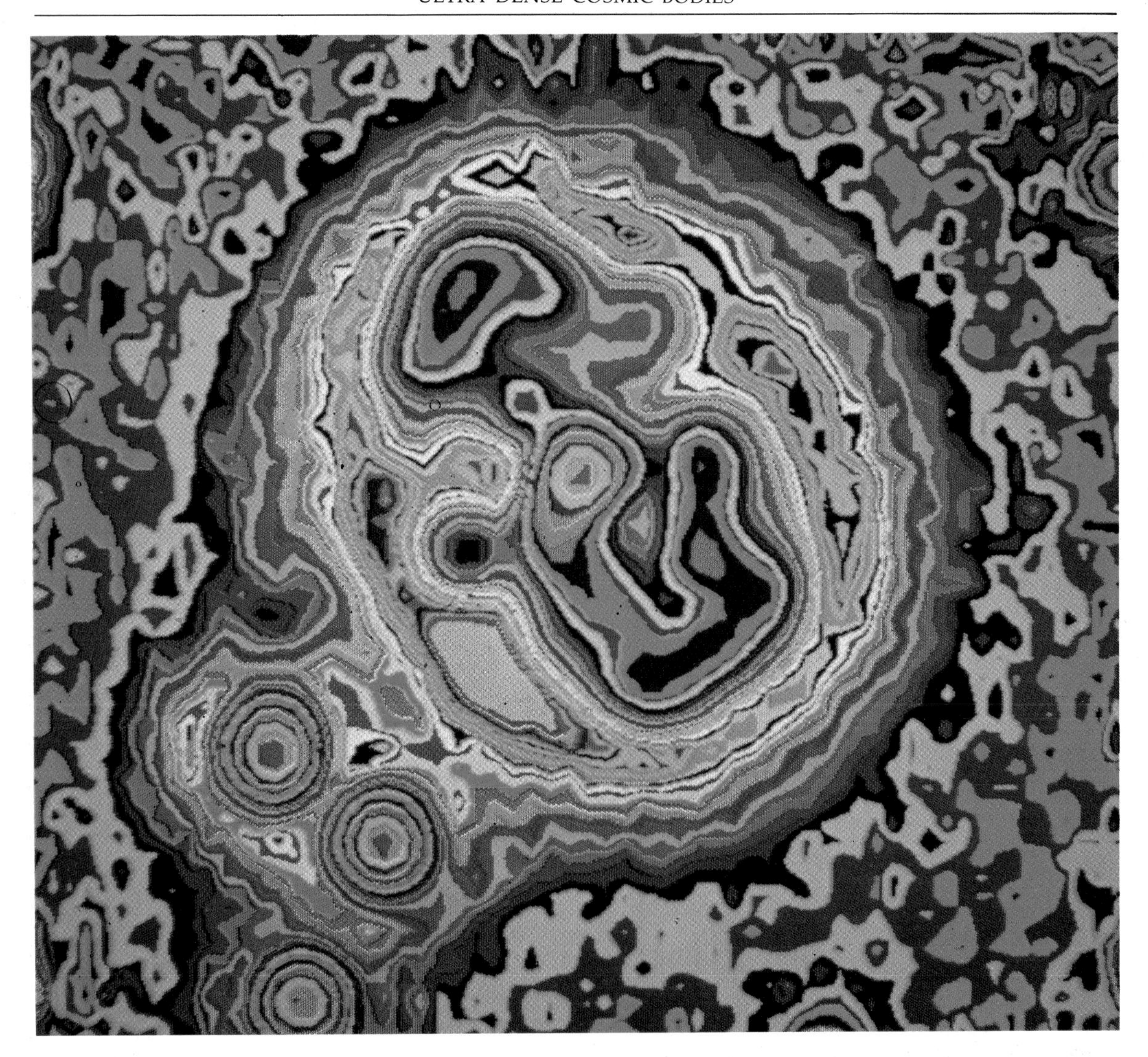

This computer-processed optical picture is of the Homunculus nebula and Eta Carinae, a massive star which may soon (astronomically speaking) become a supernova.

The Cepheids we have discussed are massive young stars (of Population I) and are referred to as **classical Cepheids**. Another type, **W-Virginis stars** (of Population II) are old, low mass stars and are intrinsically two magnitudes fainter than classical Cepheids with the same periods. They are in the helium-burning phase of their evolution and are found particularly in globular clusters.

RR-Lyrae type variables are conspicuous and very numerous members of globular clusters. These are also helium burning stars and their absolute magnitudes are all around +0·6 irrespective of their periods which range from 0·4 to 1·0 days. This is because they are all in the same phase of evolution with very similar age, mass and chemical composition. This constancy of absolute magnitude makes them ideal distance indicators for the older (Population II) regions of the Galaxy.

Ultra-dense cosmic bodies

Stars have a wide range of densities. The mean density of the Sun is $1{\cdot}4 \times 10^3$ kg per m^3 (1·4 times the density of water) and it is known that the density at the centre is over a hundred times greater. However, this is trivial in comparison with those we shall now explore. We saw that the end product of stellar evolution may be the formation of a condensed stellar remnant; perhaps a white dwarf following a planetary nebula phase of a red giant, or a neutron star via a supernova explosion. Both of these stellar bodies have incredibly high densiti , among the highest known.

There is a maximum density to which matter can be packed if it is to remain as atoms with nuclei surrounded by their electron clouds. Normal matter is mostly empty space, because, although the electron cloud occupies the volume of the atom, its mass is negligible compared with the protons and neutrons of the nucleus, each weighing roughly 2 000 times more than an electron. If the electron cloud was removed and the nuclei were squeezed together, they would repel one another because of their positive charges, and in doing so settle down to a new and stable configuration. They would then be at the density of nuclear material, about 10^{17} kg per m^3! This phenomenal density turns out to be that of a neutron star, the material of which, if packed into a match box, would weigh 2×10^{13} kg (the same as 2 000 million double decker buses). One could also achieve this density by compressing the entire Earth into a sphere of radius 200 m! Yet there is one further régime more extreme than white dwarfs and even neutron stars, and this is the realm of the black hole. We will now survey all these regions of ultra-high density in turn.

Double and Multiple Stars

Double and multiple stellar systems present many striking colour and magnitude contrasts, and many amateurs derive a great deal of satisfaction from merely seeking out and examining different systems. However, close binaries are excellent objects for testing both a telescope's resolution and an observer's acuity of eyesight, so this aspect also presents a considerable challenge.

Generally resolution is dependent upon the aperture of the equipment being used, but refractors usually resolve closer pairs than reflectors of the same aperture. (Under certain specific circumstances the optical characteristics of the reflector may make the resolution of a particular double system slightly easier, but this is rare.) Some interesting systems are listed in the table, with a note on some which are suitable for telescopic resolution tests. Many more objects are listed in some of the works given in the bibliography.

For the purposes of resolution tests, apparent doubles – i.e. those where the stars merely happen to lie on the same line of sight, and where the separation remains constant apart from any proper motion – are better objects than the true binaries, where the stars are in orbit about one another. Strictly speaking, both individuals in a binary system orbit their common centre of mass, but it is usual to regard one – usually the brighter and more massive – as being stationary and orbited by its companion. (Similar considerations apply, of course, to multiple systems.)

The study of true binaries is rather neglected by both amateur and professional astronomers, but it remains very worthwhile, partly because of this neglect. Very frequently the separations of the components of binaries and multiples are quite inaccurately quoted in textbooks, largely because many years have elapsed since the last measurements were made. The orbital motion of the stars may have converted 'easy' systems into very difficult objects, and vice versa. Even a single measurement is therefore of interest, but a series on a single object, perhaps continued over years, is of great value. If a proper orbit can be defined the orbital period and also the relative masses of the components can be found.

Generally refractors are favoured for this work, with large aperture and long focal length being great advantages. Although professional astronomers may employ very sophisticated techniques, such as photography and speckle interferometry as described on page 232–4, amateurs use one of the many forms of visual micrometer, the bifilar type being one of the best and perhaps the most readily understood and used. The measurements required are those of position angle (PA) and separation. Because of the problems encountered, most especially with close binaries and those with very unequal magnitude components, it is usual for a set of PA and separation estimates to be made at any one time, the most probable values being then derived mathematically. From these the apparent orbits may be plotted, either by continuing a series of observations over a period of years, or by combining modern measurements with those of earlier observers. The calculations required for deriving the full orbital elements and relative masses of the components are fairly complex, and are usually only applied to observations of a sufficiently high accuracy. However such studies are of great importance in the investigation of stellar characteristics and there are indications that this is now being recognized, and that more amateurs are turning to the measurement of double and multiple stars as a primary interest.

Right and centre right:
Measurement of position angle is always made from north through east (anti-clockwise in most normal telescopes). Separation is the angle between the two components of the double.

Far right:
A number of measurements made over a period of years – in some cases over many decades – can show the relative orbit of the secondary star and reveal the position of perihelion (p).

Opposite:
Double stars are conventionally shown on charts by a bar through the centre of the star. Some well known examples are shown on this chart.

By a strange chance, Gemini (the Twins) contains many double stars, including the brightest stars Castor (top) *and Pollux* (left).

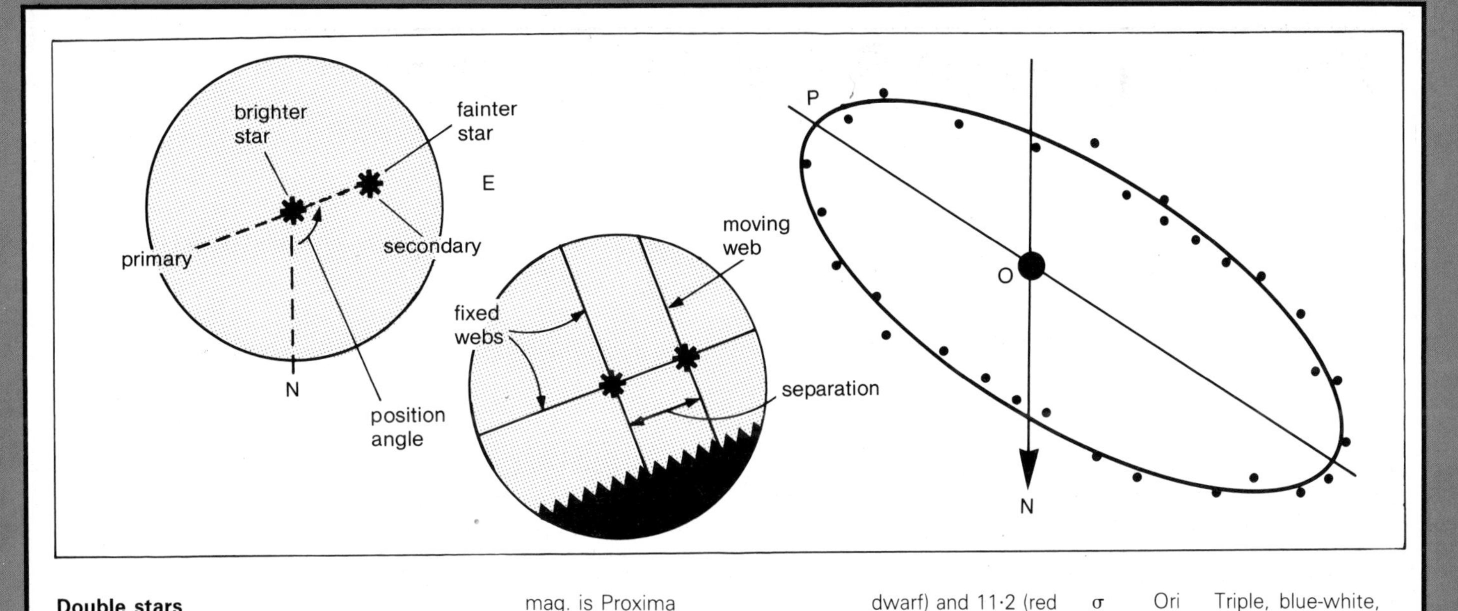

Double stars

	Name	Remarks
γ	And	Yellow and blue
ζ	Aqr	White stars in 75–100mm telescopes
γ	Ari	White 5^m stars
ε	Boo	Orange and blue-green in 75–100mm telescopes*
μ	Boo	Mags 4·3 & 6·5
ξ	Boo	Yellow and orange
ζ	Cnc	Triple system in 100mm telescopes
ι_2	Cnc	Easy pair, 4·2 and 6·6 mags
α	CVn	Easy pair
η	Cas	Yellow and red
α	Cen	Yellow stars (third component at 11 mag. is Proxima Centauri)
β	Cep	Supergiant blue mag. 3·3 star, and 8·1 companion
ζ	CrB	Blue stars
α	Cru	Blue-white pair
β	Cyg	Easy pair, yellow and blue**
o_1	Cyg	Easy pair, orange and blue, triple in good binoculars or 75mm telescope
61	Cyg	Orange stars
γ	Del	Yellow stars
υ	Dra	White 5 mag. stars
16–17	Dra	Easy blue-white pair, triple in 75mm
θ	Eri	Easy blue-white pair
o_2	Eri	Triple in 75mm telescopes, mags 4·5, 9·5 (white dwarf) and 11·2 (red dwarf)
α	Gem	(Castor) Blue-white pair and red dwarf (Each star is itself double, but only spectroscopically)
ρ	Her	Easy pair
95	Her	Yellow and white
ε	Hya	Mags 3·5 and 6·9
γ	Leo	Easy pair of yellow stars**
α	Lib	Easy pair, mags 2·9 and 5·3
β	Lyr	Yellow component famous eclipsing binary, blue companion
ε	Lyr	Famous 'double double'. Easy pair, each of double in 75 – 100mm**
β	Mon	White triple system
σ	Ori	Triple, blue-white, blue, red. Quadruple in 150mm
ζ	Pav	Red and white stars
η	Per	Orange and blue
β	Sco	Blue-white pair
ν	Sco	Triple Quadruple in150mm telescope
δ	Ser	White stars
θ	Ser	Easy pair
β	Tuc	Triple
ζ	UMa	(Mizar) Companion (Alcor) visible to good eyes, triple in small telescopes
ξ	UMa	Yellow stars
γ	Vel	Easy blue-white pair, quadruple in 75mm**
γ	Vir	Yellow-white stars**
γ	Vol	Yellowish-white and yellow stars

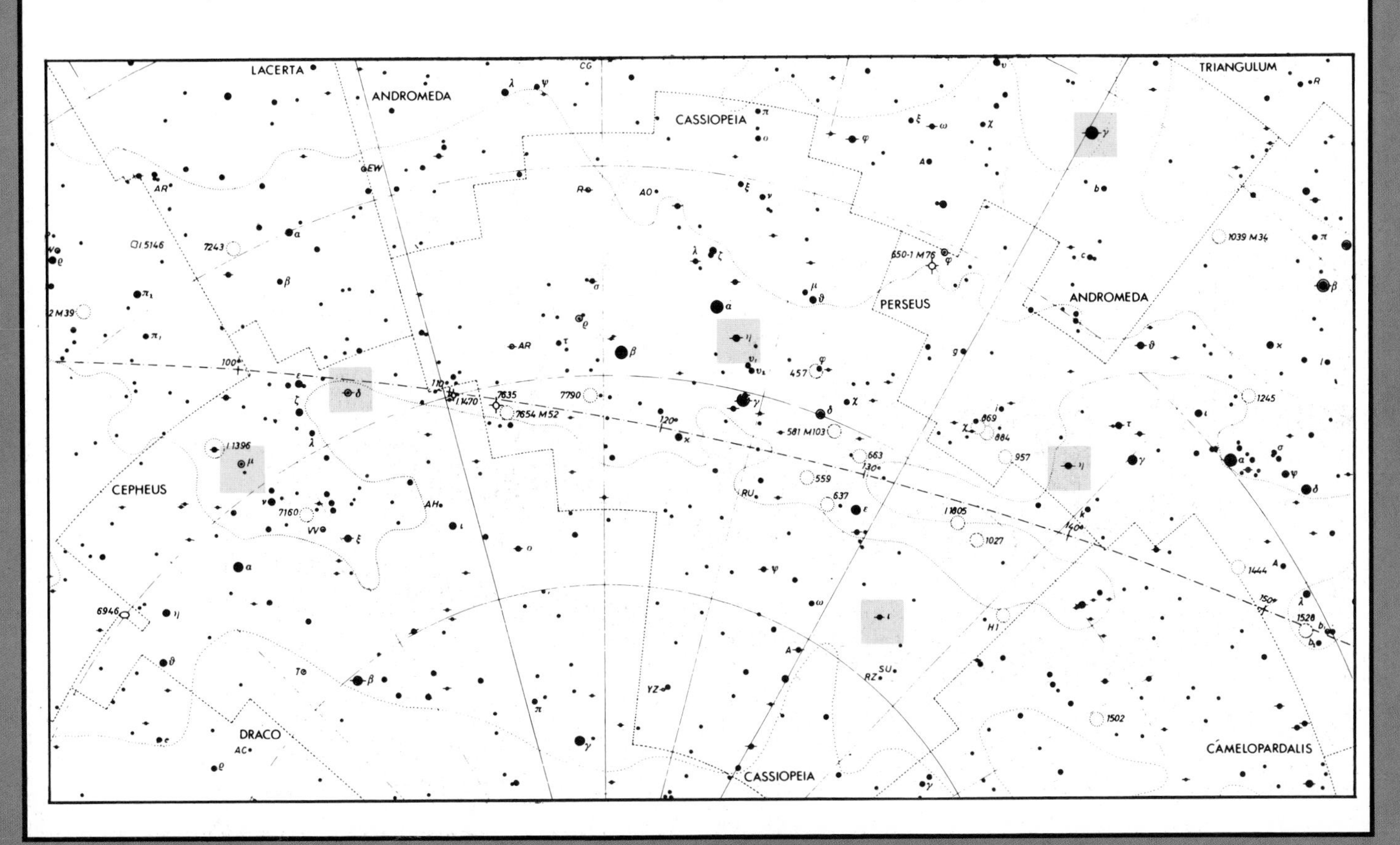

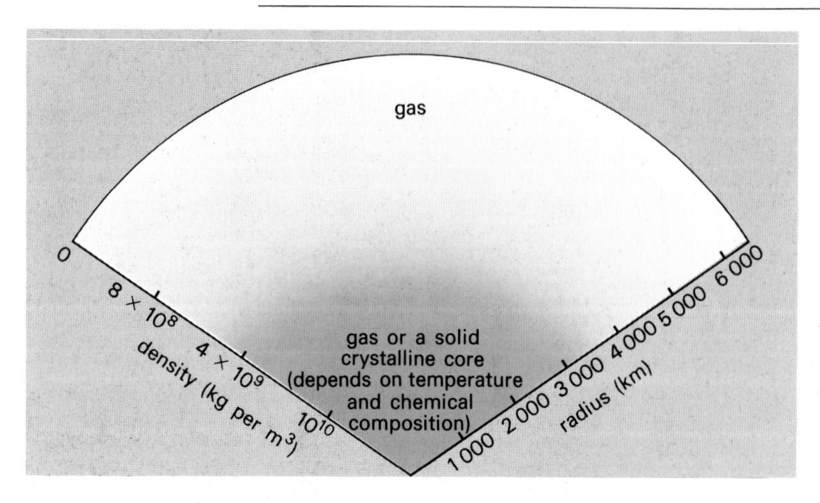

Fig. 3·23 Sector diagram of a white dwarf illustrating how the interior density varies with distance from centre. The model is for a 0·9 solar mass star and the radius is virtually the same as for the Earth. The central temperature is 10^6 K and the core is liquid if composed of carbon, but becomes solid for all heavier elements. A white dwarf composed mainly of iron would be virtually totally solid.

White dwarfs

White dwarfs are stable because their gravitational pressure is balanced by their electrons resisting being squeezed. Indeed, the behaviour of degenerate materials like this depends only on its density, and the mass of the degenerate particle (in this case the electron). Unlike normal matter, it is not dependent on temperature, with the dramatic consequences of burning we have seen for the helium flash. Another consequence is that as the mass of a degenerate body increases the radius shrinks, so the most massive white dwarfs are the smallest! However, there is a maximum mass that a white dwarf may attain – the Chandrasekhar mass limit of 1·4 $M_\odot$. Beyond this the velocities of the electrons become **relativistic** (that is, they approach the velocity of light), the degenerate material becomes unstable and a rapid collapse begins until, perhaps, the neutron star state is reached. The internal structure of a white dwarf is shown in Fig. 3·23.

The observational evidence for white dwarfs is strong. It began in 1862 with the observation of a dim companion star to Sirius, a companion whose presence had been predicted nearly 20 years earlier by Friedrich Bessel to explain a wobble in Sirius' motion. In the early 1900s, spectroscopic measurements showed that this companion had the same surface temperature as Sirius, although it was ten magnitudes fainter! This then made it a very strange object occupying a deserted region of the H–R diagram.

Now because the luminosity of the binary companion, Sirius B, was 10 000 times less than Sirius although their surface temperatures were the same, the temperature-luminosity relationship (page 45) showed that Sirius B must be 100 times smaller (that is, $R \sim R_\odot/50$). But study of the orbital motion of the pair showed that it had a mass about equal to the Sun ($M_\odot$). Therefore, with a density of nearly 10^5 times that of the Sun, Sirius B had to be a white dwarf.

Many other similar stars have now been found in the Galaxy and many more are probably present though unseen because, as they grow older, they cool until they become too faint to observe.

Fig. 3·24 Cross section of the interior of a neutron star. The core composition is unknown but will probably be a mix of exotic elementary particles. The rigid crust is a good electrical conductor and pins the enormously strong magnetic field to the surface, making it sweep out a region of space as the neutron star rotates. The mechanism for pulsar emission is believed to originate somehow with this rotating magnetic field plasma.

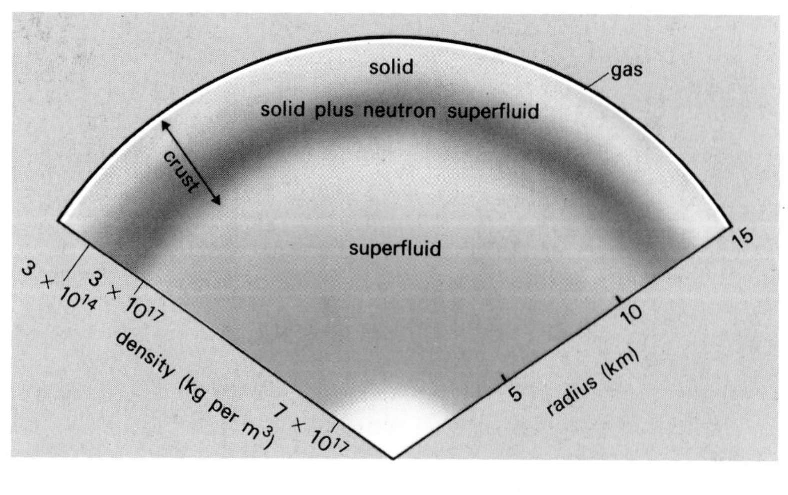

Neutron stars

Neutron stars are, by any stretch of the imagination, exotic. Like white dwarfs their internal stability comes from their atomic particles resisting squeezing, although neutrons not electrons are doing this. The high number of neutrons have mostly been formed by electrons reaching relativistic velocities and interacting with protons to produce neutrons. The entire structure of a neutron star is really bizarre; it has a mass of about one solar mass but a radius of only 10 km!

In some ways it looks more planetary than stellar, although further comparison quickly vanishes. The crust is extremely rigid, 10^{18} times more so than steel, and, because of the tremendous gravitational force, surface features are minor. The highest mountain would be measured in millimetres and climbing it would require the same energy as carrying 10^7 kg to the summit of Everest! The internal structure is shown in Fig. 3·24.

It is now thought that the exotic objects which give rise to the **X-ray bursters**, where pulses of energy lasting only 3 to 100 seconds are observed, include a neutron star as one component of a binary system. In a similar manner to the nova mechanism (page 65) these produce the X-ray pulses by accretion of material on to the neutron star. The gravitational fields around neutron stars are so strong that relativistic effects have to be taken into account, and it would seem that these X-ray bursters can only be properly explained if a modified theory of gravity proposed by J. W. Moffat, and for which there is other evidence, is accepted.

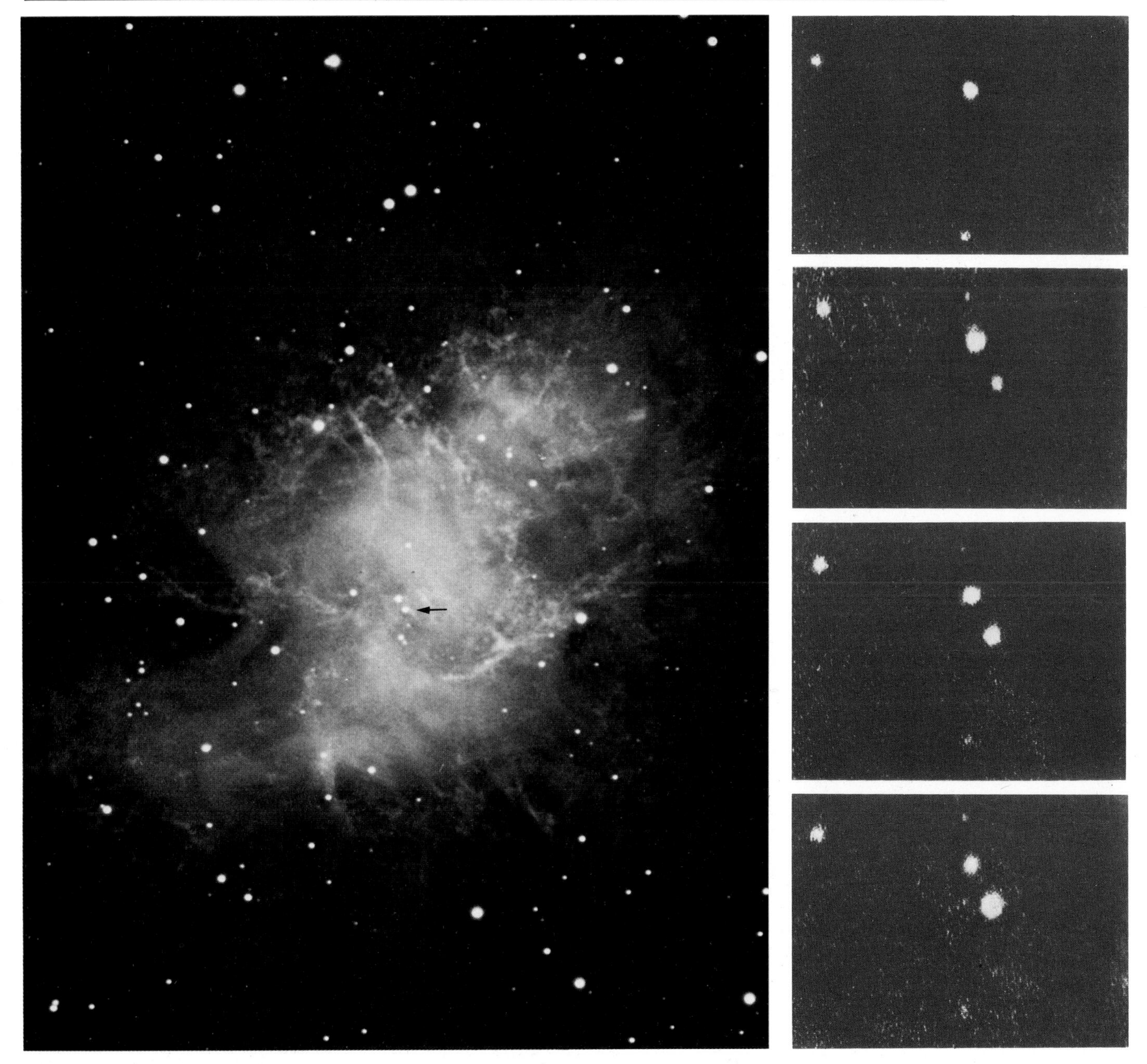

The Crab Nebula; a supernova explosion observed by Chinese astronomers in 1054 AD. The nebula is about 2 kpc distant and is observed to be expanding. One of the central stars, indicated by the arrow, is the Pulsar NP 0531, which is flashing like a lighthouse more than thirty times every second, as shown in sequence of television pictures at right.

Pulsars

In the 1930s, Walter Baade and Fritz Zwicky suggested that neutron stars might perhaps exist as the stellar remnants of a supernova explosion. This far-sighted idea lay dormant until the early 1960s when it dramatically exploded on the scene with the 1967 discovery of pulsars. These, found by a team of astronomers from Cambridge University, England, appeared as radio sources which pulsated with very short periods, of the order of seconds or less, and they maintained these periods with extreme accuracy (Fig. 3·25). Neutron stars seemed to be the most likely explanation for the generation of such short pulses and the discovery, one year later, of the pulsar NP 0531, embedded deep in the heart of the Crab Nebula supernova remnant, both confirmed Baade and Zwicky's hypothesis and proved beyond doubt that pulsars were neutron stars. Neutron stars rotate rapidly, and, although the actual mechanism of the pulsations is still unclear, it certainly seems to involve a strong magnetic field rigidly linked to the rotating surface. Somehow and somewhere close to the neutron star itself, energetic particles (probably electrons) interact with this rotating field to produce a narrow, lighthouse-like beam of radiation which we see as a pulse when it intercepts the Earth (Fig. 3·26). Pulsars have also been observed in the X-ray region as components of X-ray binary sources.

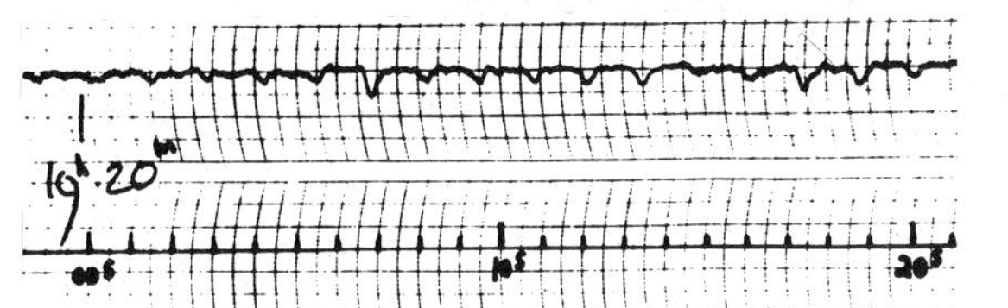

Fig. 3.25 Pulse trace (obtained on 1967 November 28) revealing the discovery of a hitherto unknown class of objects, now referred to as pulsars. The regular but variable radio emission from the first known pulsar CP 1919 is clearly visible in the upper trace. The lower trace is a timing marker. CP 1919 emits a radio pulse every 1.3373 s.

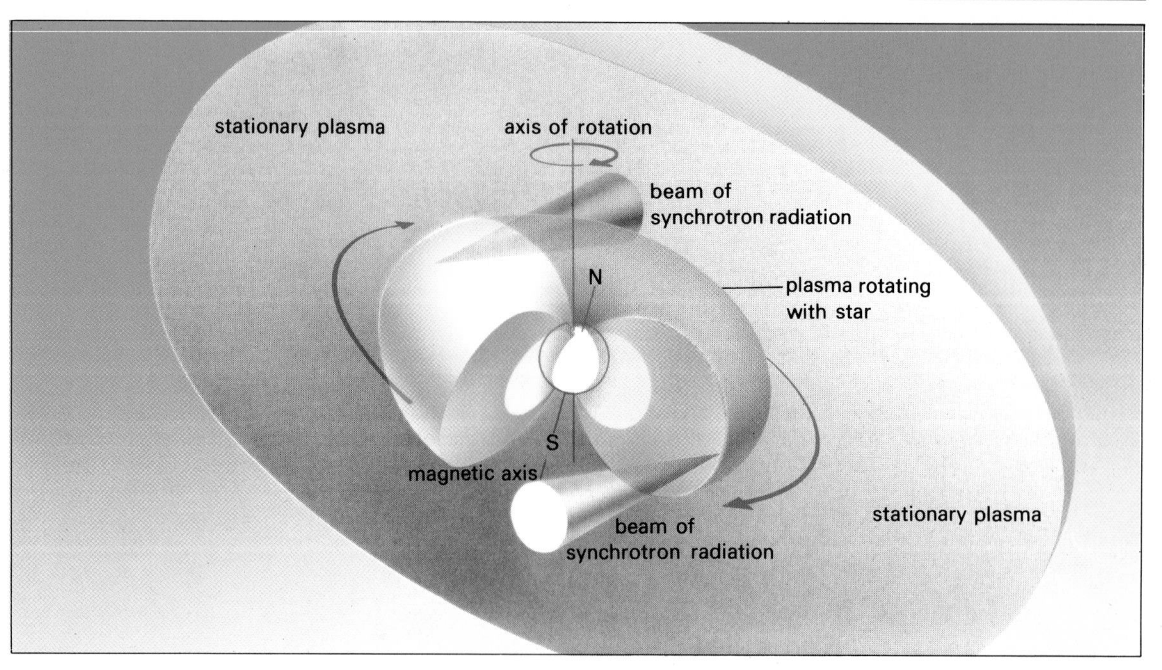

Fig. 3·26 A 'lighthouse theory' model of pulsar emission from a rapidly rotating magnetic neutron star. The magnetic field is exceptionally strong and rotates with the star. Swept up plasma is accelerated until, at the periphery where it decouples from the magnetic field, highly directional radio beams are emitted.

The Crab pulsar has a period of 33 milliseconds (ms), and, although extremely regular, it was eventually observed to be increasing gradually. This was only just perceptible, but its existence proved dramatically important in solving a mystery of the Crab Nebula, which is the remnant of a supernova explosion observed by Chinese and Arabic astronomers in 1054 AD. The mystery arose over the radio, optical and X-ray emission we now detect from it. This radiation gives every evidence of being emitted by the synchrotron process, so that its intensity must depend upon the strength of the magnetic field and the energy of the electrons. As the field in the Crab is known, the electron energies can be calculated, and it is found that up to 10^{14} electron volts are needed to explain the X-ray emission. Yet such high energy electrons would lose energy within the nebula itself in a few months. As the supernova occurred over nine hundred years ago, the mystery was how the electrons could obtain their high energies. The existence of a spinning neutron star solved the problem: the loss of energy it suffers by its observed slowing down is sufficient to explain the energy required to feed the electrons to produce the observed radiation. In fact, the neutron star acts as a powerhouse.

Observations of a few pulsars, of which the Crab is one, have shown that the regular periods occasionally go haywire and suffer a hiccup, known in the jargon as a **glitch**. These are now believed to be starquakes, caused by stresses as the neutron star undergoes its spin down. Eventually these are sufficient to cause the crust to deform and, being extremely rigid, it cracks. This causes an internal readjustment which manifests itself to us as a glitch. It is now generally suspected that neutron stars are born spinning exceedingly quickly but gradually lose energy and rotate more slowly until, after a few thousand years, their periods settle down to the order of seconds.

One pulsar (known as PSR 1937 +215) has a period of only 1·55 milliseconds, one-twentieth of that of the Crab pulsar. This is so fast that it must be spinning at about 90% of its disruption speed, if current theories are correct. It has been tentatively identified with a 22nd magnitude object, and if this is verified will be only the third pulsar to be optically confirmed, the Crab and the Vela pulsar (known to be a star of 25th magnitude), being the others. One other 'millisecond pulsar' is now known, and their high rate of spin suggests that they are very young objects.

The discovery that the pulsar 1913 +16 is one component of a binary system with an orbital period of 7·75 hours, offered the possibility of an independent test for the various gravitational theories which have been proposed. Studies have shown that both stars are compact objects with masses of about 1.4 $M_{\odot}$ (in accordance with the theoretical predictions for neutron stars), and that they are in a very close orbit, which is also highly eccentric. Einstein's general relativity theory (and one or two other versions, especially the Brans-Dicke theory) predict that when ultra-dense objects like these are in such close orbits, they should lose energy by the emission of gravity waves (page 244), and cause the orbital period to decrease. Such a decrease has now been detected for this object, and it proves to be of precisely the right amount. Moreover, theory also predicts that an eccentric orbit should precess, and this has also been confirmed, measurements having shown that it occurs at the enormous amount of 4·2 degrees per year (which may be compared with the 43 arc sec. per century for the precession of the perihelion of Mercury). There is just a slight shade of doubt over the nature of the pulsar's companion star, which could possibly be a white dwarf or helium star. However, the likelihood of this is very small indeed, and for the moment this object appears to provide striking confirmation that the theory of general relativity is correct.

Neutron stars are also thought to be associated with the sources of gamma rays which have been detected by certain spacecraft. The Crab pulsar, the Vela pulsar, and another object known as Geminga

– recently identified at X-ray wavelengths, and possibly in the visible region as well – are the only positively identified sources. However there are a very large number of transient bursts of gamma rays which are also observed, and these cannot yet be satisfactorily explained. These **gamma-ray bursters** appear to produce essentially instantaneous events, ranging in duration from only tenths of a second to about a minute. Their detection by various spacecraft in different parts of the Solar System has enabled the directions of the sources to be determined, but no certain identification with any object has ever been made, although their distribution suggests that they are likely to be galactic objects. One very energetic burst (and later, lesser ones) appeared to correspond with the radio source N49, thought to be a supernova remnant in the Large Magellanic Cloud. Oscillations with a period of eight seconds in the case of this burst, and four seconds in another event elsewhere, point to neutron stars as probable sources, as they could rotate at such rates. The mechanism for the production of these intense bursts of gamma rays is quite unknown; however, it has been suggested that collisions between asteroids and neutron stars might be involved, which could account for the apparent single event per source. A more 'normal' accretion mechanism in a binary system has also been proposed, but this might be expected to show frequent bursts. For the time being these objects remain a problem, although one possible candidate for an optical burst has been found.

Black holes

A neutron star, like a white dwarf, can have only a certain maximum mass. Above this limit, thought to be about 4 $M_\odot$, it will suffer a rapid and unrestrained collapse into a very much higher density régime – a black hole. Current theory predicts no limit to the collapse, so the density goes to infinity as the radius shrinks to zero. This mind-boggling object of zero radius is known as a **singularity**. It is surrounded by a region, the boundary of which is called the Schwarzschild radius (R_s), named after its discoverer Karl Schwarzschild. This is the outward distance at which the **escape velocity** (the velocity required to escape from the gravitational pull of the collapsed body) equals the velocity of light. Any

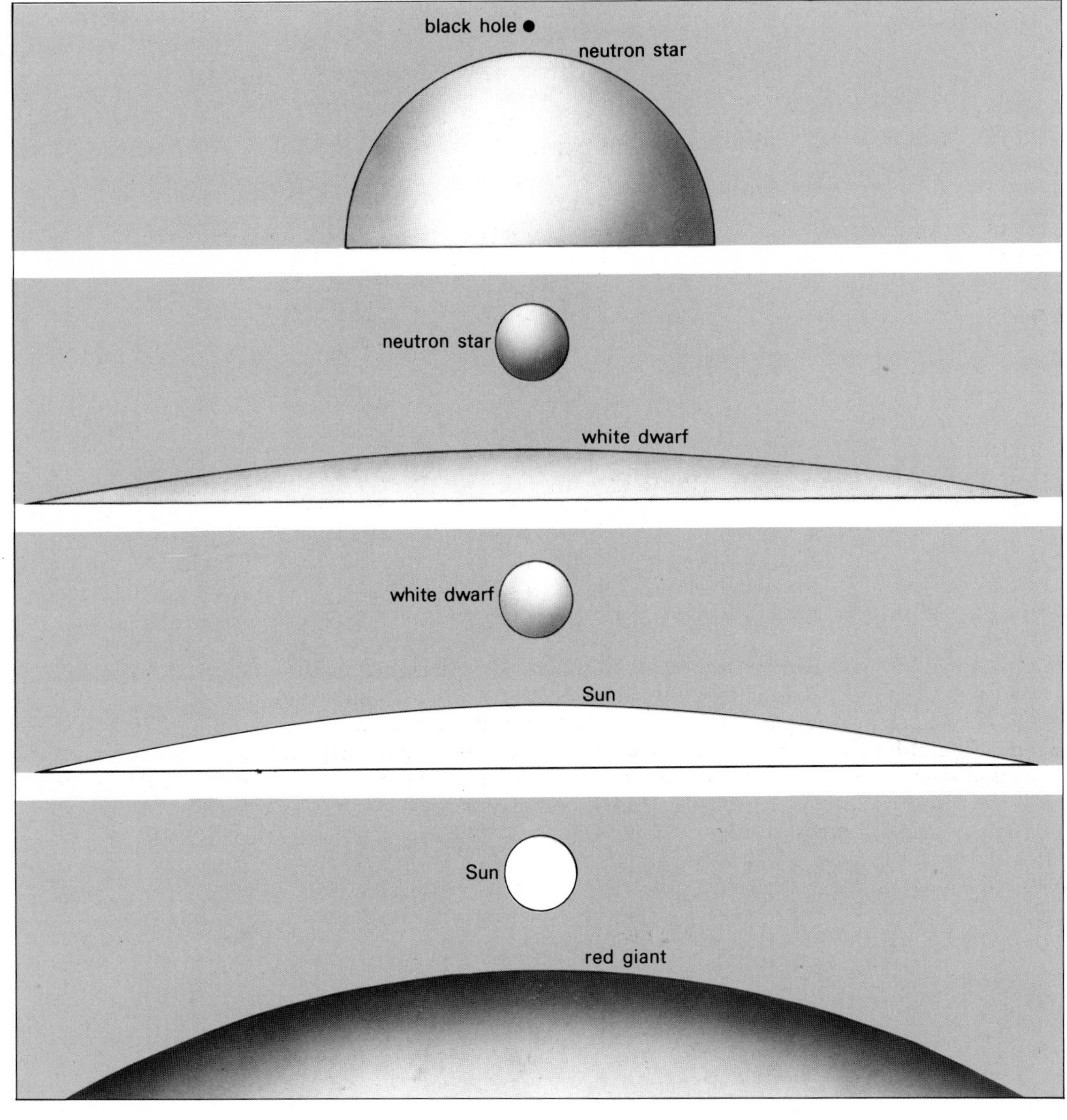

Fig. 3·27
Relative sizes of stellar bodies, all of one solar mass, showing the enormous range of density spanned by stars in different phases of evolution.

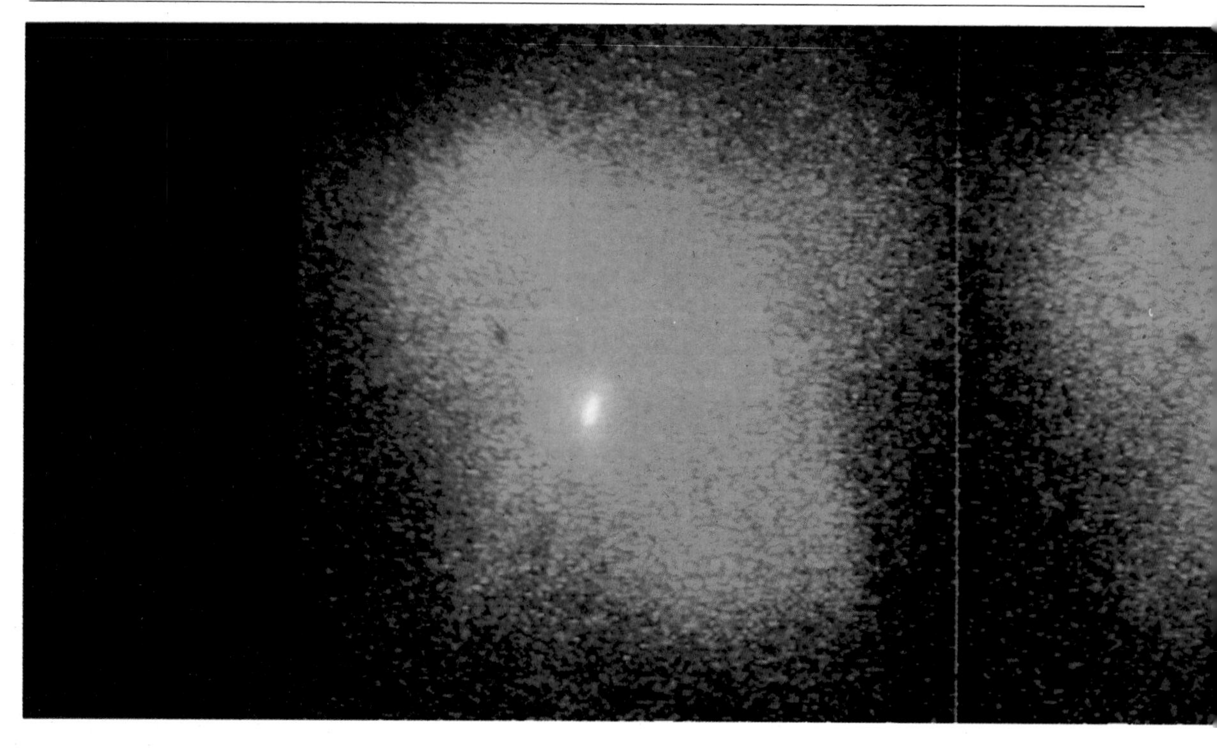

This pair of X-ray observations of the Crab pulsar by the Einstein satellite shows 'on' (left) *and 'off' phases, just like those seen at optical wavelengths (page 71).*

body, nuclear particle or photon of radiation, inside this boundary is forever trapped and isolated from the rest of the universe. Other names for this boundary are **event horizon** and 'speed of light surface'. This then is the concept of a black hole; events within this horizon are completely unknown to an external observer, and the theory ensures 'cosmic censorship' by stating that naked singularities can never exist, they will always be clothed by a black hole.

The size of the Schwarzschild radius R_s, is given by a simple expression:

$$R_s = 2GM/c^2,$$

M being the mass of the body. For the Sun, R_s turns out to be 3 km. Because the Sun is much larger than this we need not worry, but if it were to shrink to this size, then it would indeed become a black hole. A neutron star has a mass equal to the Sun and a radius of 10 km, therefore if it were to shrink by only a factor of three, it would turn into a black hole. How can it be made to shrink? Simply by adding mass until its critical mass is exceeded and it will then immediately collapse into a black hole. All further 'information' would then be lost (another form of 'censorship') because we can never know what happens in the depths of a black hole. Anything can go in but nothing, not even light, can escape. Black holes could act as the ultimate refuse disposal agent for the Galaxy, swallowing everything and spewing out nothing.

Do black holes exist? How can they form? These two most tantalizing questions can be partially answered if we consider the evolution of a massive star. As it evolves it eventually exhausts its nuclear fuel and explodes as a supernova. One of three events can then ensue: the core can be completely destroyed in the supernova explosion; or if something of the core remains and is sufficiently small, it will form a neutron star; if, though, it is too massive the core will collapse to form a black hole. For stars whose masses greatly exceed 10 $M_\odot$, their fate is either complete disruption or the formation of a black hole. Because we know such massive stars exist in the Galaxy, we suspect the existence of black holes as one natural end-product of stellar evolution. Other mechanisms for black-hole formation involve the influx of material on to a neutron star which is a member of a binary system, thus easing it over its critical mass so it will collapse; or, more speculatively perhaps, the collapse of the massive, dense, central regions of globular clusters and giant galaxies. The sizes of black holes and other stars are shown in Fig. 3·27.

If we were to witness a massive star collapsing tc a black hole (albeit from a safe distance so that we were not swallowed inside the Schwarzschild radius) we would again have a surprise. Someone in a space suit orbiting the star would be trapped with the collapse and would plunge into the singularity to be crushed out of existence as the gravitational forces increased. Indeed, if they fell in feet first they would be stretched like a piece of spaghetti because the gravitational force on the feet would vastly exceed that on the head. To us watching from a distance, however, the events of the plummet into oblivion would seem to take an infinite time to reach the point of no return – the Schwarzschild radius. These and other more peculiar effects are a consequence of the theory of relativity and the extreme distortions of space and time by the singularity (see Chapter 8).

Black holes are masses which do not radiate, hence the name. However, their gravitational fields are still present and can influence any close passers-by. This gives us a clue for black hole detection. As a member of a binary system, a black hole might visibly perturb the orbit of an observable primary. If this mass can be found then the mass of the secondary might be deduced, and if it turns out to be too great for a white dwaf or a neutron star, then it must be a black

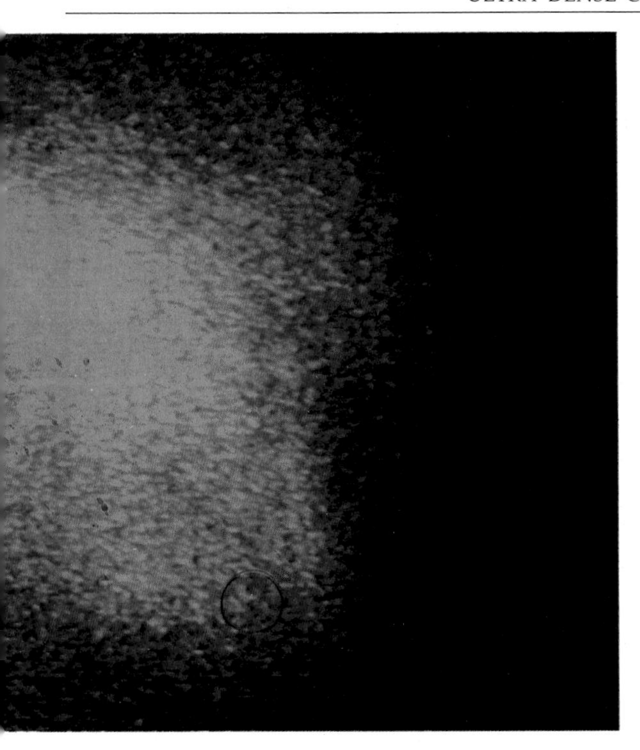

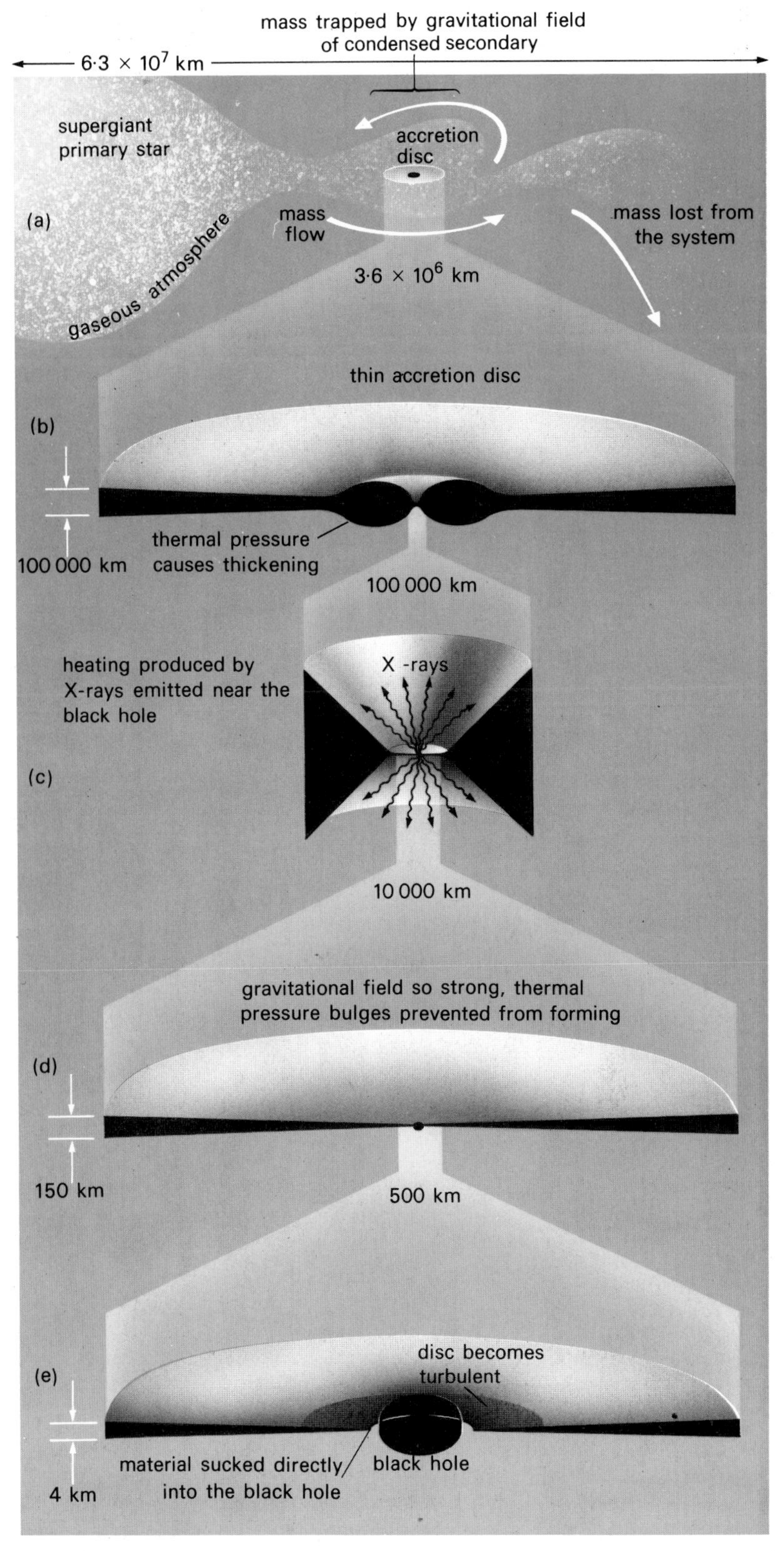

Fig. 3·28 A model of an accretion disc surrounding a black hole in Cygnus X-1. The gas pulled off the primary does not fall directly into the black hole but, instead, forms a circling accretion disc around it. This very flat disc is large in extent compared with the Schwarzschild radius of the black hole. The X-rays we observe probably emanate from only the innermost part (e) of the disc, where temperatures rapidly increase as material is eventually sucked into the black hole itself.

hole. Alternatively, the black hole could have a surrounding accretion disc as it consumes mass from its companion due to its gravitational strength. As the matter spirals towards the black hole, but well outside the event horizon, it will become heated and emit energetic radiation.

Examples of accretion disc processes are believed to account for various X-ray stars. Most have neutron stars as the collapsed component, but one particular source remains a mystery. This is Cygnus X-1, a strong and highly variable X-ray source (Fig. 3·28). The variations are non-periodic and range from flickerings of tenths of a second to flarings on a monthly scale. Optically, the object appears as a spectroscopic binary B star with a period of 5·6 days. When the mass of the unseen companion is calculated the answer turns out to be at least 6 M$_\odot$. If this is correct then it seems that the unseen companion must be a black hole. There remain some problems in assigning the mass however, and although Cygnus X-1 remains a good candidate for a black hole, another object known as LMC X-3 seems to be even more definite. Here, determination of the orbital velocity and period, and the perfectly normal appearance of the visible star (a B3 star of about 6 solar masses), suggest that the invisible component must have a mass of between 6 and 14 solar masses, most probably 9–10. A neutron star cannot exceed 4 solar masses, so that a black hole remains the only possibility.

Alternative means of detecting black holes lie in finding systems whose structure indicates the presence of a large amount of mass in a non-luminous region. Globular clusters, galactic nuclei and clusters of galaxies spring to mind as possible occupation regions. Because black holes are so bizarre and there seems every reason, in theory, to suspect that they do exist, it is tantalizing for astronomers to be still thwarted in attempts to observe conclusively even a single such object.

CHAPTER FOUR

The Sun

Although the Sun is a typical star it requires special comment for a number of reasons. It is unique to us among the stars because it provides the light and heat necessary to sustain life on Earth. For the astronomer it is the only star whose surface can be observed in detail, and it acts as a giant nearby astrophysical laboratory. Scientific studies of specific conditions found in the Sun and its atmosphere, but not on the Earth, have led to advances in the fields of plasma, nuclear and atomic physics. Moreover, by studying the Sun astronomers can investigate physical conditions typical of most stars. Biologists, climatologists and meteorologists are also interested in it because its radiant energy has significant repercussions for life on the Earth. Table 4·1 (page 91) lists the overall properties of the Sun.

The visible disc

We know that the Sun is an incandescent sphere of gas, yet when we view the **photosphere**, the highly luminous surface, its edge or limb appears sharp, as if it were a solid body. It does not gradually merge into the blackness of space as we would expect. The interpretation of this basic observation is that the region from which most of the visible radiation is coming must be thin compared with the radius of the Sun, and is only a few hundred kilometres deep. In this zone, the gas becomes more opaque with depth, with completely opaque gas below it and a transparent solar atmosphere above. Photographs of the Sun reveal that the disc is perceptibly fainter at its rim than the centre, a phenomenon known as **limb-darkening**. The explanation for this is very simple. A line of sight to the disc centre penetrates to a greater vertical depth before reaching an opaque layer. We therefore see slightly deeper into the Sun where the regions are hotter and in consequence more luminous. A line of sight to the limb on the other hand passes obliquely into the solar atmosphere and becomes opaque at a slightly higher and cooler level. The way this effect depends on magnitude and wavelength is very important in assisting astronomers to define the structure of the solar atmosphere.

The black body temperature of the Sun is obtained by comparing the continuous spectrum with that for black bodies and turns out to be 6 000 K. The effective temperature derived from the luminosity of the photosphere (using the relation ($L \propto R^2 T^4_{eff}$)) is 5 800 K, although limb-darkening means that the centre and limb temperature will straddle this average value. The solar surface also reveals fine structure brightness variations, referred to as granulation, the granules being bright patches with a dark border about 1 000 km in size. This irregular mosaic is continually changing on a time scale of minutes; high speed photography shows that the solar surface resembles a pan of simmering soup. By observing the Doppler shifts from these granules we know that their hot centres are rising whilst the cooler, darker boundaries are sinking.

This is evidence of convective motion taking place just below the visible surface and is believed to happen because hydrogen is undergoing a change from being completely ionized in the deep interior to neutral at the surface. In this intermediate zone, radiation flowing out from the core meets a sudden increase in opacity; energy is then transported to the surface not by radiation but primarily by turbulent convective currents of heated gas. The transition region begins at about 0·85 of the distance from the centre to the surface. In the lower convective zone three main layers are thought to exist, the deepest forming giant cells, with supergranular cells next and finally granular cells above, the tops of these forming the Sun's visible surface.

Chromosphere and corona

Above the photosphere the atmosphere of the Sun rapidly thins and the temperature drops to 4 000 K in a 500-km-thick layer. This thinner, cooler gas is transparent to most wavelengths of the continuous photospheric spectrum but it asborbs radiation at wavelengths characteristic of the atoms in this layer. This is the zone which produces the solar spectrum absorption lines first studied by Fraunhofer. By analysing this incredibly complex spectrum, solar astrophysicists have been able to deduce the abundances of chemical elements and also their relative states of ionization. This cooler zone, occasionally referred to as the **reversing layer**, is the lower portion of a much larger zone extending a few thousand kilometres from the surface and called the chromosphere. These outer zones have no sharp boundaries and gradually merge into one another; indeed, the outermost region, the corona, extends for many solar diameters before it merges into the general interplanetary medium and solar wind.

The chromosphere is only optically visible for a few seconds before and after a total eclipse of the

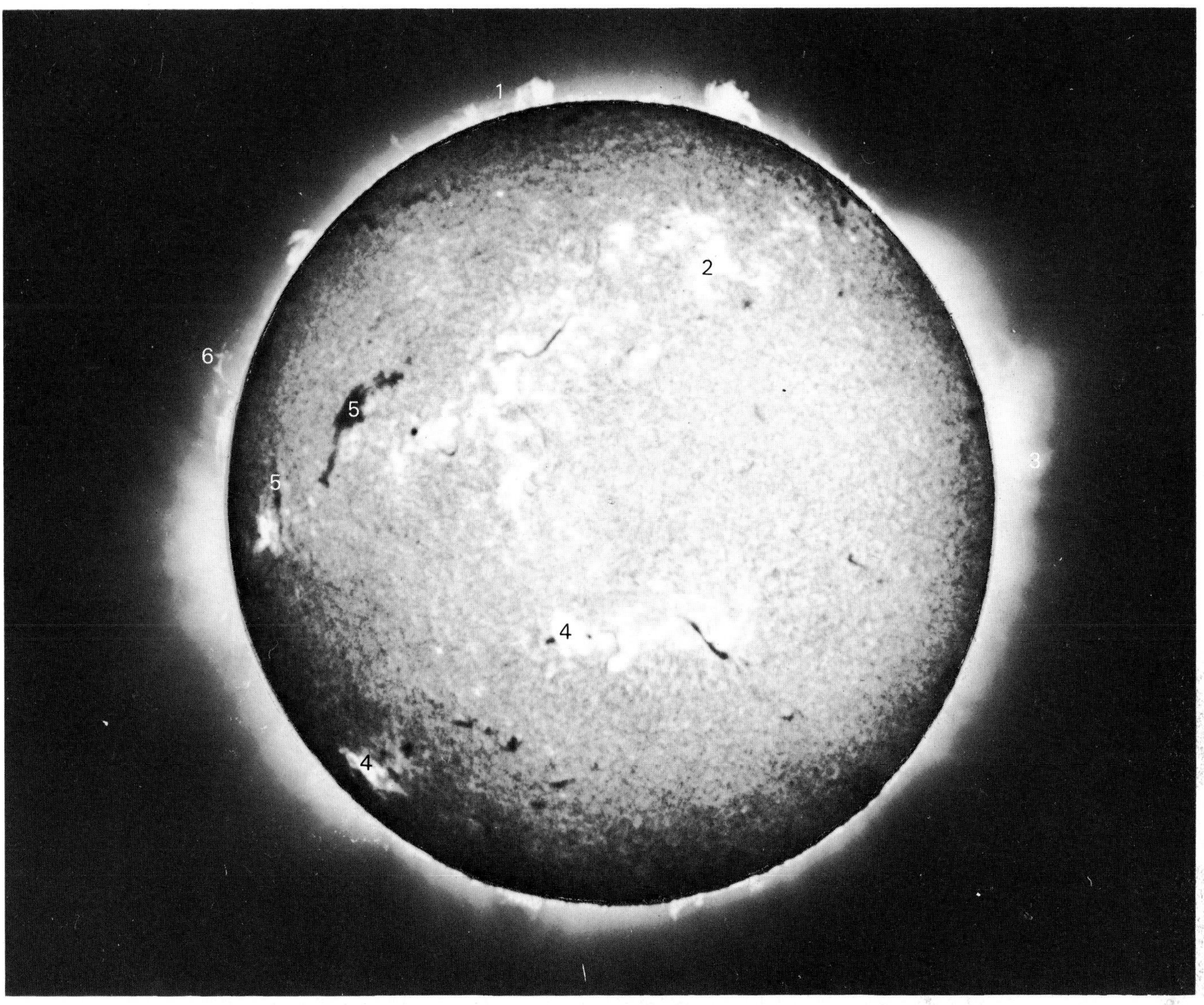

Sun. When thus observed it has a reddish colour, the emission of the dominant atomic element, hydrogen. (The reason we see the chromosphere in emission and not asborption is because the hotter photosphere is blocked off during an eclipse.) Thousands of emission lines are visible in a spectrum of the chromosphere recorded during an eclipse. In 1865, the element helium was first discovered there – indeed its very name derives from the Greek word *helios* meaning Sun. It was not until 30 years later that helium was isolated in the laboratory! Because total eclipses are very rare, special instruments have been developed for observing the outer solar atmosphere without waiting for an eclipse. These instruments do, however, need very good sites with a minimum of atmospheric scattering and are usually found at high altitude observatories. Of these the chief are the **coronagraph** and **spectroheliograph** which are narrow-band filter devices enabling the outer solar atmosphere to be studied in the light of one particular element. The most common and useful are the 656·3 nm Hα line and the 393·4 nm Ca line.

From such studies the chromosphere has been shown to have a very complex structure. Supergranular types of cells are observed along with networks of spikes resembling jets of flame and called **spicules**. These are temporary features with lifetimes of the order of minutes and extending from the base of the chromosphere to altitudes of 10 000 km. Even more spectacular are **prominences**, giant streamers of luminous gas extending well into the corona. Sometimes the prominences form loops and arches and are among the most spectacular of solar features. They are intricately connected with the Sun's strong magnetic field lines, as are all so-called 'active regions'.

As one moves outwards through the chromosphere the temperature rises until it reaches about 10^6 K where the chromosphere gradually merges into the corona. This is a region of extremely low density, which extends outwards for two or three solar radii until it blends into interplanetary space. Without specialized equipment, the corona, like the chromosphere, can only be seen during the brief few minutes of a total solar eclipse, but it is a spectacular sight. Glowing with a pearly coloured hue, its illumination is due mostly to sunlight from the photosphere being scattered by the free electrons of the highly ionized corona PLASMA. Observation of emission lines from very highly ionized elements there show that the coronal temperature exceeds 10^6 K. Recent data from ultraviolet spectrometers on board orbiting satellites

A composite photograph of the Sun with a white-light photograph of the corona and prominences superimposed on a filtergram of the disc taken in the light of a particular line of ionized hydrogen, Hα. Features marked are: quiescent prominence 1, plage 2, coronal plumes 3, sunspot groups 4, filaments 5, active prominence 6.

The full splendour of the very hot solar corona revealed during a total eclipse of the Sun (1976 October 23).

have greatly assisted the study of the high temperature plasma of the corona, while Skylab X-ray photographs have revealed the full splendour of hot active regions of the solar disc and the existence of spectacular coronal plumes. The heating of the corona is probably caused by shock waves emanating from the photosphere, so it is maintained in a dynamic state, continually expanding into interplanetary space to be replenished from the chromosphere at its base.

Solar wind

The expansion of the corona into interplanetary space is termed the solar wind; it is composed of electrons, protons, helium nuclei and other ionized material which pass from the Sun, through the Solar System, and merge eventually into the general interstellar medium. The solar wind represents a mass loss for the Sun of about 10^9 kg per s, which although large by our standards has a negligible effect on the Sun's evolution. (Supergiant stars, on the other hand, have stellar winds which carry off significant amounts of mass, up to 10^{-5} $M_\odot$ per year. This does effect their eventual evolution and may produce variable X-ray sources if the star is surrounded by a condensed companion.)

Skylab observations revealed the existence of phenomena now known as **coronal holes**, where the density and temperature are greatly reduced and appear as dark patches in X-ray pictures, but show no associated activity in visible light. It is now well established that the coronal holes are the source of high-speed streams in the solar wind, being the site

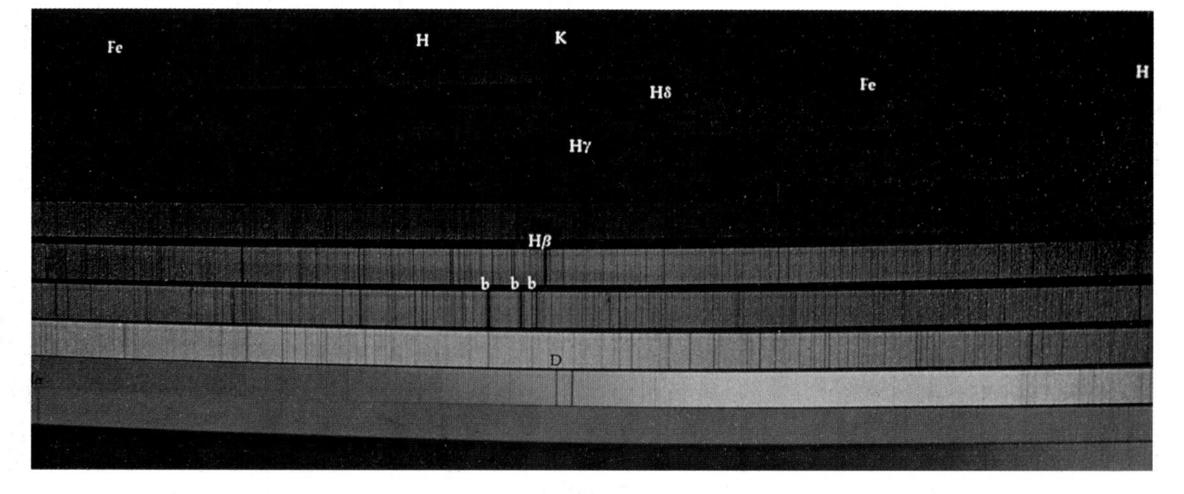

The spectrum of the Sun in visible light. Many absorption lines are seen and it was by studying such lines that the element helium was first discovered in the year 1865.

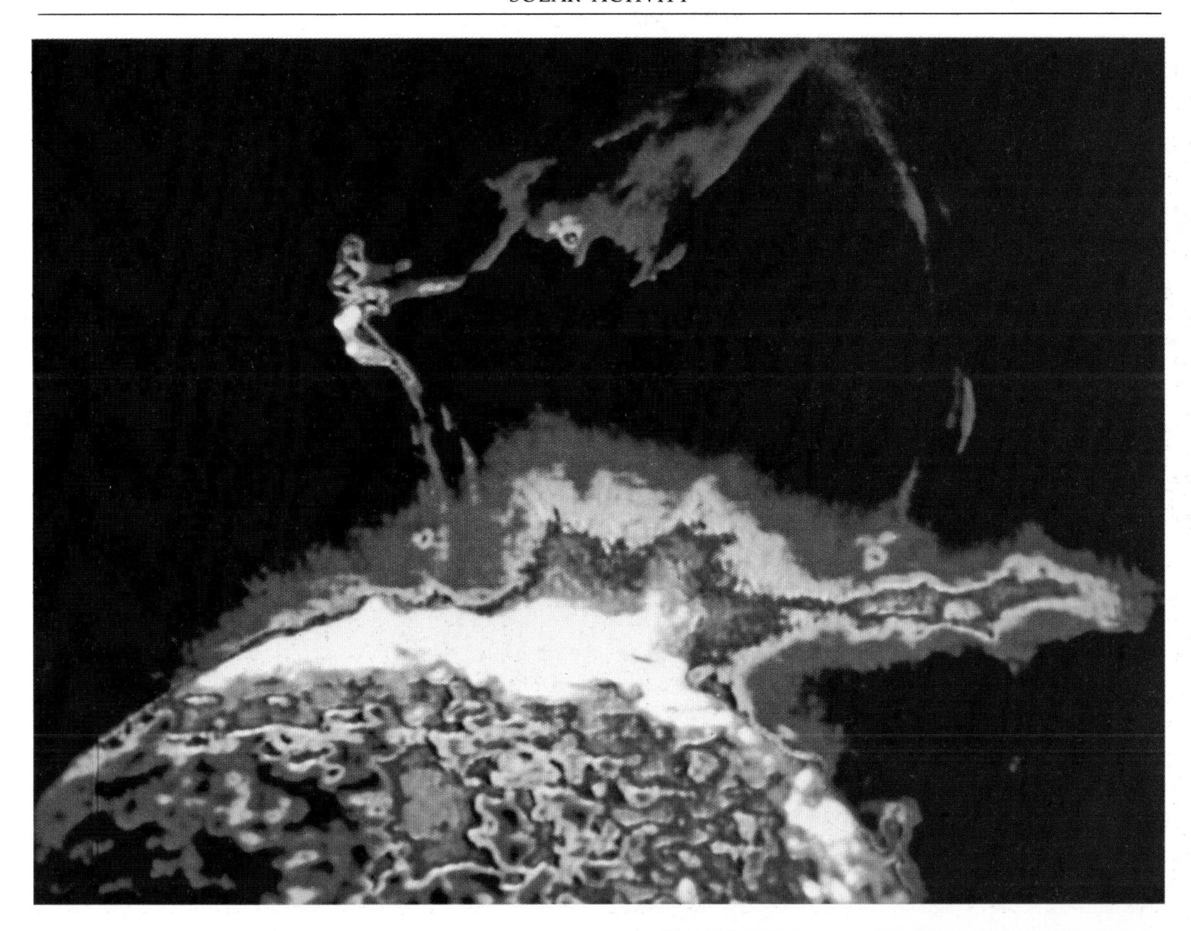

An active 'elbow' prominence is captured by the ultra-violet cameras of Skylab in 1973. The prominence extends about a million kilometres from the solar disc. The looping effect, where some of the material is raining back to the surface, is caused by interactions with magnetic fields. The picture is reproduced in false colours, the brighter colours showing the hotter regions. Ultra-violet studies where observations are made of individual highly ionized atomic species are most valuable because they show regions of specific temperatures.

of divergent magnetic field lines along which the particles may stream out into space. Similar polar plumes are also known to exist. The high-speed streams (and the coronal holes) are observed to persist over several solar rotations.

The solar wind expands out into space until its pressure is balanced by that of the interstellar gas. It is uncertain at what distance this will occur, but it is thought that this **heliopause** will be present at distances of about 50 to 100 astronomical units. The whole magnetic bubble surrounding the Solar System is itself known as the **heliosphere**, and is expected to be tear-shaped due to the flow of the interstellar gas within the Galaxy. It is hoped that data from Pioneer 10 in particular – now beyond all the known planets – will help to establish the conditions at such great distances from the Sun.

Solar activity

As the solar wind is highly ionized, it interacts strongly with magnetic fields, distorting the Earth's magnetic field as it streams past us with a velocity of 500 km per s. This distortion is only one effect the Sun has on our planet. The X-ray and ultraviolet radiation from the Sun ionizes the atoms of our outer atmosphere producing a region called the **ionosphere**. This is particularly useful to us because radio waves can be reflected from this layer to provide over-the-horizon communication on our planet. However, disturbances on the Sun can alter the balance of the ionosphere, upsetting communications and also producing the beautiful aurorae observed in polar latitudes. These are the end products of

Below:
A total eclipse of the Sun (1977 October 12) showing prominences on the limb.

The time development of activity. The small flare on the limb has an associated flare-spray. The adjacent filament, visible as a prominence on the limb, is triggered by the spray with the resulting spectacular development.

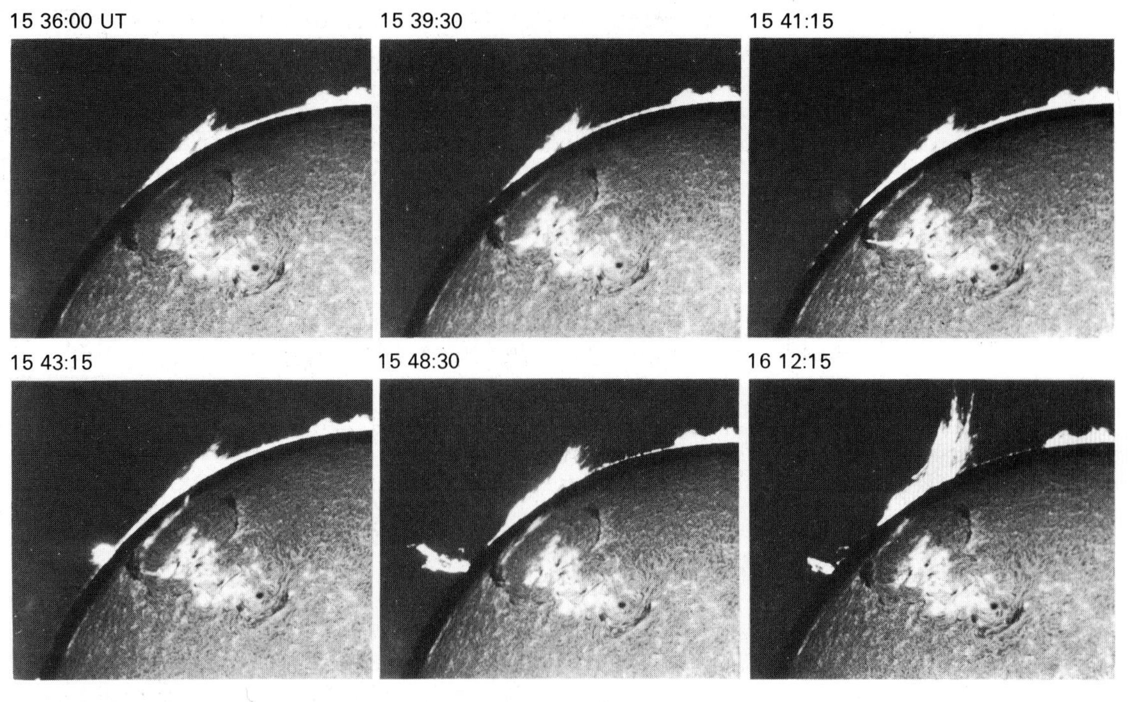

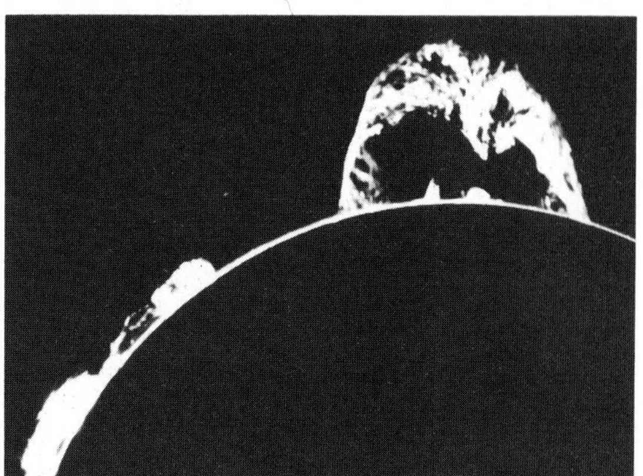

Left: *An ascending quiescent prominence photographed in the light of Hα emission. The top of the erupting loop of material is many thousands of kilometres above the solar surface. The loop will develop by probable fragmentation and the filamentary type prominence will be seen descending back to the surface region.*

Below left: *The solar disc seen in white light revealing the pockmarked surface, sunspot features and also the phenomenon of limb darkening. The largest sunspot shown is about 10 000 kilometres in diameter and is shown in greater detail right.*

Below right: *Sun spot photograph obtained under the best possible atmospheric conditions for observation. The dark circular umbra and radial streaming penumbra clearly demonstrate the appearance of a classical sunspot. Granulations, the seething tops of the convective zone, each about 1 000 kilometres across are also clearly visible. The black circle represents a scale of 5 arc secs.*

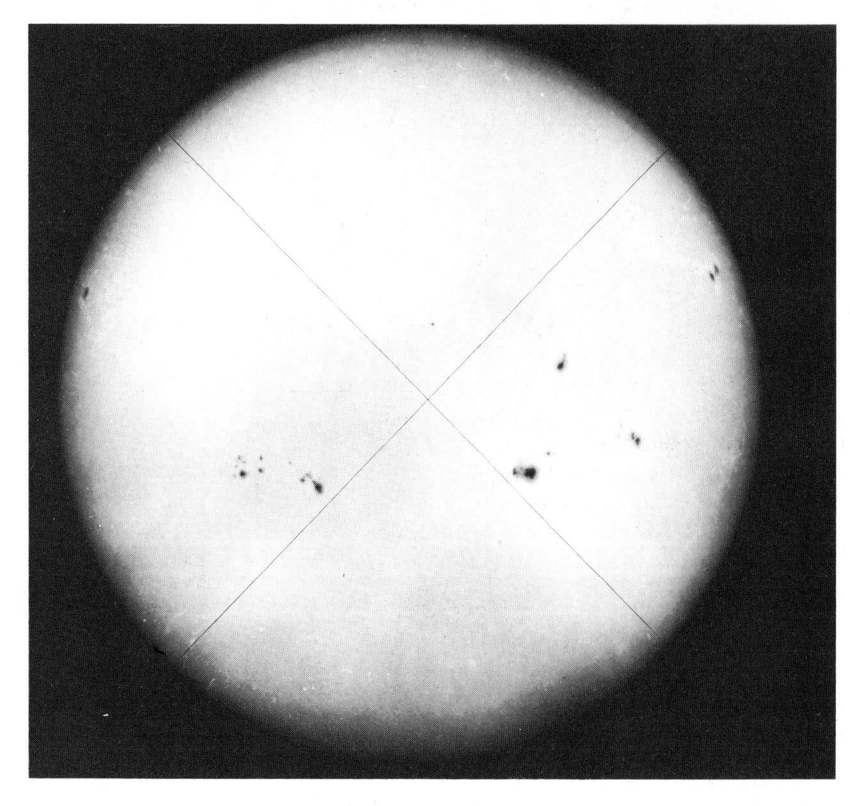

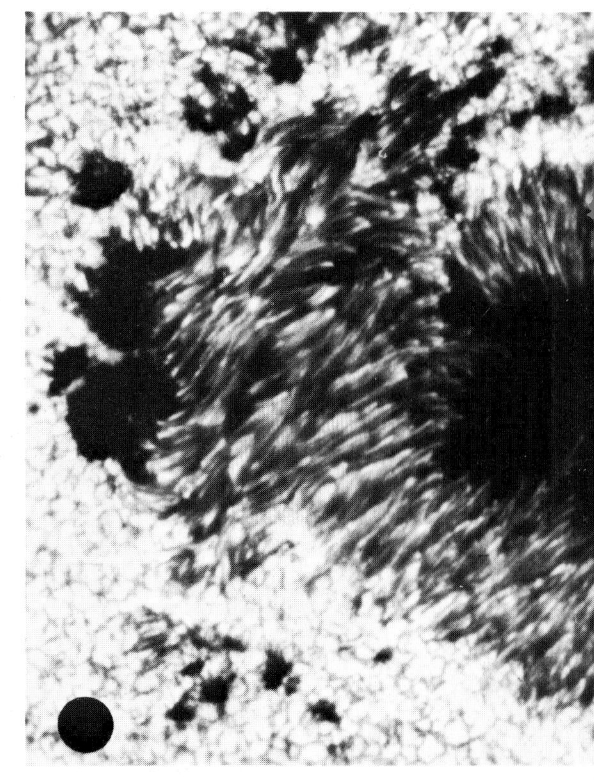

magnetic storms originating on the surface of the Sun, producing such dramatic and spectacular features as eruptive, loop and quiescent prominences, plages, plumes and **solar flares**. The latter are brilliant outbursts of light in the atmosphere of the Sun and emanate from a concentrated burst of energy, usually associated with the disruption of the solar magnetic field in the proximity of an active region. The flares emit extremely intense X-rays and ultraviolet radiation and a blast of high energy particles. It is the interaction of the latter with the magnetic field and the atmosphere of the Earth which, two days later, causes the appearance of our aurorae. Thus, we immediately have direct evidence of solar effects on our atmosphere.

What of climatic variations? This intriguing question leads us back to the solar surface to a study of the well-observed phenomenon of **sunspots**, the only form of solar activity which may occasionally be seen with the naked eye. Sunspots appear as dark blemishes against the bright solar disc, not because they are black but because they are about 2 000° cooler than the rest of the photosphere. The black central region is termed the umbra while the brighter periphery is the penumbra. Spots come in a range of sizes, the largest being around 100 000 km across but a more usual size is 10 000 km. They frequently appear in groups or clusters which may persist on the disc for a week or more. By studying the motion of sunspots across the disc, the rotation rate of the solar surface may be found. Galileo was the first to do this in 1611 and obtained 26 days for the equatorial rotation, a figure which we still believe is correct. Because the Sun is gaseous, it does not rotate like a rigid body and the polar regions rotate much more slowly, taking 37 days to make a complete revolution.

Records of the sunspots' occurrence rate have led to very useful and intriguing conclusions. We now know that sunspots are yet another form of solar magnetic activity, and it is believed that they are the result of magnetic fields generated by some form of circulating electric current in the solar interior. What seems to happen is that while solar magnetic fields are uniform and lie parallel to the solar surface, because of the turbulence of the gas in the convection zone, the field can become tangled. When this occurs the field bursts through the photosphere surface forming a sunspot in which, as observations show, the magnetic field emerges vertically. Sometimes the field forms a loop and then the two linked sunspots are termed bipolar. The number of sunspots is directly related to the general activity of the Sun.

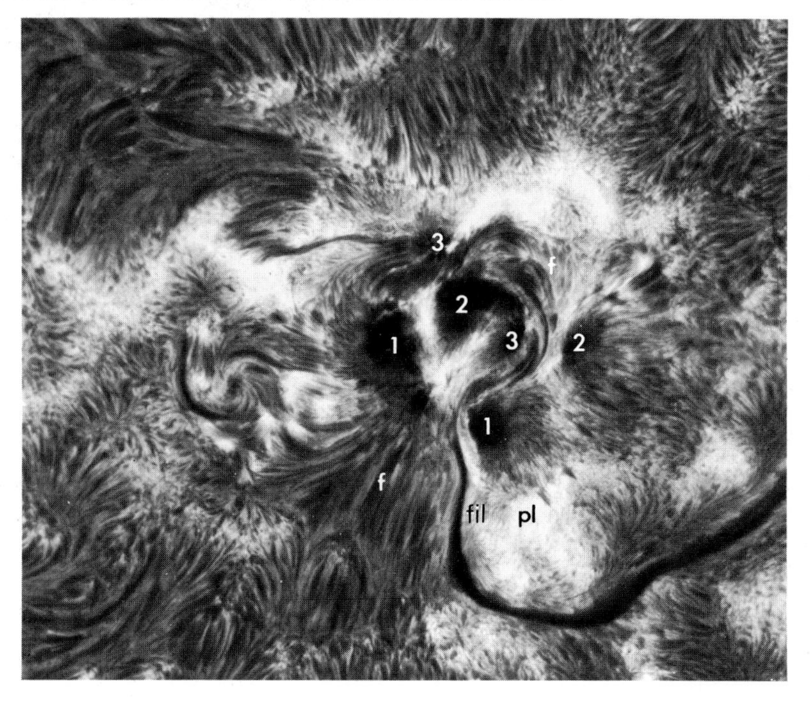

An active region observed in August 1972, just after the maximum in the 11 year solar cycle. Features which may be identified are a plage (pl), surrounded by the ribbon-like filament (fil). Three bipolar sets of sunspots (1–3) are visible and long fibrils (f) separate the polarities and follow the magnetic field pattern. The photograph was taken in the light of Hα.

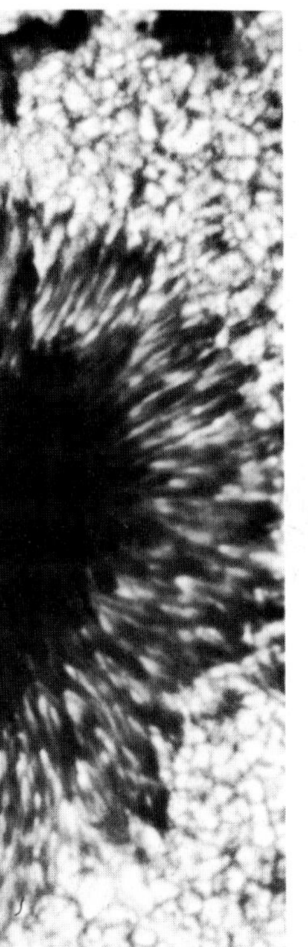

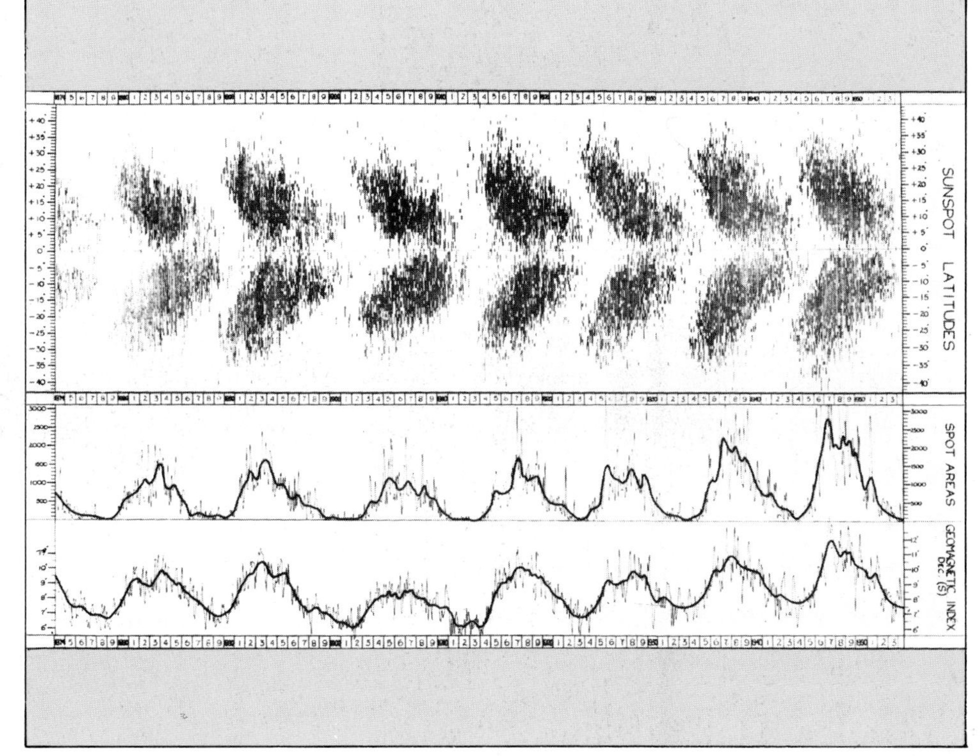

Fig. 4·1 Sunspots appear to occupy lower latitudes on the Sun as the phase of the 11 year cycle of activity progresses from one minimum to the next. A plot such as this is referred to as a Maunder butterfly diagram.

Observing the Sun

It is most important always to bear in mind that **the Sun should never be looked at through binoculars or a telescope.** The brilliant disk is so bright and hot that to observe it directly would result in permanent blindness. Even prolonged staring at the Sun with the naked eye can damage the sensitive retina. The dark 'Sun filters' supplied with some telescopes should never be used as **they are not safe:** they usually transmit harmful infrared and ultraviolet radiation, and have also been known to crack under concentrated solar heat in a telescope. Even when the Sun is low on the horizon at sunrise and sunset its infrared radiation may still be sufficiently great for it to be dangerous to use any optical equipment.

The Sun may be studied in perfect safety by adopting proper methods or by using special equipment. The simplest means is to project the solar image on to a white screen, preferably by using a proper projection box mounted on the telescope as illustrated on page 85: by preventing interference from stray light this makes the details much more easily seen. This method can be used with any telescope, even with one side of a pair of binoculars, but it is important to remember to cover the objective of any finder on a telescope, or the other half of the binoculars, to avoid any possibility of an accident. The shadow of a vane cast on to a suitably positioned target is a quite adequate, and safe, method of pointing the telescope in the right direction.

Of the special equipment which is used for observing the Sun brief mention may be made of the Herschel wedge, which reduces the intensity by a factor of between 10^3 and 10^4 – still not enough for safe viewing without filters or other devices – and of the use of specially prepared, uncoated mirrors which reduce the amount of light and heat. The intense heat of the Sun imposes a limitation on the size of telescope that can be successfully used, and those of 150mm aperture or more often have to be stopped down at the objective.

The greatest success and safety in direct viewing, and the greatest convenience, is achieved by the use of special reflecting solar filters. These consist of glass, or more commonly Mylar plastic film, coated with chromium or aluminium to reflect away the unwanted light and heat and only pass a small fraction into the telescope. As they are used over the objective, the whole telescope remains cool – a great advantage; and furthermore, quite large apertures may be used with the consequent gain in resolution. The coated film, which is somewhat similar to the thermal blankets used on spacecraft, must be of the correct type; sources of supply will be given by organizations which co-ordinate the study of the Sun.

Apart from examining the Sun in white light many amateurs study its features in the light of a single spectral line, usually Hα at 6536 A, either by using special – and expensive – filters or by means of a complex instrument known as a spectrohelioscope (illustrated on page 88). Essentially, this spreads the light into a spectrum and then uses a second slit to reject all light except one narrow spectral band. Spectrohelioscopes have the advantages over filters that they may examine any spectral region, not just one, as well as the fact that, although complex, they do not require accurate temperature control, which must be strictly observed in the case of filters.

The simplest and most obvious form of observation is to note or draw the distribution of sunspots on the solar disk. If the projection method is used it is simple to plot the positions of any sunspots or faculae on to a suitable sheet of paper. It is usual to arrange for the magnification provided by the optical train to be such that a standard-sized blank is used, with a fixed solar diameter. Made on a daily basis, such disk drawings enable the evolution of individual sunspot groups to be followed, and the adoption of a standard diameter means that records from different observers may be easily compared. An assessment of the level of sunspot activity can be competently made by the amateur and is usually expressed in terms of the Zürich relative sunspot number R, which is derived from the formula $R = k(f + 10g)$, where g represents the number of sunspot groups, f the total number of their component spots, and k is a constant which depends upon the estimated efficiency of a particular observer and the equipment he uses.

Hydrogen-alpha observations allow many fine

Opposite:
A composite of two photographs taken through a H alpha filter, one of the solar disc and the other of the prominences, taken by R. J. Poole, Canada, on 1980 August 16.

Below left:
The large sunspot of 1972 May 20, drawn by Harold Hill over the period 06.45 to 07.10 UT, using a 75 mm objective.

Below and opposite:
A fine series of drawings by Harold Hill, observing in hydrogen-α light showing the development of a large loop prominence.

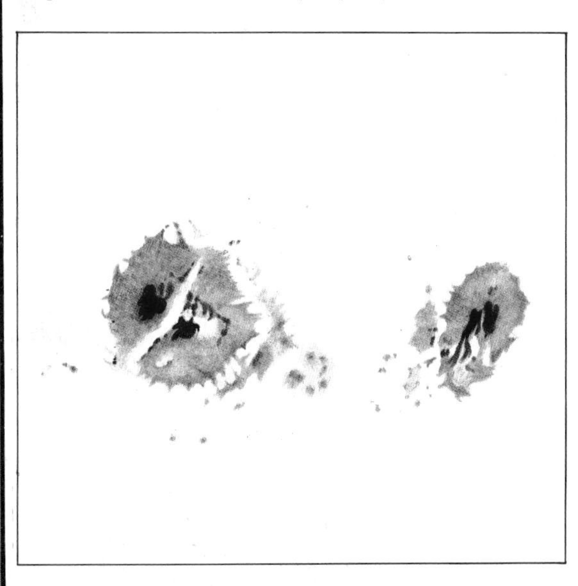

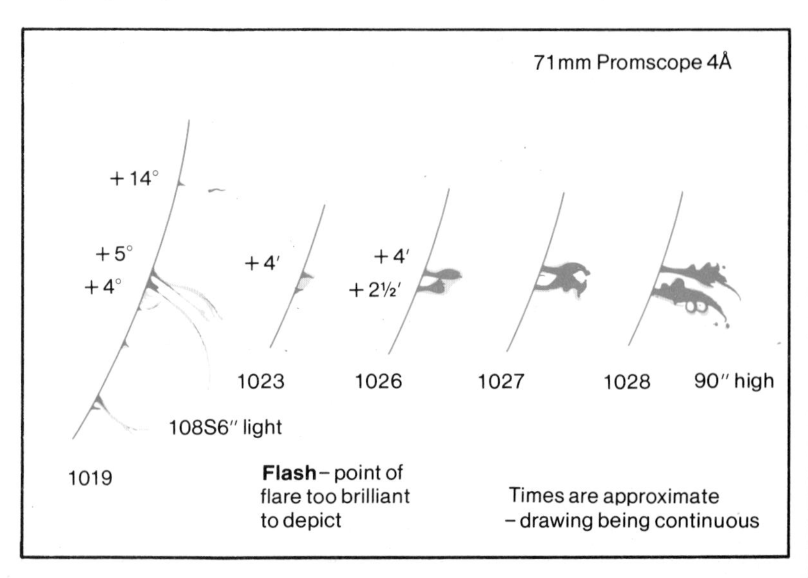

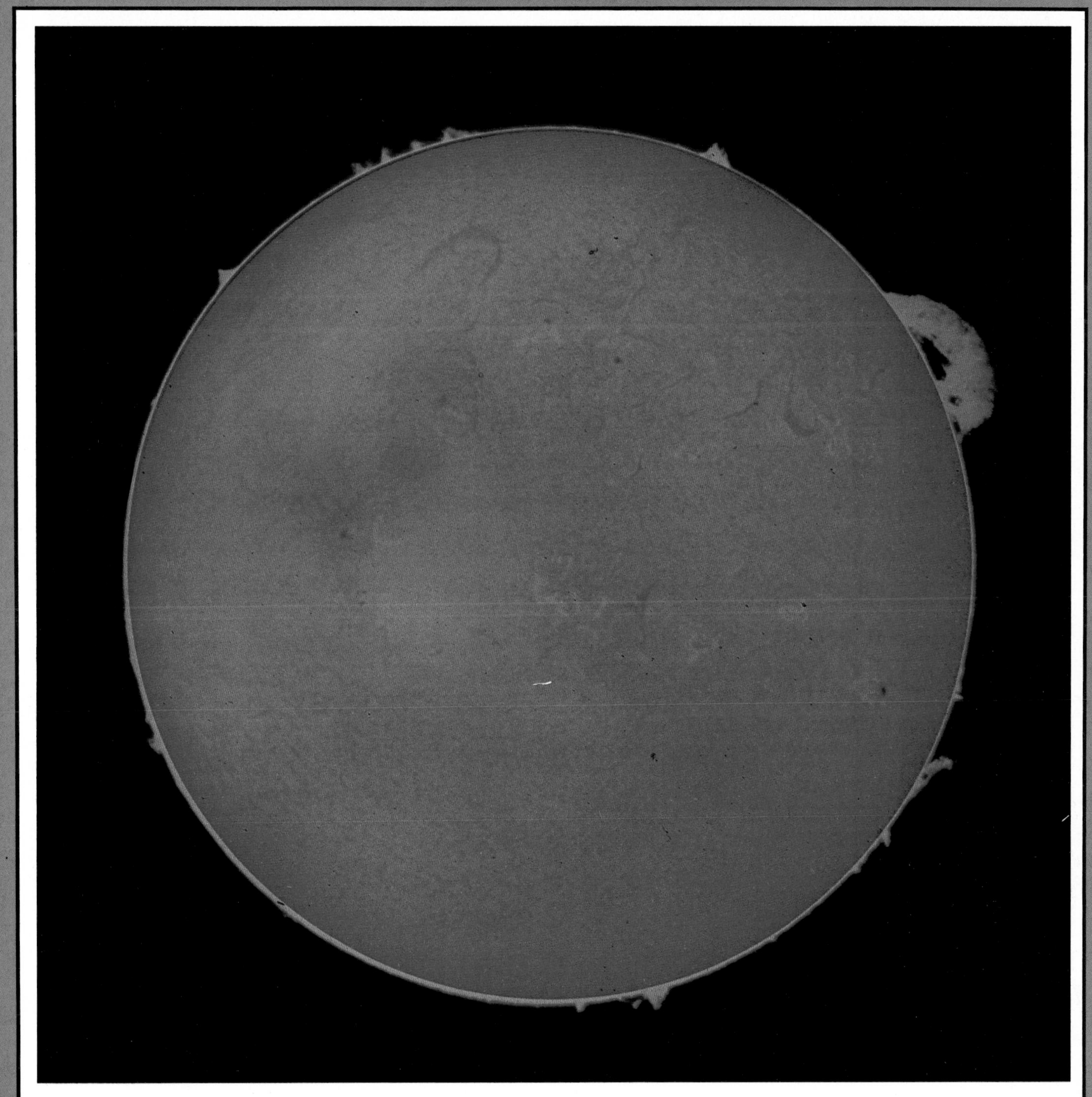

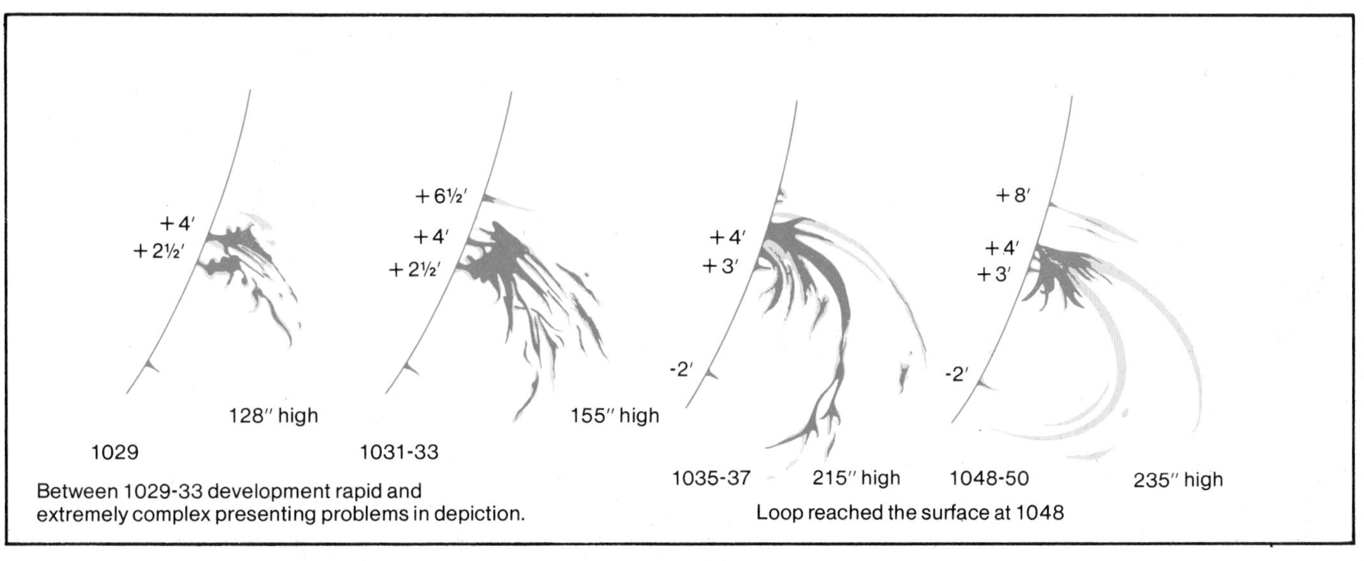

Between 1029-33 development rapid and extremely complex presenting problems in depiction.

Loop reached the surface at 1048

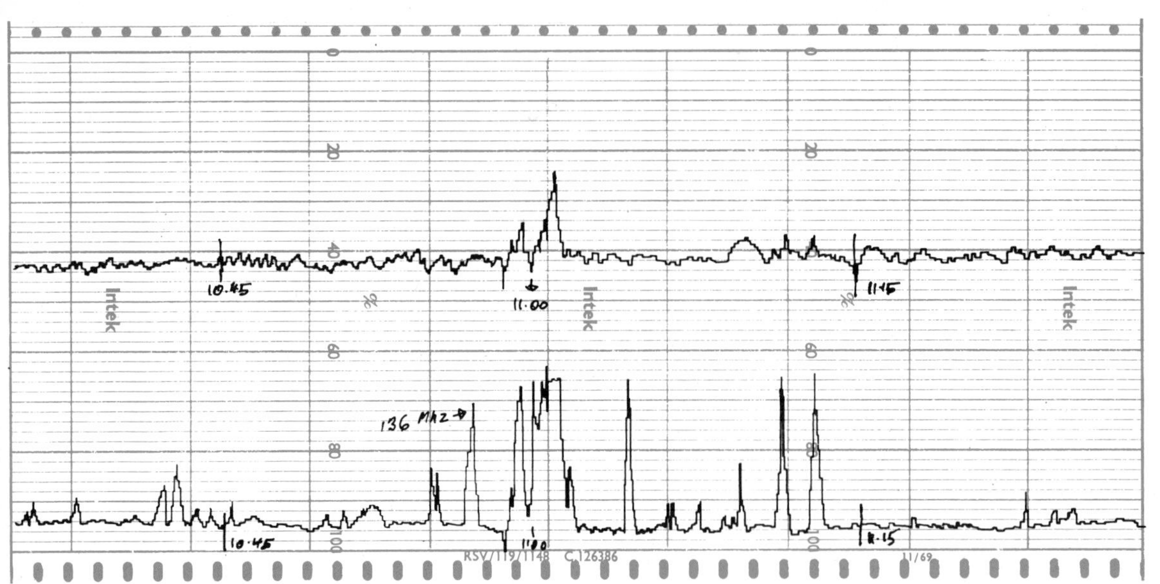

A trace recorded by H. Hatfield, showing the course of a radio flare in two different wavelengths.

details to be recorded, e.g. prominences (called filaments when seen against the disk), plages and sunspots. Flares are also more readily seen than in white light, and should be fully recorded. Detailed structure is best recorded by photography, either in white light or in a narrow spectral band, and whole-disk photographs are a useful record of solar activity. However, once again, when attempting white-light photography, precautions must be taken to ensure that the shutter of the camera – and the observer's eye – are not subjected to the concentrated heat of the primary image.

It has just been mentioned that solar flares may be detected and recorded visually and photographically, but there is also the possibility of making radio observations. Some observers monitor the Sun for activity throughout the day, and every day, using nothing more sophisticated than a fairly good communications receiver and a small Yagi aerial driven to track the Sun. Such equipment can be connected to a suitable chart recorder to give a permanent trace of the activity, and also arranged to provide an automatic warning when a flare occurs. Any such records can be correlated with those of other observers, and also with subsequent auroral activity. As an extension of this general field many amateurs are now observing and recording the changes in the Earth's magnetic field which occur with solar activity and auroral displays.

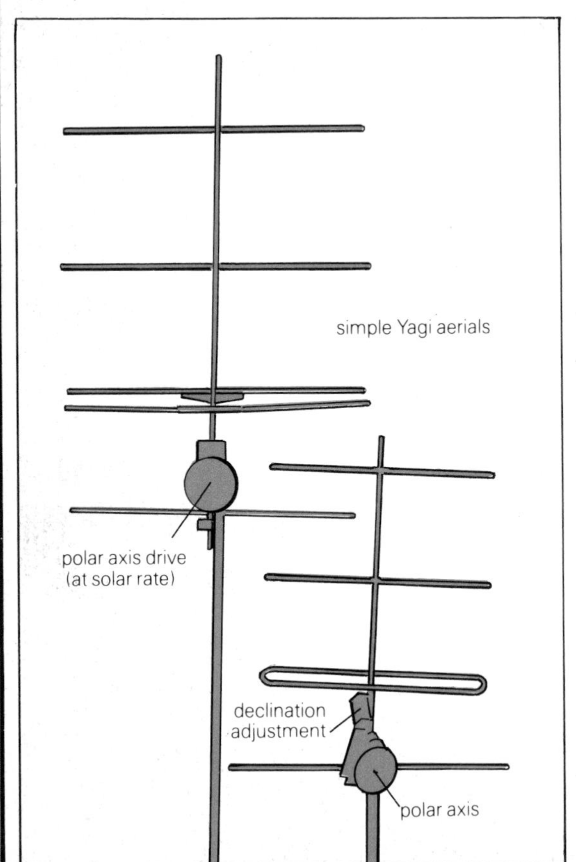

Bottom left:
These two simple radio aerials (136 and 197 MHz) are mounted so that they are easily driven to follow the sun throughout the day.

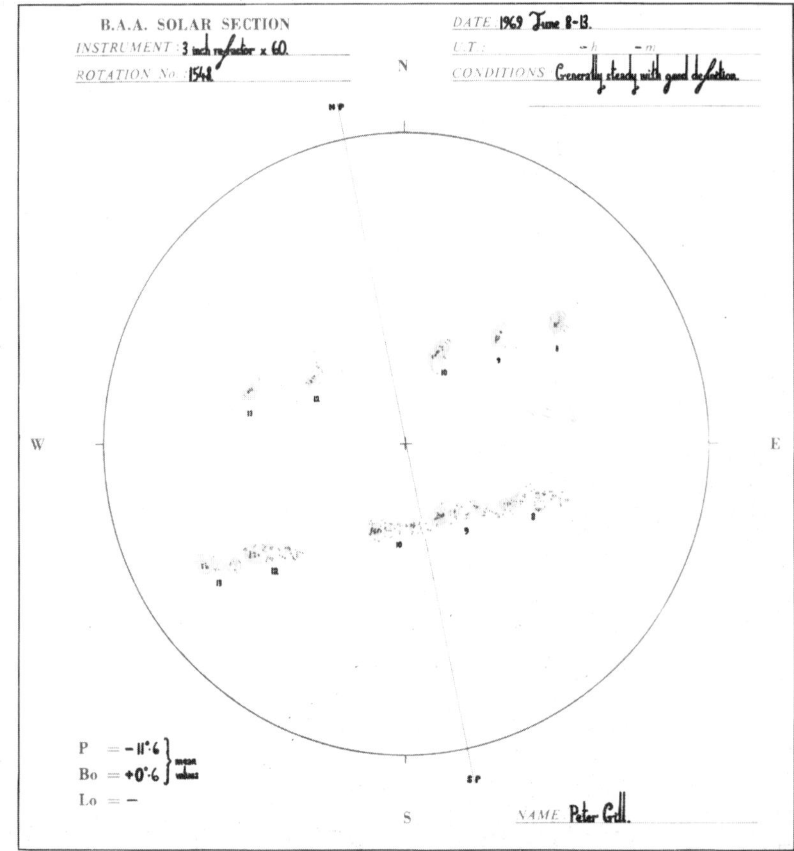

A composite drawing from observations made using the equipment shown opposite, which shows the day to day changes in structure of two large sunspot groups that were visible during the first half of June 1969.

Right:
A telescope with an aperture-reduction mask and a special filter which reflects away most of the light may be used for the safe observation of the sun.

Far right:
A proper projection box is the simplest and safest method of observing the sun telescopically.

Below:
A white light whole-disc photograph of the Sun, showing a great amount of detail.

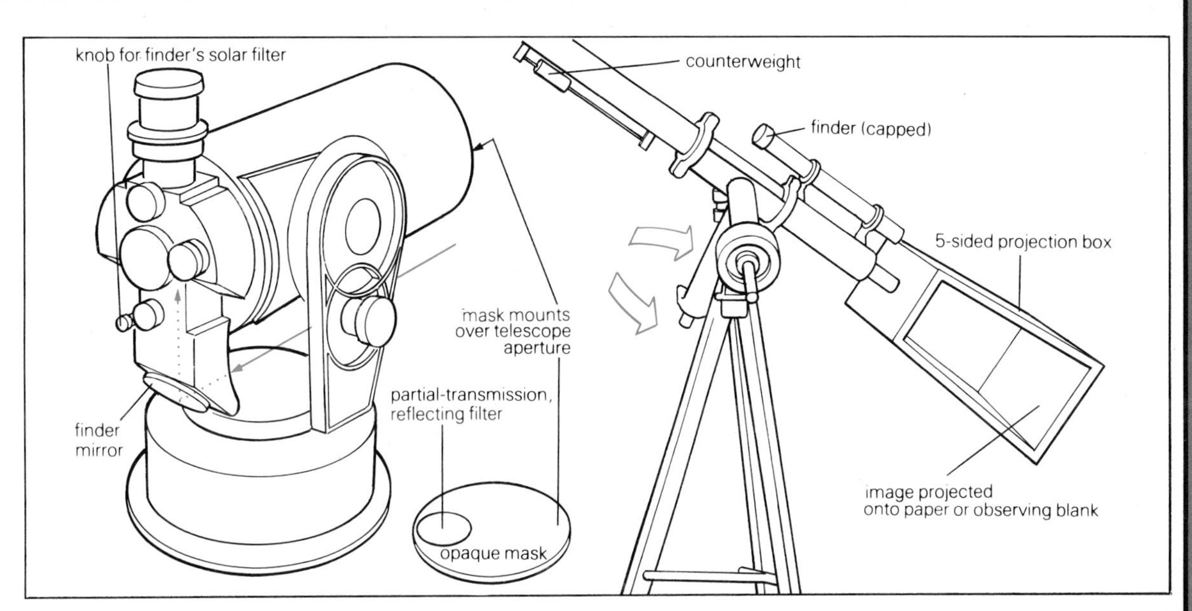

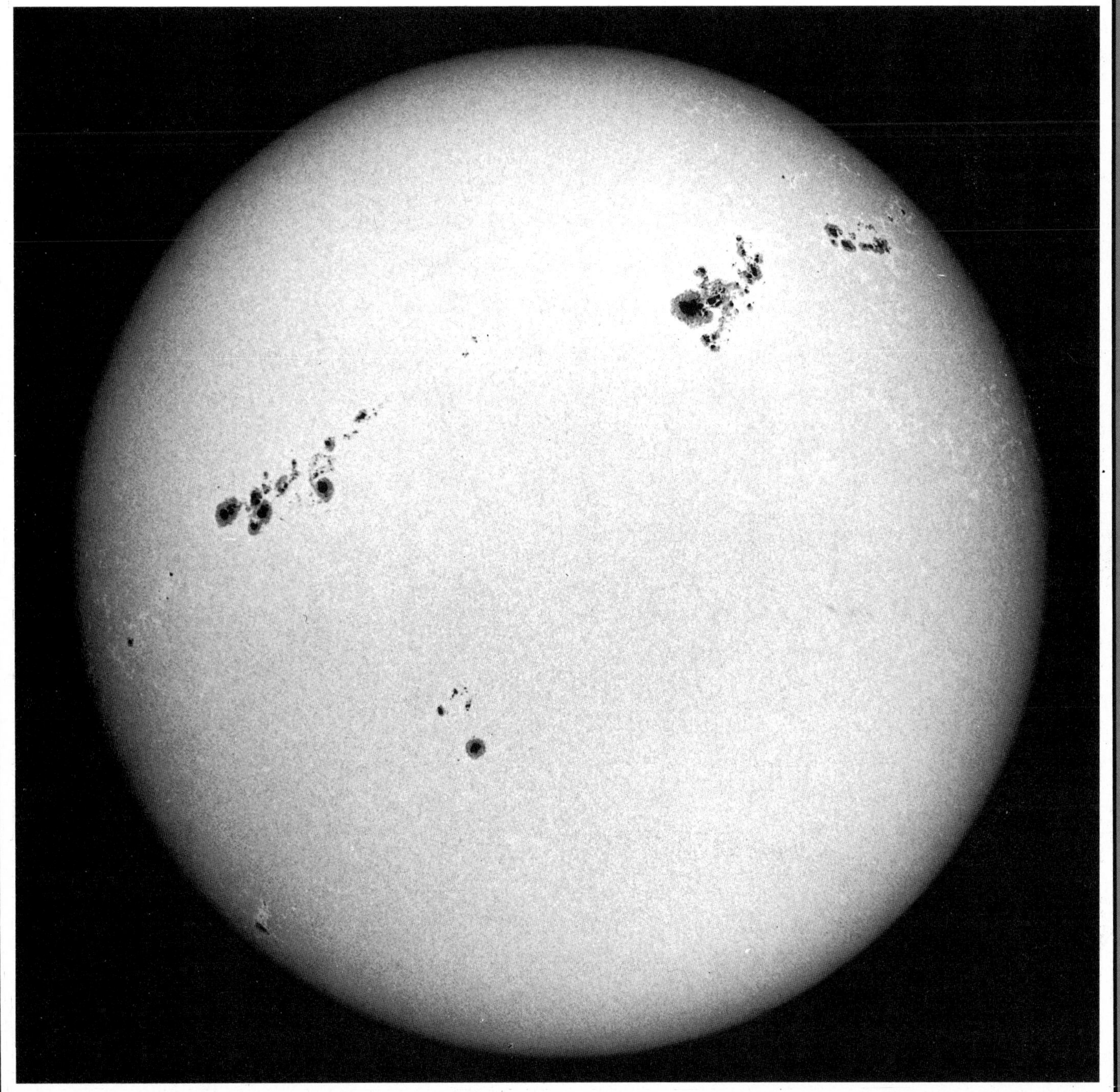

A solar eclipse is widely regarded as one of the most magnificent natural spectacles, and astronomers have always had a tendency to rush off to far-away places to catch sight of one. They still do, even though coronagraphs (page 77) and satellite-borne instruments are available to professionals, who can thus study the corona at any time. Few astronomers can hope to have an eclipse occur close to their homes, but in recent years 'eclipse chasing' has become a very popular – if rather expensive – hobby for a lot of people.

Some astronomers find the experience so awe-inspiring that they quite deliberately do not try to make any 'proper' observations, perhaps contenting themselves with taking one or two photographs. others may carry vast amounts of complicated equipment halfway round the world and have such a busy programme that they leave no time actually to look up and enjoy the sight. However, at any eclipse the duration of the total phase is so short – the longest possible is only 7½ minutes, and it is usually much less – that even attempts at simple photography need to be carefully planned beforehand.

It should be emphasized once again that all the usual precautions for viewing the Sun (page 82) *must* be followed during the partial phases, and that it is particularly important for astronomers to remember that they may be encouraging ordinary bystanders to look at the Sun, so their safety should also be borne in mind. The projection method is probably the best one to adopt, especially as many people can view the one image. The equipment need not be at all elaborate, and one side of a pair of binoculars, hand-held, and a piece of card, will frequently suffice. It is still useful to be able to take a quick look through a suitable filter to see the progress of the partial phase, and some astronomers make a habit of carrying with them a number of simple, cheap filters which they can give to anyone for viewing the Sun. These filters consist of two layers of fully exposed, fully developed black-and-white film, firmly fixed in a card (or other) holder. (Colour film is not suitable.) Such filters are quite safe, but it should be noted that photographic neutral density filters, polarizing filters (even with polarizers crossed), and most other forms of filter – except some of the very dense welding glasses – are *not* safe as they may allow invisible, yet still harmful, infrared and ultraviolet radiation through to the eye. The filters made from exposed film are worth having in any case – you can use them at any time to look for naked-eye sunspots.

The only time that it is ever safe to use ordinary unfiltered optical equipment, such as binoculars, on the Sun is when the hot photosphere is completely covered during the total phase. Then, indeed, removing the filters is necessary to photograph, or to observe visually, the detail of the inner corona or prominences. Again for reasons of safety it is a good idea to arrange for a timer (or a very dedicated timekeeper) to warn of the approach of the end of totality, so that the proper filters may be replaced. Such a warning can also serve to prepare the observers for their fleeting glimpse of Baily's Beads or the diamond-ring effect.

The accurate timing of solar eclipses is a form of observation which can be carried out relatively easily and which has many applications. The equipment used is more or less the same as that required for work on lunar occultations (page 109), although it is normally necessary to rely upon radio time signals

Below:
Total solar eclipses are typically plotted on diagrams such as this one for the event on 1983 June 11.

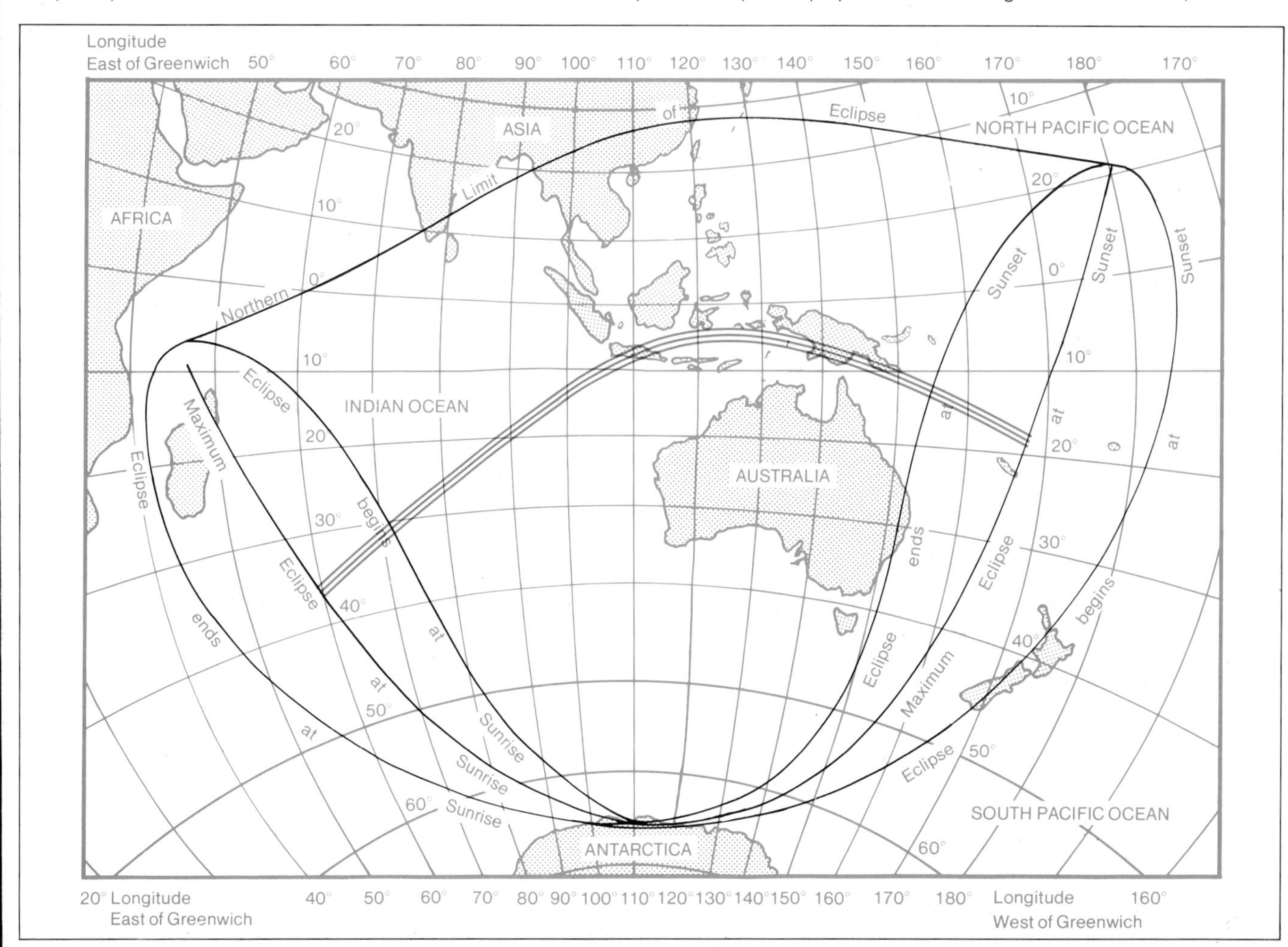

Most eclipse tracks cross large expanses of the oceans, and sometimes a ship may offer more favourable conditions than occur onland. The picture above was taken on 1973 June 30.

for precise information, as a telephone service may be neither convenient, nor even available. A more difficult problem in many countries is that of obtaining the accurate latitude and longitude, and the exact height above sea-level, of the observing site. The lack of this precise information can greatly reduce the value of the observations, lessening their accuracy for further analysis; it may be better to select a site for which the positional co-ordinates are well established but which is somewhat away from the eclipse central line (although still within the zone of totality), rather than choose a position which is theoretically ideal, but for which the position is poorly known.

Generally the times of the first and last contacts, at the beginning and end of the partial phases, are not very precisely established because of the difficulty of deciding exactly when the Moon's limb touches and leaves the photosphere. The second and third contacts, marking the start and end of totality, are much more definite and easy to observe, giving a correspondingly greater accuracy. The timings may not only be used to check the predictions and the accuracy of the theory of lunar motion – one of the most complex problems in dynamical astronomy – but also to establish the exact diameter of the Sun. In conjunction with historical observations, this information is being used in modern studies of the possible variations in the size – and presumably also in the energy output – of the Sun, which some scientists believe take place over the course of time (page 90–91).

Just as with grazing occultations (page 109), the use of observers at the northern and southern edges of the total eclipse track makes it possible to establish where a true total eclipse occurs, as opposed to a deep partial phase. This is a rather different method of obtaining the solar diameter, and again has been used in connection with historical reports. Outside the zone of totality, i.e. where only a partial eclipse is seen, timings are still useful as they may be employed to measure the lengths of chords across the solar disk, and thus also establish the diameter. The method is less precise than the others mentioned, but it can be used to gain some useful information.

Observation of annular eclipses is valuable in an even more direct manner, especially when the apparent size of the sun is only just greater than that of the Moon. This can give rise to a more or less complete circle of Baily's Beads where the photosphere is visible through depressions of the Moon's limb. As the Moon's diameter is accurately known, this simple observation allows a very accurate determination of the size of the solar disk to be made.

The main eclipse activity is of course photography, and this is usually carried out with telephoto lenses or some form of camera/telescope combination. Excellent results have been obtained using just binoculars and an ordinary camera focused on infinity, but properly combined equipment will naturally give the best results. The size of film to be used will govern the image scale required, and thus the focal length needed, while the effect to be photographed will also have to be taken into account. For example, photographs of prominences around the Sun's limb will require a large image scale and are very suitable for a square film format. Baily's Beads, the diamond ring effect and the outer corona will require much wider fields. Exposures for the different effects will also vary widely. For this reason, and to overcome the problems encountered in hurrying to change lenses or telescopes during the progress of the

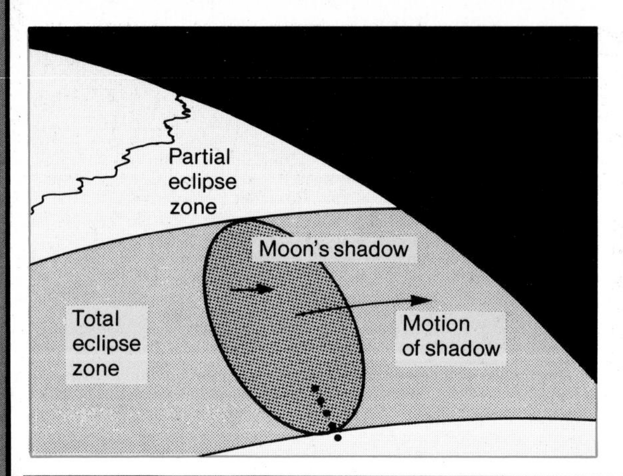

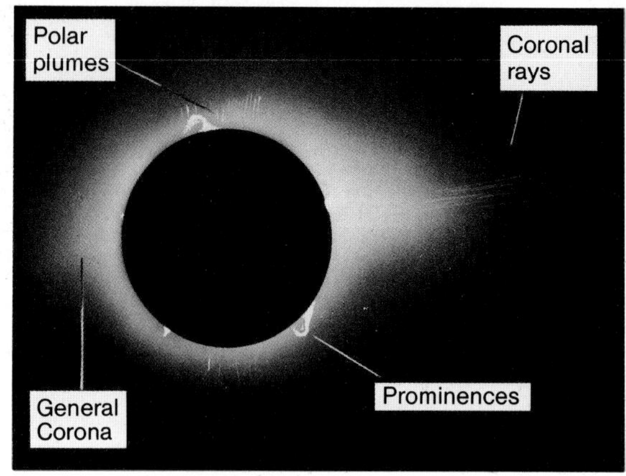

Far left: *Teams of observers located near the edge of the band of totality can provide important information about the exact size of the Moon's shadow, and hence of the Sun itself.*

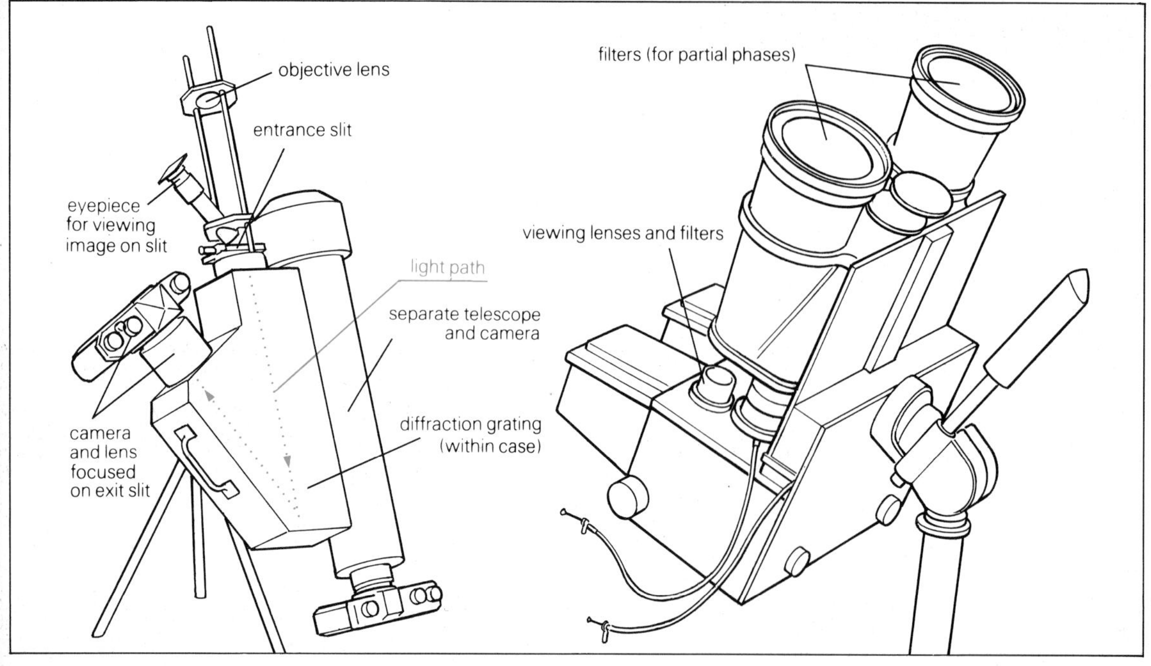

Far left: *A portable spectrograph; the instrument illustrated was specially designed for use at solar eclipses.*

Left: *Even the simplest camera equipment may be adapted for eclipse photography, as in this case where a pair of binoculars (divided into two halves) is used in conjunction with two basic twin-lens reflex cameras.*

eclipse, most observers will try to use more than one camera. In any case, as with all aspects of eclipse observation, plenty of practice is needed beforehand to take advantage of those few moments of totality.

Another factor which must be borne in mind is that of the sensitivity of the film used to hydrogen-alpha light, this being essential for prominence photography. It should not be assumed that colour films will not suffer from a lack of response, as some, most annoyingly, have red sensitivity curves which drop to zero just short of the hydrogen-alpha wavelength.

It is usual for amateurs to photograph the corona in white light, although sometimes various colour filters are used. The great difference between the brightness of the inner and outer coronal regions means that a series of exposure times is really necessary to record detail at various distances from the solar limb. To record the great extent of the faint outer corona a driven telescope mount will be required, but naturally in such pictures the inner corona will be greatly overexposed. Great success has been achieved by astronomers who have made special radial-density filters, i.e. neutral filters in which the density decreases outwards from the centre. These enable a single exposure to capture detail throughout the corona, but unfortunately they are not readily available. The ray structure of the corona is also made more pronounced by suitably orientated polarizing filters, and these are well worth trying.

Ciné-photography of both the partial and total phases can be very successful, but here a well-driven mount is essential, and care must be taken at the appropriate times to remove and replace the filters used for the partial phases.

It is interesting to note the other objects which can be seen during the total phase. These may include bright stars, planets and, if one is exceptionally lucky, comets. In a few cases comets are only known from observations made during eclipses; and in any event *any* details which can be observed and recorded are very valuable indeed.

There are a number of other effects of solar eclipses which are also often observed, although they may not be strictly astronomical in nature. They include the ripple-like shadows known as shadow bands, caused by varying refraction in the atmosphere, and only seen for a few seconds when the Sun is very nearly completely covered. The appearance of the surrounding landscape and of the sky during the period around totality is also worth recording, perhaps by ciné-photography. Continuous records of the brightness of the sky at the zenith and of the changing air temperatures are of considerable value. The drop in temperature during the eclipse is the cause of the 'eclipse wind' which may, on occasions, be of considerable force. Finally, a solar eclipse has a significant effect upon the ionosphere and thus alters conditions for radio propagation, the changes in which can be studied even by those not fortunate enough to be within the zone of totality.

Opposite page, top left: *Projection of the Sun's image – in this case by means of a right-angle telescope – is the best way of enabling several observers to see the partial phases.*

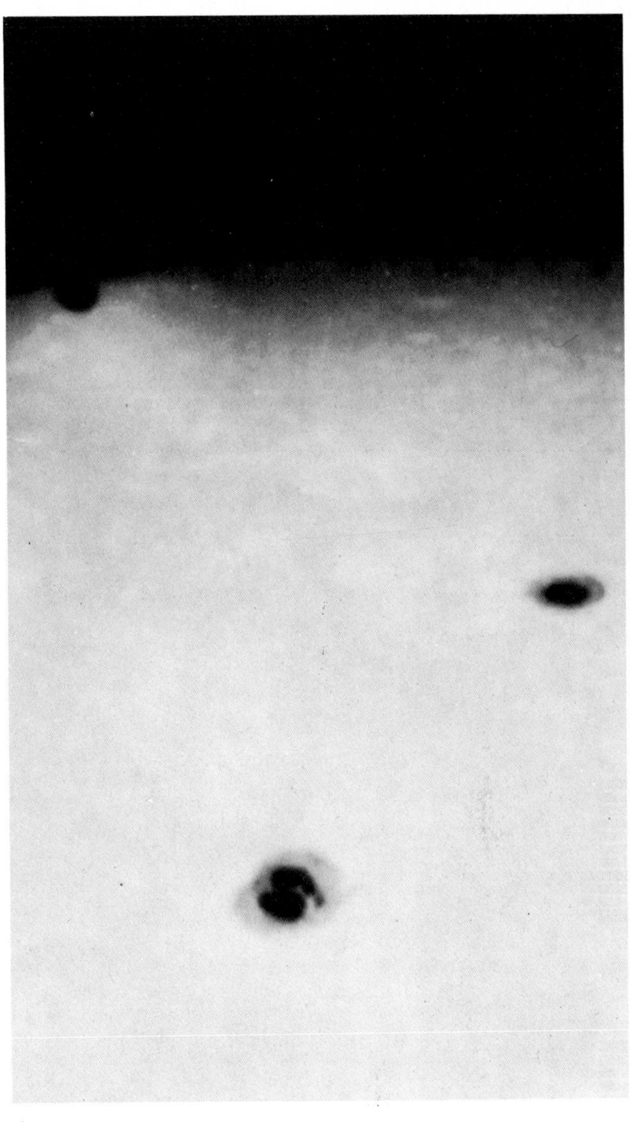

Above right:
Observing transits of Mercury or Venus requires the same methods as viewing the Sun or partial eclipses. In this photograph by J. S. Korintus, Mercury is just leaving the solar disk after the transit on 1970 May 9.

Solar Eclipses 1984–1999

Date	Type of Eclipse	Maxium Duration	Track
1984 May 30	Annular		Atlantic, Equatorial Africa, Somalia
1984 Nov. 22–23	Total	1m 59s	E. Indies, S. Pacific
1985 May 19	Partial		Arctic
1985 Nov. 12	Total	1m 55s	S. Pacific, Antarctica
1986 Apr. 9	Partial		Antarctic
1986 Oct. 3	Annular/ Total	0m 1s	N. Atlantic
1987 Mar. 29	Annular/ Total	0m 56s	Argentina, Atlantic, Congo, Indian Ocean
1987 Sept. 23	Annular		USSR, China, Pacific
1988 Mar. 11	Total	3m 46s	Indian Ocean, E. Indies, Pacific
1988 Sept. 11	Annular		Indian Ocean, S. of Australia, Antarctic
1989 Mar. 7	Partial		Arctic
1989 Aug. 31	Partial		Antarctic
1990 Jan. 26	Annular		Antarctica
1990 July 22	Total	2m 33s	Finland, USSR, Pacific
1991 Jan. 15–16	Annular		Australia, New Zealand, Pacific
1991 July 11	Total	6m 54s	Pacific, Central America, Brazil
1992 Jan. 4–5	Annular		Central Pacific
1992 Dec. 24	Partial		Arctic
1993 May 21	Partial		Arctic
1994 May 10	Annular		Pacific, Mexico, USA, Canada, Atlantic
1994 Nov. 3	Total	4m 23s	Peru, Brazil, S. Atlantic
1995 Apr. 29	Annular		S. Pacific, Peru, Brazil, S. Atlantic
1995 Oct. 24	Total	2m 5s	Iran, India, E. Indies, Pacific
1996 Apr. 17	Partial		Antarctic
1996 Oct. 12	Partial		Arctic
1997 Mar. 9	Total	2m 50s	USSR, Arctic
1997 Sept. 2	Partial		Antarctic
1998 Feb. 26	Total	3m 56s	Pacific, S. of Panama, Atlantic
1998 Aug. 22	Annular		Indian Ocean, E. Indies, Pacific
1999 Feb. 16	Annular		Indian Ocean, Australia, Pacific
1999 Aug. 11	Total	2m 23s	Atlantic, England, France, Central Europe, Turkey, India

A false-colour image of the 1973 June 30 solar eclipse, which has been prepared to show the areas of equal brightness within the corona.

Detailed records of sunspot numbers began in the seventeenth century and a mere glance at the data reveals a conspicuous cycle of solar activity whereby sunspots disappear and reappear with a period of 11 years (Fig. 4·1). The reason is not completely understood but this very regular repetition was at least comforting, suggesting a stable Sun. Recently, however, a careful study of data such as frequency of aurorae, the lack of coronal plumes during eclipses, and the radioisotopes in tree ring samples, showed strong confirmation of earlier work by Edward Maunder questioning this stable situation (Fig. 4·2). He suggested sunspot records revealed that for a period of 70 years ending in 1715 virtually no sunspots were visible on the solar disc. This **Maunder minimum** is now thought to be one dip in a successive pattern. Research by the solar astronomer John Eddy suggests quite persuasively that the epochs of solar lassitude correspond to climatic conditions on the Earth such as the severity of winters in the northern hemisphere, average temperature depressions and glacial advance. It would appear that changes on the Sun are the dominant agent of climatic excursions lasting between fifty and a few hundred years. However, although controversy has long raged over the question of whether the solar luminosity varies and was the cause of the ice ages, the present consensus is that these are caused not by a change in the Sun but rather by a small and extremely long period wobble in the orbit of the Earth.

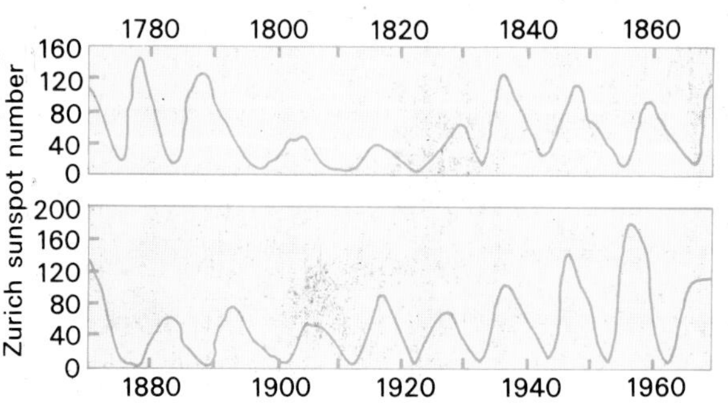

Fig. 4·2 right: *The eleven year cycle of solar activity is well demonstrated by this plot of sunspot numbers in the two hundred years since 1770. Notice how the degree of activity is irregular over many cycles, indicating perhaps ever longer overall cycles such as have been suggested by the American astronomer John Eddy.*

A considerable amount of recent research has concentrated upon the question of whether the Sun is altering in size, and conflicting opinions have been advanced. Changes in the overall diameter may be expected to occur over very long periods of time, but evidence has been presented that some shrinkage has occurred in just a few hundred years. Other discussions argue that little alteration has taken place, and only further research into historical records and careful determination of the current diameter of the Sun can settle the issue. There have been suggestions that there might be oscillations of the Sun's radius, but these are as yet unconfirmed. However, recent investigations have apparently established that the Sun is very slightly oblate. This in itself may be in accord with some recent proposals for amendments to theories of gravitation – more specifically, to Einstein's theory.

The solar neutrino problem

Solar astronomers believed they knew the temperature at the core of the Sun and the general properties of its interior. Nuclear physicists believed they understood the nuclear reaction processes and so between them they could predict the number of neutrinos emitted from a particular step of the proton-proton chain. An elaborate experiment was set up deep in a gold mine in the USA in an attempt to confirm these ideas and detect the neutrinos. Surprisingly, though, very few neutrinos have been detected and, because the apparatus did not appear to be at fault, immediate suspicion fell on our so-called knowledge of the solar interior. Speculative solutions to the problem include a cooler core so reducing the neutrino yield, a switched-off core where the Sun is 'coasting' between phases of nuclear burning, rapid core rotation, and very low heavy element abundance in the core. Yet another suggestion is that there may be a very considerable core of heavy elements such as iron, and that this core has existed from the time of the Sun's formation, the material being supplied by the initial solar nebula. Even more ingenious experiments are being mounted to try to determine both low and high-energy neutrino fluxes, and the results from these may throw some light upon this vexed question. One thing is certain; the discrepancy between theory and observation is uncomfortably large, showing that we do not even understand the details of the internal mechanics of our own Sun, let alone being in a position to solve many of the riddles the cosmos presents to us. A sobering thought, but a challenge which spurs astronomers to devise better instruments and more ingenious theories.

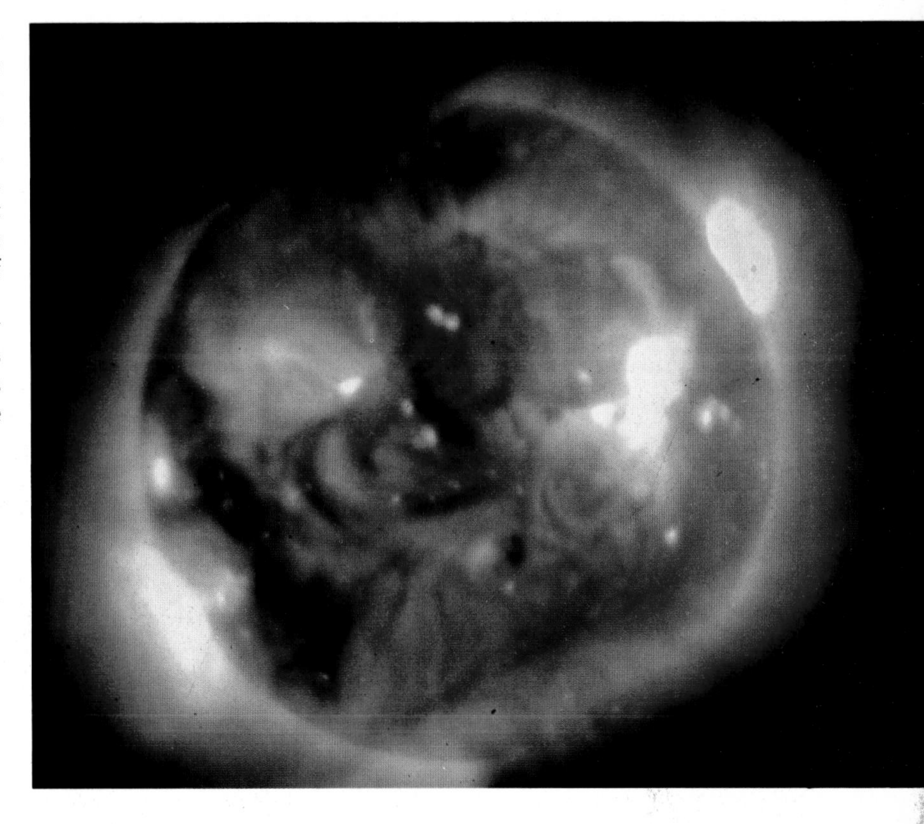

The Sun as it appears in X-ray light. This photograph was taken by astronauts aboard Skylab and reveals the hot bright inner corona and some flare hotspots across the disc. A new and previously unnoticed phenomenon of coronal holes was discovered. These are regions where the coronal density and temperature are reduced and appear as dark patches on X-ray pictures, but have no associated activity in visible light. It is speculated that these coronal holes may be the source of the high-speed ionized particles of the solar wind. Note: *The colours in this photograph are not real; they are computer generated to provide contrasts so that detail may be detected more easily.*

Table 4·1 **Some general properties of the Sun**

property	value
angular diameter in the sky	31·99 minutes of arc
mean Earth-Sun distance	$1{\cdot}496 \times 10^8$ km
radius $R_\odot$	$6{\cdot}96 \times 10^5$ km
mass $M_\odot$	$1{\cdot}99 \times 10^{30}$ kg
mean density	$1{\cdot}41 \times 10^3$ kg
effective surface temperature	5 800 K
spectral type	G2V
apparent magnitude m_v	−26·74
absolute magnitude M_v	+ 4·83
luminosity $L_\odot$	$3{\cdot}83 \times 10^{26}$ W
equatorial rotation period	26 days

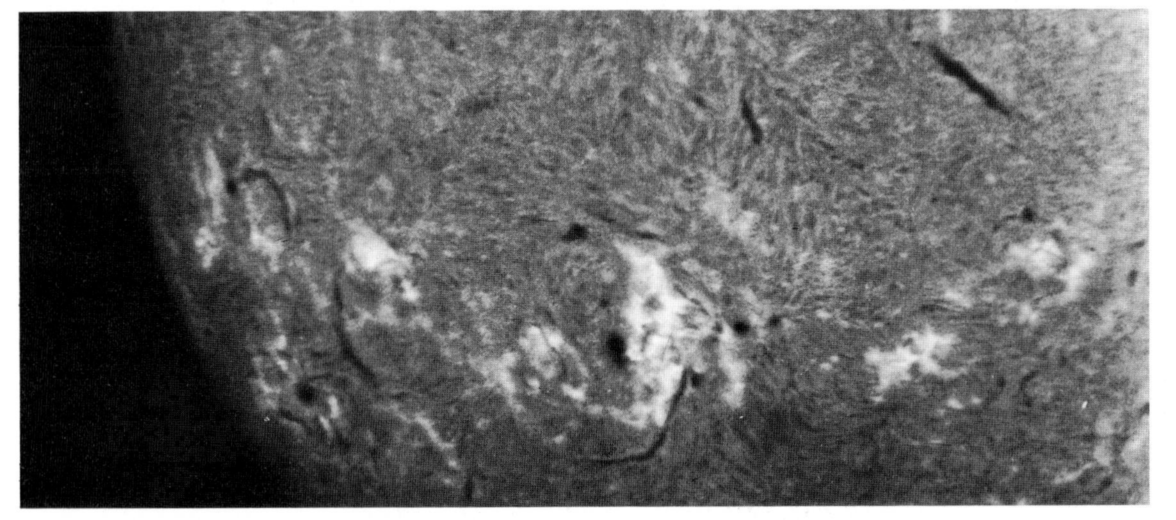

This photograph by Brian Manning taken in hydrogen-α light, shows considerable detail upon the Sun's disk, including bright faculae and dark filaments.

CHAPTER FIVE

The Solar System

The Earth

The number and diversity of objects within the solar system is not always appreciated. Besides the Sun there are nine planets, tens of known satellites, a few thousand minor planets and a host of comets numbering perhaps millions. The range of sizes is also vast. Comets may have haloes of tenuous gas which extend 3×10^7 km into space; the Sun, a far denser body, has a diameter of almost $1{\cdot}4 \times 10^6$ km, but all the other bodies are far smaller. Even the giant planet Jupiter is only a tenth the size of the Sun, while Mercury is little more than a thirtieth as big as Jupiter and Pluto less than one fortieth as big. Smaller still are their satellites and the many thousands of minor planets, some of which are only a few hundred metres across. In addition, there are the tiny bodies which make up the rings of Jupiter, Saturn and Uranus, and the innumerable meteors and particles of interplanetary dust.

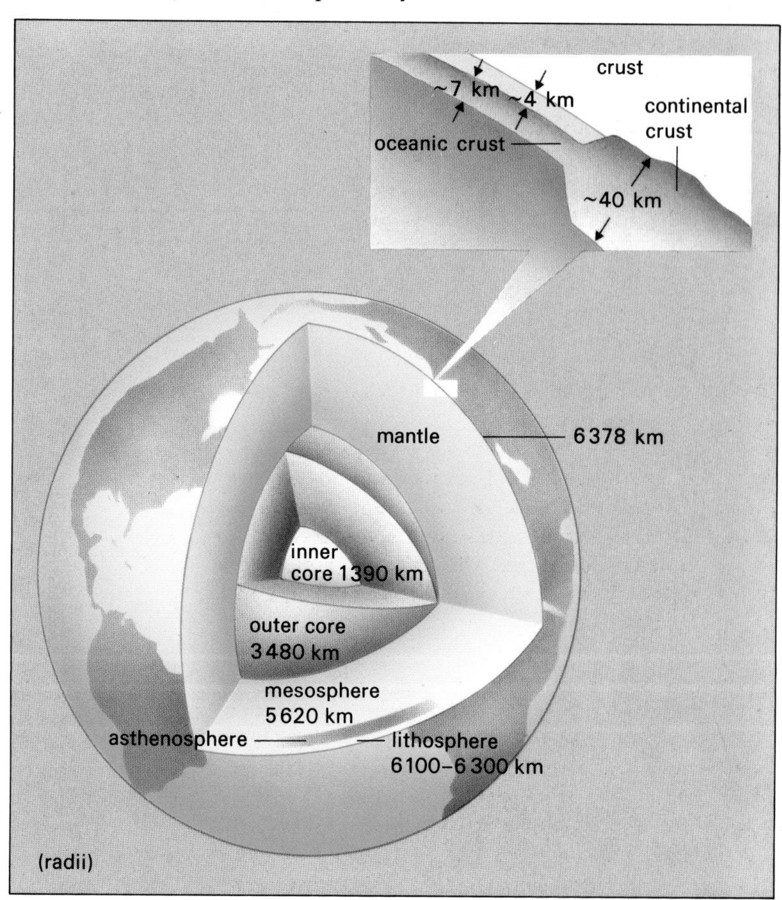

Fig. 5·1 The interior of the Earth. Average figures are given for crustal thicknesses and the radii of the various layers. The asthenosphere may be regarded as plastic and the outer core is liquid.

Although in the past it was natural to try to interpret the planets in the light of conditions on Earth, one of the most important results of space research is that it is now possible to examine our planet with new insights into the processes which are at work, and to determine how it resembles, or differs from, all the other planets. We now see that as a planet, it is comparatively undistinguished except for the fact that it is the only one with large quantities of liquid water and an oxygen-rich atmosphere. Geologically it is very active, being subjected to endless changes due to mountain-building and erosion, and in these respects it is somewhat similar to the planets Venus and Mars.

It is the largest of the inner planets with an equatorial diameter of 12 756 km, but rotation has caused a flattening at the poles, reducing the polar diameter to 12 714 km. The accurate tracking of Earth-orbiting spacecraft which, obviously, are affected by gravitational irregularities, has revealed that it is uneven and slightly pear-shaped, with a bulge in the southern hemisphere. In addition, such tracking gives information which reveals the distribution of mass within a planetary body and provides a guide to the densities of the Earth within the various internal layers at various depths.

Interior

The structure of the Earth has been principally determined by the study of seismic waves produced by both earthquakes and man-made explosions. From a study of the arrival times of the various waves which are produced, it is possible to determine the nature and depth of the different internal layers in which they have travelled. The general structure is shown in Fig. 5·1, the major divisions being the **crust**, the **mantle** and the **core**. The mantle itself consists of three regions, the highest and deepest of which are rigid, while the intervening one is weak and composed of material which is able to flow. One form of seismic wave is unable to travel through liquids and its absence from certain records indicates that the majority of the core is in liquid form. Waves which do traverse it, however, reveal the presence of a solid inner core.

The densities which have been found are given in Table 5·1 and suggest that the mantle is principally composed of silicates, while the predominant material in both the inner and outer core is iron, although there may be a difference in detailed composition between these two regions. It seems that lighter elements such as silicon are present, and that there could be a small proportion of nickel, perhaps 6 per cent, as found in metallic meteorites. The crust, which does not greatly affect the Earth's overall density because of its relative thinness (7–40 km), is very light with large quantities of silicon and aluminium compounds and a high concentration of sodium, potassium and the two radioactive elements uranium and thorium.

It is known that there is a flow of heat from the interior of the Earth, but that this is only about 0·04 per cent of the heat received from the Sun. Part of this internal heat is probably a remnant of the heating produced when the Earth was originally formed, and the rest has come from the decay of radioactive elements. It is difficult to calculate the temperature which exists at the centre, but there is general agreement that this must be in the region of 4 000–5 000 K to be consistent with seismic results. This heat is thought to produce convection within the liquid outer core (which is probably of importance with regard to the Earth's magnetic field), and to be a contributory factor in causing the flow in the middle layer of the mantle, which is suggested by plate tectonic studies.

Table 5·1 **Major subdivisions of the Earth**

	thickness or radius (km)	mass (kg)	mean density (kg per m³)
oceanic crust	7	$7{\cdot}0 \times 10^{21}$	2 800
continental crust	40	$1{\cdot}6 \times 10^{22}$	2 800
mantle	2 870	$4{\cdot}08 \times 10^{24}$	4 600
core	3 480	$1{\cdot}87 \times 10^{24}$	10 600
oceans	4	$1{\cdot}39 \times 10^{21}$	1 000
atmosphere	—	$5{\cdot}1 \times 10^{18}$	—

The distribution of density within the core is difficult to determine, there being a range of possibilities, but the probable mean densities are approximately 10 000 kg per m³ for the outer core, and 13 000 kg per m³ for the inner.

Plate tectonics

The most obvious division of the Earth's surface is into continents (40 per cent) and ocean floors (60 per cent). With this in mind, and using geological evidence from the continents as well as noting the way the coastlines of European, African, and North and South American continents fit together, Alfred Wegener proposed in 1915 a theory of continental drift. The continents were supposed to have formed one large land mass and then to have drifted apart. Increasing information about the oceans, the sea-beds and rock magnetism has led to continental drift being superseded by the theory of plate tectonics. This states that the crust is carried upon the outer layer of the mantle, which is in the form of a number of relatively rigid, thin (100–200 km) plates which are in motion both with respect to one another and to the interior of the Earth. This movement, which is naturally very slow, is permitted by a flow in the underlying weak layer. A circulation of material exists, rising towards the surface (generally in the centre of the oceanic areas), flowing horizontally over considerable distances and then descending again into the interior (Fig. 5·2).

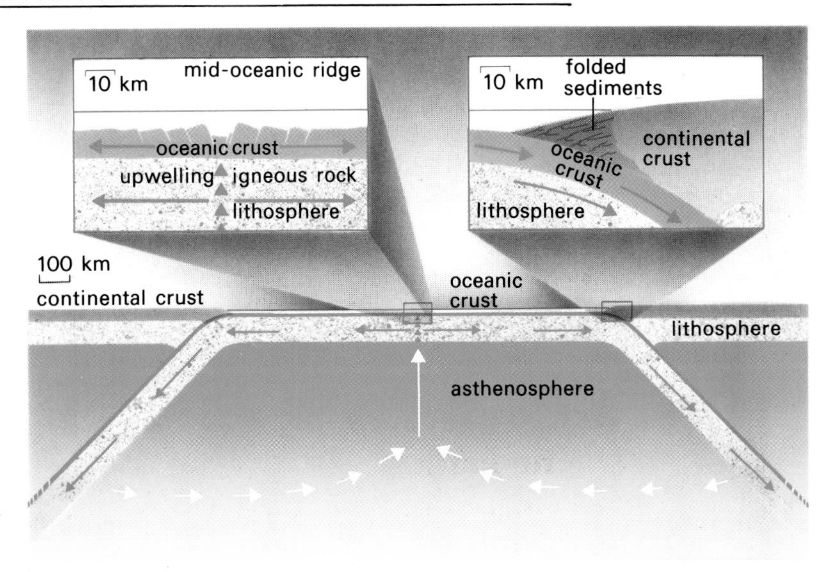

Fig. 5·2
A schematic diagram of plate tectonic movements. Material rises from the mantle under the mid-oceanic ridges and, spreading sideways, carries plates of both oceanic and continental crust. Mountains are formed where crustal regions 'collide'.

The upwelling is particularly noticeable at the mid-oceanic ridges which rise about 2 km above the general level of the ocean floors and where the injection of IGNEOUS ROCK adds to the edges of the separating areas of crust. On cooling such MAGMAS are magnetized by the effect of the Earth's general magnetic field. One of the main factors leading to the acceptance of the plate tectonic theory was the discovery that due to periodic reversals in the Earth's magnetic field, the rocks had been magnetized in opposite directions to give a series of magnetic stripes, and that these were repeated on the other side of the mid-oceanic ridge.

A further effect of the separation of the plates is the production of rift valleys, or **graben**, along the centre of the oceanic ridges. The regions where the plates descend into the interior are often associated with ocean trenches. By determining the depths at which earthquakes occur, it is found that the plates descend at an angle of 45° down to about 700 km where the increasing temperature and pressure cause them to lose their identity.

It is the mobility of the plates which has resulted in the long, curved mountain-building belts which are so characteristic of the Earth and which are absent from the other planets. The average thickness of the continental crust is 40 km, compared with the 7 km of the oceanic crust, and where these two types meet, the oceanic one descends beneath the continental region. Large quantities of volcanic rocks are produced and the resulting mountains are like those of the Andean Cordillera. The other conspicuous type of mountain belt, such as the Himalayan and Alpine chains, is formed by the interaction of two continental regions resulting in the folding, over-thrusting and uplift of thick layers of SEDIMENTARY ROCKS.

The mountain belts surround regions of continental rocks which have been geologically stable for long periods of time. By RADIOISOTOPE DATING methods it has been established that these shield areas, or **cratons**, are generally $2{\cdot}2$–$2{\cdot}7 \times 10^9$ years old and that the very oldest Earth rocks have ages of

$3{\cdot}8 \times 10^9$ years. (The age of the Solar System itself is calculated at $4{\cdot}65 \times 10^9$ years, and that of the earliest primitive life forms on Earth at $3{\cdot}4 \times 10^9$ years.)

Atmosphere

The origin and evolution of the Earth and the other planets will be discussed later, but the formation of the present atmosphere may profitably be mentioned here. Its composition is given in Table 5·2 and is largely the result of gases emerging from the interior. These materials eventually replaced the original atmosphere which probably contained considerable amounts of hydrogen and helium. Large quantities of carbon dioxide once existed, but the greater part of this is now dissolved in the oceans and locked up in carbonate rocks such as limestone. By photosynthesis in plants, a further fraction has been converted into the atmospheric oxygen essential to animal life forms, which could not arise until oxygen became an appreciable fraction of the atmosphere, that is, about $0{\cdot}6 \times 10^9$ years ago. The vast amount of terrestrial water was also produced from the interior, but atmo-

The North African coast of Morocco at the end of the Anti-Atlas mountains which run down from top right. False-colour photography brings out the mountains' structural features. A small storm is out over the Atlantic to the north-west.

The mouth of the Colorado river, showing sediment being carried out into the Gulf of California. Low-tide flats are visible and a tear-drop shaped depositional island. The false-colour photography accentuates the pattern of the cultivated fields in the lower Colorado valley.

spheric water vapour is omitted from the table as it is present in such variable amounts, being about 4 per cent near the surface, but absent above about 12 km. The noble gases argon and helium are predominantly decay products of potassium and uranium respectively.

Part of the energy that the Earth receives from the Sun is absorbed by atmospheric molecules, principally of oxygen and water vapour. These filter certain wavelengths of radiation from space and so restrict the wavelengths available to Earth-based astronomers. Another part is reflected back without heating the surface at all, and this portion is termed the **albedo**; it varies greatly according to the type of surface on which the solar energy is falling (Table 5·3). Estimation of albedo is very important in determining the nature of the surface of many of the planetary bodies, and especially minor planets (see p. 154). The Earth's overall albedo is approximately 40 per cent and the fraction absorbed gives rise to the mean temperature of about 283 K for the surface, and 250 K for the atmosphere. Re-radiation to space at infrared wavelengths is also affected by atmospheric absorption, primarily by water and carbon dioxide, which thus trap radiation in the so-called **greenhouse effect**, which causes a warming of the air.

Table 5·2 **Earth atmospheric composition** (mean dry atmosphere below 25 km)

component	symbol	percentage volume
nitrogen	N_2	78·08
oxygen	O_2	20·94
argon	Ar	0·93
carbon dioxide	CO_2	0·03*
neon	Ne	0·0018
helium	He	0·0005
ozone	O_3	0·00006
hydrogen	H	0·00005
krypton	Kr	trace
xenon	Xe	trace
methane	CH_4	trace

(*very variable)

There is obviously an excess of energy received in the equatorial regions compared with that at the poles, and this is transported poleward partly by ocean currents, but predominantly by horizontal atmospheric motion, thus driving the Earth's vigorous weather systems.

Apollo 7 photograph of part of the Himalyas and Tibet, showing the mountain ranges so characteristic of the Earth. The area covered includes Annapurna in the far distance, and Everest and Makalu (casting large shadows) in the foreground.

Tracking Artificial Satellites

Observations of artificial satellites can be used to provide information about two aspects of the Earth. They are affected by any irregularities in the Earth's gravitational field, and precise determination of their orbits since the first satellites were launched in 1957 has gradually resulted in a most detailed knowledge of gravitational variations from place to place. (Studies of the motion of natural satellites has similarly given some information about the gravitational fields of other planets. However, this knowledge has been greatly increased in those cases where spacecraft have passed close by, or even better, orbited the body. The discovery of the lunar mascons described on page 117 is a case in point.) Information can also be gained about conditions in the upper atmosphere, especially its density. The orbits of satellites are greatly affected by the density of the regions through which they are passing, particularly at perigee. Strong solar flares, by causing heating and expansion of the upper layers of the atmosphere, greatly increase their density and thus have a marked effect on satellite orbits. In the final stages of a satellite's lifetime, such changes can precipitate decay and re-entry into the atmosphere.

All that is required for these studies is a number of accurate positional determinations at known times. Naturally this is more easily said than done, but although special large cameras are used by some professional teams, visual observations are quite capable of yielding the required degree of accuracy. Timing usually presents few problems, and stop watches or the simultaneous recording of time signals and the observer's event markers are quite suitable. As with lunar occultations (page 109), telephone or radio time signals should be used.

The measurement of positions is rather more difficult and the precise means that are adopted will depend upon the equipment employed. The purpose is to obtain a satellite's right ascension and declination at a determined instant. Some instruments such as theodolites – not usually found in amateur hands – will give readings in altitude and azimuth, but these will, essentially, be converted into RA and Dec. when analysed.

In the early days of studies of artificial satellites, teams of observers used 'fixed' telescopes to establish a 'fence' of overlapping fields in the sky. It was then possible for the particular observer (or observers) who saw the satellite to obtain the time of its transit across a crosswire in the field. Nowadays it is more common to find individual observers with binoculars or similar small-aperture equipment. The method then consists of observing the satellite's path against the stellar background, and of choosing a pair of suitable reference stars and noting its position relative to them at a certain instant. The stars chosen should be readily identifiable, and given in a good atlas and catalogue. As the positional information regarding the fainter stars (below about 9th magnitude) is not very reliable, bright objects should be chosen whenever possible. If possible more than one observation should be obtained during the satellite's pass, but this may not be practicable on some occasions as the satellite may run into eclipse. From the positions and times (and additional information such as the observer's exact latitude, longitude and altitude) the orbit may be derived.

With so many objects in orbit, it is essential that the proper identity of the satellites should be known, and it is normal for 'Look Data' predictions to be prepared for each observer, giving information on where individual satellites may be seen. Some observers prepare such predictions for themselves – especially if they have computers – but the normal method is for a national centre to be responsible.

One of the earliest and brightest artifical satellites, Echo 1, is shown in this photograph taken looking towards the galactic centre.

Far right:
Large binoculars, mounted on a specially made but simple stand, are particularly suitable equipment for observation of artificial satellites.

Magnitude observations are also of interest (although less vital scientifically) and may be made by methods similar to those used for variable stars (pages 56–57). As satellites, rockets and debris are of all sizes and reflective properties, as well as being at various distances from the Earth, a whole range of magnitudes is encountered, the brightest being visible to the naked eye. Although satellites move more slowly than meteors (one way in which re-entries can be distinguished from fireballs), they also suffer from the effect that moving objects must be rather brighter than fixed ones for them to be perceived by the eye. This, as well as the aperture of the telescope equipment, must be taken into account when choosing objects to observe. If tumbling accidentally, or having been set spinning deliberately, satellites also show regular magnitude variations, while large reflective surfaces such as solar panels can give rise to bright flashes. All these facts are useful pieces of information which should be recorded. Quite apart from the magnitude variations mentioned, of course, all objects will fade as they pass into the Earth's penumbra, or disappear completely within the umbral cone.

Satellite re-entries give rise to phenomena very similar to fireballs (pages 162–3) and they should be observed and reported in just the same way. They can be very bright, and on occasions can exceed the brightness of the full Moon. As with fireballs, some are thus visible in daylight, when it will usually be necessary to record the track in terms of altitude and azimuth rather than in celestial co-ordinates. If a bright re-entry is expected, it may be possible to photograph it deliberately, as happened when Skylab burnt up over Western Australia. Normally, however, the best chance of securing a photograph must lie with one of the regular fireball patrols.

Although some very experienced observers may use large telescopes and specialized techniques to provide an exceptionally high degree of accuracy, the observation of artificial satellites is yet another example of how amateur observation with comparatively simple techniques and equipment can give results of great scientific value.

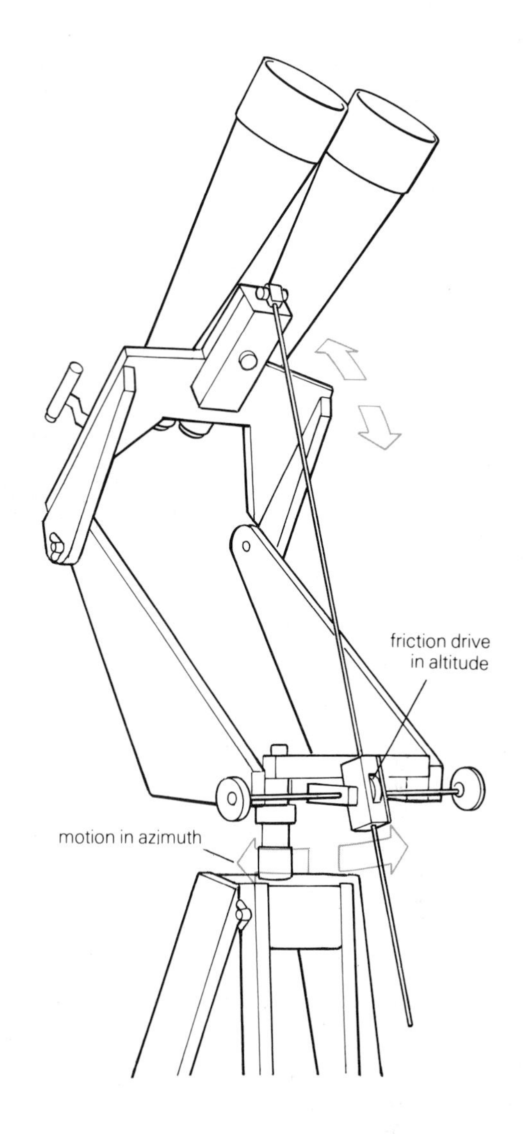

Satellite observation consists of estimating the position in terms of the distance between suitable stars (such as those marked A and B on this chart), by one or other of the methods shown here, whichever is most appropriate at the time.

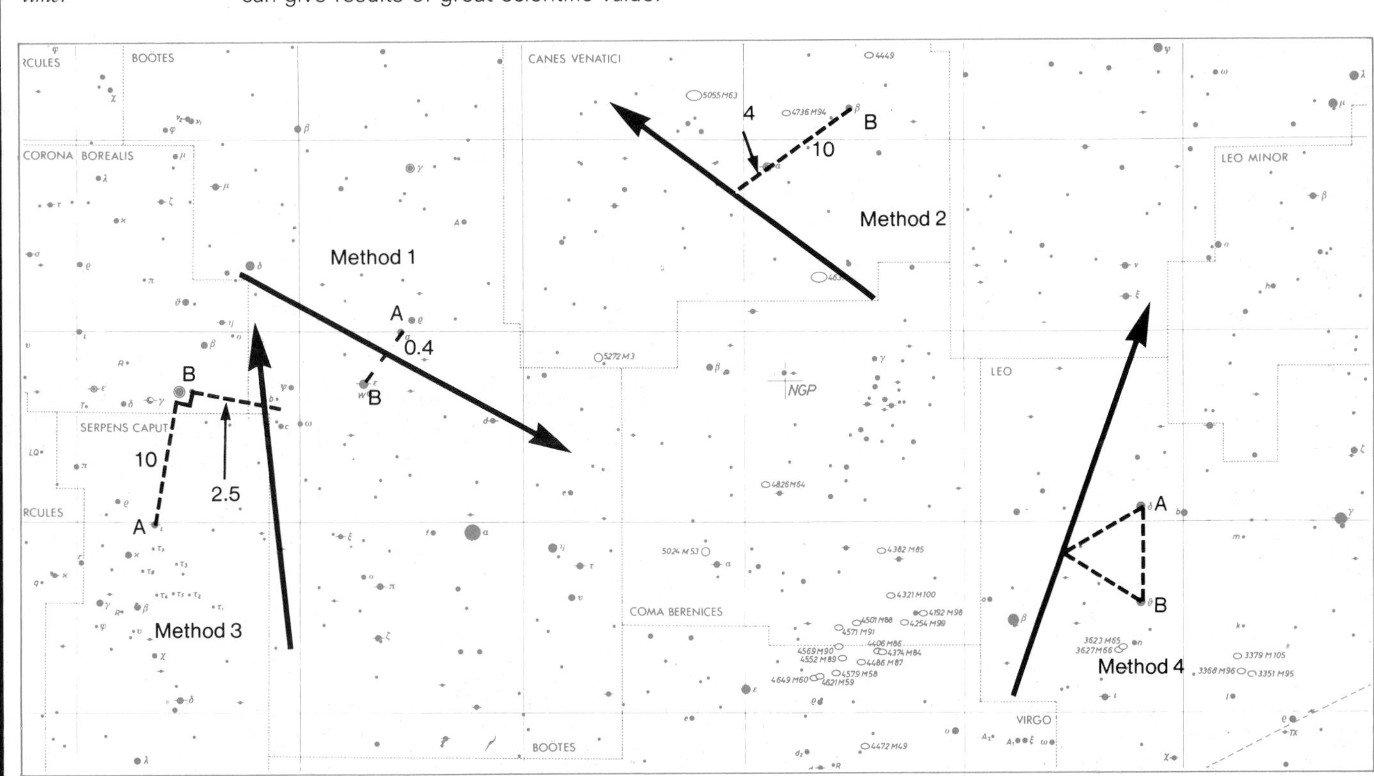

An infrared image of the Earth, obtained by the European geostationary meteorological satellite, Meteosat, on 1978 February 12 at 18·25 UT. The eastern Mediterranean, Red Sea and Iberian peninsula can be readily distinguished, while most of Europe lies beneath broken cloud cover.

The generalized atmospheric structure is shown in Fig. 5·3, together with temperature and pressure distributions. The lowest layer, or troposphere, contains 75 per cent of the total mass and is the zone where turbulence and weather systems are the most marked.

Ultraviolet and X-ray radiations from the Sun cause increasing ionization above 100 km, resulting in the various regions known as **ionospheric layers**, which are so important because they reflect radio waves of various wavelengths, particularly those used in radio communication.

Temperature rises rapidly with height due to absorption of solar radiation until in the exosphere, which mainly consists of oxygen, hydrogen and helium, the atoms are able to acquire sufficient velocity to escape into space.

Geomagnetism

The metallic core is responsible for the Earth's magnetic field. The exact method by which this takes place is unknown, but it seems that the liquid outer core is largely responsible, rather than the solid inner region. The shape of the MAGNETIC LINES OF FORCE is very similar to that of a common bar magnet, but the axis of the lines does not pass through the centre of the Earth; it is tilted by 11·5° to the Earth's rotational axis.

The geomagnetic lines of force extend out into space and define the region of the Earth's magnetic influence which is known as the **magnetosphere**. In an undisturbed situation the field would extend to a distance of at least 100 Earth radii, but the outer lines of force are strongly affected by interaction with the

The Earth-Moon system. The reversal of phases means that as seen from the Moon, the Earth is 'new' (as here) when the Moon appears nearly full from Earth.

ionized particles from the Sun, which are collectively known as the solar wind (page 78).

The boundary of the Earth's magnetic field where this interacts with the solar wind is termed the **magnetopause**, and the pressure exerted by the particles is so great that the field lines are compressed down to a distance of about 10 Earth radii on the sunward side (Fig. 5·4), although it may be even less at times of intense solar activity. On the opposite side the outer lines of force are not closed and stretch out into space to form the **magnetotail**. The full extent of this has yet to be determined by space probes, but it is certainly present far outside the orbit of the Moon (60 Earth radii) and probably beyond 1 000 Earth radii.

The solar wind particles have a very high velocity (250 to 400 km per s) and have an effect upon the number of cosmic-ray particles which are detected on Earth. At times of high solar activity, the cosmic-ray count drops and this is due to the solar wind sweeping the particles away from the Earth. Furthermore, low-energy cosmic ray particles cannot approach the Earth's middle latitudes due to the effects of the geomagnetic field.

The majority of particles within the magnetosphere are trapped by the field in a region which, beginning about 3 000 km above the Earth, extends out to 4 Earth radii.

At one time it was thought that this **Van Allen radiation belt** was formed of two separate regions with differing particle populations, but it is now known that high-energy protons are concentrated close to the Earth, with electrons and lower-energy protons at greater distances.

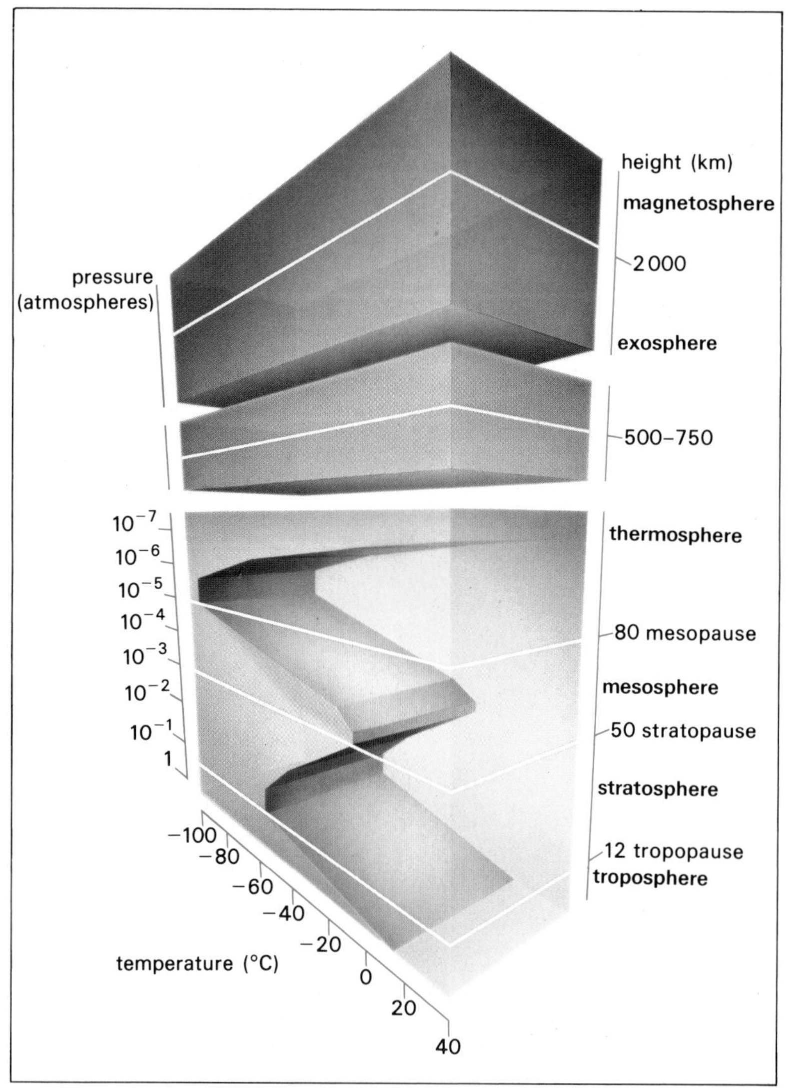

The aurorae

The large quantities of energetic particles ejected by solar flares can reach low levels in the polar regions, where they excite the atoms in the upper atmosphere, resulting in the emission of visible light and producing the polar aurorae. The base of auroral displays is usually at a height of approximately 100 km, although emission has been noted as high as 1 000 km, and they are normally overhead roughly 15° to 30° from the magnetic poles, although at times of intense solar activity, they have been observed from as far south as the geomagnetic equator. The aurorae can assume various forms, but the most striking are the rayed structures and the curtain-like sheets which can be observed when the bands of light are overhead. These display the alignment of the magnetic lines of force along which the particles are entering the atmosphere. The changes in the numbers of electrons in the upper atmosphere which are associated with auroral activity affect radio communication and are known as **ionospheric storms**. They also produce similar disturbances of the Earth's magnetic field.

The Earth-Moon system

When compared with the other satellites in the Solar System, the Moon is exceptionally large relative to its primary, with the Earth-Moon mass ratio being 81·3 : 1. (This ratio is only exceeded by that of Pluto: Charon, both of which are small bodies of probably similar composition, and where the ratio is of the order of 20:1. The next largest is that of Neptune: Triton which is approximately 500:1.) As a result, both bodies are actually in orbit about the centre of mass of the system, which is about 1 000 km from

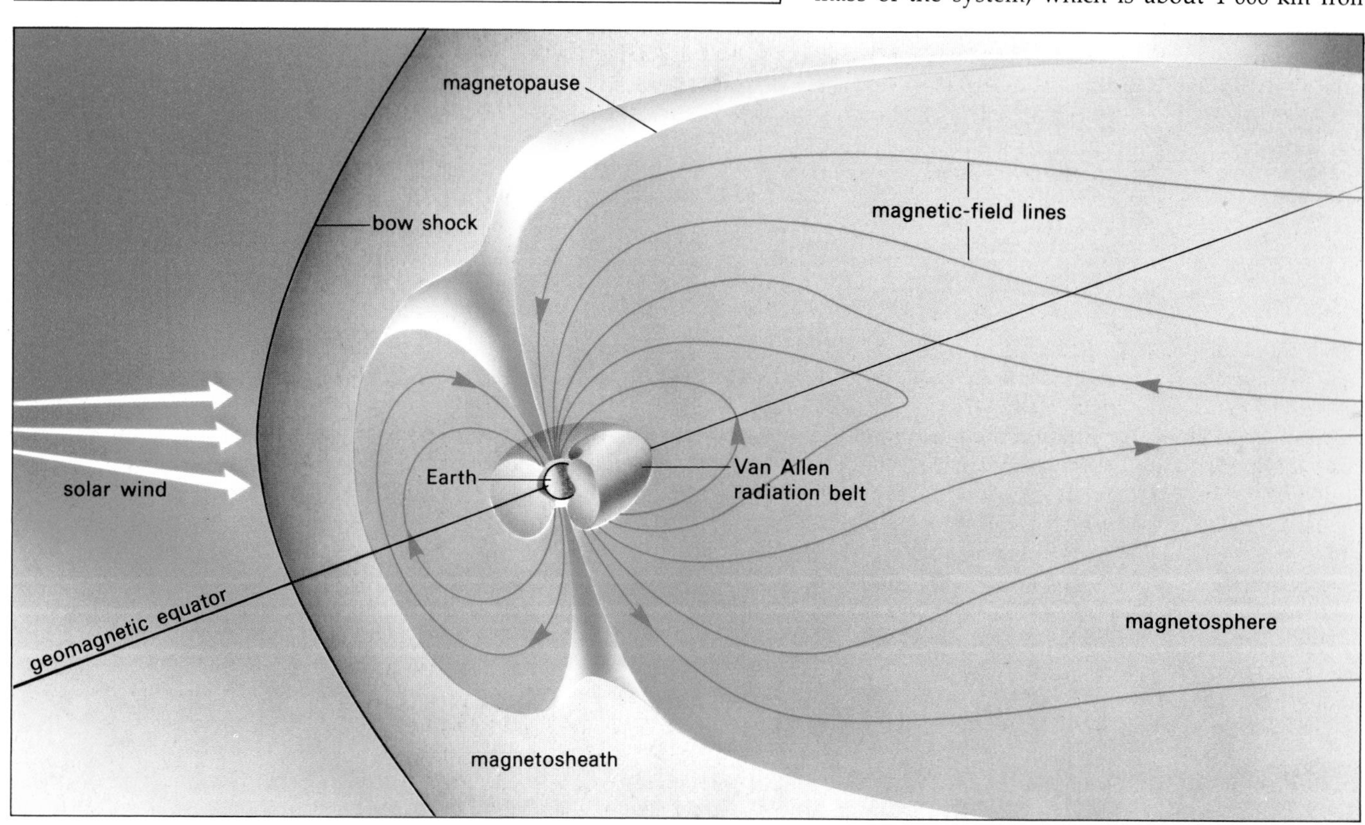

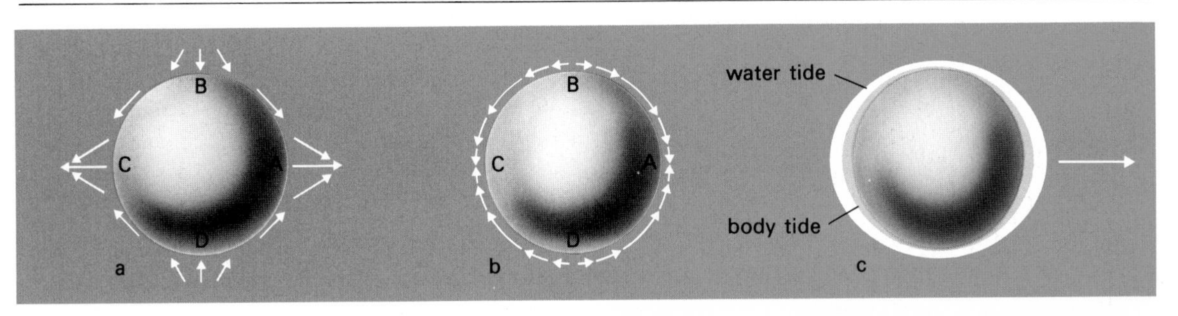

Fig. 5·5 left: *The tidal forces produced on the Earth are shown in a. (Notice that these act inwards at B and D.) The forces acting around the Earth are the most important in raising the oceanic tides (b). Both the Earth's body and water tidal bulges are shown in c.*

Table 5·3 **Typical ranges of albedo**

	(per cent)
clouds	44–80
land surfaces	8–40
water surfaces	4–>50
vegetation	9–25
snow and ice	up to 85

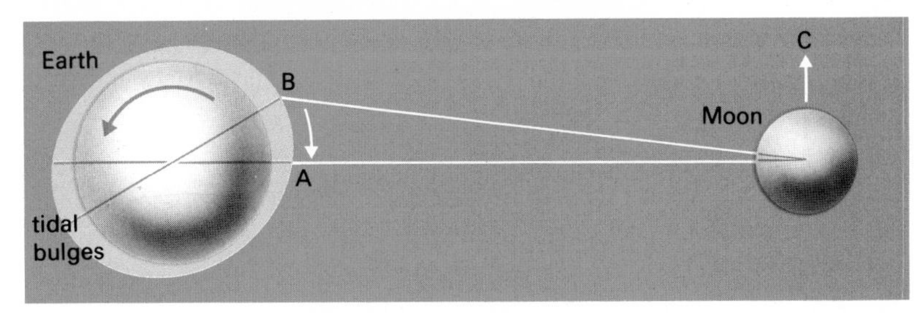

Fig. 5·6 above: *Due to the effect of friction, the Earth's tidal bulges lie ahead of the Earth-Moon line. As a result, the Earth is retarded from B towards A, and the Moon accelerated towards C.*

the centre of the Earth. Similar terms are used to describe the orbit of the Moon as for planetary or cometary orbits, with the closest and furthest points being known as perigee and apogee respectively. The motion of the Moon in its orbit is actually very complex. It has already been noted above that the synodic period (for example, from New Moon to New Moon) 29·53059 days, differs from the sidereal period of 27·32166 days, but due to various perturbations the orientation of the orbital plane in space and of the actual orbit on that plane change continually. Consequently the periods as measured from node to node (the **draconitic month**) and from (say) perigee to perigee (the **anomalistic month**) are 27·12222 days and 27·55455 days respectively. The first of these periods is of significance in calculating the occurrence of eclipses, as they may only occur when the Moon is very close to a node when the three bodies (Earth, Moon and Sun) are in line. The second is used in predicting tidal heights, as, naturally, the gravitational effect of the Moon is greatest when at perigee. There is even a fifth lunar period (the **tropical month**) measured from equinox to equinox, which differs slightly from the sidereal period, being 27·32158 days. (Similar considerations apply to the Earth to give tropical, sidereal and anomalistic years.)

The Moon always turns the same face towards the Earth, its sidereal periods of axial and orbital rotation being the same (said to be in a 1 : 1 **resonance**). However due to the inclination of the orbit, the varying speed along its path and the position of the observer on the Earth, **librations** in latitude and longitude occur. These mean that the Moon's visible face is not always precisely the same and that a total of about 59 per cent of the surface can in fact be seen from the Earth over a period of time.

The major effect which the Moon has upon the Earth is that of the tides. The actual explanation of the mechanisms involved is complex, but the combined effects of the motion of the Earth and Moon in their orbits about the centre of mass, the variation in the gravitational attraction of the Moon with distance, and the angle at which this is exerted, result in objects on the surface of the Earth being subjected to the forces shown in Fig. 5·5a. As far as the ocean tides are concerned, the vertical component of this force is less important than the horizontal (Fig. 5·5b), which effectively heaps the water up on opposite sides of the Earth (Fig. 5·5c).

The Sun also exerts tide-raising forces which are about 0·46 of those of the Moon, to give **spring tides** when the forces of Sun and Moon are aligned (close to Full and New Moon) and **neap tides** when acting at right-angles (near the Moon's 1st and 3rd quarter). There are numerous other effects such as the variation in the distance of the Moon from the Earth, of the Earth from the Sun, of the Moon's declination, as well as the shape and volume of the seas, which greatly complicate the task of actually predicting the times and heights of the tides at any place on Earth. It may be noted in passing that tides also occur within the bodies of the Earth and the Moon (the effects on the Moon will be mentioned later) and in the atmosphere, although in both cases the variations primarily have periods of a month.

Due to the effect of friction, both the body and water tidal bulges of the Earth lie ahead of the Earth-Moon line. As a result, the Moon is subjected to a small force which tends to accelerate it in its orbit, and the Earth's rotation is correspondingly slowed down (Fig. 5·6). As a consequence, the Moon's orbit is expanding and the Earth's day is becoming longer. This latter fact is confirmed by a study of certain fossil corals which have daily growth bands and where yearly variations can also be seen. These show that approximately 4×10^8 years ago, the 'day' was approximately 22 hours long. At that time the distance of the Moon was about 58 Earth radii, rather than the present 60. The effect will continue, until in the far distant future the Earth's axial period and the Moon's orbital period are equal at about 60 present Earth days.

Quite apart from this slowing effect by the Moon, the length of the Earth's day is also subject to slight, sudden irregularities which are caused by major earthquakes. These not only affect the rate of rotation, but also produce small alterations in the inclination of the axis of rotation.

Fig. 5·3 facing page, top: *The structure of the Earth's atmosphere. The temperature rises very rapidly with height in the thermosphere. Ionization layers occur at about 60 km, 110 km and also 150 km (day) or 250 km (night).*

Fig. 5·4 facing page, bottom: *The Earth's magnetosphere is enveloped by the stream of particles forming the solar wind. Due to the inclination of the magnetic and rotational axes, the position of the magnetopause and the orientation of the field lines vary both daily and over a year.*

Aurorae

The observation of aurorae is quite extensively covered by amateur astronomers and needs very little in the way of equipment in its simpler forms. As aurorae occur along the auroral ovals surrounding the north and south magnetic poles, observers tend to be concentrated at fairly high latitudes. However, it is useful for astronomers at low latitudes to be aware of the fact that they may be able to see aurorae – even if only on rare occasions – and that they may sometimes be more favourably placed than people at high latitudes. They should take every opportunity to check for the presence of aurorae, and record their form. A very valuable contribution to auroral studies is made by officers of ships at sea, who submit reports of sightings, and who thus provide help to fill the otherwise very considerable gaps in coverage which would result from relying upon land observers alone.

The restricted population (and land areas) in the Southern Hemisphere means that many southern auroral displays are poorly covered, Antarctic observers being sometimes *too* far south. It is therefore difficult to make correlations between activity seen in the two hemispheres, although this can occasionally be tried. Auroral activity does, of course, vary with the sunspot cycle, reaching a maximum one or two years after sunspot maximum, when in general the auroral ovals expand and move down to lower latitudes.

Auroral observations are reasonably straightforward and consist of recording which of the main types are present: arcs (arches with smooth lower borders), bands (irregular lower borders or folds), patches (resembling isolated clouds), veils (widely spread, evenly illuminated areas) and rays (streaks of light extending upwards into the sky). The basic forms may be described as homogeneous, striated (with bands roughly parallel to the lower border) or rayed (appearing to be formed of many individual rays). Further information to record includes details of behaviour (e.g. quiescent or active), brightness, colour and form (e.g. multiple or, if seen at the zenith, coronal). When arcs or bands are seen it is most important to record the elevation above the horizon of the highest point of the lower border. If several observers report this information it is then possible to derive the altitude and position of the display.

Although most people are fascinated by the constantly changing forms in an auroral display, and will watch continuously, efforts should be made to note full details at the standard times: every hour, and at 15, 30 and 45 minutes past each hour. This same procedure should be followed with photography as well, although naturally one might wish to take other photographs at intermediate times in addition.

Photography is of course of very great value. Exposures will usually be fairly short – perhaps of the order of half a minute, although this depends upon the speed of the film. Photographs which record the stellar background are particularly valuable, as they enable easy determination of the extent and precise position of a display. There is a lack of

Opposite page, top: *The frequency with which aurorae can be seen varies very greatly with the observer's location on the Earth.*

Opposite, centre right: *Measurements of the angle from the horizon to the bottom of auroral arcs are most important.*

Multiple rayed bands (below) *and a homogenous single band covering the V of the Hyades* (facing page, below). *Both auroral displays were photographed in Canada by W. Cobley.* Opposite: *an aurora photographed from Fort Augustus, Scotland, by D. Gavine.*

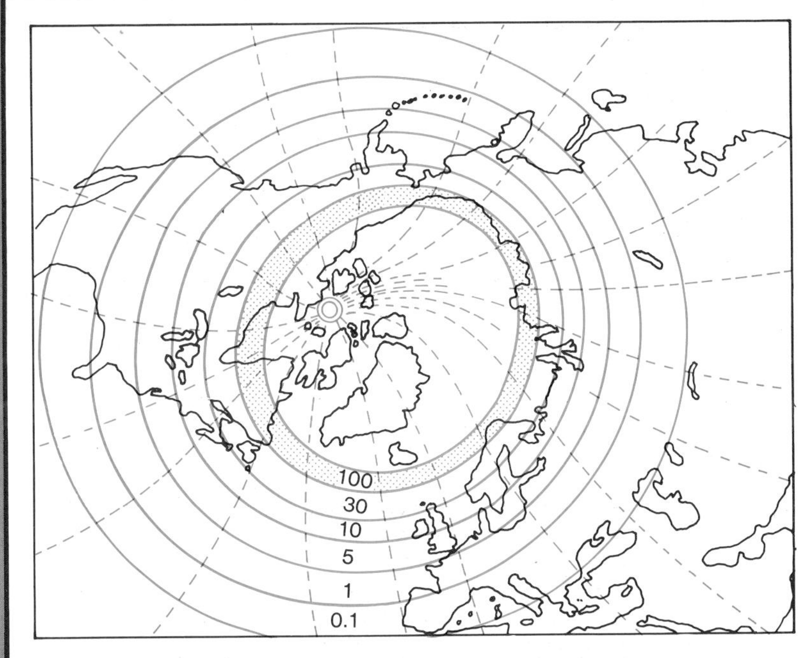

knowledge about the precise heights at which some of the colours are present, but photographs taken on modern fast colour films should be capable of being used to determine this. However, care needs to be taken to ensure that consistent results are obtained.

The magnetic disturbances associated with aurorae can be very marked, and it is quite possible to construct simple magnetometers – usually of the suspended-magnet type – which will give an indication of the changes taking place in the Earth's magnetic field. Some amateurs have constructed more sophisticated devices, including those known as fluxgate magnetometers, which permit continuous recording of the magnetic disturbances. The changes in the ionosphere with the occurrence of aurorae have always been known to affect radio reception, and the detection of such anomalous conditions is yet another aspect which can be followed with comparatively simple equipment. Both magnetic and radio studies have the great additional advantage, of course, that they can proceed whatever the weather and the cloud conditions.

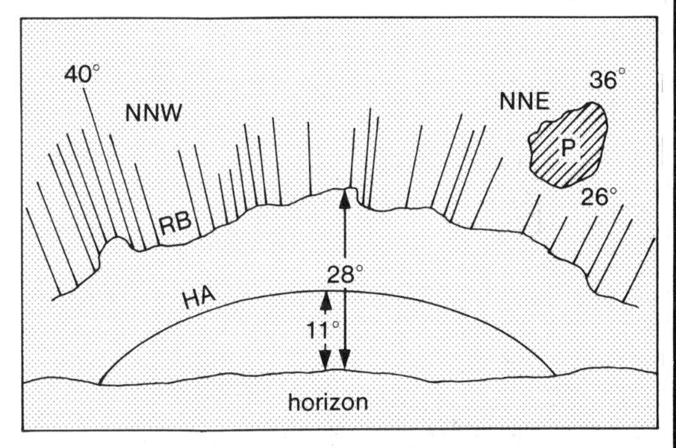

The Moon

Although the Moon was mapped through telescopes before its exploration by manned and unmanned spacecraft, we now have factual evidence to help us interpret its features. Indeed, due to photographic missions in lunar orbit, higher quality maps are available for practically the whole surface than for some parts of the Earth. Moreover, the samples returned from the Apollo and Luna landing sites have completely changed ideas about the origin and early evolution of the Solar System, as well as answering many questions about the Moon itself.

The Moon, which has a diameter of 3 476 km, orbits the Earth at an average distance of 384 402 km. The relative sizes of these two bodies are shown in Fig. 5·7, while Table 5·4 gives additional data.

Tracking orbiting spacecraft has shown that the Moon is not perfectly spherical, but is slightly elongated towards the Earth. The Earth's tidal forces have locked on to this distortion and caused the Moon always to turn the same face towards us. This near side is divided into light and dark coloured areas, called by the early investigators **terrae** and **maria**, from the Latin words for lands and seas respectively, since they mistakenly supposed the surface to be similar to that of the Earth. Use of these terms has persisted despite more accurate knowledge, although the terrae are now more frequently referred to as **highlands**.

Around the edges of the maria these highland areas may form conspicuous mountain chains stretching for hundreds of kilometres, with the Apennines, for example, reaching 7 km above the nearby plains. The maria are either circular or irregular in shape and we now know that they are concentrated on the near side with only a few minor examples elsewhere.

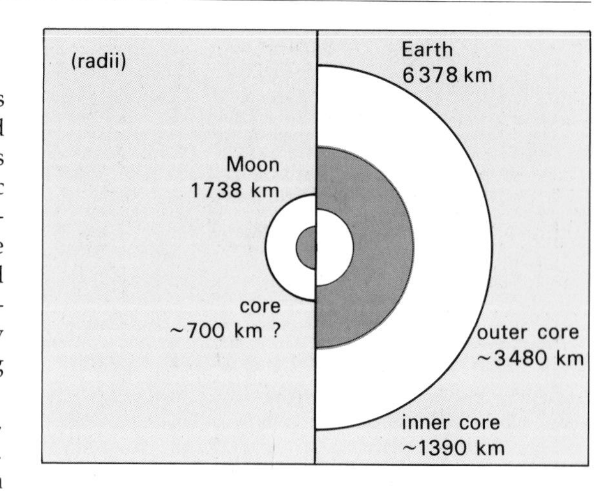

Fig. 5·7 far right: *Comparative sizes of the Earth and the Moon. The size of the Moon's core (if any) is uncertain.*

The lunar farside crater Tsiolkovskii, photographed from Apollo 15 in lunar orbit. Its diameter is 180 km and the slump terracing and central peak are typical of craters of this size, but the lava flooding of the floor is unusual.

Table 5·4 **Moon-Earth comparative data**

	Moon	Earth
equatorial diameter (km)	3 476	12 756
sidereal period of axial rotation	27·322d	23_h 56_m 04_s
inclination to ecliptic	1^0 32′	23^0 27′
density (kg per m^3)	3 340	5 517
mass (Earth = 1)	0·0123	1·0000
surface gravity (Earth = 1)	0·1653	1·0000
escape velocity (km per s)	2·37	11·2
albedo	0·07	0·36
mean Earth-Moon distance	384 402 km	

Craters

Craters have been found on all four inner planets, the Moon and the satellites of Mars, Jupiter and Saturn. In the case of the Moon, although present all over the surface, they are particularly numerous in the highlands. They are circular, or approximately circular features with raised walls and range from large multi-ringed structures with diameters of hundreds of kilometres, all the way down to microscopic pits on the surface. The Moon is exceptionally rich

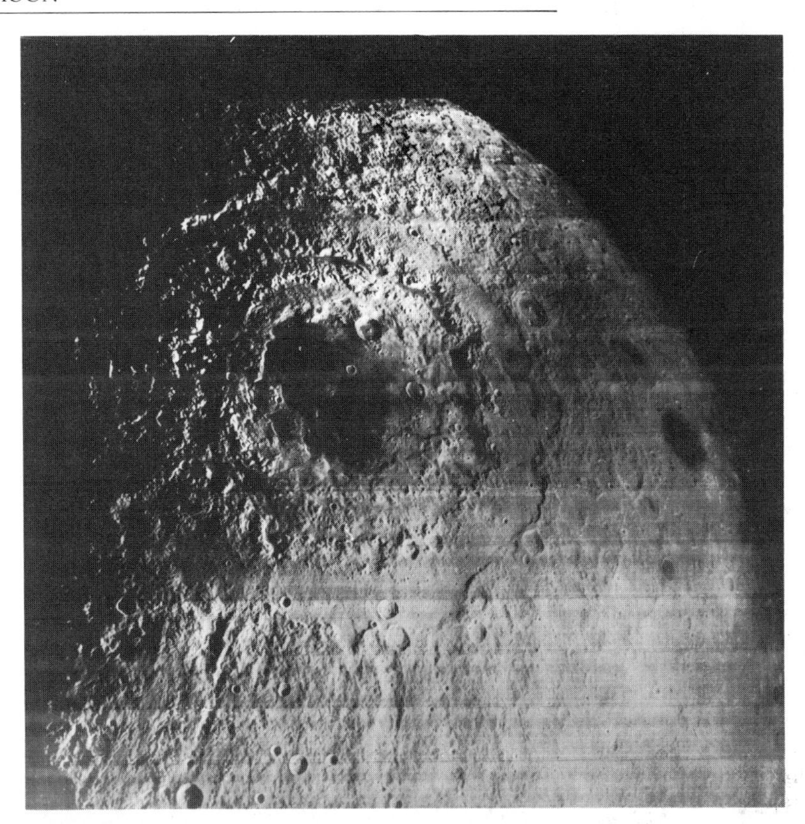

Above:
Lunar Orbiter IV photograph of the Mare Orientale basin, and its multiple mountain rings. The outer ring, the Montes Cordilliera, has a diameter of 900 km.

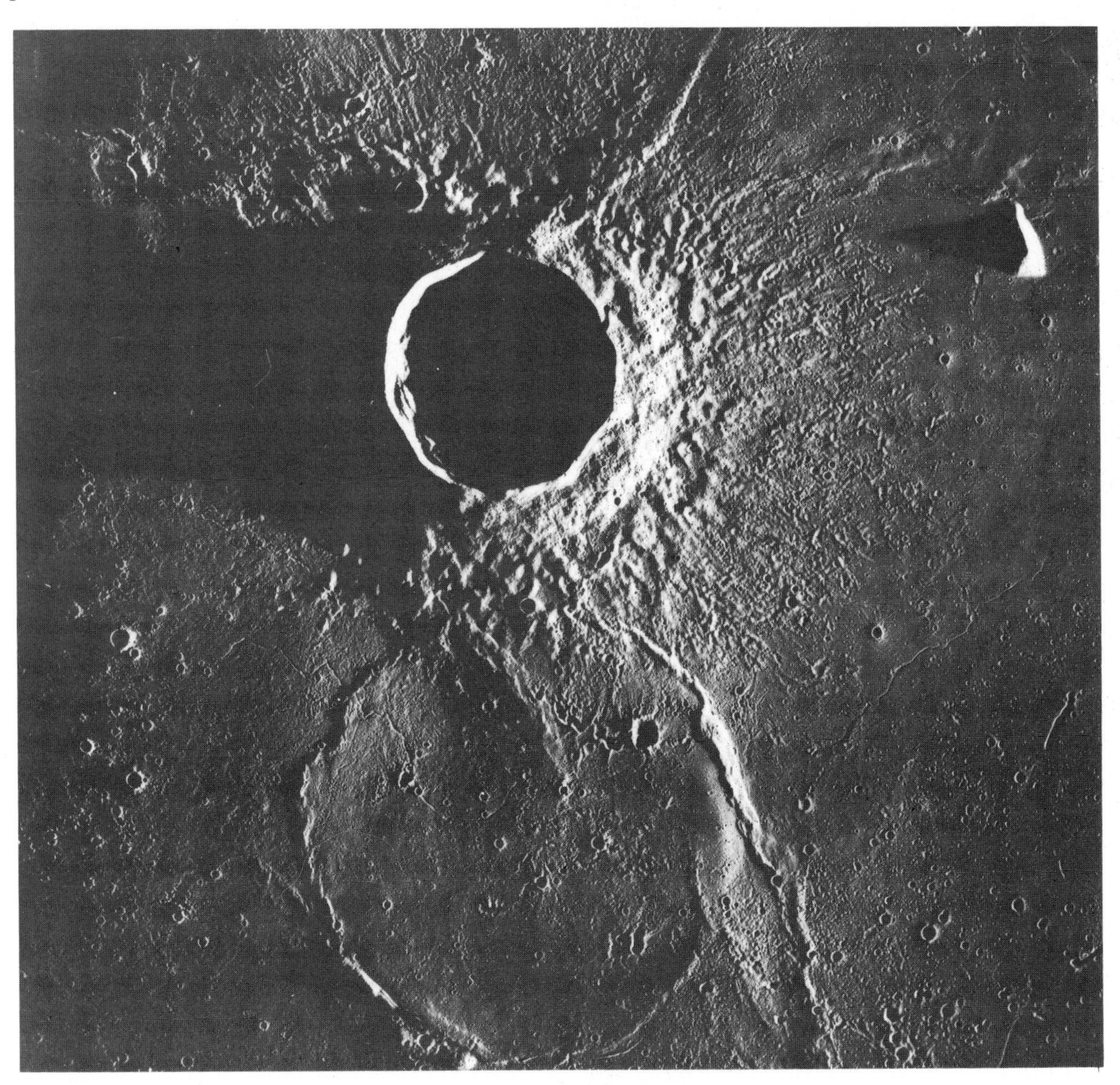

The crater Lambert in the Mare Imbrium, showing the radial structure of the ejecta blanket and also many secondary craters. The mare ring structure to the south has a diameter of about 50 km and seems to have been caused by lava flows covering, and then subsiding onto an earlier crater wall.

The large (94km diameter) ruined walled plain Fra Mauro is marked by the spacecraft boom. Apollo 14 landed in the rugged terrain just outside the ring and towards the top.

in craters with diameters of 20–50 km and these, together with the larger sizes, are very shallow in relation to their diameter, with depths of just a few kilometres. Smaller craters are often clean-cut and bowl shaped, but the larger they are, the more likely it is that they will be partially filled by material which has fallen from the walls. In the very largest cases there has usually been extensive slumping to form internal terraces and, in some instances, flooding by dark mare material has produced a level floor. Big craters frequently show a central peak, or even a ring of peaks around the centre. Such features show many similarities to the multi-ringed basins, one of the youngest of which is the 900-km-diameter Mare Orientale.

The origin of all the craters has long been a subject for debate between those who favour a volcanic origin and those who believe they were formed by impact. However, the evidence from spacecraft data and from the rocks gathered by the astronauts makes it seem highly likely that the majority are impact features. The impact of a high-velocity meteorite vaporizes both the object itself and part of the underlying rock down to as much as a few kilometres in depth. The explosion caused by this pocket of hot gas can transport debris to very great distances, but

the form of the crater which is produced is always circular, regardless of the direction of the impacting body. Explosion craters have been studied on Earth and they show rims of uplifted bedrock which correspond to the raised walls of lunar craters, while measurements of the diameters and depths show complete agreement between the two bodies. Elastic rebound of the rocks can cause the central peaks and peak rings which are seen in the larger, naturally formed craters on both the Earth and Moon. Débris from the explosions can be seen as characteristic **ejecta blankets** and the larger fragments can themselves produce further secondary craters. Under Earth's high gravity conditions, the ejecta fall close to the main crater, but because the Moon's gravity is only about one-sixth of that of the Earth, relatively young craters on the Moon often show bright rays which may stretch for hundreds of kilometres. These and the smaller, bright haloes are composed of fine dust and glass beads flung out by the impact.

The large multi-ringed structures are thought to have been caused by the impact of very large bodies, perhaps tens of kilometres in diameter, although there is still considerable discussion over the exact way in which the surrounding mountain rings were formed.

The vast ejecta blankets from these impacts can be traced over wide areas of the Moon's surface, while some of the earlier structures have been almost obliterated by large numbers of smaller, later craters.

At the other end of the scale, low-velocity impacts of small bodies do not cause explosive cratering, but merely excavate pits by throwing out loose materials. Such activity is important in the uppermost layers of the surface, termed the **regolith**, which varies in thickness from 4–5 m on the maria to 10 m and more in the highlands. The uppermost surface layer is composed of the finest dust, but the regolith's composition ranges from this to large blocks several metres across.

In the airless and waterless lunar environment it is impacts which have fractured and powdered this material, and which are responsible for the very gradual erosion and obliteration of craters and other features such as ray systems.

A type of rock which is very common in lunar samples is a **breccia**, in which fragments of rock have been welded together by the heat of later impacts, before being broken up yet again. In some highland samples, as many as four different generations may be recognized, showing the results of repeated impacts.

A composite of two photographs taken on the lunar surface at the Apollo 17 landing site in the Taurus Mountains near the crater Littrow. The valley is thought to be a graben flooded with basalt flows totalling about 1 400 m. Astronaut Schmidt and the Lunar Rover give an idea of the scale of the surface features, while South Massif on the right is about 8 km distant and reaches a height of 2 500 m. The large broken boulder is a breccia, and has rolled about 1·5 km down the slope.

Observing the Moon

The Moon is usually the first object to which a newly acquired telescope is turned. Its surface exhibits a wealth of fascinating detail and no description can convey the excitement of observing it for the first time. In addition to the well-known craters and extensive maria there are mountain ranges, valleys, domes, rilles and ray systems. All of these display dramatic alterations in their appearance in the course of a lunation as the Moon's phase changes – the rays, for example, are very conspicuous near Full Moon when most of the other features are less apparent, lacking shadows under the nearly vertical illumination.

Visual observation of the Moon is not as significant today as it was twenty-five or so years ago, when the amateur was at the forefront of lunar research. A decade of intensive exploration by spacecraft has caused a change of emphasis and most work now is concerned with the elusive obscurations and colorations known collectively as transient lunar phenomena, or TLP. (The alternative term lunar transient phenomena, or LTP is also sometimes used.) The reality of these events has been established almost exclusively by amateur observations made in the past twenty years, and only amateurs continue to monitor them. It has been found that they are more pronounced when viewed through certain colour filters, and simple devices have been constructed to allow the filters to be alternated rapidly, giving rise to a 'blink' which can be easily detected by the observer, just as blink comparators work by presenting two photographs alternately to the person examining the plates. Attempts have been made to use photoelectric devices for this task of detecting changes on the lunar surface, but they have proved to have little success when compared with the human eye. When an event is detected it is usual for other observers to be alerted and given just a general description of the area in which the TLP is occurring, so that they may search independently. Needless to say, familiarity with the lunar surface is a prerequisite for anyone wishing to undertake a regular patrol, and the best way of achieving this is by drawing specific areas of the Moon under all conditions of illumination.

Drawing the lunar surface is great fun and can be highly recommended. Although now of little scientific value, with the existence of highly accurate maps made from lunar orbit – although some of the polar regions remain poorly covered – it can still give great personal satisfaction. A series of drawings of a particular crater and its surroundings, for example, under different conditions of illumination, will show striking changes in appearance and prove a very instructive exercise. A medium-power eyepiece should be used, i.e. one which is sufficient to show fine detail without causing the definition to deteriorate, and a soft pencil should be employed to sketch in the relative positions of the major features and their associated shadows, followed by careful positioning of the more delicate features. No alteration should be made to the drawing after leaving the telescope, although the shadows may be painted over with Indian ink and the very bright areas highlighted with a white chinagraph pencil. There is no inherent difficulty is precisely positioning features by eye and almost anyone can become proficient after a little practice – even those who would not feel that they were particularly 'artistic'.

The Moon is also the ideal subject for photography, offering as it does an abundance of suitable subjects, plenty of light, and a large-sized image. Quite satisfactory results can be obtained by simply applying the lens of a camera, focused at infinity, to the eyepiece of a telescope and making the exposures. Some experimentation may be needed to find the best exposure duration to suit the film in use, but times of between 1/25 and 1/2 second are

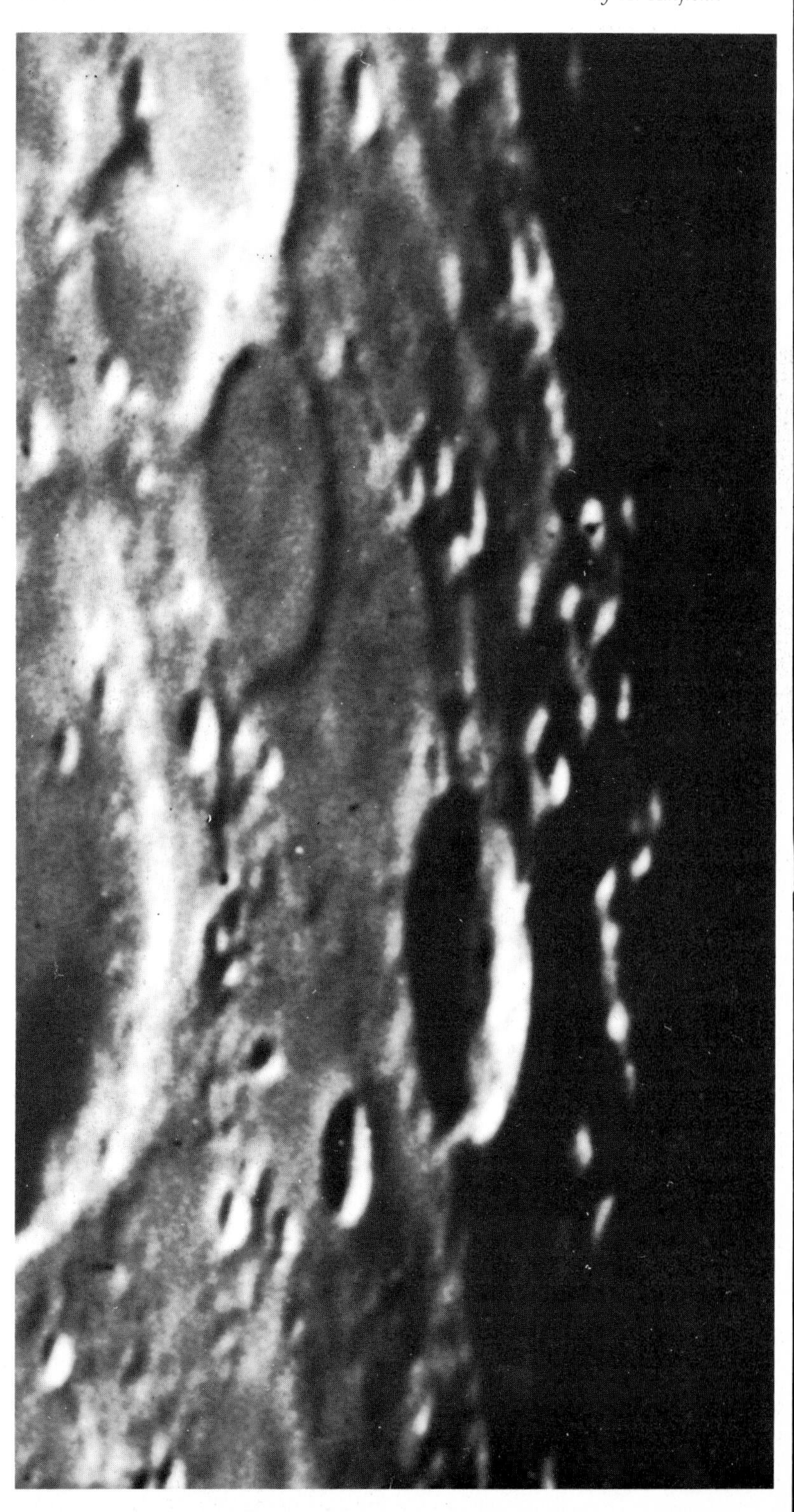

The lunar crater Inghirami and its environs, on 1966 December 25; a 2-second exposure on Ilford G30 Chromatic Plate (ASA 10) taken by H. Hatfield.

feasible with this method. For more serious work a driven telescope is desirable, and better still, one capable of being driven at the lunar rate rather than the usual sidereal rate. Many amateurs then project the image through a high-quality eyepiece directly on to the film. The sensitivity of a photographic emulsion is no match for the acuity of the human eye, and more detail can be seen on the Moon and planets by an observer than can be photographed through the same instrument. Photography has the advantage, however, that an exposure can be made in a fraction of the time it would take to make a drawing and the area covered is usually considerably greater. The effects of libration (page 101), mean that some features at the lunar limbs are easily visible at some times and quite out of sight at others. It may therefore take a considerable time to see or to be able to photograph some particular formation under just the right conditions of illumination.

A drawing by H. Hill of the same crater and made at the same time as the photograph opposite.

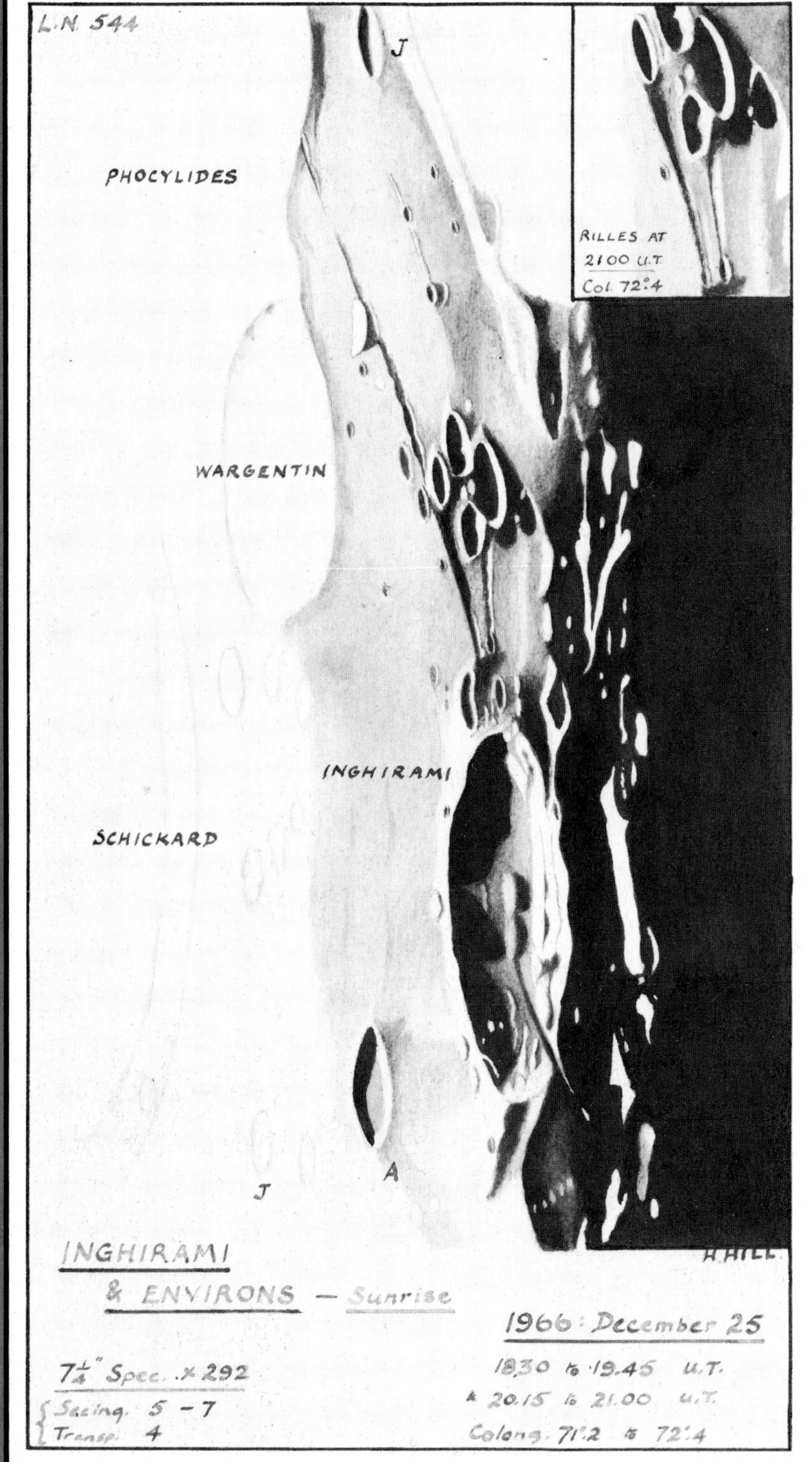

The appreciable motion of the Moon against the background stars – about its own diameter in one hour – causes it to pass between an observer on the Earth and a number of stars during the course of a month. The various lunar motions cause it to trace out a path in a wide band each side of the ecliptic, and there is a large number of such occultations which can occur. The disappearance of a star behind the dark limb of the Moon is a very striking phenomenon, and indeed, as the Moon is airless, is so sudden that it can come as quite a shock to an observer. However, the accurate timing of these events is very important indeed for determining the position of the Moon. Even simple visual methods giving an accuracy of perhaps ± 0.3 seconds of time enable the position of the Moon to be determined to an accuracy of about ± 0.15 arc sec. (about 300m), which is several times better than results obtainable with a transit circle. Depending upon the observer and the precise methods used, even higher accuracies can be achieved. It was an analysis of such observations that first indicated that the Earth's rate of rotation was not constant, and this led to the concept of ephemeris time (page 17).

Predictions for the various occultations are produced by several professional observatories and published in various yearbooks, or sent directly to observers who have established their seriousness and accuracy in this type of work. In a similar manner professional astronomers are prepared to analyse accurate timings, and so occultations provide an excellent opportunity for observers with small telescopes to contribute to astronomical studies. The only basic ancillary equipment is a good stop-watch, preferably of the 'split-action' type, which can be stopped against the first available time signal. For the latter most amateur astronomers use the telephone service's 'speaking clock'. From the recorded time and other details, such as the observer's geographic latitude and longitude, the appropriate calculations may be made. Many observers use rather more sophisticated methods of timing, for example picking up the continuously broadcast time signals available on certain radio frequencies, recording these on a tape recorder, and superimposing a signal at the time of the event. Photoelectric detection of occultations is yet another technique, and one which will yield highly accurate times.

The first of these more sophisticated methods is required to make the most of the events known as grazing occultations. In these the star appears just to brush the Moon's limb, and because of the irregularity of its surface may disappear and reappear several times. The observers' position on the Earth is all-important here, as just a few metres may make all the difference between seeing an occultation and missing it entirely. Because of the problems of exact prediction, grazes are best observed by a team, the members of which are able to place themselves at intervals along a line, ideally at right angles to the predicted graze limit. A team of observers has the further advantage that a series of timings is obtained, thus allowing a detailed limb profile to be determined. This may be compared with the known profile, or in cases where it was previously unknown, serve for future reference. Under any circumstances the result is highly accurate knowledge of the Moon's position at that moment of time.

Disappearances at the bright limb and

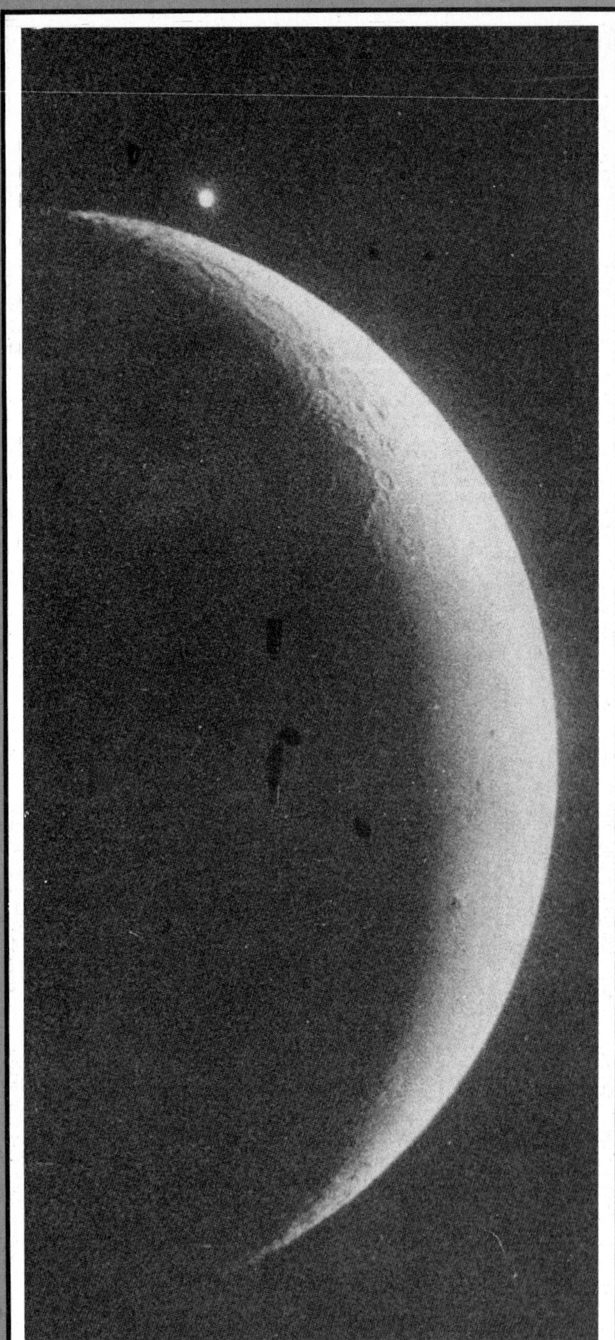

reappearances may also be observed and timed, but there is a tendency for these to be less accurate, mainly because bright-limb disappearances are more difficult to follow, and because the position of reappearances is usually a little uncertain.

Sometimes stars appear to fade gradually, or rather in two stages, instead of disappearing instantly, and this has been shown to be related – at least in some cases – to the fact that they are actually binary systems, unresolved by the telescope. Indeed in a few events the binary nature of previously unknown systems has been established.

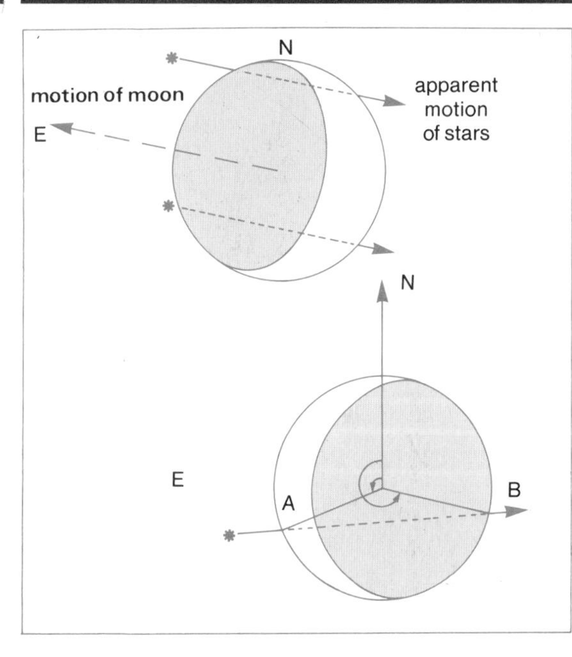

Venus, photographed (left) *just before occultation by the Moon, and just after* (right), *21 minutes later.*

Far left: *Observations of a grazing occultation may be used to provide a very accurate profile of that part of the Moon.*

Left: *Occultation predictions include position angles (see page 68), but bright limb disappearances (A) and all reappearances – even at the dark limb (B) – can be difficult to observe with accuracy.*

The total lunar eclipse of 1982 January 9 in a 5-minute exposure through a 150 mm f/19 refractor, taken by D. Buczynski.

LUNAR ECLIPSES 1984–200

Date	Type of Eclipse	Duration*	Region of visibility*
1984 (No eclipses)			
1985 May 4	Total	70m	Europe, Africa, Asia, Australia
1985 Oct. 28	Total	42m	Europe, Africa, Asia, Australia
1986 Apr. 24	Total	68m	E. Asia, Australia, Antarctic
1986 Oct. 17	Total	74m	Asia, Europe, Africa, Australia
1987 Oct. 7	Partial	–	–
1988 Mar. 3	Partial	–	E. Europe, E. Africa, Asia, Australia
1988 Aug. 27	Partial	–	E. Asia, Australia, Antarctic, N. America
1989 Feb. 20	Total	76m	Europe, Asia, Australia, N. America
1989 Aug. 17	Total	98m	N. & S. America, Europe, Africa
1990 Feb. 9	Total	46m	E. Atlantic, Europe, Africa, Asia, Australia
1990 Aug. 6	Partial	–	Australia, Asia, Antarctic
1991 Dec. 21	Partial	–	Asia, N. Australia, N. America
1992 June 15	Partial	–	N. & S. America, Antarctic, W. Africa
1992 Dec. 10	Total	74m	N. & S. America, Europe, Africa W. Asia
1993 June 4	Total	98m	E. Asia, Australia, Antarctic
1993 Nov. 29	Total	50m	N. & S. America, W. Europe, NE, Asia.
1994 May 25	Partial	–	N. & S. America, W. Europe, Africa
1995 Apr. 15	Partial	–	E. Asia, Australia, Pacific, N. America
1996 Apr. 4	Total	84m	N. & S. America, Europe, Africa, W. Asia
1996 Sept. 27	Total	72m	N. & S. America, Europe, Africa
1997 Mar. 24	Partial	–	N. & S. America, W. Europe, W. Africa
1997 Sept. 16	Total	66m	Europe, Africa, Asia, Australia, Antarctic
1998 (no eclipses)			
1999 July 28	Partial	–	E. Asia, Australia, Pacific, N. America
2000 Jan. 21	Total	84m	N. & S. America, Europe, W. Africa
2000 July 16	Total	102m	E. Asia, Australia, Antarctic

Astronomers are usually only interested in observing partial or total lunar eclipses, and in fact these are the only ones listed in most handbooks. Those eclipses when the Moon only enters the penumbral shadow offer little scope for any proper observations, quite apart from the fact that the reduction in the brightness of the Moon is usually so small that without prior information one would not generally realize that any eclipse was taking place. When the Moon is partially or completely immersed in the Earth's umbral shadow, however, it is far more interesting. Due to the refraction of the Sun's light within the Earth's atmosphere, the boundary between the penumbra and the umbra is never completely sharp; neither is the umbra absolutely dark, the Moon's surface usually being illuminated to a greater or lesser degree by light refracted into the umbral cone. This atmospheric refraction also gives rise to a dispersion of colours, so that occasionally the edge of the umbra appears to show an indistinct spectrum with the blue tint furthest away from the centre of the umbra, which may still be receiving some long-wavelength, deep red light.

There are very great variations in the overall brightness of the Moon when in the umbra, and also in the general tint of the surface. There are a few occasions on record when the eclipses have been very dark indeed and the Moon has completely disappeared for a while in mid-eclipse, but these events are infrequent. The visibility of individual lunar features such as the maria and craters can vary very considerably from eclipse to eclipse, and is not simply linked to the overall darkness of the eclipse.

There seem to be three primary factors which cause these effects: (i) changes in the clarity of the Earth's atmosphere, (ii) fluctuations in the level of solar activity, and (iii) the variation in the luminescence of the lunar surface from place to place.

The most important factor in the Earth's atmosphere seems to be the presence of volcanic dust injected into the higher layers by particularly violent explosive eruptions. This obviously absorbs light directly, and thus leads to a darker eclipse. Similarly, dust from particularly strong meteor showers is thought to have caused dark eclipses, but positive confirmation of this must await a greater number of occurrences.

The luminescence of the lunar surface material will obviously vary with the composition at any given point; however, as luminescence is caused by the flux of energetic particles from the Sun, it is also related to the number of solar-wind particles arriving at any particular area of the Moon at the time of the eclipse. A relationship with solar activity (or more properly, with the latitude of the most active areas on the Sun's surface) was demonstrated by the French astronomer André Danjon, who established that the brightest eclipses (presumably those with the greatest luminescence) occurred just before sunspot minimum – that is, when the active areas were closest to the solar equator. It is now known that the supply of the most energetic solar wind particles is largely controlled by the existence of

Copernicus, seen here in an Apollo photograph, is one of the brightest ray craters on the Moon, and can frequently be seen from Earth during lunar eclipses.

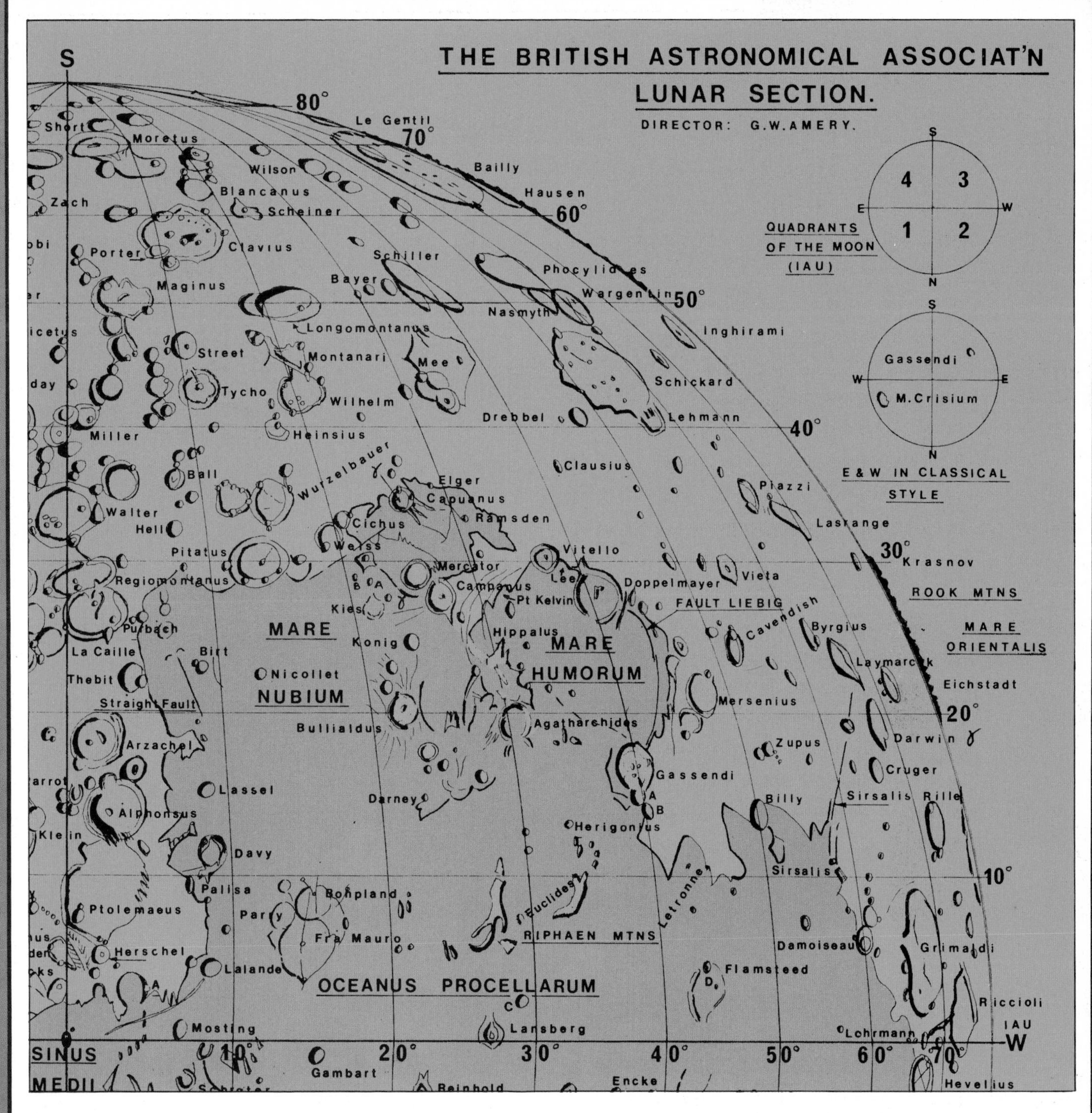

Part of an outline map of the Moon: such a map is very useful for learning the positions of the major features.

coronal holes (pages 78–79), but as yet no more precise correlation has been established between the darkness of lunar eclipses and the conditions in the outer layers of the solar atmosphere.

Observation of lunar eclipses, then, offers some insights into processes involving all three bodies – Sun, Moon and Earth – although not necessarily simultaneously.

One effect which has yet to be mentioned, and which is present to a variable degree at every eclipse, is that the Earth's shadow is larger than would be expected on purely geometrical grounds. The umbral shadow is always enlarged and flattened to a considerably greater extent than can be accounted for by the polar flattening of the Earth. Various reasons for this have been suggested, but none of them are fully accepted, although it does seem that the height and opacity of the atmospheric layers play a considerable part. Further observations which will help to establish the correct mechanism are quite easy to make, as they only involve the accurate timing of events. These may either be the four instants at which the limbs of the Moon enter and leave the umbra, or the times when individual features are crossed by the edge of the umbra. From the known positions of the Moon and of features upon it, the exact extent of the umbral shadow may then be established. Small telescopes and binoculars are quite sufficient for this work, which is easy to undertake.

The overall visibility and colour of the eclipse can be easily estimated and recorded using the naked eye alone, although specialized photometers have been constructed to record the brightness of the whole Moon by comparing it with stars of known magnitude. Such experiments are unusual however, and it is more normal for observers to estimate the visibility of certain specific features by the use of various coloured filters. Some observers use photoelectric photometers for the same purpose,

and are consequently able to construct even more accurate light-curves of the changes which take place in the course of an eclipse.

Photography of lunar eclipses is a fascinating exercise and it is not particularly difficult to obtain a good record of the whole event. Colour photographs will provide at least some indication of the intensity of the colour present in the umbra, but it must be borne in mind that the colour bias of photographic film varies slightly from one brand to another, and in any case depends upon the length of the exposure, so it may not be a truly accurate record. In general, the fairly long exposures required, even with fast films (black and white or colour) during the mid-eclipse phases, do not allow photographs to be used for 'scientific' purposes. It is still worth attempting to take them, especially if some form of driven mount is available.

When the Moon is full the bright ray craters, especially Tycho, Copernicus and Kepler, can be seen to best advantage.

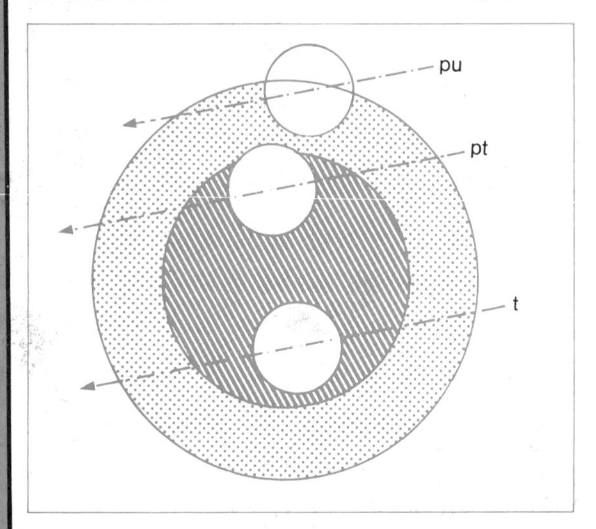

Far left:
Lunar eclipses may be total (t), partial (pt) or penumbral (pu), the latter usually being far too faint to be observed without special equipment.

Observing conditions, whether from Earth or as in this case, from one of the Lunar Orbiters, are usually best when the area is close to the terminator, which divides light and dark portions of the Moon.

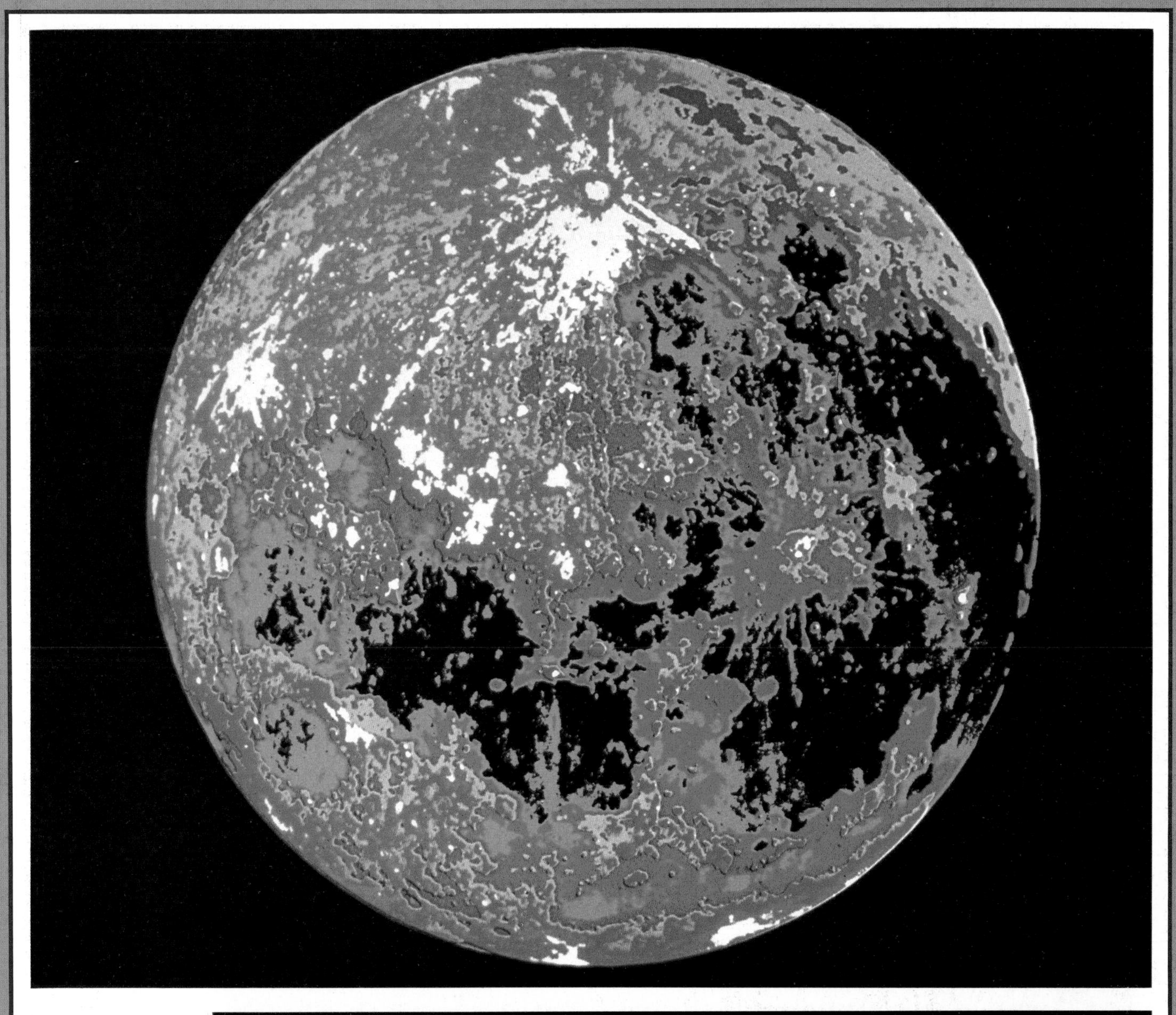

This colour composite picture has been produced to accentuate the range of albedo of the Moon's surface between the darkest maria and the brilliant ray systems.

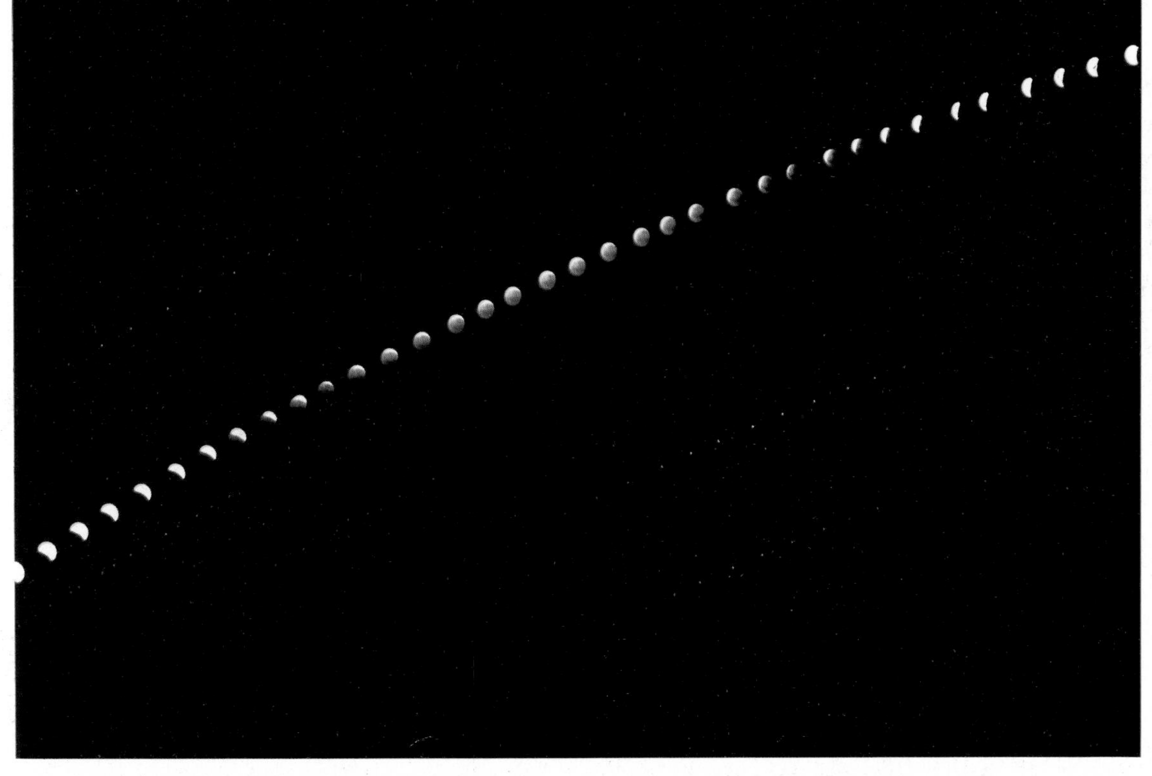

A series of exposures made over the duration of the lunar eclipse of 1982 January 9 by P. Parviainen, Finland.

Lava flows in the Mare Imbrium which are here about 35 m high and 10–25 km wide. The lava source was off the picture to the lower left and some flows are about 1 200 km long.

Facing page, top right: *The Appenines, part of the rim of the Imbrium basin, with a portion of Mare Serenitatis in the distance. Hadley Rille, which is typical of the sinuous type, begins at an elongated fracture, runs approximately parallel to the base of the Appenines and turns at the Apollo 15 landing site through a right angle before gradually fading out into Palus Putredinis.*

The Maria

There are good reasons for believing that there were many large impacts in the early part of the Moon's history and some of these excavated basins 20–25 km deep. At a later time these basins were flooded by vast quantities of lava which more or less completely filled them to form the circular maria. The irregular maria, on the other hand, have been produced merely by flooding of low-lying terrain and the lava infill is much thinner. The flooding has been in the form of a series of thin flows, the latest of which can be easily recognized, and which in some cases are known to extend for as much as 1 200 km, showing that the material was very fluid. Similar, although not so extensive flooding is known on Earth, in the Deccan Traps in India and the Colombia River region in North America in particular. The very fluid lava is of the type known as basalt, and is similar on the Earth and the Moon.

In many cases the lunar basalts have more or less completely submerged old craters, sometimes leaving only the faintest ghost rings to be seen. Later craters which have formed in the mare material itself, provide excellent examples of the typical crater shape and deposits.

In the mare areas there are a few domed areas and low arches which have probably been produced by upwelling lavas, but the numerous mare **wrinkle ridges** have almost certainly been caused by compression of the surface layers when the lava flows cooled.

Generally, but not exclusively, associated with the edges of the maria are valley-like **rilles**. The straight and bow-shaped types are troughs or graben due to faulting, which may cut indiscriminately across mare and craters alike. The bow-shaped rilles are usually concentric with the mare basins, and both types are probably caused by movement on ancient underlying lines of weakness. The sinuous and meandering rilles are very different and at first sight look like river valleys. However, unlike water channels they are deepest where they are widest, and in fact they show points of resemblance to collapsed lava tunnels.

Part of Mare Serenitatis, showing mare wrinkle ridges and many graben which are generally concentric to the mare basin.

Volcanism

Apart from the mare domes which have been mentioned, there are a few other features which are thought to be due to volcanic action. Some domes on the edges of the highlands seem to have been formed before the maria were filled and may be very ancient, while there are a few low cones which resemble cinder cones produced by low-energy eruptions. Moreover, a few dark areas are apparently covered in cinders and ashes from more energetic eruptions. In a number of crater chains the pits greatly resemble the formations known on Earth as volcanic maars, where explosive release of gas has bored a hole in overlying rocks, although there has been no major ejection of volcanic materials. All these features are of minor importance, however, and the few large craters which on the grounds of their positions and associated features may be volcanic, are greatly outnumbered by the impact formations. Indeed, although the Apollo 16 landing site of Descartes was chosen because the surface rocks were possibly volcanic, in the event the astronauts collected large quantities of impact breccias.

The occasional obscurations and glows which are known as **transient lunar phenomena** are most frequently seen around the edges of maria and near to certain relatively fresh craters. These events are more numerous when the Moon is at perigee, suggesting that tidal forces are causing slight movement of the crust which permits gas to escape from the interior. Gas emissions have been detected from Earth, and more reliable observations have been made from orbiting spacecraft, but there are no reasons for supposing that any eruptive volcanic activity is taking place.

Interior

The seismometers installed by the Apollo missions have recorded waves from both moonquakes and from the impact of natural bodies and spent spacecraft stages. More than 3 000 natural moonquakes have been recorded per year. The majority of these occurred at very great depths (about 900 km), with just a few taking place in the topmost crustal layer. There are more moonquakes when the Moon is at perigee, once more suggesting that tidal forces are important. A 206-day period is also detectable and this can be related to orbital perturbations by the Sun.

Table 5·5 **Density of lunar rocks**

		density range (kg per m³)	
crust	highland rocks	2 750–3 000	2 950 mean
	mare basalts	3 300–3 400	
	lower crust	3 000–3 100	
lithosphere, asthenosphere and core (if any)		3 390 (bulk density)	
Moon total		3 340	

The picture of the interior which has been built up from the seismic evidence is shown in Fig. 5·8 and data on the lunar rocks and crust are given in Tables 5·5 and 5·6. In contrast to the Earth, the Moon is rigid down to a depth of about 1 000 km, below which there is a weak, possibly molten, zone. It is difficult to estimate the temperature at the centre, but it is probably in the region of 1 300 K. As the overall density of the Moon is much less than that of the Earth (3 340 kg per m³ against 5 517 kg per m³) it cannot have a very large metallic core of iron and nickel, and the centre is probably composed of a mixture of iron and sulphur, with perhaps a very little nickel. Rock samples indicate that the Moon had a magnetic field a very long time ago, although there is none now, and this could have been produced by such a composition when most of the interior was fluid.

Mascons

Analysis of the motion of spacecraft in orbit around the Moon – largely the Lunar Orbiter series of craft – showed certain positive gravity anomalies in restricted regions. These mass concentrations, or **mascons**, were found to underlie the circular maria in particular. The exact circumstances of their formation are still subject to debate but it would appear that some impact basins became the sites for a concentration of denser material either at the surface or at depth. Normally on a body like the Earth, any concentration of or reduction in the mass of the rocks at any one point is compensated by the rising or sinking of the crust and a displacement of material at great depth. This mechanism, known as **isostasy**, gives rise to an approximate equilibrium everywhere. However, on the Moon, with its thick, rigid crust and lithosphere, this has not occurred in all cases, thus giving rise to the mascons, and the less pronounced negative gravitational anomalies.

Table 5·6 **Structure of lunar crust**

	approximate thickness	
regolith	3–40 m	total crustal thickness 61 km (approx).
mare basalts	<20 km	
brecciated crust	25 km	
lower crust	35 km	

A denser sub-crustal layer may exist between depths of 60–150 km.

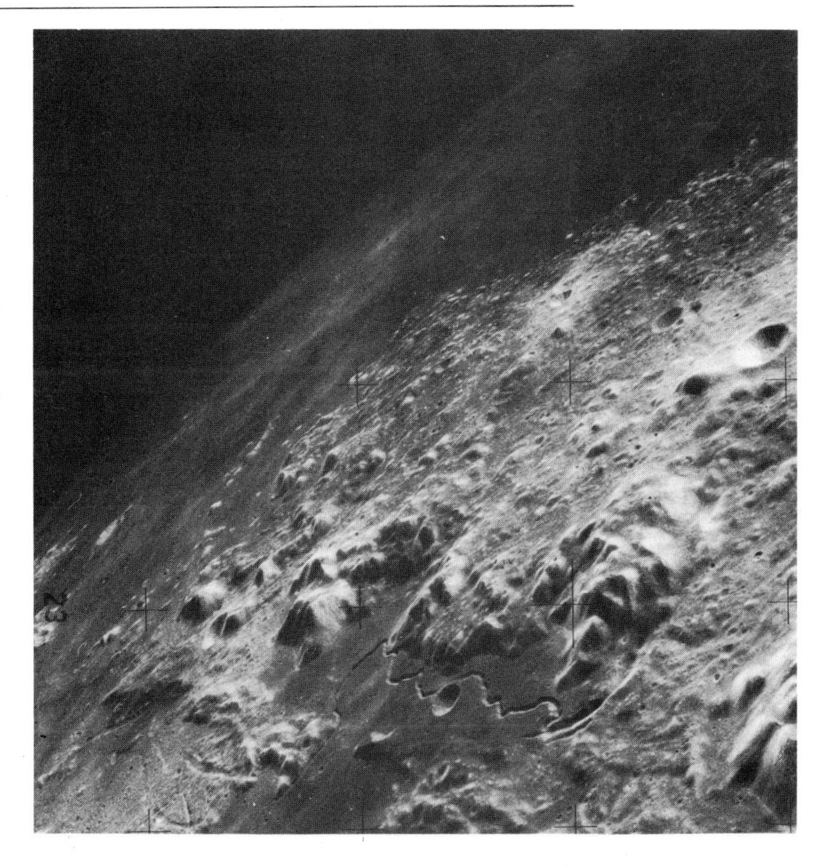

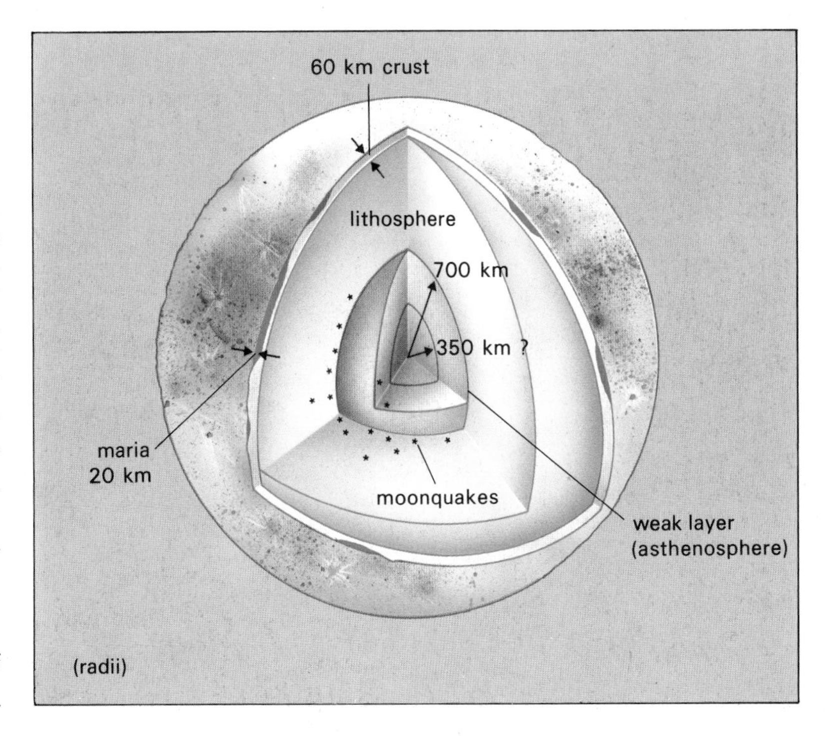

Fig. 5·8 The interior of the Moon, showing the position of the major moonquakes. The thickness and irregularity of the crust are exaggerated, as is the depth of the mare basins. The weak layer may extend to the centre, or there could be a completely solid inner region.

Age and origin of the Moon

There are various dating techniques which may be used on lunar samples and these have shown that the majority of the rocks are very old indeed. The highland materials, for example, commonly have ages of 4·3–4·0 × 10^9 years and the very oldest rock has been dated at 4·6 × 10^9 years, very close to the age estimated for the Solar System itself. The mare basalts on the other hand have ages of 3·8–3·1 × 10^9 years and are thus much younger than the highlands, although still comparable in age to the very oldest rocks known on Earth (3·8 × 10^9 years). No significantly younger rocks are known to exist on the Moon.

By counting the number of craters on identical areas, the relative ages of different surface materials may be determined. As we are now able to date some of these materials exactly from the samples which have been obtained, it is possible to estimate the rate at which craters have been formed. It seems that on the Moon the flux of meteoroids declined rapidly until about 3·0 × 10^9 years ago, since when it has remained at approximately the same level and is now similar to the number and size of meteoroids which are known to enter the Earth's atmosphere.

It is thought that the Moon (like all the planets) was formed in a very short time from smaller bodies a few hundred kilometres in diameter, which are known as planetesimals. In the intense early bombardment, the heat of the impacts caused the whole surface to become molten down to about 200 km. From this molten layer the crustal (highland) rocks formed, and these continued to be cratered after they became rigid. At a later date, the interior heated up due to radioactivity of the materials and the mare lavas escaped to the surface, filling some of the large basins which remained. The interior of the Moon gradually cooled until now only the very centre retains any heat. Being without a molten interior today, the Moon does not possess any significant magnetic field. There are, however, traces of very early magnetism, but this might have originated outside the Moon itself.

Right:
Mariner 10 photograph of part of the surface of Mercury. The prominent lobate scarp which runs diagonally across the picture is actually considerably longer, only about 300 km being shown here.

Mercury

Mercury has never been an easy object to observe from the Earth because of its closeness to the Sun. It never reaches an elongation greater than about 27° 45′, and it is also very small with a diameter of 4 878 km, thus presenting a disc never more than 11 arc sec. across. Only a few vague markings are visible with even the largest telescopes. It was long supposed that its axial rotation was the same as its orbital period of 88 days, so that one hemisphere permanently faced the Sun, leading to very high temperatures on that side while the other was very cold. In 1965, using radar echoes, it was discovered that the rotation period was approximately 59 days, suggesting that tidal interaction with the Sun has caused 2 orbital periods to equal exactly 3 axial rotations. This effect, known as **spin-orbit coupling**, has resulted in a rotation period of 58·65 days. The spacecraft Mariner 10, to which we owe practically all modern knowledge about the planet, was placed into a similar resonant orbit, making its second and third encounters with the planet 2 and 4 Mercurian 'years' (3 and 6 rotations) after its initial approach. Some

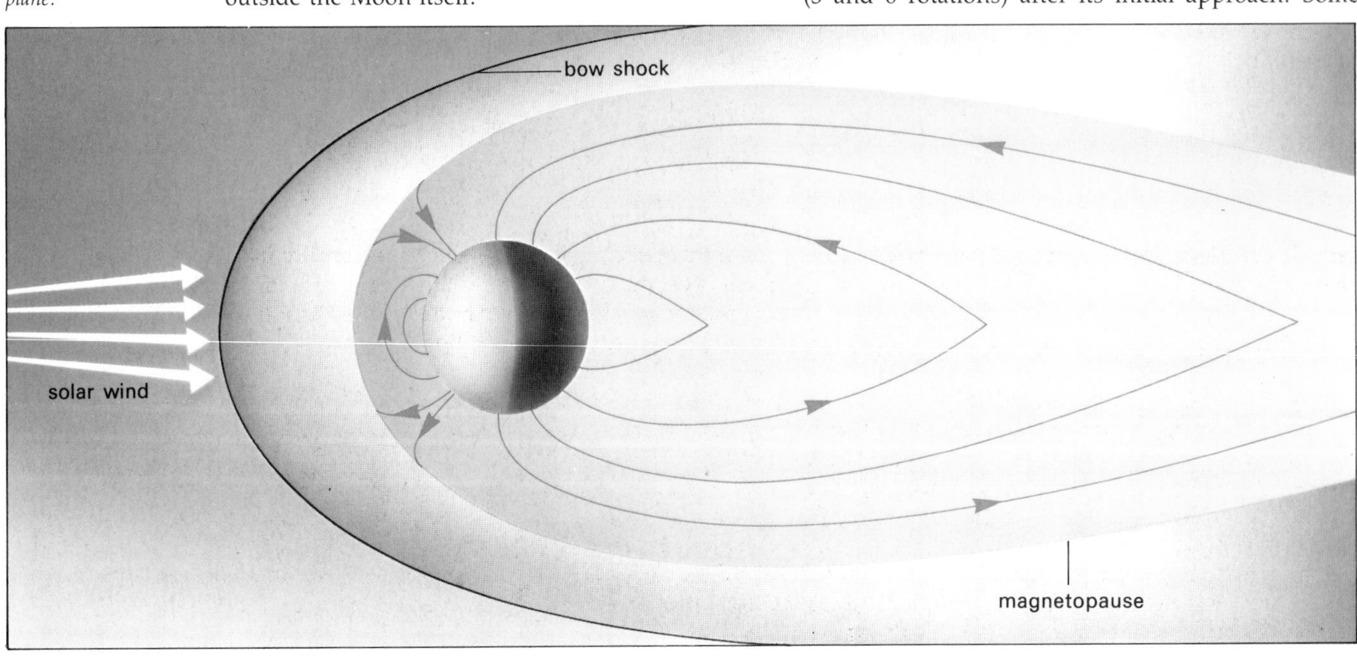

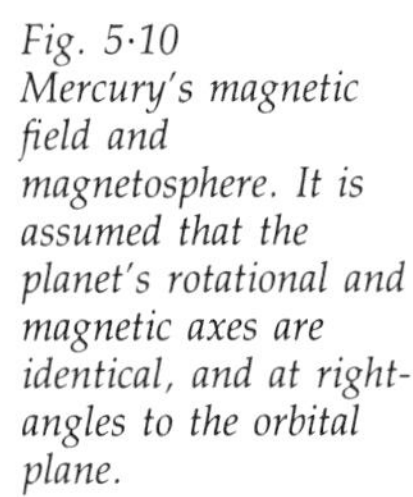

Fig. 5·10
Mercury's magnetic field and magnetosphere. It is assumed that the planet's rotational and magnetic axes are identical, and at right-angles to the orbital plane.

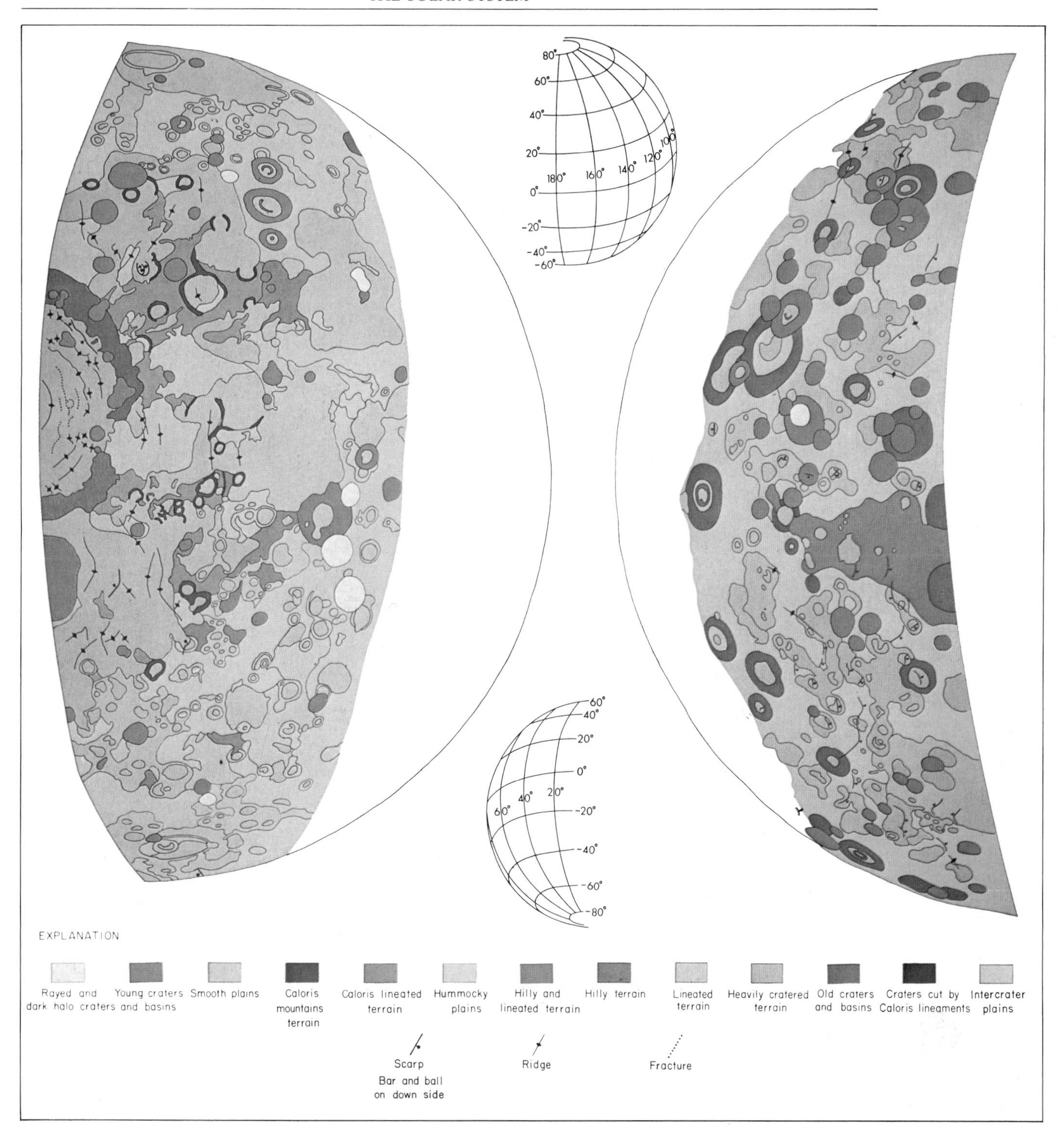

Geological map of part of the surface of Mercury prepared from Mariner 10 photographs. The most striking feature is a portion of the Great Caloris basin (left edge) which is shown in greater detail on page 120.

general data relating to the planet and its orbit are given in Table 5·7.

Interior and magnetism

One of the most remarkable facts about Mercury is its very high density, greater than any of the other planets except the Earth. This is surprising in such a small body (Fig. 5·9, page 120) and suggests the presence of a relatively large metallic iron-nickel (Fe–Ni) core, which contains about 80 per cent of the planet's mass (as compared with the Earth's 32 per cent). This was unexpectedly confirmed by the Mariner 10 observations which showed that Mercury has a magnetic field, a fact which presumably indicates that the planet has a fluid core. The field strength is much weaker than that of the Earth (just over 1 per cent) and the closed field lines are compressed to about 2 000 km from the surface by the effect of the solar wind (Fig. 5·10). The exact position of the magnetic axis is unknown, but it is thought to coincide with the rotational axis, which is probably at right-angles to the orbital plane.

The very high density of the planet remains a mystery, but although some doubts have been expressed about past measurements, the results obtained from the Mariner 10 tracking showed that the density is correct.

Table 5·7 **Mercury-Earth comparative data**

	Mercury	Earth
equatorial diameter (km)	4 878	12 756
sidereal period of axial rotation	58·65d	23_h 56_m 04_s
inclination to orbit	0°?	23° 27′
density (kg per m³)	5 500	5 517
mass (Earth = 1)	0·055	1·0000
surface gravity (Earth = 1)	0·38	1·0000
escape velocity (km per s)	4·3	11·2
albedo	0·06	0·36

mean Sun-Mercury distance 0·3870987 au

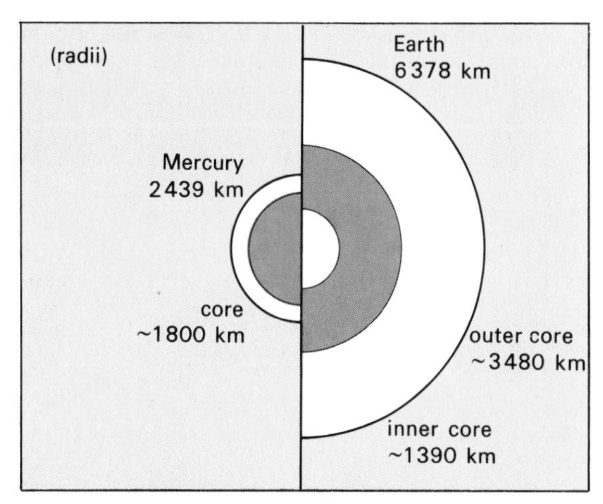

Fig. 5·9 Comparative sizes of Mercury and the Earth. Despite the relatively large core, Mercury's total mass is only about 5·5 per cent of the Earth's.

Photomosaic showing part of the great Caloris basin on Mercury (left centre). *The diameter of the mountain boundary ring is about 1 300 km, with heights up to 2 km. The floor is intensely disrupted by ridges and fractures into a wrinkled appearance. The radial pattern of the surrounding ejecta blanket is evident.*

Atmosphere and surface temperatures

Mercury has been found to have a very tenuous and transient atmosphere with a surface pressure of less than 2×10^{-12} ATMOSPHERES (compared with the Earth's 1 atmosphere). It is composed of helium atoms which the planet captures from the solar wind and retains for about 200 days before they gain sufficient energy to escape again into space.

Although the surface temperatures are not so extreme as had been thought previously, they reach 700 K at the equator of the sunward hemisphere and cool to less than 100 K on the dark side. Due to the orbital coupling either longitude 0° or 180° is towards the Sun when the planet is at its closest, at perihelion, while longitudes 90° or 270° face it at aphelion. At perihelion, the speed of the planet's movement along its orbit exceeds the small rotational velocity so that the Sun makes a small retrograde loop. As a result of the various motions and the orbit's great eccentricity, the 0° and 180° meridians receive about two-and-a-half times as much radiation as those at 90° and 270°.

Surface

Mariner 10 returned a series of pictures which showed that the surface of the planet is covered with a large number of craters. It was only possible to examine a little more than one-third of the surface, but as on the Moon there seems to be a division into highland and lower mare-like areas. The highlands are not as saturated with 20–50 km craters as the Moon and there remain extensive flatter areas which are known as **intercrater plains**. It has been suggested that these regions represent the original Mercurian surface which has undergone a lesser degree of cratering than the Moon, but close examination shows evidence of a large number of highly degraded craters and depressions within the plains. It therefore seems possible that the surface has gone through a process of heating and softening, perhaps somewhat akin to that which formed the Moon's original crust. Since the crust became completely rigid, insufficient impacts have occurred to cover the surface with craters.

The intercrater plains and some craters are broken by the highly distinctive features called **lobate scarps**, which are up to 3 km high and may run for hundreds of kilometres across the surface. These scarps have no counterpart on the Moon, despite resembling mare wrinkle ridges in some respects, but have apparently been caused by major crustal compression. They suggest that the planet's radius has decreased by about 1–2 km, which could have been produced by solidification of as little as 6 per cent of the planet's iron core.

The craters themselves greatly resemble those of the Moon, but secondary craters are closer to the main feature and ray systems are less extensive. This is to be expected in view of the higher surface gravity which reduces the area covered by ejected material to about one-sixth of that on the Moon. Central peaks and peak rings are also present at smaller diameters, in accordance with calculated values, and in this respect Mercury shows a great resemblance to Mars, which has an almost identical surface gravity. Multi-ringed basins are also seen and the great Caloris basin closely resembles the lunar Mare Orientale both in structure and volume of material ejected.

Crater counts on the ejecta from this basin and on the lava flooding of the floor, as well as the other extensive areas on this hemisphere, imply that the materials of the mare-like smooth plains were erupted very shortly after its formation, which again suggests that the crustal heat persisted for rather longer than on the Moon. On the opposite side of the planet to the Caloris Basin is an area of very peculiar hilly and lined terrain the formation of which is difficult to explain. It could have been produced by seismic energy focused from the Caloris impact on the other side of the planet.

In summary, it may be said that the history of Mercury seems to have been very similar to that of the Moon, with crustal heating and one or more major episodes of impact cratering. Crater density counts indicate that Mercury, the Moon and Mars have all been affected by a similar meteoroid flux in the recent past, and as this is in accordance with the current terrestrial rate, we may assume that it has also applied to Venus.

Venus

Despite the fact that Venus approaches closer to the Earth than any other major planet and that it is very similar in size and total mass to the Earth (Table 5·8), and the several Mariner, Venus Pioneer and Venera spacecraft missions, we know less about its surface features than those of any other body of the inner Solar System.

Venus has a very extensive atmosphere with the high albedo of 76 per cent, and this completely hides the surface. Even the rotation period could not be established with any confidence until 1962, when radar methods indicated a retrograde period of 243 days. This may be a result of tidal resonance with the Earth, as an axial rotation period of 243·16 days would result in the same side of the planet facing the Earth at each inferior conjunction. It is, however, difficult to establish how this can have occurred, unless Venus is asymmetrical like the Moon. The indistinct markings, sometimes visible from Earth and on the Mariner photographs, show an apparent 4-day rotation period for the upper atmosphere and this will be discussed later.

Interior and magnetism

The planet's overall density is fairly close to that of the Earth and it would be reasonable to assume that both planets had a similar composition when they were formed. The implication of this is that the core of Venus has a radius of about 3 100 km (Fig. 5·11) with a considerable (but unknown) proportion being

Table 5·8 **Venus-Earth comparative data**

	Venus	**Earth**
equatorial diameter (km)	12 104	12 756
sidereal period of axial rotation	243·16d	$23_h\ 56_m\ 04_s$
inclination to orbit	178°	23° 27′
density (kg per m^3)	5 250	5 517
mass (Earth = 1)	0·815	1·0000
surface gravity (Earth = 1)	0·903	1·0000
escape velocity (km per s)	10·36	11·2
albedo	0·76	0·36

mean Sun-Venus distance 0·7233322 au

fluid. There is expected to be a mantle and a crust which we may reasonably assume will fairly closely resemble those of the Earth. Despite the presumed fluid core, the planet has no detectable magnetic field, the axial rotation being apparently too slow to produce one.

Distinct problems have arisen since the discovery by spacecraft experiments that there are considerable anomalies in the abundances of certain gases, and of argon isotopes in particular. The amount of argon 40 (the product of radioactive decay of potassium) is much lower than on the Earth, being only about one third. Yet argon 36 and argon 38, the isotopes most likely to be present in any primordial nebula, are together 75 times as abundant as on Earth. In a similar manner the neon abundance is about 45 times as great. Other gases do not show this discrepancy;

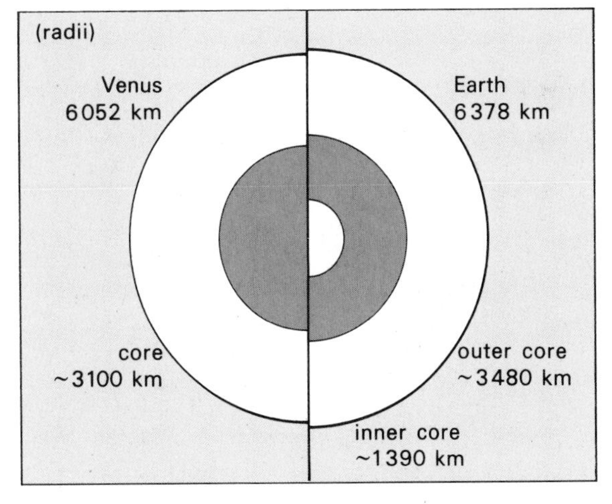

Fig. 5·11 Comparative sizes of Venus and the Earth. Venus may have a solid inner core like that of the Earth.

carbon dioxide, nitrogen and krypton are present at only about twice or three times their terrestrial levels. There is as yet no generally accepted theory to account for these differences, although various suggestions have been made. However, they do provide yet more information which can be used to obtain some idea of how the Solar System itself was

Surface features

The expected similarity of the interiors of the two planets would suggest that Venus should exhibit considerable tectonic movements and volcanism. However, for reasons to be discussed later, there is no free water on the surface or in the atmosphere,

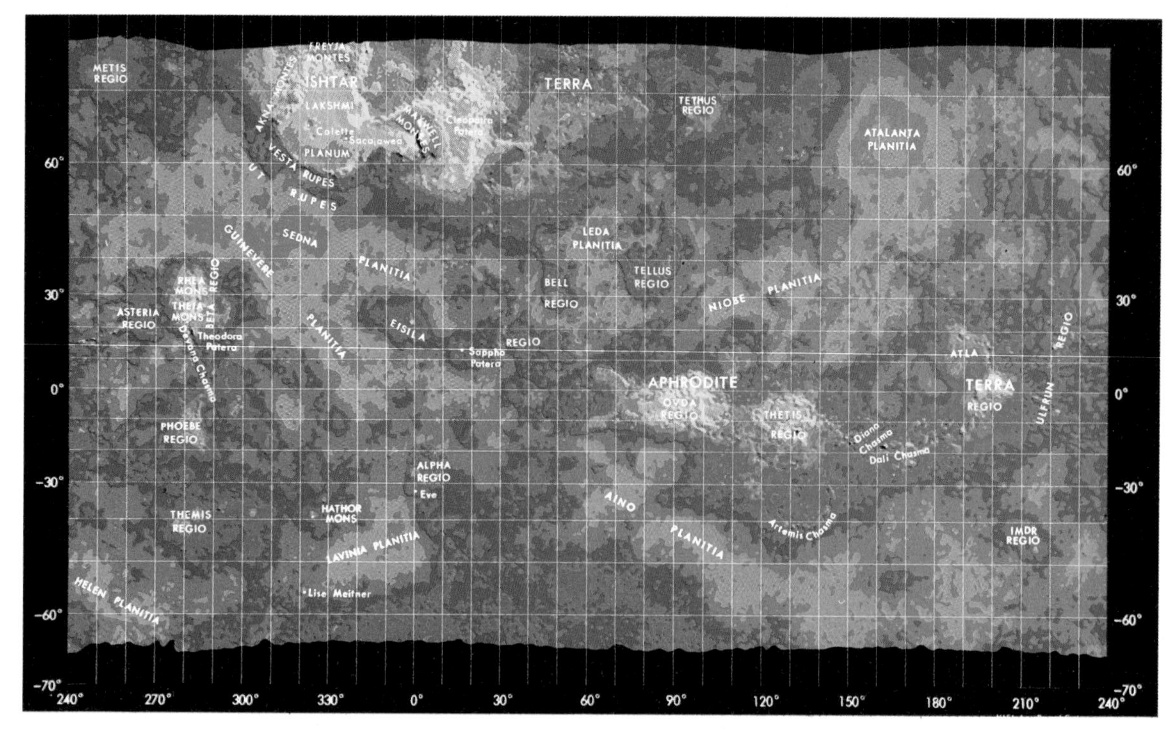

A radar map of the surface of Venus, made by the Orbiter section of the Pioneer Venus spacecraft.

so that any sedimentary rocks would have to be formed from wind-borne deposits. Radar techniques are able to penetrate the dense atmosphere, and studies from Earth (using the large Arecibo telescope) and from the Orbiter section of the Venus Pioneer mission have revealed considerable surface detail. (The Orbiter had low resolution of about 100 km, whereas Earth-based radar can approach 6 km in certain regions.) The main portion of the planet is surprisingly flat, and this has caused speculation that there is little current tectonic activity on Venus relative to that on Earth, and that it is more like Mars in this respect. The lack of water (thought to help to 'lubricate' Earth's plate tectonic movements) and the high temperature which would possibly maintain rocks in a somewhat plastic state (and thus allow them to deform, and any elevations to subside) may be contributory factors. But there are high regions, such as Aphrodite Terra, Beta Regio (Thea Mons and Rhea Mons) and especially Ishtar Terra, where Maxwell Montes rise to 12 km above the mean planetary radius of 6 051·2 km. These must be sustained somehow, and although suggestions have been made that they are relics of earlier tectonic activity and represent the planet's continental regions, it is distinctly possible that they are situated over 'hot spots' in the mantle and that volcanic activity is still occurring. Many of the landforms are similar to terrestrial shield volcanoes, and high-resolution images of Maxwell Montes appear to show a 100 km summit crater and other volcanic features. The other two regions, and especially Beta Regio, show a strong concentration of lightning-like activity. As the atmospheric processes do not appear to be able to generate lightning in a manner similar to that occurring on Earth, this is further evidence of volcanism and the accompanying electrical discharges within the cloud of ejecta.

There are indications of other considerable features, including a couple of graben-like structures, large basins and a considerable number of craters, ranging in size from about 160 km to 35 km in diameter. There are also indications of larger impact structures, including one 1 800 km in diameter, but there appears to be a deficiency of small craters. This is probably due to the presence of a very massive atmosphere which prevents small meteoroids from reaching the surface, as well as perhaps to the thermal effects mentioned earlier. Surface winds are now known to be very light, so that any erosion of

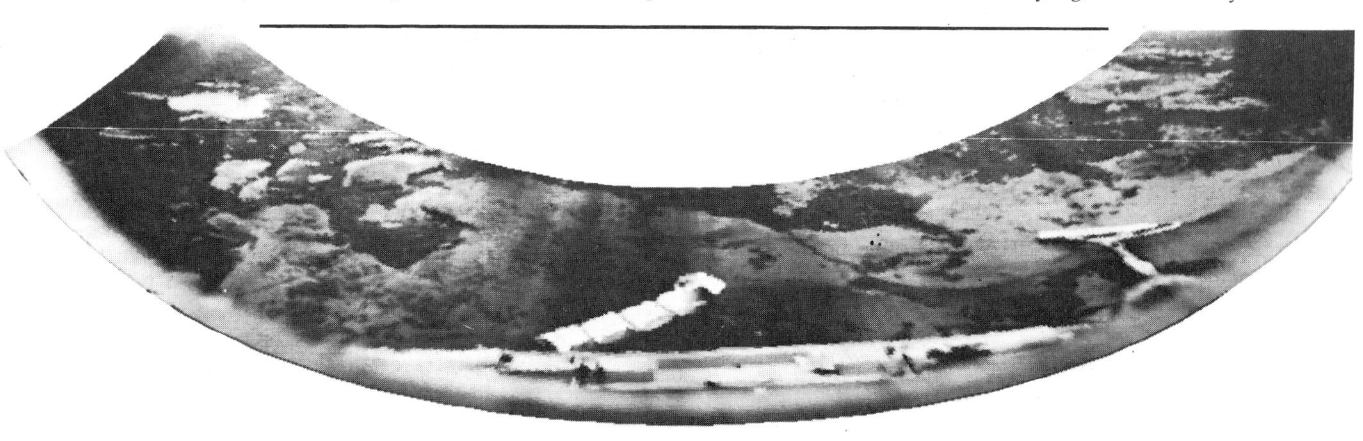

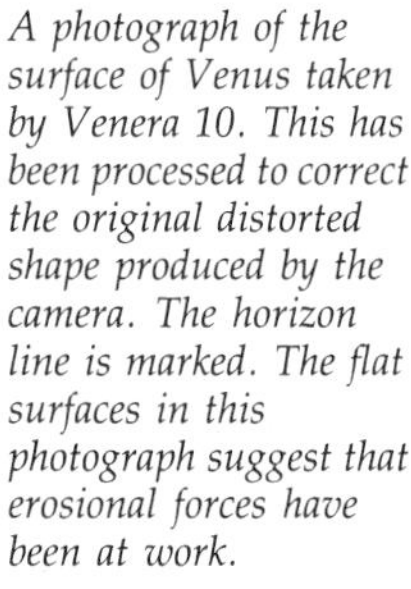

A photograph of the surface of Venus taken by Venera 10. This has been processed to correct the original distorted shape produced by the camera. The horizon line is marked. The flat surfaces in this photograph suggest that erosional forces have been at work.

surface rocks is likely to be by chemical and thermal processes rather than by wind-blown particles.

Our knowledge of the surface composition is based partly upon measurements of radioactivity made by Veneras 8, 9 and 10. The first of these indicated a similarity to terrestrial granites, and the others a close resemblance to both terrestrial and lunar basalts. The more sophisticated techniques used on Veneras 13 and 14 (using drilling and X-ray fluorescence) found an unusual high-potassium basalt, with some similarities to lunar highlands and terrestrial continental rocks, and a type similar to that of the Earth's ocean floors. Veneras 9, 10, 13 and 14 returned pictures of the areas surrounding their landing places (the last two in colour). The first, in particular, proved to be surprisingly rough; the others were generally smoother, suggesting that erosional processes are at work. Veneras 10 and 13 showed flat rock outcrops with intervening patches of darker 'soil' fragments. The view from Venera 14 was somewhat different, with just a flat, apparently layered rock expanse, but the layers in themselves are rather difficult to understand and have been proposed as either cemented fine particles or thin layers of the original lava.

Atmosphere

The dominant constituent of the atmosphere is carbon dioxide (CO_2) and this amounts to 97 per cent of the total mass. The overall quantity of carbon dioxide is approximately the same as the total held by the combined oceanic, rock and atmospheric reservoirs on Earth, implying that both planets were formed with, or have accreted, similar quantities of the gas. A considerable number of other components of the atmosphere have now been identified. Particularly significant is the very small amount of water (H_2O) and this will be discussed below.

Because of the vast amount of carbon dioxide, the atmospheric mass and surface pressure are both about ninety times that of the Earth (Fig 5·12). As a result of the carbon dioxide's greenhouse effect (page 95), the surface temperature has been raised to the very high level of 760 K, comparable to, if not hotter than, that at the surface of Mercury. However, there is now considerable evidence that these high temperatures have not prevailed throughout the planet's history.

The extreme surface temperature is, of course, far above the boiling point of water, and it was a matter of some surprise when early radio studies showed that the atmosphere contains very little water vapour. Subsequent direct measurements agree with this, and although there are some discrepancies, lie in the range of 0·1 to 0·01 per cent. This scarcity is very difficult to understand unless Venus was formed from very different material than the Earth, or accreted its atmosphere in a different way. But there is direct evidence that Venus once had a very considerable quantity of water. This information comes from a determination of the ratio of hydrogen to deuterium (heavy hydrogen). There is far more deuterium on Venus than on Earth, and this shows that a large amount of water was once present, indeed that oceans probably existed. The greenhouse effect would cause this liquid water to be evaporated into the atmosphere (incidentally contributing to a still stronger trapping of infrared radiation), and the water molecules would be dissociated by the Sun's ultraviolet radiation at the top of the atmosphere. The ordinary light hydrogen atoms would be most easily lost to space, but the heavier deuterium isotope would be retained, leading to its present high concentration. The free oxygen is likely to have been incorporated into surface rocks, and it has been suggested that the oxidation process would further increase the heating of the crustal layers. However, it does seem likely that Venus had large bodies of water at an early period in its history, and it is by no means impossible that some form of life did originate upon the planet before the runaway greenhouse effect made it too hot for any life forms.

The structure and composition of the clouds long remained a mystery, but has now been reasonably well established. The yellowish tinge is probably due to the presence of sulphur (S), while the dark markings, which show up strikingly only in ultraviolet wavelengths, are most probably caused by

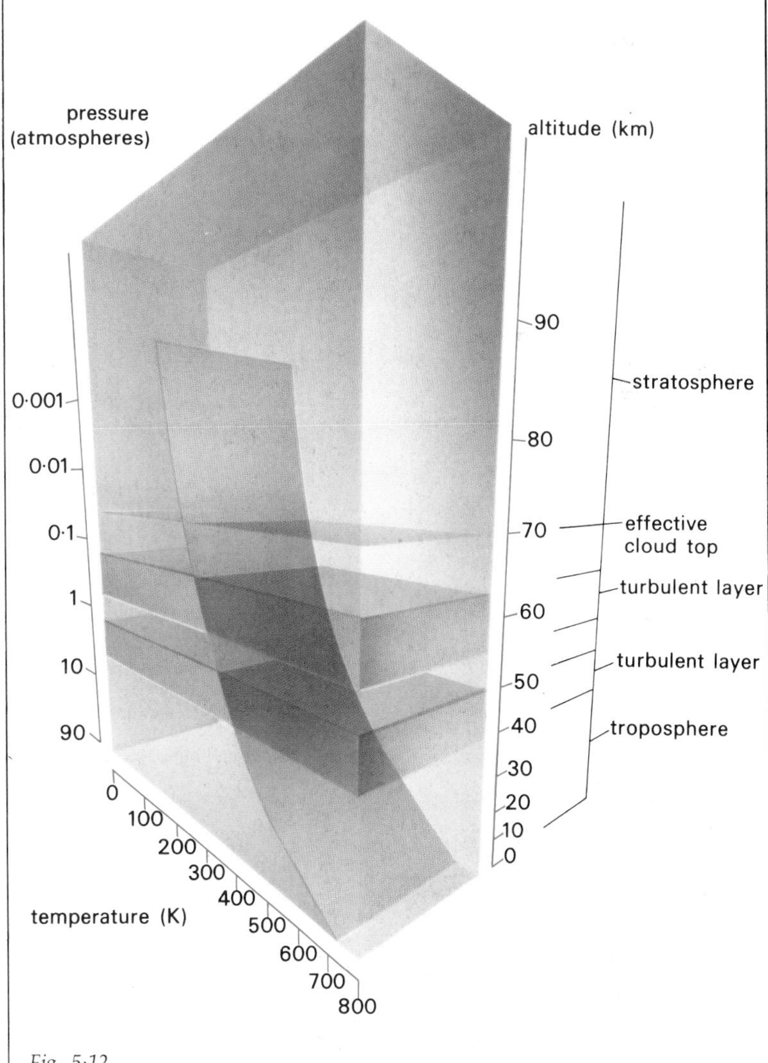

Fig. 5·12
The structure of the atmosphere of Venus. Atmospheric particles (cloud droplets) reach a peak between 40–50 km and are largely absent between 10–30 km.

absorption by sulphur dioxide. The clouds exhibit a strongly layered structure, the densest layers occurring between about 70 and 47·5 km, with hazes both above this (extending to about 90 km) and below (down to about 31 km). The lower atmosphere is essentially clear, which is one of the reasons why conventional lightning processes are unlikely to occur.

There are various sizes of cloud particle present, but the majority appear to be droplets of sulphuric acid (H_2SO_4), some tiny and some larger, the latter probably accumulating on nuclei of small particles. The very largest cloud particles are quite possibly fairly large crystals, most likely to be some form of chloride, although this is still uncertain. A number of other components must be present to account for the optical properties of the clouds.

The atmospheric circulation of the planet is unusual. There is a very high-speed circulation of the upper and middle layers around the planet, in the opposite direction to the axial rotation. The velocities are very high at about 100 m per second, giving rise to an apparent rotation period of about 4 days for the cloud tops.

At high levels the night-side temperature is very low (unlike on Earth, where both day and night sides have high temperatures). The steep temperature gradient and consequent pressure difference is the cause of the high-speed winds.

Superimposed upon this rotation of the cloud around the planet is slower motion between the equator and poles. Most of the absorption of solar heat occurs at fairly high levels in the cloud layer, and this drives a fairly slow circulation in a layer of restricted depth. By friction this layer drives a circulation in the stratosphere above it, and one in the lower atmosphere (down to about 40 km from the surface), so that the circulation is rather like three counter-rotating conveyor belts running between equator and poles. Yet another circulation cell appears to exist at the surface, driven by the small amount of heat which penetrates to such a low level, and there may also be smaller intermediate eddies contributing to the overall stack of counter-rotating layers.

At the poles the general circulation of the layers appears to form a pair of very large counter-rotating eddies which allow the lower, hotter layers of the atmosphere to be seen through holes in the cloud layer. The general atmospheric circulation is quite unlike that of the Earth, which essentially consists of three circulation cells between equator and poles, driven by heat absorbed at the surface.

The Venera landers and the four Venus Pioneer probes (one of which, although not designed to survive impact with the surface, did return 67 minutes of data) show that information can be gained slowly about the nature of the surface, as well as about the atmosphere through which they descend. Further advances in understanding should come with the proposed orbiting vehicle carrying a radar experiment, as well as the balloon probes now under development, which would remain in the atmospheric layers for some considerable time.

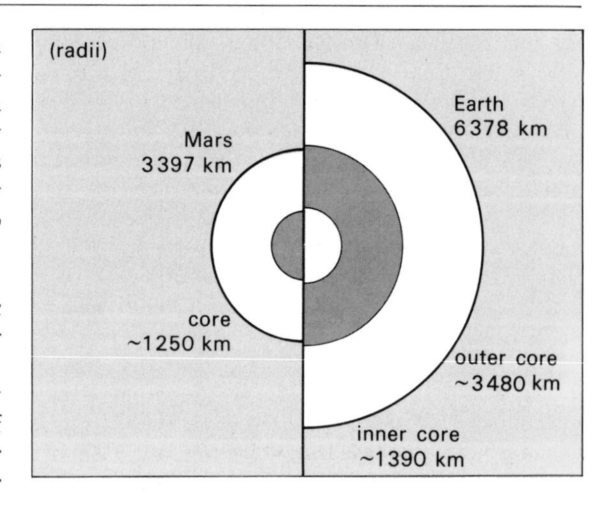

Fig. 5·13 far right: Comparative sizes of Mars and the Earth. The planet's total mass is about 11 per cent of the Earth's (about twice that of Mercury).

Mars

Mars, unlike Venus, has only a thin atmosphere and the planet has long been extensively studied from Earth. Detailed maps have been drawn of its surface markings which were thought probably to be related to high and low areas, perhaps somewhat similar to the division on the Moon. Apart from these features it shows clouds, brilliant polar caps which alter in size with the seasons, similar seasonal changes in some of the surface markings and occasional vast dust storms which may obscure the whole planet. Rather ironically, the markings which were thought to indicate the nature of the surface have now been shown by spacecraft pictures to bear little relation to the actual features, whereas the other characteristics have been fully confirmed.

Mars has an equatorial diameter of 6 794 km, an axial period of $24_h37_m23_s$ and is inclined to its orbit by 24° 46′, the last two factors being very similar to those of the Earth (Table 5·9). It has a lower density than any of the other inner planets (although higher than the Moon) and has a surface gravity almost exactly equal to that of Mercury, despite being more than 3 800 km greater in diameter. The crustal and mantle materials of the inner planets appear to be very similar and this, together with spacecraft tracking results, suggests that the core of Mars is smaller (Fig. 5·13) and less dense than those of the other planets, with the most probable material being iron sulphide (FeS). The nature of the core may be clarified by future information about the planet's magnetic field. Results from the Soviet Mars probes and Viking spacecraft appear to indicate the presence of a weak magnetic field, although nothing had been detected by the three Mariner missions.

Craters

Both the Mariner 4 and 7 spacecraft showed that the surface of Mars was cratered, but it was Mariner 9 which revealed the true distribution of the craters and discovered numerous other interesting features. The Soviet Mars 5 probe also returned high resolution pictures, but the highest quality images have naturally been obtained by the later Viking Orbiters. The most striking fact which has been revealed is that the surface is divided into two approximate hemispheres, one of which, the southern, is generally high, heavily cratered and ancient, while the northern is formed of low-lying, relatively featureless plains which have far less, and generally fresher craters. These plains appear to be predominantly formed of volcanic materials and this general impression was confirmed by Viking Lander 2. The boundary between the two regions is marked by boundary scarps which are remarkably uniform in height at 1–2 km, and areas of the so-called **fretted terrain** where the highlands are being eroded.

Craters in the northern hemisphere only rarely exceed 10 km in diameter, but in the southern they include several multi-ringed structures which range in size up to that of Hellas, which is about 2 000 km across and some 4 km deep. The number of central peaks and peak rings is similar to that of Mercury, as expected from the equal surface gravities. The effect of gravity can also be seen in the fact that Martian craters are shallower than those on the Moon. Distinct blankets of ejecta can be seen around some of the younger impact craters.

Table 5·9 **Mars-Earth comparative data**

	Mars	**Earth**
equatorial diameter (km)	6 794	12 756
sidereal period of axial rotation	$24_h37_m23_s$	$23_h56_m04_s$
inclination to orbit	24^0 46′	23^0 27′
density (kg per m^3)	3 933	5 517
mass (Earth=1)	0.107	1.0000
surface gravity (Earth=1)	0.38	1.0000
esape velocity (km per s)	5.03	11.2
albedo	0.16	0.36
mean Sun-Mars distance 1.5236915 au		

Volcanism

Although there are no signs of curved mountain chains, a prominent part of the surface is that in the three volcanic regions in the Tharsis, Elysium and Hellas areas. One of these, the Tharsis volcanic province, is especially notable as it contains extensive volcanic plains, four exceptionally large volcanoes and numerous other features. The largest volcano, Olympus Mons, has a diameter of 600 km and a height of 26 km, making it twice as high as the large feature on Venus and much bigger than Hawaii, which forms the largest such volcano on Earth (Fig. 5·14). It has a complex central **caldera** or crater-like depression, gives evidence of extensive lava flows, and a boundary cliff which is up to 4 km high in places.

The other three volcanoes in the same area are very large with diameters of about 400 km and heights of 19 km, but there are also smaller domes which were probably formed by more viscous lava. To the north of the region there is a considerably older, degraded structure, Alba Patera, which seems to have had a diameter of about 1 600 km, making it the largest volcanic feature on the planet.

The Elysium area contains volcanoes and volcanic plains, but the structures in the Hellas region are much less distinct and certainly very much older,

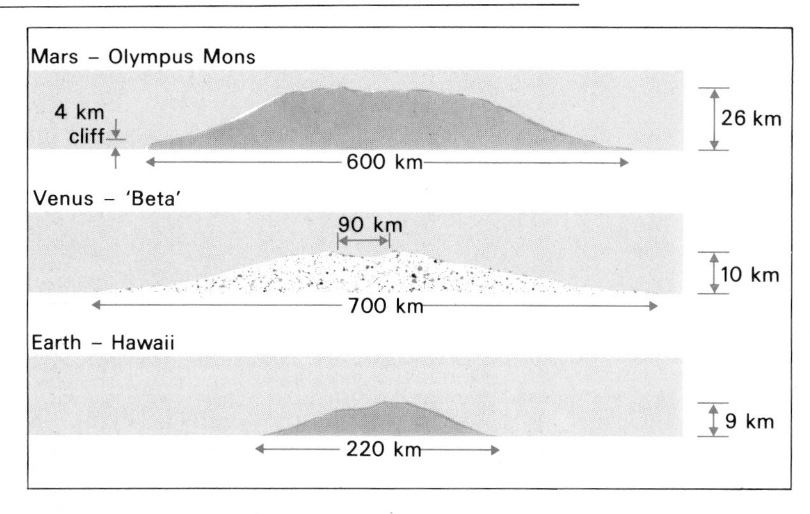

Fig. 5·14 above: A comparison of the largest recent volcanic structures on Earth, Venus and Mars, all of which have central calderas. The highest point on the island of Hawaii is Mauna Kea, but only Mauna Loa is now active. The details of the object on Venus are naturally only tentative.

being highly degraded. Crater counting and other techniques have enabled the features to be dated and this shows that Mars has been volcanically active practically since its formation until the present. The Hellas features are the oldest, with ages of $3{\cdot}5$–$4{\cdot}0 \times 10^9$ years, and the Tharsis area the youngest, at perhaps $2{\cdot}0 \times 10^8$ years. It is quite possible that a small amount of activity could take place at the present time, although there is no actual evidence for this. The exceptionally long period of activity in just a few regions argues against major horizontal plate tectonic movement having occurred as on Earth, and it may be shown that, as on the Moon, the molten zone has been slowly migrating towards the centre of the planet, with the magma for Olympus Mons now originating at a depth of about 200 km, some four times as deep as on Earth.

Rifts, troughs and valleys

The Tharsis volcanoes lie on part of a very large uplifted area, which is about 5 000 km across and 7 km high. The exact causes of this bulge are obscure,

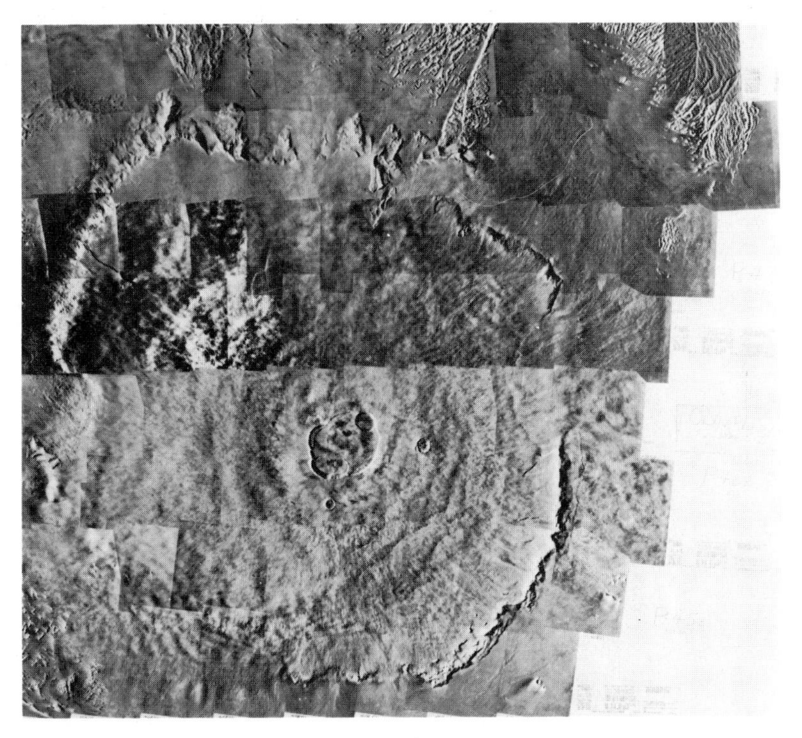

Photomosaic of Olympus Mons, which has a diameter of about 600 km and height of 27 km. The complex central caldera has a diameter of about 70 km.

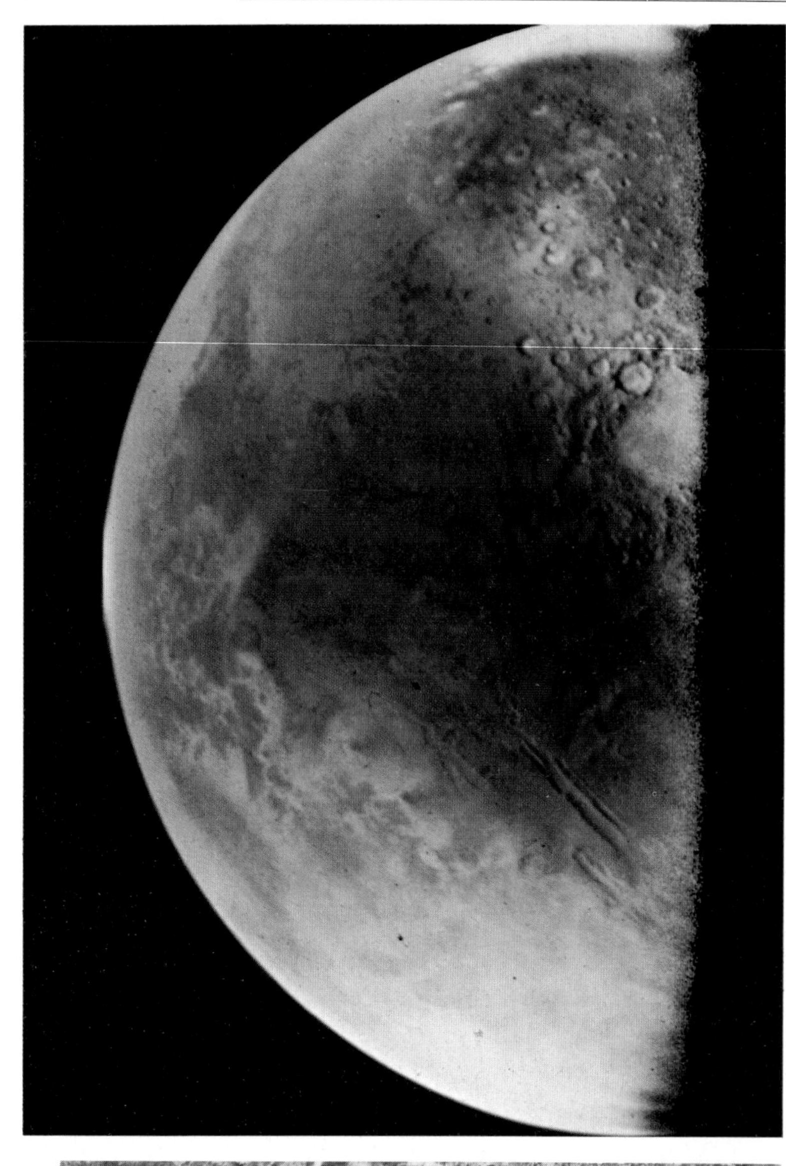

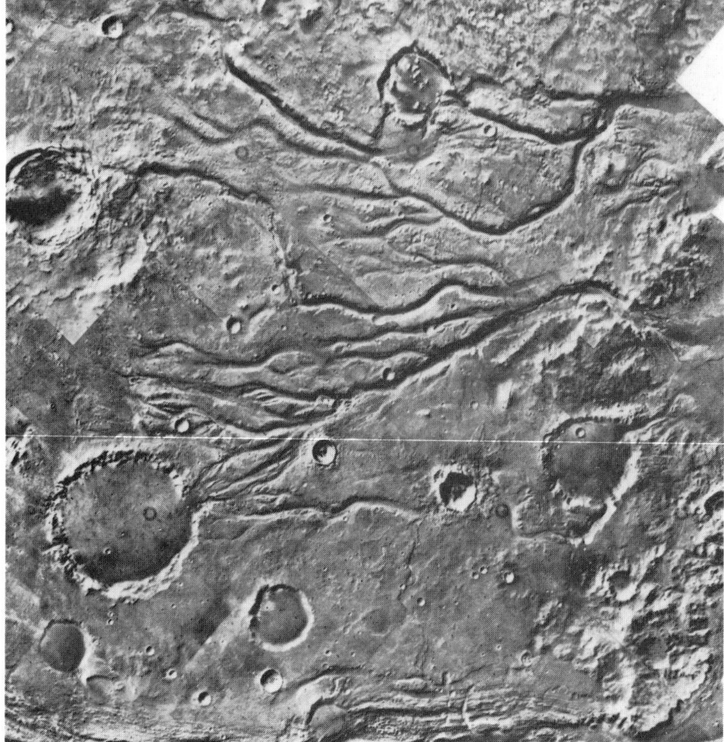

but it has been suggested that due to the probable chemical composition of the mantle, Mars may have expanded slightly as the interior cooled. The top of this uplifted region is marked by numerous intersecting graben-like troughs (in the Labyrinthus Noctis area) and is the centre of a vast system of radial fractures, some of which extend for thousands of kilometres. These are particularly important on the eastern side, where they have apparently led to the formation of the immense Vallis Marineris canyon system, which is several kilometres deep and extends for about 4 000 km across the planet, roughly parallel to the equator. At its eastern end it joins the Chryse Trough which runs from south to north across the cratered upland terrain.

The Chryse Trough contains the major part of the so-called **chaotic terrain** where irregular depressions are filled with an apparently haphazard arrangement of blocks ranging in size from tens of kilometres to a few hundreds of metres. In this area there is no distinct drainage, but further downslope, sinuous river-like channels run down on to the Chryse Plain (the landing site of Viking 1) and thus reach the northern plains (*below left*). These channels greatly resemble those formed by intense flooding of arid areas on Earth and are found in other parts of the cratered terrain, close to the boundary with the northern plains.

Water and wind action

Present conditions on Mars do not permit rainfall or areas of liquid water to occur and it is a matter of debate whether they could ever have been present. It is possible, however, that the early, fairly dense atmosphere contained considerable quantities of water vapour and that as the planet cooled this was taken up by the surface materials to form kilometre-thick layers of ground ice. Subsequently, tectonic movements and fracturing, impact cratering or volcanic heat could either expose the ice and cause it to sublime (turn directly to water vapour) or, in some cases, produce the short-lived but vigorous flooding which led to the formation of the channels. Slumping of surface materials following removal of ground ice would lead to the typical chaotic terrain, and study of the Vallis Marineris area shows that fracturing initiates the formation of chains of small pits. These gradually grow and eventually form the vast canyons that we see.

Destruction of ground ice would lead to a smaller volume of solid debris than might otherwise be expected, but the almost complete lack of sedimentary deposits associated with the various depressions almost certainly indicates removal of the small particles by wind action. There is ample evidence for such transport of fine materials in the numerous light and dark streaks which can be seen, as well as in the changes of some of the surface markings. Moreover, the recurrent alterations in the dark features appear to be due to seasonal variation in the pattern of surface winds.

Areas of dunes are visible and similar small features are shown by Viking Lander 1 pictures (pages 130–131).

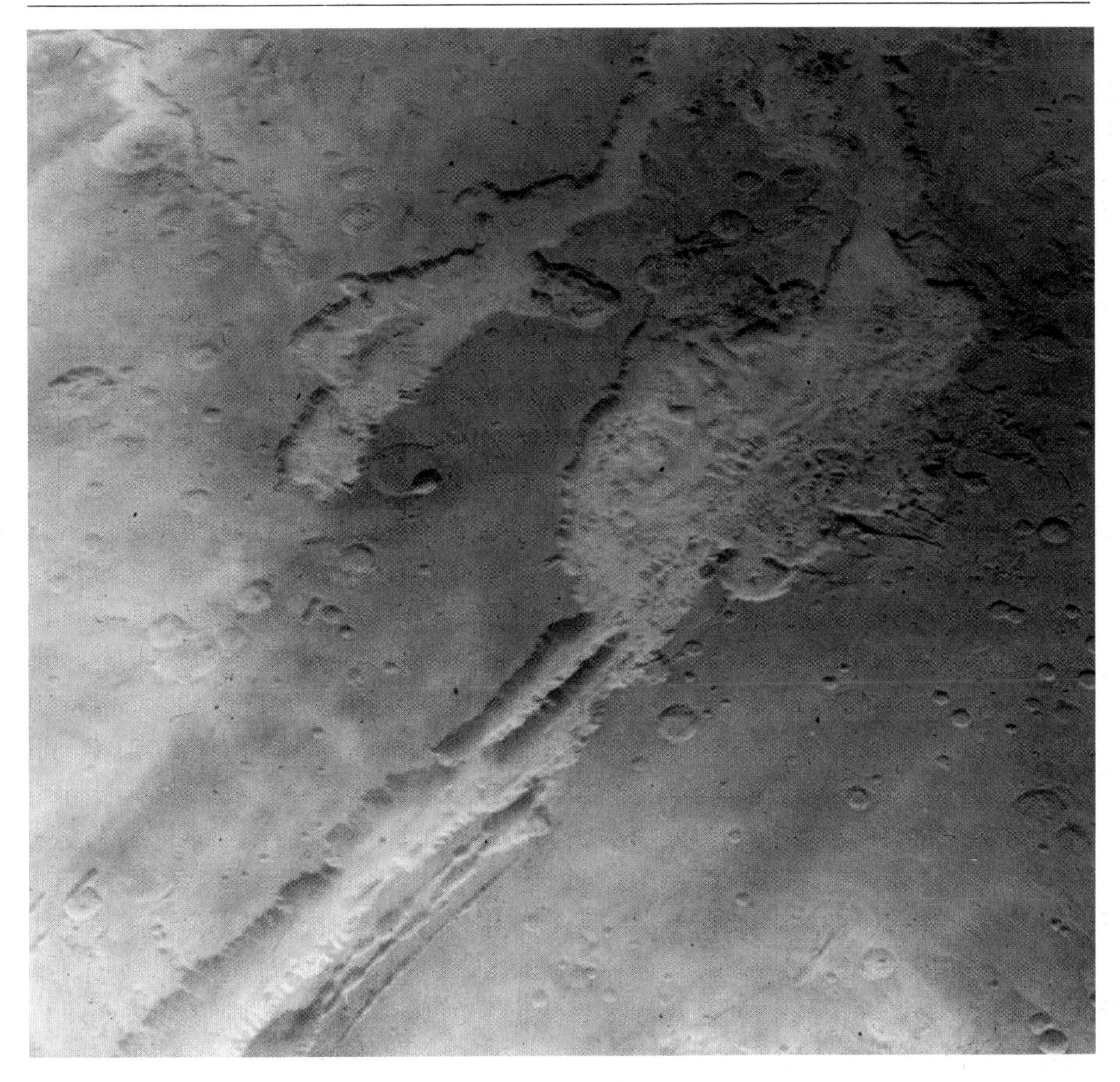

Opposite page, top: *A composite of three single-colour photographs of Mars made by Viking Orbiter 1, with south at top to assist recognition of surface relief. Part of the south polar cap can be seen, while frost covers the area between it and Argyre basin, which is itself beneath frost or haze. Vallis Marineris is recognizable towards the bottom, but water ice clouds cover the Tharsis volcanic region.*

Left: *Viking Orbiter 1 photomosaic of Vallis Marineris, the area covered being about 1 800 km by 2 000 km. The canyon system is 4 000 km long and up to 120 km wide and 6 km deep. Note how the depressions are extending along lines of structural weakness (bottom centre). General levels trend downwards towards the Chryse Trough which is out of the picture to the upper right.*

Due to the lesser density of the atmosphere, higher wind speeds are needed on Mars to transport particles (30–60 m per s) than on Earth (less than 10 m per s). The Viking Landers actually recorded speeds of only 10–20 m per s on the surface, but found much higher velocities higher in the atmosphere (100 m per s at about 10 km).

The major planet-wide dust storms all seem to originate in the southern hemisphere, but there is a strong tendency for the fine particles to be removed from the high cratered terrain and to be deposited on the lowlands, particularly on the northern plains. The erosive action of wind-borne particles can be seen in large areas of grooved and fluted surfaces all over the planet.

Winds have transported material towards the poles and formed two major sedimentary deposits, probably composed of mixed ice and dust, which cover older pitted terrain. One deposit is smooth and unlayered, but the other, on which the polar ice caps themselves are resting, is composed of numerous layers about 30 m thick, giving it a highly distinctive appearance (page 131, top left), and is known as **laminated terrain**.

Both types of deposit are apparently being eroded at the present time.

Atmosphere

The Viking Landers have shown that the average surface pressure is $7{\cdot}4 \times 10^{-3}$ atmospheres and that the mean temperature is 230 K, with a summer maximum of about 265 K and a winter minimum close to 150 K.

Details of the atmospheric composition are given in Table 5·10, where it will be seen that the main constituent is carbon dioxide amounting to about 96 per cent. It is difficult to establish the amount of carbon dioxide and water which have been produced by loss of trapped gas on Mars, but the quantities seem to have been many, possibly hundreds of times less than on Earth.

Water ice forms the permanent polar ice caps and the majority of the clouds and fogs which are observed.

There has long been a debate concerning whether temperatures are low enough for carbon dioxide to freeze, but spacecraft measurements indicate temperatures as low as 125 K, so that the seasonal caps are probably formed of both water and carbon dioxide ices. Carbon dioxide ice is also present in some clouds at high altitudes and in the winter polar regions.

Opposite page, bottom: *Photomosaic of the channelled terrain west of the Viking Lander 1 site in Chryse Planitia. The slope is from west to east (left to right). Note that the channels cut some craters, but also have other, later craters superimposed upon them.*

Watching the Planets

Observation of the planets has long been a field of special interest to amateur observers, and although professional patrols and spacecraft missions have provided a vast amount of new information, the latter in particular are naturally limited in their coverage. Amateurs are able to maintain a watch upon the various planets and ensure continuity by following the changes which occur over a period of time. It is just this sort of continuity which has made earlier observations so valuable to workers investigating planetary atmospheres, in particular, in the light of modern knowledge.

As planetary disks are so small, detailed observation requires large apertures and long focal lengths capable of giving high magnifications. Refractors or catadioptric systems (pages 230–1) have certain advantages over reflectors for this sort of work, but few persons can now afford the large-size refractors favoured by wealthy amateurs in the late nineteenth and early twentieth centuries. Reflectors are frequently used, therefore, although a growing number of Maksutov and Schmidt-Cassegrain systems are also employed. Visual techniques for drawing the various features are generally the same as those for the Moon (pages 108–109), while Mars, Jupiter and Saturn offer the opportunity to try working in colour.

Of the inferior planets, **Mercury** is the more difficult to observe as it is always close to the Sun and is at best visible only low in the sky before sunrise or after sunset. For most amateurs, especially those at high latitudes where it remains low in the sky, there is always a sense of achievement in spotting the planet low in the evening or morning twilight. Very little detail can be seen on the surface, even with the largest telescopes, and this is limited to faint albedo markings. Its study therefore is likely to be undertaken by only the most dedicated observers.

Venus can be quite well placed on occasions and its phases discerned even through a fairly small telescope, but the intense brilliance of the planet and the low contrast of the atmospheric features mean that these can only be seen with difficulty, although they can be accentuated by the use of suitable filters. At some time in the future it may be possible to correlate the observed distribution of bright and dark patches (usually very indistinct) with atmospheric activity or features, but this has yet to take place with any degree of certainty. However, it may be noted that amateur observations indicated the 4-day rotation period for the upper atmosphere even before the first spacecraft observations.

Telescopes with setting circles may be used to find the planet in the daytime sky, when observation is sometimes easier because of the reduced contrast between the brilliant planet and the background of the sky. Despite the difficulties in observing the planet, more work needs to be done on certain unresolved problems such as the 'Schroter Effect', a discrepancy between the observed and calculated times of dichotomy (half-phase), and the existence of the 'Ashen Light', when the unilluminated portion of the disk appears, near inferior conjunction, to be faintly luminous.

Mars is not easy to observe as the high eccentricity of its orbit causes its maximum apparent diameter to vary from 25 to only 14 arc sec. as its minimum distance from the Earth varies from 5.6×10^7 km at a perihelic opposition to 10^8 km at an aphelic opposition. In addition, the relative motion of the two planets is such that the period most favourable for telescopic work lasts only a few months and occurs at intervals of more than two years. Detailed observation of the planet requires a telescope of at least 200mm aperture, but the more prominent surface features can be distinguished through a smaller instrument. Careful drawings may be used for the study of several effects, such as the seasonal changes which occur – particularly the advance and retreat of the polar caps and the changes which take place from year to year. Similarly the variations in the extent and intensity of the dark markings are a measure of the transport of material over the surface of the planet, while atmospheric conditions are monitored by observing the hazes and clouds which occasionally obscure the surface, as well as the sometimes planet-wide dust-storms.

Jupiter undoubtedly offers the best prospect for the planetary observer. Its apparent diameter is rather more than 46 arc sec. at its mean opposition distance and is never less than 30 arc sec., so the planet may be profitably observed for several months around the time of opposition, which occurs at intervals of about 13 months. Even through a small telescope the disk of the planet is seen to be crossed by dark belts separated by light zones, interspersed with streaks and oval features of which the most prominent (at present) is the Great Red Spot. Drawings of the relative positions and prominence of the numerous features can help to determine the various circulations which are taking place in the atmosphere, and which with modern spacecraft measurements are leading to an understanding of the processes at work. Due to the short rotation period of less than 10 hours, many

The identification of the various belts, zones and regions on Jupiter. (A similar system is used for Saturn, although details are more rarely seen on that planet.) The abbreviations are: E equatorial; GRS Great Red Spot; N north; P polar; R region; S south; T temperate; Tr tropical; Z zone.

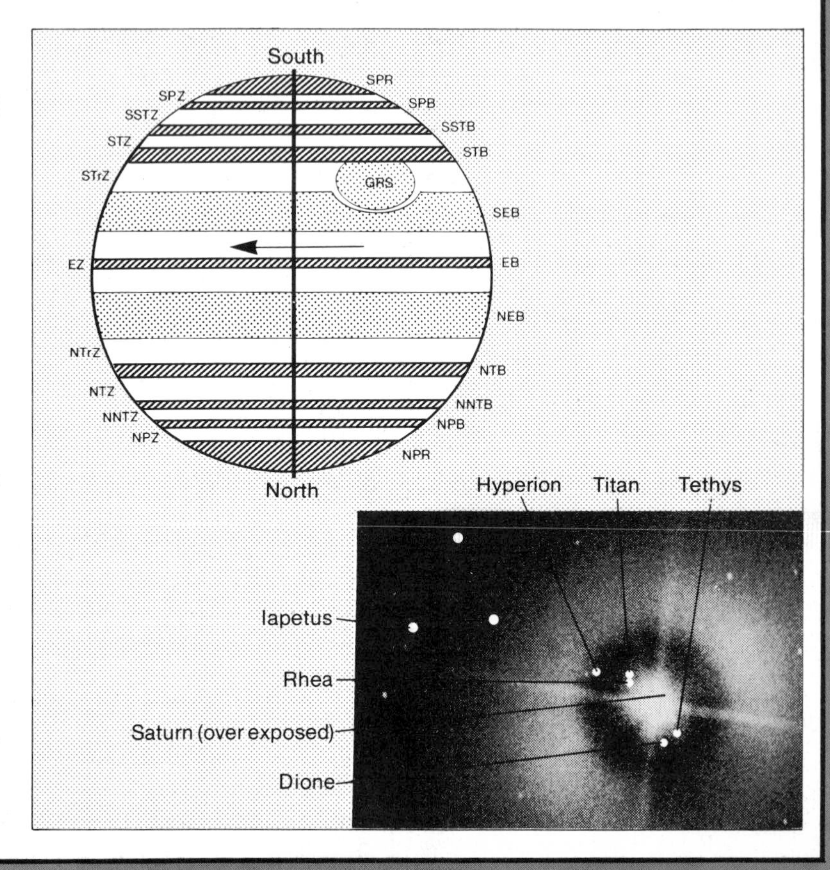

different features may be observed during a single, fairly short observational session. This leads to a very valuable method of determining the position of a feature by timing its passage across the central meridian, from which its longitude may be derived.

Although the four Galilean satellites can be seen with only limited optical aid, somewhat larger equipment is required to observe the various forms of satellite phenomena which arise from their small orbital inclinations to the equatorial plane of Jupiter. As a result of the similar orbital inclinations of Jupiter and the Earth, transits of the satellites and their shadows across the planet's disk, eclipses of the satellites by the shadow-cone of the planet and occultations of the satellites by Jupiter itself are regularly visible from the Earth. Apart from being interesting to observe, a comparison of the actual times of their occurrence with those predicted is important for an understanding of the gravitational interaction of the satellites and the planet.

Saturn with its system of rings is an awe-inspiring sight even through quite a small telescope (except when the rings are presented edge-on to the Earth, as happens every 15 years). However, a large aperture is really required for detailed observation of the planet, which generally shows less distinct phenomena than Jupiter. Long-lived features are only very rarely seen – and cause appropriate excitement when they do appear – so that routine amateur work concentrates upon the estimation of the relative intensities of the rings and the rather faint belts and zones. When any distinct details can be seen the method of Central Meridian transits can be applied (as for Jupiter). Phenomena involving Saturn's satellites are rare when compared with Jupiter's, as they can be observed only during the five successive apparitions that occur when the Earth is passing through the plane of the rings. In general, passage through the ring-plane is eagerly awaited by observers, as they also make observations of the visibility of the rings themselves as they fade towards disappearance and subsequently reappear. Around this period bright 'knots' may sometimes be seen in the rings, and these probably represent clumping of the ring particles or undulations in the general surface.

The magnitudes of Saturn's satellites are poorly known and estimates may therefore be attempted by amateurs. Variations in brightness (and apparent colour) are detectable, and these are dependent upon the positions of the satellites in their orbits (as would be expected with synchronized rotation). However, this work is very difficult to carry out on account of the considerable interference from the glare of the planet itself, and, as in the case of minor planets and comets, the lack of suitable comparison stars. It is usual with this kind of work to take one satellite as the standard against which the others may be compared, even though it is itself almost certainly variable. With so many problems, a proper photometer – preferably photoelectric – is really required.

The remaining planets – Uranus, Neptune and Pluto – can be located using progressively larger telescopes, Uranus being visible as a faint point of light in binoculars, and Pluto needing about 300mm aperture. A telescope of this size shows Uranus as a disk, but no detail can be seen and none of the three planets offers much for most amateurs.

General photography of the planets is difficult because of the small sizes of the disks. Nevertheless some amateurs, especially those with fairly large-aperture equipment, have achieved excellent results, even though these cannot compare for detail with visual observations. The main requirements are patience, perseverance and a telescope with a long focal length. Some success has been achieved by the use of the multiple-image technique, where a series of exposures, obtained over a short period of time on a single occasion, are combined in the darkroom to produce a single image with enhanced detail. However, this method and those requiring hypersensitized film are only applied by the most advanced amateurs.

A remarkable photograph showing several of the satellites of Saturn, taken by K. Kaila, Finland. The image of Saturn itself is, of course, greatly overexposed; its brightness makes photography of the satellites extremely difficult.

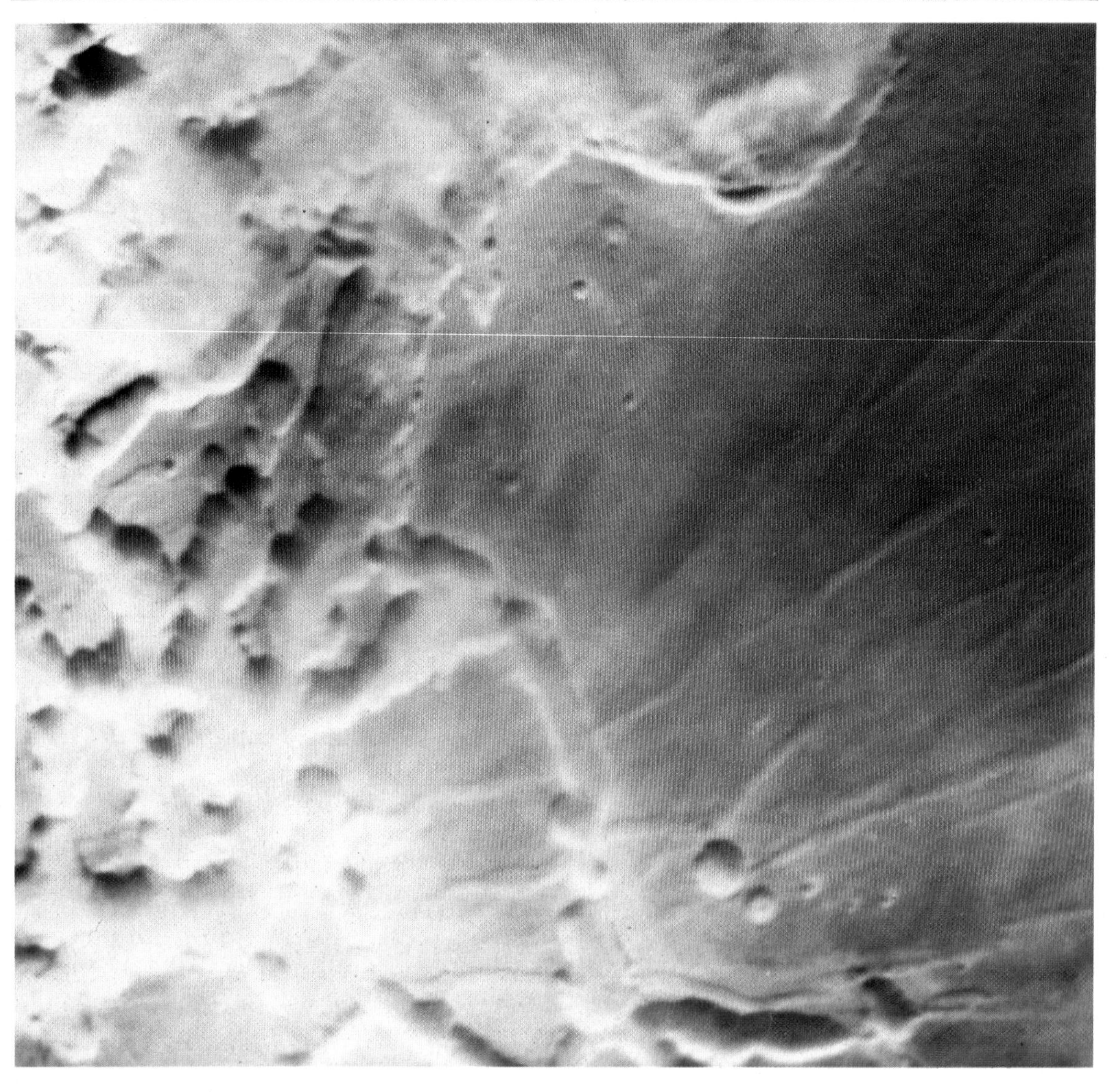

Above: *Part of the Noctis Labyrinthus area, with early-morning clouds of water ice within the canyons. The clouds are generally confined to the valleys, only rarely covering parts of the plateau surface. The area shown in this photograph is about 10 000 km² and the large partial crater is known as Oudemans.*

Below: *Early morning photograph taken by Viking Lander 1, covering about 100° from roughly north-east to south-east. The small dune field indicates winds from the upper left, that is northerly winds. The large boulder at the left is about 3 m long and 1 m high.*

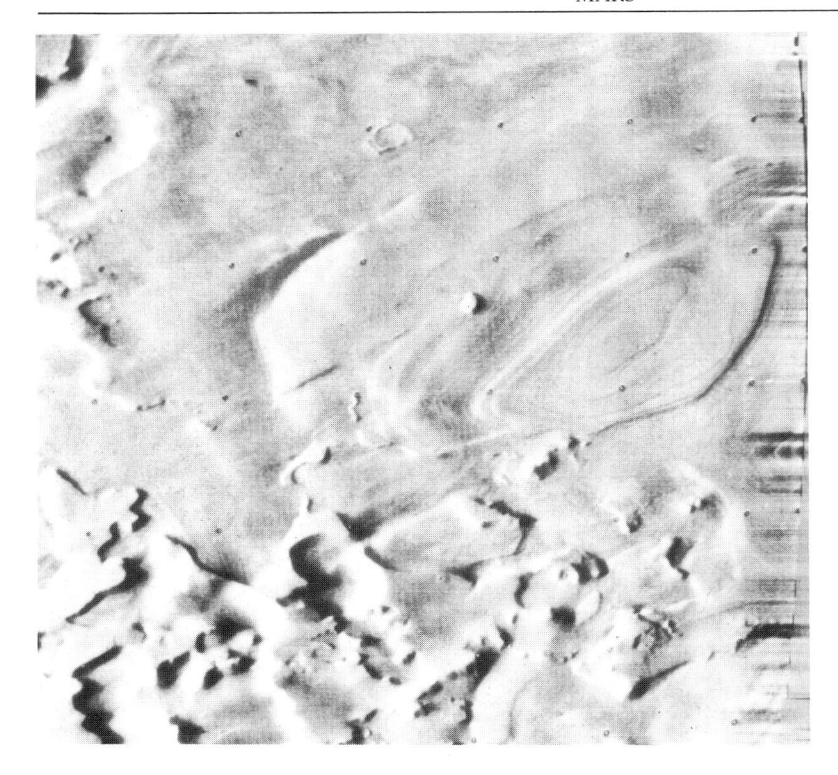

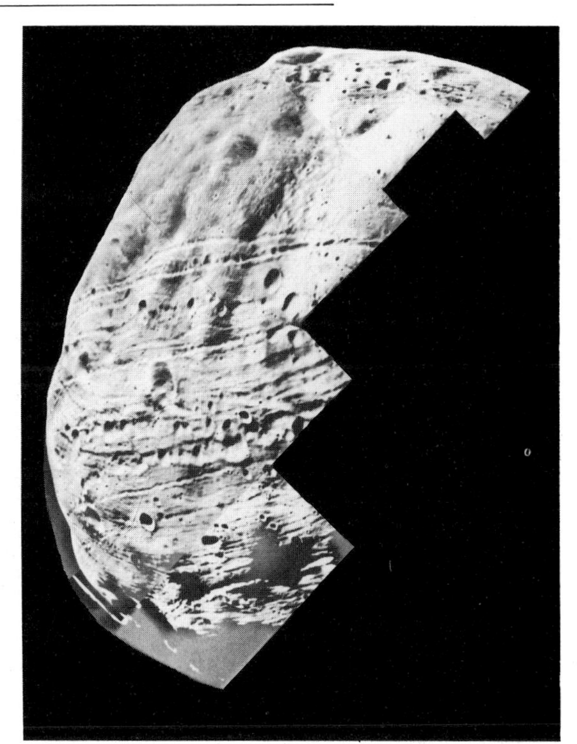

Table 5·10 **Mars: atmospheric composition**

component		**percentage volume**
carbon dioxide	CO_2	about 96
nitrogen	N_2	2.5
argon	Ar	1.5
oxygen	O_2	0.1
krypton	Kr	trace
xenon	Xe	trace
helium (probable)	He	trace
neon (probable)	Ne	trace
water	H_2O	—

Chemical species probably formed within the amosphere include: carbon monoxide (CO), atomic oxygen (O), atomic hydrogen (H), molecular hydrogen (H_2), the hydroxyl radical (OH), and hydrogen peroxide (H_2O_2).

The search for life

At one time there was fairly widespread belief in the existence of higher life forms on Mars, due to the supposition that the planet possessed an artificial canal system. However it is now known that these canals were an optical illusion, and the environment seems obviously unsuitable for any complex plant or animal species. One of the major tasks for the Viking Landers was to search for possible signs of life which might have formed in earlier times when the atmosphere was denser and warmer, and liquid water could have existed on the surface. Micro-organisms are known to survive in very harsh environments on Earth and also under simulated Martian conditions, so they could have persisted until the present time. Apart from the cameras which could obviously have

Above lett: *Layered terrain and etched terrain near the south pole of Mars. The laminations are probably caused by alternating layers of ice and dust deposits.*

Top right: *Photomosaic of Phobos, taken by Viking Orbiter 1 when about 300 km from the satellite. The striking linear grooves are about 500 m wide. The crater Stickney is indistinctly seen on the limb at top left.*

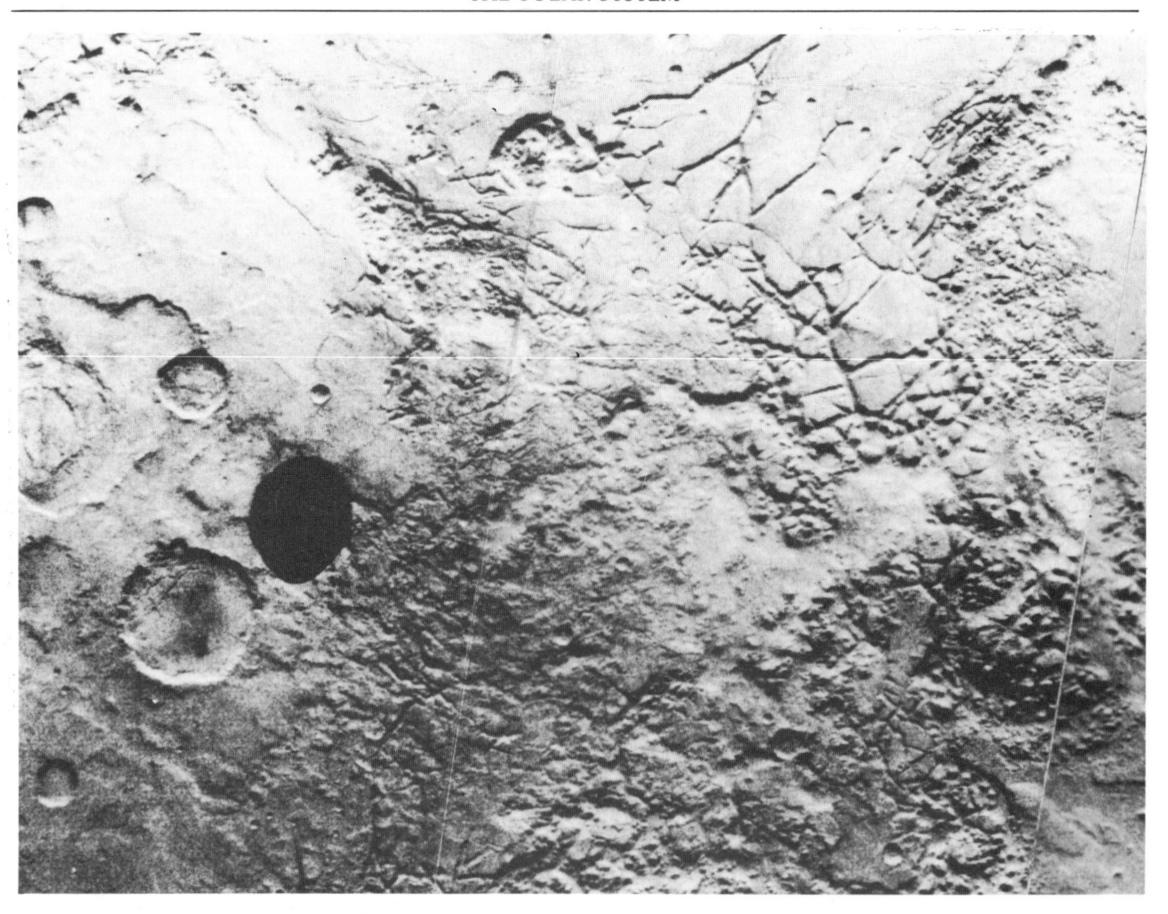

Phobos over the Margaritifer Sinus region of Mars, from Viking Orbiter 1 when about 13 700 km above Mars and 6 700 km from the satellite, which appears dark due to its very low albedo. Chaotic terrain can be seen near the head of Ares Vallis, which is a major channel running into the Chryse Basin.

detected any larger life forms, each Lander carried equipment to perform three types of biological test on soil samples. Basically the experiments were designed to establish whether there was an exchange of gases within the atmosphere and to detect the production of gaseous carbon compounds or complex organic (carbon-based) substances. Some of the results can just possibly be taken to indicate biological activity, but the general opinion is that the observed changes have been caused by unexpected but purely chemical reactions.

Satellites

Prior to the Mariner and Viking Orbiter observations little was known about the satellites of Mars apart from their orbital periods and distances from the planet. Phobos, the larger, orbits the planet in 7·65 hours at a distance of 9 350 km (2·75 Mars radii). Although the orbit is direct, the same as Mars' rotation, the period is so much shorter that it would appear to rise in the west and set in the east. The second moon, Deimos, orbits in 30·3 hours at 23 490 km (6·9 Mars radii) and would appear to take about 60 hours to cross the sky.

The spacecraft pictures show that both satellites are irregular bodies which are shaped like potatoes, with approximate diameters of 27, 21 and 19 km for Phobos and 15, 12 and 11 km for Deimos. Tidal forces acting on these irregular bodies have pulled both satellites into synchronous rotation so that the same faces are always turned towards Mars.

Both objects are seen to be covered in craters, with the largest on each being Stickney (diameter 10 km) on Phobos, and Voltaire (diameter 2 km) on Deimos. The craters show uplifted rims but no ejecta blankets or central peaks, as is to be expected when the very low surface gravity is taken into account. The sharpness of some of the features suggests that both bodies are solid, rather than loose blocks bound together by gravitational forces. Other results indicate that they are covered with rubble generated by meteoric impacts.

A suggestion that they might be formed of basalt has been disproved by tracking of Viking Orbiter 2, which has shown that Phobos has a very low density close to 2 000 kg per m^3. This, together with studies of the spectral characteristics of the surface materials, indicate that the satellites may closely resemble a certain type of meteorite known as a carbonaceous chondrite, which will be discussed later (see p. 150). Indeed, they may well be remnants of the material from which Mars itself was originally formed. Phobos exhibits some remarkable parallel grooves (page 131; top right). These have been shown to be associated with Stickney, suggesting that they are fracture lines produced by the impact which formed the crater. Future investigations may enable some of the outstanding problems about these tiny bodies to be settled, but certainly they show many features which are expected to apply to bodies such as the minor planets.

Jupiter

Ever since Galileo Galilei turned his primitive telescope towards Jupiter in 1609 and noted its four major satellites, the planet has been under constant study. It has proved to be a world of superlatives: it is by far the largest of the planets with an equatorial diameter of 142 796 km (11·2 times that of the Earth); it is more massive than all of the other planets put together (318 times the Earth's mass); it has the shortest rotation period; it has a vast magnetic field and is a powerful source of radio waves. In keeping with its importance it has at least seventeen satellites and exerts great influence on the orbits of the minor planets and comets by its gravitational perturbations. However, despite its apparent size, it is small when compared with the Sun which has a mass 1 047 times that of Jupiter. Table 5·11 gives some of the details of this planet and its orbit.

Table 5·11 **Jupiter-Earth comparative data**

	Jupiter	Earth
equatorial diameter (km)	142 796	12 756
sidereal period of axial rotation	$9_h55_m30_s$	$23_h56_m04_s$
inclination to orbit	3° 04′	23° 27′
density (kg per m³)	1 330	5 517
mass (Earth = 1)	1318·7	1·0000
surface gravity (Earth = 1)	2·643	1·0000
escape velocity (km per s)	60·22	11·2
albedo	0·73	0·36

mean Sun-Jupiter distance 5·2028039 au

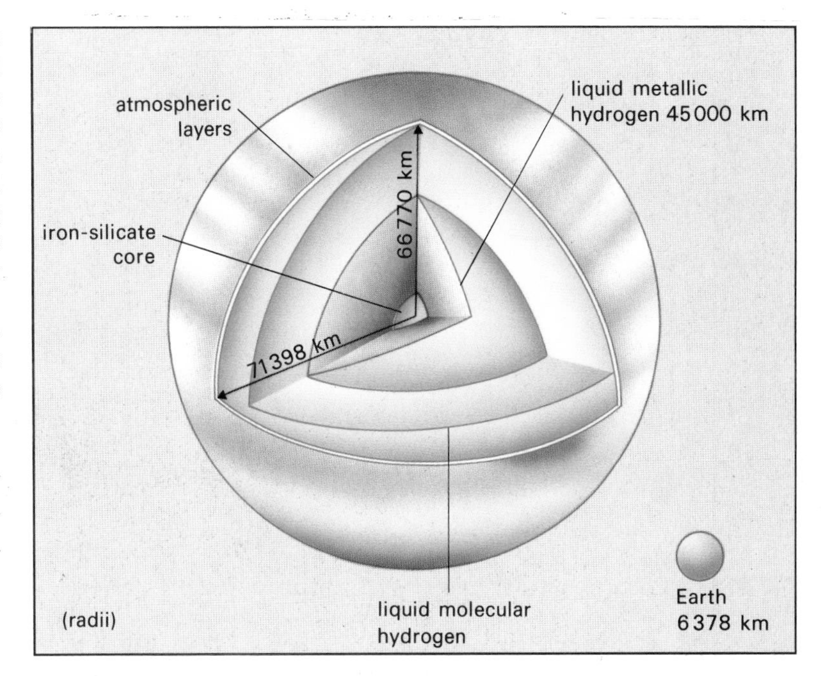

Fig. 5·15: Comparative sizes of Jupiter and the Earth. The size of Jupiter's probable rocky core is uncertain. The thickness of the planet's atmospheric layers is about 1 000 km.

Interior

Jupiter's density is only 1 330 kg per m³ (less than a quarter of the Earth's) and this indicates that it is almost entirely composed of very light elements, notably hydrogen and helium. On theoretical grounds it is expected that the ratio of these two elements should be about the same as that found in the Sun, and this has been confirmed by the Pioneer and Voyager spacecraft measurements. A minor amount (perhaps 10–20 Earth masses) of heavier elements should be present, and this may be expected to be concentrated into a small iron-silicate core, although the presence of such a core is not evident either from study of the orbits of its satellites or spacecraft tracking. Tracking has, however, shown that the body of Jupiter is in fact liquid, and this is in agreement with calculations of the planet's internal temperature, which is expected to reach 20 000–30 000 K at the centre. This heat is the remains of that produced by release of gravitational energy when the planet originally formed. The combined effect of the temperature and the immense internal pressure is that hydrogen will be liquid throughout the planet. In an outer layer the hydrogen is in the molecular (H_2) form, but at about 40 000–50 000 km from the centre the molecules are separated into individual atoms (H), causing the material to become an electrical conductor and it is described as being in the **metallic state**. Figure 5·15 indicates the probable internal structure of the planet.

Observation of Jupiter's radio emission has established that the rotation period of the body of the planet is $9_h55_m29{\cdot}75_s$. (This is known as the System III rotation period to distinguish it from the other two periods which are found from atmospheric features and which will be mentioned later.)

As a result of this rapid rotation, the liquid state and the low overall density, the globe is flattened by about 6 per cent with the polar diameter being 133 540 km.

Atmosphere

Above the liquid molecular hydrogen layer there is a thick and complex atmosphere and it is clouds within this which are responsible for all the visible features. The most immediately obvious are a series of light-coloured zones and dark belts which encircle the planet parallel to the equator. Apart from the banded structure, however, there is a wealth of streaks, irregular patches, ovals and spots of all kinds, as well as the famous Great Red Spot, which is about 25 000 km long by 10 000 km wide and roughly equal in area to the whole of the Earth's surface. Observations from Earth showed that markings within the approximately 20°-wide equatorial zone have a rotation period of $9_h50_m30_s$ (System I), whereas areas to the north and south of this rotate in $9_h55_m41_s$ (System II), much closer to the period of the interior. Voyager images showed that this difference is largely due to the very strong eastwards flow within the equatorial zone, which may reach velocities as high as 150 m per second. The whole atmospheric circulation at low and middle latitudes is strongly zonal with both eastwards and westwards currents at various latitudes. Considerable local turbulence and small-scale features occur with the strong wind shear which is present in the regions between the opposing currents. It is now known that the colours of the various belts and zones are very variable, but the dominant winds are remarkably constant, having been traced from Earth-based observations made over many decades, and including a very substantial contribution from amateur astronomers.

Jupiter, as photographed by the spacecraft Voyager 1 on 1979 January 29, from a distance of 35·6 million kilometres. This composite has been assembled from images obtained through three colour filters. Great detail is shown in the area surrounding the Great Red Spot and in most of the belts and zones, including tiny red spots in the northern hemisphere.

The driving mechanism for this activity is uncertain as the temperature difference between equator and poles due to solar heating is very low (only about 3 K), and in any case little poleward flow occurs, unlike on Venus and the Earth. It is thought that the persistent circulation pattern could well be the result of deep-seated processes occurring at lower layers of the atmosphere or in the planetary interior, and the strong zonal flows are certainly related to the high speed of rotation of the planet. The observed average atmospheric temperature of 125 K is higher than could be produced by solar radiation alone (about 105 K) and has long been known to be due to a flow of heat from Jupiter's interior. The interior heat source contributes about 1·7 times the heating produced by the Sun, and is likely to be mainly due to residual heat from the time of the planet's formation (when the temperature in the deep interior is thought to have been around 50 000 K).

It has now been established that the white zones are, in general, the cold tops of rising air masses, the clouds being formed of ammonia crystals at temperatures of about 141 K. The darker belts are descending currents with a higher temperature – reaching as much as 149 K. However, there are significant exceptions to the dependence of colour upon height, the Great Red Spot being the obvious example as this feature is both high and cold. The Great Red Spot has much in common with the more numerous white ovals and they can all be likened to the persistent high-pressure areas which occur on Earth. The exceptionally long lifetime of the Jovian features (the Great

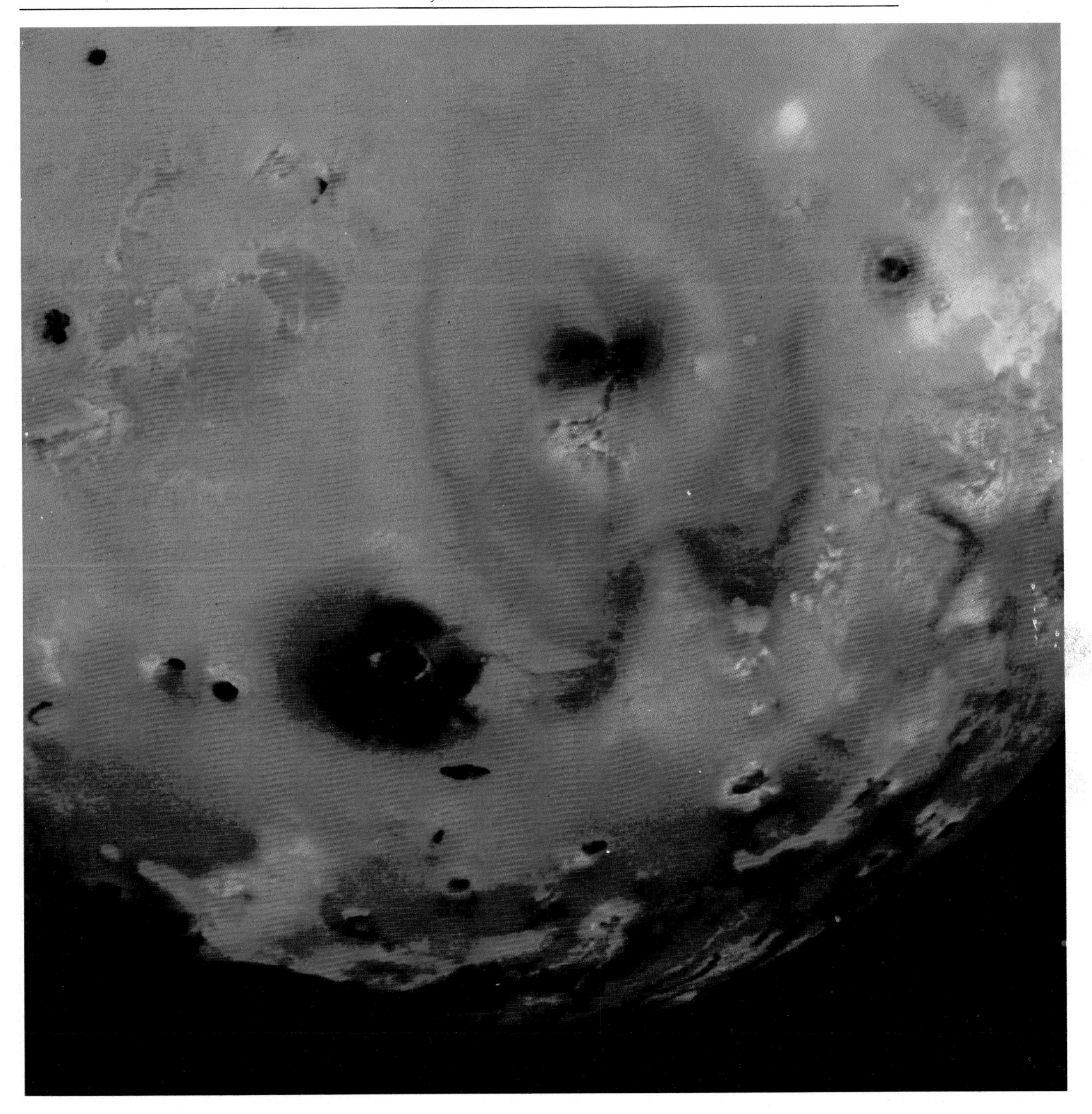

Red Spot has been observed for more than 300 years) is largely related to the fact that temperatures are so low on the planet that very little energy is lost by these systems.

A large number of gases and compounds are known to exist in the atmosphere and these are listed in Table 5·12.

The outermost region consists of a hydrogen haze, below which are the visible cloud-tops of ammonia cirrus.

Even deeper it is generally agreed that the clouds are formed of ammonium hydrosulphide crystals (NH_4SH), even though this cannot be observed directly, at a likely temperature of 260 K. Further down there are layers of water ice and water droplets. (See Fig. 5·16.)

Table 5·12 **Jupiter: atmospheric composition**

known components			
hydrogen	H_2	carbon monoxide	CO
helium	He	hydrogen sulphide	H_2S
methane	CH_4	hydrogen cyanide	HCN
ammonia	NH_3	germanium hydride	GeH^4
water	H_2O		
ethane	C_2H_6		
acetylene	C_2H_2		
phosphine	PH_3		
ammonium hydrosulphide (probable)	NH_4SH		

The surface of Io, the innermost of the Galilean satellites, showing the giant volcano Pele (the heart-shaped area). The dark areas are other volcanic craters; the white areas are sulphur dioxide 'snow', while the overall reddish coloration is due to sulphur, ejected in molten form from the volcanoes.

Fig. 5·16
The atmospheric structure of Jupiter. The height of clouds observed in the zones is approximately 120 km. The temperature is rather uncertain above the stratopause.

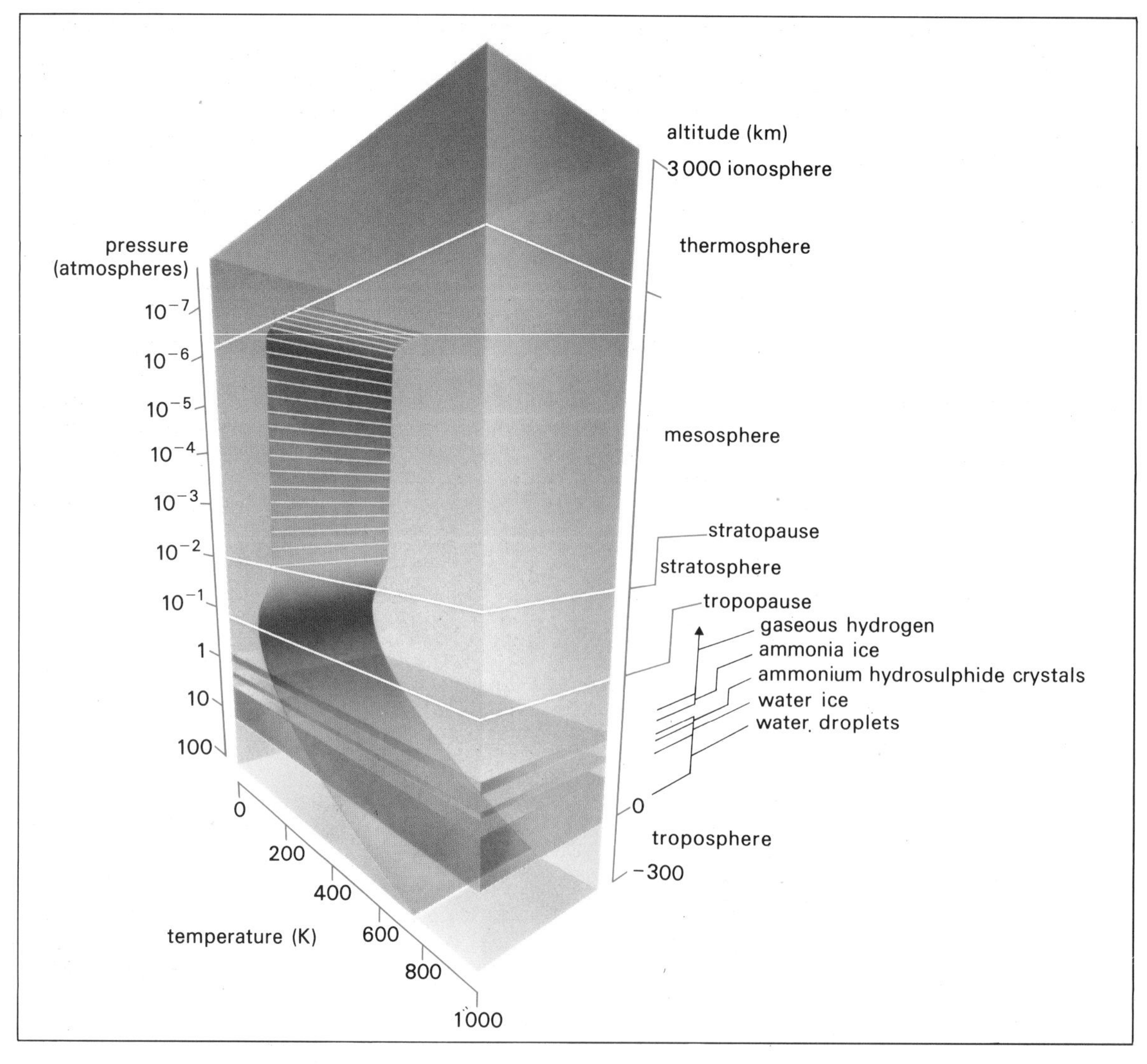

The cause of the distinct brown and reddish colouring of the clouds remains uncertain. It is most likely that it is related to the presence of phosphine (PH_3), which may be broken down by ultraviolet radiation from the Sun to give red phosphorus (P_4). This process should be most pronounced at the highest levels, and would thus nicely account for the colour of the Great Red Spot, which is the highest feature on the planet. However, some of the smaller red spots are very much lower, so the matter is unresolved. Other colouring agents almost certainly exist, such as complex hydrocarbon compounds produced from the methane (CH_4) and ammonia (NH_3) present in the atmosphere.

Magnetic field

Jupiter's strong magnetic field is probably generated as a result of the internal heat producing convection currents inside the liquid metallic hydrogen. Close to the planet the field is extremely complex and possibly related to circulation within this liquid metallic hydrogen. From about three Jupiter radii, however, the simple bar-magnet type of dipole field is like the Earth's, and the magnetic axis is tilted by about 10–11° from the axis of rotation and displaced by about 7 000 km from the centre of the planet. Between about 20 and 50 Jupiter radii there is a sheet of low-energy trapped particles which themselves produce a magnetic field, influencing the overall lines of force (Fig. 5·17). Further out the dipole field is weak; it ends at the magnetopause, but the position of this depends strongly on solar activity, and may range from 100 to 50 Jupiter radii. In the other direction the spacecraft measurements indicate that the magnetic tail extends out beyond the orbit of Saturn, having been detected at about 7×10^8 km (more than 4·5 au) from Jupiter.

The immensely strong magnetic field has given rise to a vast radiation belt, rotating with the planet, with a trapped plasma of highly energetic electrons and protons. By means of a process known as synchrotron radiation, the trapped electrons produce powerful radio emission at wavelengths between 5 and 300 cm. Although all the inner satellites are within this belt and are bombarded by the charged particles, the density reaches a maximum in a ring-shaped torus around the orbit of Io. This satellite is connected to Jupiter by a flux tube of magnetic field lines which carries an immense current of about 5 million amps. Electrons from this flux tube sometimes precipitate into the upper atmosphere, giving rise to the aurorae observed by Voyager 1. Similar discharges are responsible for the bursts of radio emission between about 7·5 and 670 m which are observed from Earth. These are linked with specific longitudes on the planet and also with Io's position in its orbit.

Quite apart from the aurorae detected by Voyager 1, both Voyagers recorded lightning discharges on the planet, as well as picking up the accompanying radio emissions.

The rings

One of the most surprising discoveries made by Voyager was the existence of a system of rings around Jupiter. These are quite unlike the extensive rings of Saturn, which accounts for the fact that they had not been discovered previously, although they have since been detected from the Earth. They are very tenuous – Pioneer II even passed through them at the time of its Jupiter encounter in 1974 without suffering any damage – and the total thickness is probably less than 1 km. The main rings are quite narrow and occur in two bands, the outer, brighter one lying between about 52 200 and 53 000 km from the top of the atmosphere. The fainter inner belt is about 5 000 km wide and extends down to about 47 000 km. However there is also a very tenuous population of particles which extends all the way down to the cloud tops. All the particles are very dark, and they are thought to be very small rocky grains. Their origin is unknown, but they must be continually renewed, perhaps from the small satellites found to be orbiting just outside the rings, as their orbits must slowly decay and bring them down into the atmosphere.

Satellites

The two small satellites orbiting at the edge of the rings were not the only objects discovered by Voyager. One other small body (since named Thebe) is certain, and there are at least another three suspected objects. But the four large satellites, which prior to the Voyager missions were poorly known, having only the vaguest discernible markings, have now been revealed as distinctive objects in their own right. It has long been known that they were the size of small planets, but what was surprising was the fact that they are all different.

It was certainly not expected that the innermost, Io, would prove to be the most volcanically active body in the Solar System, with as many as eight vents active at any one time. Despite the fact that Io should undergo more intense meteoritic bombardment than any of the other satellites, due to the 'focusing' effect of Jupiter's gravitational field, not one impact crater can be found on any of the images returned by the Voyager cameras. This in itself would be enough to show that the surface is very young and must be geologically active, but this was confirmed by the detection of the volcanic plumes rising to between 70 and 280 km above the surface. Further examination showed the volcanic vents themselves and extensive lava flows. However, the ejected material and the 'lava' is quite unlike that of the volcanoes on Earth, as it consists of molten sulphur and sulphur dioxide. It is largely the former material which is responsible for the striking red and orange colour of the satellite, with the extensive white areas of the surface being comprised of sulphur dioxide snow.

The exceptionally high ejection velocities from the volcanoes (in some cases probably exceeding 1 000 m per s – several times greater than any found on Earth), together with the intense bombardment of the surface by energetic particles trapped in Jupiter's magnetosphere, are thought to contribute to the loss of material into space. Considerable quantities of both neutral and ionized sodium (Na), and ionized sulphur, potassium (K) and oxygen have been found in Io's torus around the planet.

A small body such as Io, which is only slightly larger than the Moon, should have long ago lost the

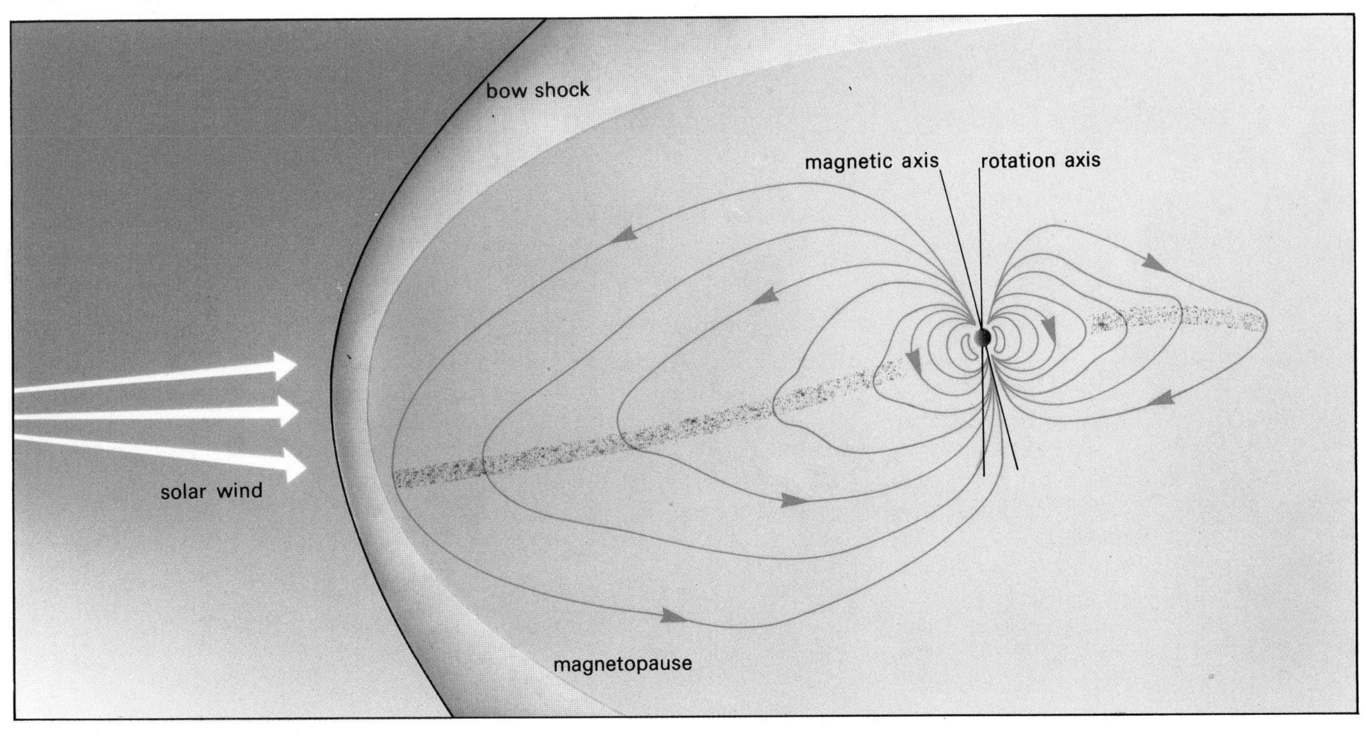

Fig. 5·17 Jupiter's magnetosphere, showing the sheet of trapped particles (stippled). This diagram is approximately to scale for a 'quiet' state of the solar wind.

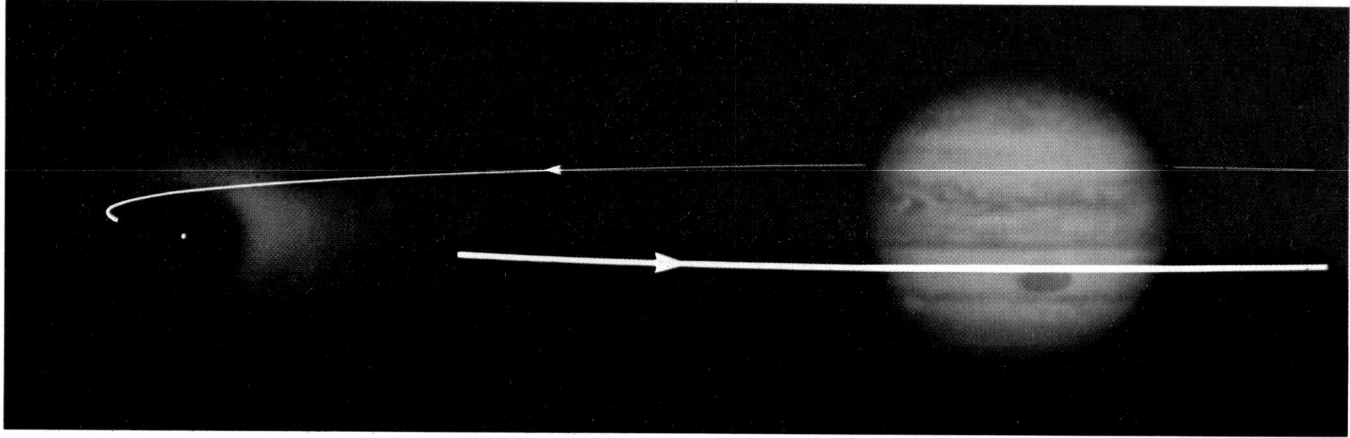

Key *Diameter of Io: 3640 km* ⊢—⊣ *10 arc sec* ⊢—⊣ *30 000 km*

heat generated during its accretion, and also that from radioactive decay processes. However, the other Galilean satellites cause perturbations in Io's orbit so that its distance from Jupiter varies slightly. Jupiter's gravitational field is so strong that even these small changes cause great tidal distortions of Io, and thus produce sufficient heating of the interior to account for all the volcanic activity.

The satellite appears to consist of a molten silicate interior, just possibly with a solid core, overlain by a layer of liquid sulphur several kilometres deep. Above this is a layer consisting of a mixture of solid sulphur and liquid sulphur dioxide (SO_2) covered by a solid crust of sulphur and sulphur dioxide.

Europa

The next satellite out from the planet, Europa, is quite similar in size and density to Io (and the Moon), and again was expected to be covered in craters. However, it also proved to be unique in the Solar System, for a completely different reason. Only a few small impact craters have been found; the rest of the surface is incredibly smooth. A network of straight, curved or irregular dark markings covers the whole surface, and these range from less than 10 km to about 70 km in width. There are also randomly located dark spots, but all these markings appear to have quite negligible vertical height, so that the satellite has been described as 'a billiard ball covered in scribbles from a felt-tipped pen'. Even stranger is yet another network of markings, this time faint and light-coloured, quite independent of the dark ones, and also covering the whole satellite. These are only about 10 km wide, and they do show some vertical relief, although this is less than a few hundred metres. But the most surprising thing about these ridges is that they are not straight: they run across the surface in a regular series of curves or scallops, ranging from about 100 to 300–400 km across.

The first impression that the surface is cracked is quite possibly correct, even though the lack of vertical relief seems surprising. Studies show that parts of the surface are covered with apparently freshwater frost, as well as traces of sulphur (almost certainly derived from Io). However, there is less sulphur than would be expected, which may well indicate that some has been buried beneath fresh frost deposits. These considerations, together with the lack of impact craters, suggest that processes are still acting to smooth out the surface.

Europa, like Io, is subject to tidal forces which could well maintain a certain degree of heating in the interior.

It would seem that a solid rocky core is covered by a thick layer of water and ice (perhaps about 100 km deep). Liquid water could escape to the surface through the cracks and give rise to the frost deposits before the cracks themselves freeze over once again, perhaps after a few years. The low rigidity of the icy crust would account for the lack of impact craters. The darker markings could well have been formed when the underlying water layer contained some mixture of other substances at an earlier period in the body's history.

Ganymede

Ganymede, the largest satellite in the Solar System, and Callisto both have lower densities than Io and Europa – about 1 900 km per m³. This suggests that they are both comprised of roughly half rock and half ice. Once again they are thought to have rocky cores surrounded by water or icy layers with icy crusts.

The surface of Genymede is very varied. The oldest regions consist of dark plains, one of which, Regio Galileo, is as much as 4 000 km across and preserves signs of a major impact in a series of low ridges (about 100 m high) spaced about 50 km apart. All this old terrain appears to have been fractured into separate blocks, some of which have been displaced, and some completely replaced by younger, lighter-coloured material consisting of long parallel lines of valleys and ridges about 15 km across and 1 km high. This 'grooved terrain' is highly complex in appearance, not only cutting into the old plains, but also intersecting older areas of the same type of surface, suggesting many mountain-building episodes. Still other regions show rough mountainous terrain, and Ganymede's surface seems to be the one place in the Solar System to have undergone geological changes like those produced by plate tectonics on Earth.

Some craters appear relatively fresh, with bright haloes, presumably from ice or water ejected by the impact, but most of the surface is actually very old. Crater counts suggest that the dark plains date back to about 4×10^9 years, and even the most recent grooved terrain seems to be about $3{\cdot}5 \times 10^9$ years old - roughly the same as the lunar highlands. The low relief probably results from a time when the interior of the satellite was rather warmer and the crust more plastic.

Callisto

Callisto seems to possess an even thicker icy crust than Ganymede, and it is very heavily cratered. However, all the craters are shallower than similar-sized ones on any of the terrestrial planets, and many of the later ones which might have been expected seem to have disappeared completely. There are remnants of large impacts, but they all have very little vertical relief. One, Valhalla, has a bright central region, about 600 km across, probably representing the original impact crater, and is surrounded by an immense set of 'ripples' which makes its overall diameter nearly 3 000 km – far larger than any feature such as Mare Orientale on the Moon or the Caloris Basin on Mercury. It seems certain that flow has occurred in the icy surface to obliterate many of the very old impact scars, and to reduce the height of the remainder. Apart from this, however, there appears to have been very little true geological activity on Callisto, and certainly not to the extent found on the other Jovian satellites.

The other satellites

Very little is known about any of the other satellites except for Amalthea, of which Voyager returned images showing it to be an irregular rocky body

Opposite page, top: *Voyager 2 photograph of Jupiter taken on 1979 June 28. One of the dark 'barges' can be seen in the North Equatorial Belt. In the lower half of the picture, south of the Equatorial Zone, a chaotic region of whiter clouds is visible, lying west of the Great Red Spot, which is out of the picture to the right.*

Opposite page, bottom: *A composite picture of a ground-based photograph of Jupiter and a three-hour exposure of Io's sodium cloud. The dark circular area was caused by the occulting disc used to prevent interference by light from Io itself. The size of Io is shown by the white dot.*

Table 5·13 **Satellites of Jupiter**

number	name	distance (km)	sidereal period (d)	inclination	eccentricity	diameter (km)	magnitude
XIV	Adrastea**	128 000	0·295	0·0	·0	40	17
XVI	Metis**	128 000	0·295	0·0	·0	40	17
V	Amalthea	180 900	0·489	0·4	·003	270 × 170 × 150	14·1
XV	Thebe	221 000	0·670	0·0	·0	80	16
I	Io	421 700	1·769	0·0	·0	3632	5·0
II	Europa	671 000	3·551	0·0	·0001	3126	5·3
III	Ganymede	1 070 400	7·155	0·2	·001	5276	4·6
IV	Callisto	1 882 600	16·689	0·2	·007	4820	5·6
XIII	Leda	11 110 000	240	26·7	·146	10	20
VI	Himalia	11 470 000	251	27·6	·158	180	14·7
X	Lysithea	11 710 000	260	29·0	·130	20	18·6
VII	Elara	11 740 000	260	24·8	·207	80	16·0
XII	Ananke	20 700 000	617	147	·17	20	18·8
XI	Carme	22 350 000	692	164	·21	30	18·1
VIII	Pasiphae	23 300 000	735	145	·38	40	18·8
IX	Sinope	23 700 000	758	153	·28	30	18·3

Much of the information for the smaller satellites is uncertain, in particular the diameters, which are subject to errors of 5–10%. The eccentricity of satellites I, II and III is variable.
** Yet to receive official confirmation.

covered in a reddish material – possibly derived from Io. Even less is known about the remaining objects except for their orbits, and details about these have yet to be fully confirmed in the cases of Adrastea, Metis and Thebe. The satellites lying beyond the four large Galileans fall into two groups, and although their origin is obscure it has been suggested that they might have originated from the break-up of two larger bodies. In the case of the outer group, which has retrograde motion, they are very likely to be captured minor planets.

Saturn

The most spectacular object in the Solar System as seen through a telescope is probably Saturn with its magnificent system of rings. Apart from these, however, Saturn shows many similarities with Jupiter (Table 5·15),and is a very sizeable planet with an equatorial diameter of 120 000 km (nearly nine-and-a-half times that of the Earth). It is particularly remarkable for its very low density of only 706 kg per m^3, less than that of any other known planetary or satellite body, which indicates that, like Jupiter, it is primarily composed of hydrogen and helium. It, too, is expected to have a core of silicate materials, relatively larger than Jupiter's, surrounded by layers of metallic and molecular hydrogen. This core probably has a diameter of about 20 000 km and contains some 3–4 Earth masses, while it may be surrounded by a 5 000-km-thick layer of ice. Unlike Jupiter, the layer of metallic hydrogen is probably fairly small at about 8 000 km thick and the major portion of the planet is formed of liquid molecular hydrogen (Fig. 5·18).

Observations show that Saturn, like Jupiter, has an internal source of heat, and that the amount of energy is relatively more important, being about two to three times the radiation received from the Sun. The original heat of accretion is thought not to be sufficient to account for this and it is proposed that an additional source is the separation of helium from the hydrogen-helium mixture in the interior. Helium is probably 'condensing' and 'raining' down towards the centre, releasing heat as it does so.

Saturn has an even greater polar flattening than Jupiter (about 10 per cent compared with 6 per cent) and the polar diameter is about 108 600 km. The atmospheric rotation period was rather difficult to establish as the markings visible from Earth were never so distinct as on Jupiter. The generally accepted period is 10_h14_m (System I) with greater speeds towards the poles. Voyager experiments detected radio bursts at very long wavelengths which were shown to have a period of 10_h40_m, this now being accepted as the sidereal period of the planet (System III).

As with the other planets known to possess metallic cores and have rapid rotation rates (Jupiter and the Earth), a planetary magnetic field is generated in the interior. Before the Pioneer 11 and Voyager missions nothing certain was known about Saturn's magnetic field, but it has now been established that it is intermediate in strength between those of Jupiter and the Earth. It was something of a surprise when the geomagnetic axis was found almost to coincide with the rotational axis – they deviate by only 0·7 degree – as one theory for the generation of planetary magnetism required a much greater divergence. Just as Io is the primary source of the charged particles and atoms within the Jovian magnetosphere (with perhaps some contribution from Europa), so Titan provides most of the material in Saturn's system, with some gas also possibly derived from the inner satellites Enceladus, Tethys and Dione.

Titan orbits near the position of the magnetopause, although apparently it is usually inside it. It contributes large amounts of nitrogen from its atmosphere, as well as methane, which is broken down and provides the basic source for the large hydrogen torus found to encompass the orbits of Titan and Rhea.

All the satellites within the radiation belt affect the population of charged particles at their respective distances, but Titan in particular causes strong magnetic and plasma effects behind it. The number of charged particles drops dramatically at the edge of the rings, which essentially sweep up all particles within that radius.

The atmosphere

The atmospheric composition is roughly the same as that of Jupiter, but the overlying hydrogen haze is very much thicker, so that the planet appears comparatively featureless. Sufficient details have been observed, however, for it to be determined that the overall structure is very similar to Jupiter's, with strongly zonal flow at low latitudes and greater convective activity towards the poles. The equatorial wind speeds are exceptionally great, reaching 500 m per second (nearly 1 500 km per hour). To the north and south of this eastward jet stream, easterly and westerly flows alternate in a very regular fashion. There was a particularly turbulent westerly jet at about 47° north at the time of the Voyager 2 encounter.

Below the hydrogen haze the highest clouds are again ammonia cirrus, with a temperature of about 95 K at the top of the troposphere, where the pressure is about 0·07 atmosphere. Unlike Jupiter, breaks in the cloud cover appear to be very rare indeed on Saturn – if they occur at all. There is actually quite a number of features within the ammonia clouds, however, the largest being oval spots quite similar to those on Jupiter but very much smaller. Like the Great Red Spot and the white ovals on Jupiter, they are high-pressure areas which can certainly persist for about a year or more. Smaller-scale features may only last for a few days. There is probably little seasonal variation on Saturn, despite its axial tilt of nearly 27°, due to the large mass of the atmosphere.

Saturn also experiences aurorae at high latitudes where energetic particles from the magnetosphere are precipitated into the upper atmosphere. Radio discharges similar to those produced by lightning were initially thought to be associated in some way with the ring system, but are now believed to be truly the result of lightning discharges in the atmosphere.

The rings

Saturn's ring system, impressive even from Earth, has been shown by the Voyager missions to be of quite amazing complexity. The details of the main features are given in Table 5·14. It should be realized that there are really no distinct boundaries between the main rings, although the Cassini Division comes closest to being so. There are instead definite changes in the density of particles found at a given distance from the planet. The three main rings, A, B and C, have been known for a long time, the last being sometimes known as the Crêpe Ring on account of its tenuous appearance. Although discovery of the D Ring from Earth has been announced in the past, it does not really correspond with the features found on the Voyager images.

The whole system of rings consists of many thousands of individual ringlets, and doubtless with higher resolution even more would be shown. Understanding the dynamics of all these ringlets is a major problem, and there are probably several factors which contribute, or which may predominate in particular regions of the whole system. It was long thought that the major divisions were caused by gravitational resonances with certain of the satellites outside the rings, particularly Mimas; however, except in the case of the Cassini Division – or more especially the Huygens Gap – this cannot account for the system's complexity. Current theories favour the idea that much of the structure is governed by density waves somewhat similar to those suggested to control the distribution of matter in spiral galaxies. The distribution of ringlets certainly changes with time, although it is expected that the overall structure will probably be reasonably constant. Many of the individual ringlets are eccentric, with the centre of Saturn at one focus of the ellipse, and it is suspected, although unconfirmed, that they may be controlled either by a single satellite or by a pair of 'shepherd'

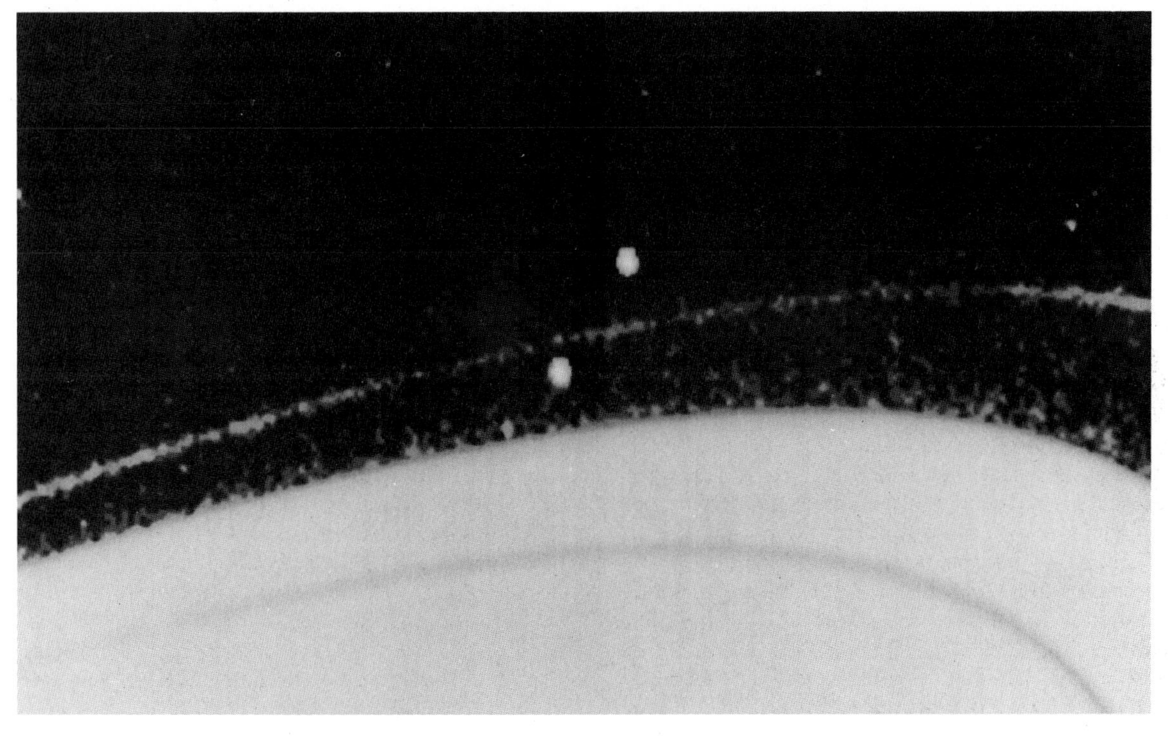

Opposite:
A Voyager 2 picture showing the thin F Ring and its two 'shepherd' satellites, and part of the much wider A Ring in the foreground. The two satellites confine the F Ring to its narrow band, while another satellite (not shown) orbits the very edge of the A ring.

This computer-enhanced photograph of Titan shows some of the haze layers overlying the otherwise featureless, thick, nitrogen-rich atmosphere.

Opposite page, top:
A composite Voyager picture (in true colour) of Saturn also showing three of the satellites – Tethys, Dione and Rhea. The black spot on the southern hemisphere of the planet is the shadow of Tethys. Some of the 'spokes' can be seen (particularly left) on the bright B Ring.

Opposite page, bottom left:
A close-up view in false colour of two of Saturn's rings – the C Ring (blue in this picture) and the B Ring (orange). Seen from this close, the two rings are clearly made up of many smaller ringlets, of which more than 60 are evident in this frame.

Opposite page, bottom right:
Unlike Jupiter, Saturn has a thick overlying haze which hides most of the atmospheric features. Complete computer-processing and image-enhancement was required before any details could be seen, as in the case of this Voyager 2 picture.

satellites (as described below for the F ring) orbiting within the ring system.

The particles comprising the whole ring system must number many thousands of millions, the largest being of the order of 10 m in diameter, while the smallest are only micron-sized (10^{-6} m). The brightest ring, the B Ring, showed remarkable dark irregular 'spokes', which formed, apparently rotated with the ring, and then dissipated. These are now thought to be due to tiny particles like smoke, out of the general plane of the rings, perhaps levitated by electrostatic forces. (The appearances described are those of the ring system as seen from the sunlit side; these change completely when viewed from 'beneath' – that is from the 'shadow' side. Then the densest parts of the rings – the B Ring in particular – are opaque and appear black, whereas the Gaps and Divisions and some of the less dense portions of the various Rings appear bright with scattered sunlight.)

It has long been established that the ring particles mainly consist of water ice, but subtle variations in the colour in various parts of the system suggest that other substances could be present. It is unlikely that the rings could be of recent origin; they probably consist of material remaining after the formation of Saturn and the larger satellites. It is possible that they were formed from larger bodies which were fragmented by collision or tidal forces.

The A Ring appears to be bounded by the small satellite which orbits at its outer edge, in the same way as Jupiter's rings seem to be restricted by the two satellites orbiting just outside them. The narrow F Ring, which is a few hundred kilometres across, is gravitationally limited by the two small 'shepherd' satellites (XV and XVI) which orbit on either side.

This ring appears to consist of several strands with various kinks as well as an apparently 'twisted' or 'braided' structure. Gravitational interactions with the confining satellites probably produce the kinks, but the twists are more likely to be illusory – the

Table 5·14 **Saturn's rings**

		distance from Saturn's centre		
Cloud tops		60 300		
D Ring	inner edge	67 000		
D Ring	outer edge	73 200		
C Ring	inner edge			
		87 500	Maxwell Division	
C Ring	outer edge	92 200		
B Ring	inner edge			
B Ring	outer edge	117 500	Huygens Gap	Cassini Division
		119 250	centre	
A Ring	inner edge	121 000	outer edge	
		135 500	Encke Division	
A Ring	outer edge	136 500	Keeler Gap	
F Ring		140 600		
G Ring		170 000		
E Ring	inner edge	210 000	(approx.)	
E Ring	outer edge	300 000	(approx.)	

result of the displacement of some of the strands from the general plane of the rings. The Encke Division also contains a ringlet which exhibits kinks, but the cause in this case is quite unknown.

Further out from the planet, well beyond the main rings, and outside the orbits of satellites X and XI (provisionally called Janus and Epimetheus) there exists the very tenuous G ring, about which little is known. Neither it nor the final ring, the broad E Ring, show any of the fine structure of the main belt.

Satellites

The system of satellites around Saturn is quite remarkable, not only for their number (seventeen confirmed and a further six suspected), but also for the extraordinary orbital characteristics of several of

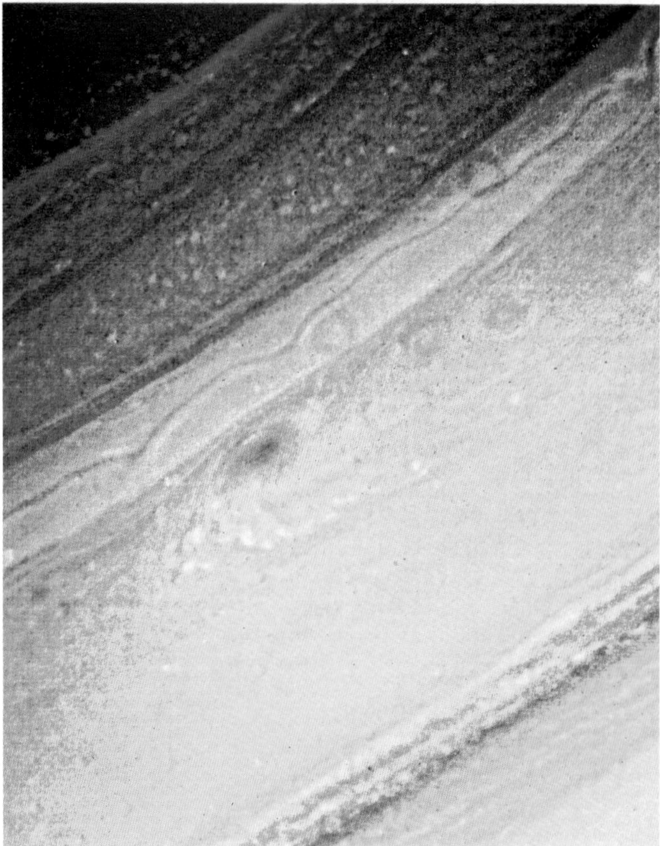

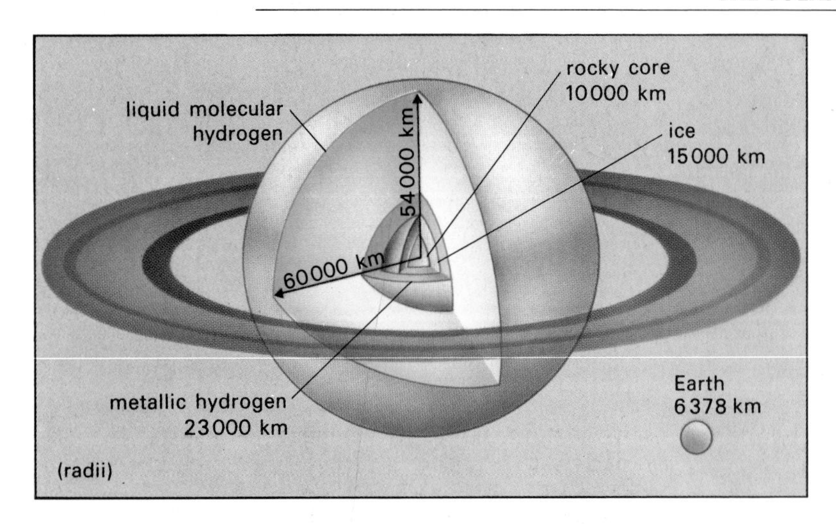

Fig. 5·18
Comparative sizes of Saturn and the Earth. Unlike Jupiter, Saturn probably has a layer of ice surrounding the rocky core.

Table 5·15 **Saturn-Earth comparative data**

	Saturn	Earth
equatorial diameter (km)	120 660	12 756
sidereal period of axial rotation	$10_h 40_m$	$23_h\ 56_m\ 04_s$
inclination to orbit	26° 44′	23°27′
density (kg per m^3)	706	5 517
mass (Earth = 1)	743·6	1·0000
surface gravity (Earth = 1)	1·159	1·0000
escape velocity (km per s)	36·26	11·2
albedo	0·76	0·36

mean Sun-Saturn distance 9·5388437 au

them. One, Titan, is a planetary-sized body in its own right, and is the only satellite in the Solar System with a dense atmosphere. The next six satellites, in descending order of size, do not show the extreme diversity of Jupiter's Galilean satellites, but they do possess some remarkable features, and can conveniently be discussed in pairs of similar sizes.

Titan

Although methane had been detected from Earth as early as 1944, the atmosphere of Titan has proved to be truly extraordinary, with more compounds being discovered than on Saturn itself. Most surprisingly, nitrogen forms the bulk of the atmosphere, and this is even more massive than the Earth's, with a surface pressure of 1·6 atmospheres. There are thin haze layers in the outer atmosphere, but the surface is completely hidden beneath a dense blanket of photochemically produced clouds about 200-300 km above the solid surface. It is quite probable that large complex organic molecules are being formed from the compounds present, and that they aggregate to form 'smog' particles which sink down to the surface.

As with Ganymede and Callisto, Titan's density of about 1 900 kg per m^3 suggests a composition of about half water ice and half silicate material, which is probably differentiated into a rocky core and a surface ice layer about 100 km thick. Conditions at the surface must be remarkable. The temperature is about 95 K, and is almost certainly raised by some form of greenhouse effect. This temperature is close to the temperature at which methane is either solid or liquid, so that methane clouds in a nitrogen/methane atmosphere may be raining methane down on to a surface covered with methane ice! Quite apart from this, complex organic molecules must precipitate out on to the surface, possibly in sufficient quantity to have formed a layer 100 m thick over the lifetime of the satellite.

The atmospheric conditions on Titan are so unusual, and its chemistry must be so similar in many ways to that of the primitive Earth, that the satellite is probably a more important target for future space probes than Saturn itself. The techniques used in such a mission would have to resemble those employed in examining Venus with its thick atmosphere, or in the forthcoming Galileo probe to Jupiter.

Mimas and Enceladus

The two large satellites closest to Saturn, Mimas and Enceladus, are both icy bodies. Mimas, however, is fairly strongly cratered, and has one enormous crater (named Herschel) which is about 130 km in diameter, as much as 10 km deep, and with walls rising 5 km above surface level. A high central peak was created by the rebound from the impact, which probably came very close to disrupting completely the 400 km diameter satellite.

Enceladus is very different, with an exceptionally high albedo of well over 90 per cent. Although some parts of the surface have a moderately high density of craters, other portions are relatively smooth, suggesting that some process has been acting to reju-

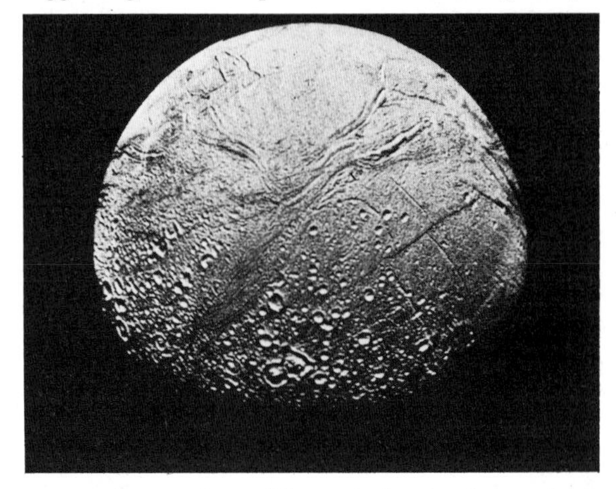

Far right:
The smoother portions of the surface of Enceladus appear to indicate that some 'resurfacing' of the icy crust has taken place, perhaps by water 'volcanism'.

venate the surface. Water 'volcanism' has been proposed, and it has even been suggested that this has been the source of the material forming the E Ring. But such a small body as Enceladus could not be expected to have, or to retain, any internal heat, so it must be assumed that some form of gravitational tides are acting to heat the interior. It is as yet uncertain, but it is possible that Dione may act in this way, in the form of an orbital resonance, just as Europa acts upon Io in the Jovian system.

Tethys and Dione

Both Tethys and Dione are slightly larger than 1 000 km in diameter and are quite heavily cratered. Tethys has one large impact scar, Odysseus, which is 400 km across. Unlike Herschel on Mimas, Odysseus has become subdued in relief, presumably due to flow in the crust. Even more remarkable is the long system of valleys and troughs (named Ithaca Chasma) which reaches for 2 000 km – about three-quarters of the way around Tethys. Although just possibly related to the giant impact, it is more likely to be due to cracking of the surface as a liquid water layer froze and expanded.

Dione also shows some signs of geological activity, and many areas appear to have been resurfaced. Its density is higher than that of Tethys, so it probably had a greater source of interior heat from its rocky materials. The trailing hemisphere (all the Saturnian satellites appear to have synchronous rotation) is covered in streaks which seem to consist of surface ice deposits. These could have formed by venting of water through cracks early in the history of the satellite, perhaps being destroyed on the leading hemisphere as this would be more subject to meteoritic impacts.

Rhea and Iapetus

Rhea also shows similar wispy markings on its trailing hemisphere, presumably caused by the same process as those on Dione. The surface is heavily cratered, but the distribution of sizes suggests that there were distinct episodes of cratering, the later one with many smaller bodies and perhaps being associated with the fragmentation of another satellite (or satellites), or even with the formation of the rings.

Iapetus has always been known to show a remarkable change in brightness around its orbit. We now know that the leading hemisphere is generally reddish-black – as black as pitch! – while the trailing hemisphere is white. It cannot be decided if the material is of internal or of external origin, since there is no evidence for white spots on the leading hemisphere as might be expected from impacts breaking through a thin surface layer deposited from outside; yet the very orientation suggests just such a source, perhaps from Phoebe, the outermost satellite. On the trailing side the dark material is concentrated on crater floors, suggesting an internal origin, although whether ammonia- or methane-volcanism can account for this distribution is an open question.

The smaller satellites

Hyperion is irregular in shape, suggesting that it is a result of fragmentation, and it is distinctly darker than most of the other satellites, possibly indicating that it has retained some of its original dirty ice crust.

Little is known of the outermost satellite, Phoebe, which has a retrograde orbit at nearly four times the distance of Iapetus. Its albedo is low, comparable with that of the carbonaceous minor planets, and this together with its orbit suggests that it has been captured, perhaps at a relatively recent time.

Between the orbits of Mimas and the 'shepherd' satellites controlling the F Ring, two bodies circle the planet. Their behaviour is quite unlike that of any other objects in the Solar System. These 'co-orbital satellites' have orbits which are separated by less than the size of the objects themselves. Every four years the inner one catches up with the outer, but finds no room to pass. However, as they approach they attract one another gravitationally, swing round one another, and exchange orbits! Four years later they go through the same performance, which can apparently be repeated indefinitely.

Of the remaining small satellites, two, Telesto and Calypso, occupy Lagrangian positions in the orbit of Tethys, just as the Trojan minor planets do in Jupiter's orbit around the Sun (page 153), and one a similar situation in the orbit of Dione. Two more are suspected in the orbit of Tethys, and another in that of Dione. A further suspected body may be in a special form of interaction with Mimas, placing it into

Table 5·16 **Satellites of Saturn**

number	name	distance (km)	sidereal period (d)	inclination	eccentricity	diameter (km)	magnitude
XVII*	–	137 000	0·602	0·3	·002	40 × 20 × ?	18
XVI*	–	139 400	0·613	0·0	·004	140 × 100 × 80	16
XV*	–	141 700	0·629	0·1	·004	110 × 90 × 70	16
X*	–	151 400	0·694	0·3	·009	220 × 200 × 160	15
XI*	–	151 500	0·695	0·1	·007	140 × 120 × 100	16
I	Mimas	186 000	0·942	1·5	·020	392	12.1
II	Enceladus	238 000	1·370	0·0	·004	510	11.8
III	Tethys	295 000	1·888	1·1	·000	1060	10.3
XIII*	–	295 000	1·888	–	–	34 × 28 × 26	19
XIV*	–	295 000	1·888	–	–	34 × 22 × 22	20
IV	Dione	377 000	2·737	0·0	·002	1120	10.4
XII*	–	377 000	2·737	0·2	·005	36 × 32 × 30	19
V	Rhea	527 000	4·518	0·4	·001	1530	9.7
VI	Titan	1 222 000	15·95	0·3	·029	5150	8.4
VII	Hyperion	1 481 000	21·28	0·4	·104	410 × 260 × 220	14.2
VIII	Iapetus	3 561 000	79·33	14·7	·028	1460	11.0
IX	Phoebe	12 954 000	550	150	·163	220	16.5

*Unofficial number. The satellites also have the following informal designations:
XVII A ring shepherd; XVI Inner F ring shepherd; XV Outer F ring shepherd;
X & XI Co-orbital satellites; XIII & XIV Tethys Lagrangian; XII Dione Lagrangian.
In addition, at least six further objects are unconfirmed, among them those with the informal designations: Mimas Horseshoe; Tethys Lagrangian (2); Dione Lagrangian. The remaining two possible objects are situated between the orbits of Tethys and Dione, and between Dione and Rhea.

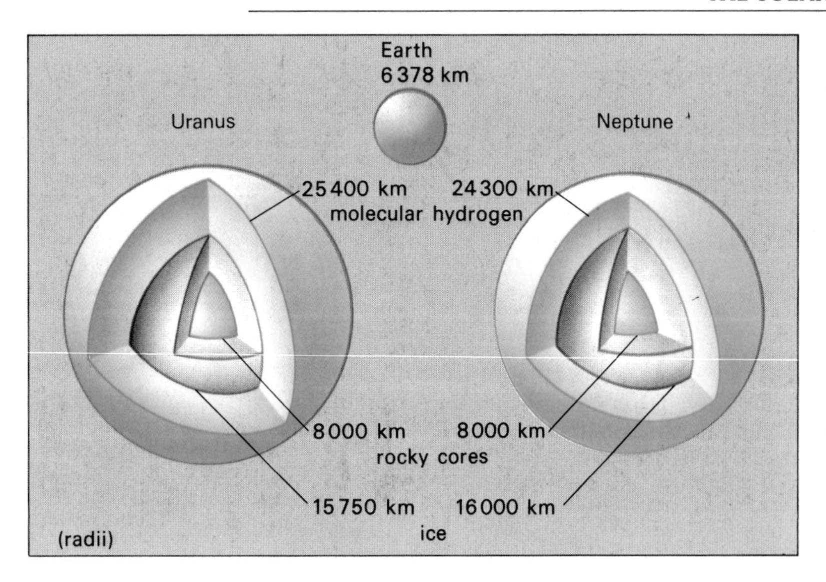

Fig. 5·19 Comparative sizes of Uranus, Neptune and the Earth. The sizes of the internal layers for the two gas giants are only tentative.

a 'horseshoe' orbit where it alternately approaches and recedes from the larger body.

Investigation of the Saturnian system has provided a great deal of insight into the characteristics of bodies which are composed largely of ice, and which probably bear a considerable resemblance to some of the primitive planetesimals and to some of the minor planets. The study of the dynamics of the rings and of some of the minor satellites remains poorly understood, but certain aspects may become clearer when Voyager examines Uranus, a different kind of system.

Uranus and Neptune

Although small in comparison with Jupiter and Saturn, the next two planets are still very large bodies, and the four planets are sometimes known as the 'gas giants'. Uranus and Neptune are very similar in size (equatorial diameters of 50 800 km and 48 600 km respectively) and mass, with Neptune being about 15 per cent more massive than Uranus (Table 5·17). Their characteristics are so similar that they may be conveniently discussed together. The densities of both planets are low and they probably have identical internal structures, with rocky cores surrounded by layers of ice and molecular hydrogen (Fig. 5·19). Unlike Jupiter and Saturn, they have measured temperatures which agree with those calculated for their distances from the Sun, 57 K and 45 K approximately, indicating that they have no major internal sources of heat.

Rotation

There is considerable doubt about the rotation periods of both planets. The cloud layers, which might give an indication of the correct periods, show no distinct features at all and spectroscopic measurements have given apparently contradictory answers. For a long time, periods of 10_h 50_m and 15_h 48_m have been quoted for Uranus and Neptune respectively but recent results suggest that these may have to be revised to approximately 16_h and 18_h.

Uranus is unique among the planets in that its equatorial plane (as shown by the orbits of the satellites) is almost perpendicular to the orbital plane. The axis has apparently been tilted by an angle of about 98°, so that, strictly speaking, the rotation and the orbits of the satellites are retrograde. The cause of this is completely unknown, but it has been suggested that the impact of a very large body could have been responsible. The effect of the axial inclination is that for very long periods of time (Uranus has an orbital period of about 84 years) the poles will face towards the Sun, and calculations indicate that over an orbital period they actually receive more solar radiation than the equator.

Atmospheres

Spectroscopic examination has shown the presence of only two gases in the atmospheres of these planets, methane and molecular hydrogen, although both ammonia and helium are also expected to be present, with ammonia ice forming the principal cloud particles. Neptune has recently been found to show unexpected changes of brightness at infrared wavelengths, and these are thought possibly to be due to alteration in the amount of high clouds, which may be composed of particles of frozen methane or argon.

Although with the apparent detection of radio bursts there had been some suggestion that Uranus had a magnetosphere, this remained uncertain. The discovery that the planet is unusually bright at various ultraviolet wavelengths now suggests that there is some auroral activity, and that most of the radiation comes from excited hydrogen atoms. The only likely mechanism for such excitation is the acceleration of charged particles in a magnetic field, which therefore seems likely to exist on Uranus. The planet's peculiar axial orientation should lead to some unusual features in its magnetotail, and these may be detectable when Voyager 2 encounters the planet in 1986.

Satellites

Uranus has five fairly small satellites, while Neptune has two or possibly three, one of which, Triton, is exceptionally large and has a retrograde orbit very close to the planet. Details of the orbits and possible sizes are given in Table 5·18, although it should be mentioned that the diameters are only rough estimates based upon the apparent magnitudes.

In 1977, observation of the occultation of a star by Uranus led to the discovery that within the orbit of Miranda, the innermost satellite, there were five narrow 'rings'. Subsequent observations, including further occultations, have established that there are actually nine rings located between about 42 000 km and 51 000 km from the centre of the planet. At least two of the rings, and possibly three, appear to be double, or have some form of complicated structure. However, they are all very narrow (unlike e.g. Saturn's rings) with the widest being no more than about 60 km. There are some suggestions that a minor satellite body has been detected in association with one of the rings. Such satellites have been proposed as one means of ensuring the dynamical

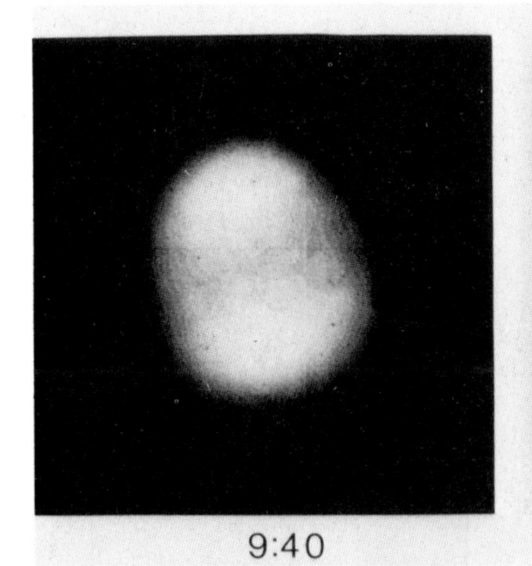

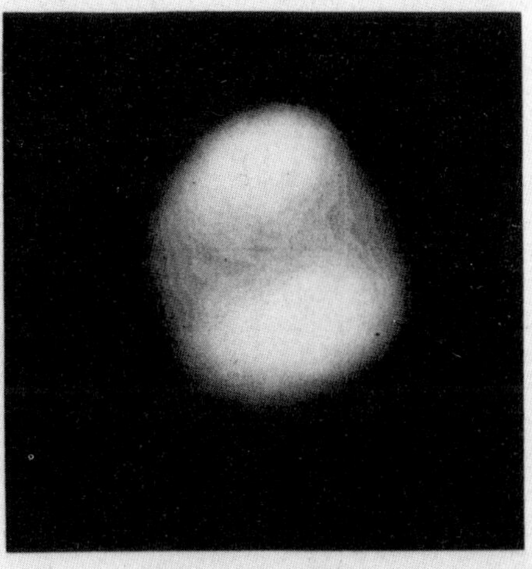

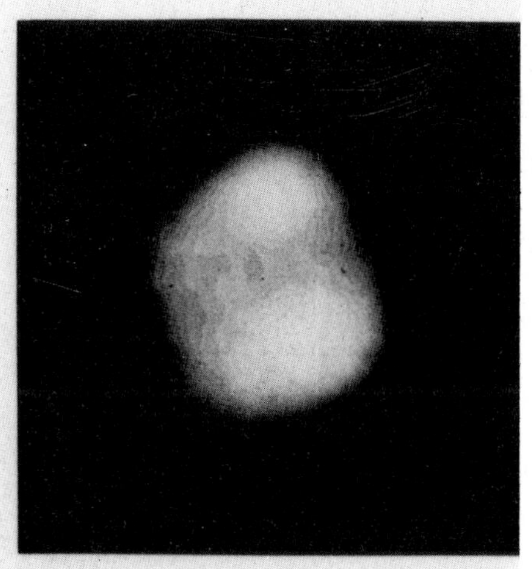

Very few details can ever be detected on Neptune, but these methane-band photographs, taken with CCD equipment (see page 235), do appear to show slight changes in the high cloud cover.

stability of such narrow, low-mass rings. Measurements show that the particles are very dark, having an albedo of less than 5 per cent, so that they are presumably neither formed of ice, nor ice-covered (unlike Saturn's ring particles). Similar measurements have established that the larger satellites of Uranus, although probably composed mainly of icy materials, are also fairly dark, and may thus have been covered in low-albedo material.

The discovery of rings around Jupiter and Uranus as well as Saturn has intensified the search for comparable features around Neptune. As yet there is only rather tenuous evidence that a ring could exist rather close to the planet (between 28 500 and 32 500 km from the planet's centre). However, experiments have produced evidence that Neptune may have a third satellite, perhaps 180 km across and orbiting about 50 000 km from the planet. Confirmation is likely to be difficult as the expected brightness is only around magnitude 20, which is too faint for direct detection so close to the planet. Unlike Saturn's rings which are thousands of kilometres wide, the width of the four inner rings probably does not exceed 10 km. The outer ring may be rather wider, or possibly even be double, and may not lie in the same place as the others. Later results suggest that there may be as many as nine rings around the planet.

Neptune's large satellite Triton also poses some problems. Its highly inclined retrograde orbit is very hard to explain if it is assumed to have formed at the same period as the planet. It would be expected to have a fairly eccentric elliptical, rather than perfectly circular orbit, if it had been captured later. It has been suggested that Pluto could be an escaped satellite of Neptune and that this could have affected Triton's orbit, but this is now considered to be unlikely, especially since the discovery that Pluto has a large satellite. There are conflicting reports about the existence of a methane atmosphere on Triton, although as it is certainly comparable in size with Pluto it seems possible that such an atmosphere could exist. Its confirmation, or evidence for its non-existence, will have to await further research.

Table 5·17 **Uranus – Neptune – Earth comparative data**

	Uranus	**Neptune**	**Earth**
equatorial diameter (km)	50 800	48 600	12 756
sidereal period of axial rotation	16_h?	$18_h 12_m \pm 24_m$	$23_h\ 56_m\ 04_s$
inclination to orbit	97° 53′	28° 48′	23° 27′
density (kg per m^3)	1 270	1 700	5 517
mass (Earth = 1)	14·6	17·2	1·0000
surface gravity (Earth = 1)	1·11	1·21	1·0000
escape velocity (km per s)	22·5	23·9	11·2
albedo	0·93	0·84	0·36
mean distance from Sun	19·181843 au	30·057984 au	

Table 5·18 **Satellites of Uranus and Neptune**

number	name	distance (km)	sidereal period (d)	inclination	eccentricity	diameter (km)	magnitude
Uranus							
V	Miranda	130 400	1·41349	0·0°	0·00	300	16·5
I	Ariel	191 700	2·520384	0·0°	0·0028	800	14·4
II	Umbriel	267 100	4·144183	0·0°	0·0035	600	15·3
III	Titania	438 300	8·705876	0·0°	0·0024	1 100	14·0
IV	Oberon	586 200	13·463262	0·0°	0·0007	1 000	14·2
Neptune							
I	Triton	355 200	5·876844	159·9°	0·000	3 700	13·5
II	Nereid	5 562 000	359·881	27·7°	0·7493	300	18·7

Watching Minor Planets

It might at first be thought that the minor planets offer little scope for amateur study since most of them are quite faint and no detail can ever be seen even in the very largest telescopes. However, although they will never be as widely observed as the major planets, they do have their devoted band of followers. There is a certain challenge in locating and following a faint speck of light as it makes its way against the background stars, and this exercise provides valuable experience in finding one's way around the sky. The positions of some of the brighter objects are given in certain yearly publications, so that the paths may be plotted on suitable charts. ('Suitable' charts in this case implies that they show stars fainter than the expected brightness.

The next stage is to photograph the minor planets, and in its simplest form photography records the change in position from night to night, which can be easily seen on comparing the photographs. Some of the objects which make close approaches to the Earth may have so great a motion that an exposure driven at sidereal rate will show a trail even after a fairly short exposure. However, most such objects are fairly faint, so that moderate apertures are required.

The majority of the orbits of minor planets are reasonably well-known and they are less subject than comets to planetary perturbations. Nevertheless good photographs, especially of some of the fainter Earth-crossing objects are required for positional measurements. Unfortunately the equipment needed to carry out the measurement of 'plates' – usually film nowadays – is complex, and not readily constructed by most amateurs. However a few have built their own measuring engines and usually undertake work on suitable objects for other observers in the national groups, normally doing cometary work as well.

The measurement of magnitudes of these objects is not an easy task as they do not stay still. Locating suitable comparison stars becomes very difficult, especially as, apart from variable star fields, the magnitudes of stars in the general field are poorly determined, even around magnitudes 8–9. Some form of photometry is usually regarded as being most successful, and this can be undertaken with suitable photographs, using variable star techniques. However, more amateurs are turning to photoelectric photometry and achieving notable success, even managing to obtain rotational light-curves.

A completely different study, but of the greatest importance, is the observation of stellar occultations by minor planets. These are not easy to predict, given the uncertainties in the orbits and the inaccuracies in catalogued stellar positions. After a considerable period without success, professional astronomers, particularly Gordon Taylor at the Royal Greenwich Observatory at Herstmonceux, have successfully predicted a number of occultations in recent years. In nearly every case it has proved necessary to secure a photographic plate showing both the star and minor planet in order to make the final corrections, sometimes only hours before the event, to the predicted track of the occultation across the surface of the Earth. Even minor errors in the known positions may cause the tracks to shift by hundreds of kilometres, and, as some of the objects are so small, to miss particular observers completely. True enthusiasts may then pack up their telescopes and rush across the country.

The techniques of observing an event are the same as those used for lunar occultations, consisting of timing the disappearance and reappearance of the star as accurately as possible. If more than one observer has been successful these timings can indicate the size (and shape) of the body, which under favourable circumstances can be very accurate. Indeed this method is the best means of determining the sizes of these small bodies, and has also been used to detect the presence of satellite bodies, both certain and unconfirmed.

Opposite page, top: *The motion of the minor planet Iris* (bottom centre) *is shown on these two photographs taken by H. B. Ridley, West Chinnock, on 1980 September 5* (left) *and September 8.*

Below left: *The observation of an occultation by Pallas on 1983 May 29 accurately determined the minor planet's shape. The length of each line represents the time the star was hidden for one particular observer. (After D. W. Dunham and M. Marr, Maryland.)*

Below: *The photoelectric light-curve of the minor planet Thyra, obtained by R. Miles of Mouldsworth, Cheshire, from observations on 4 nights in 1983 February.*

Minor planets may prove difficult to identify unless their complicated paths are plotted on a suitable atlas.

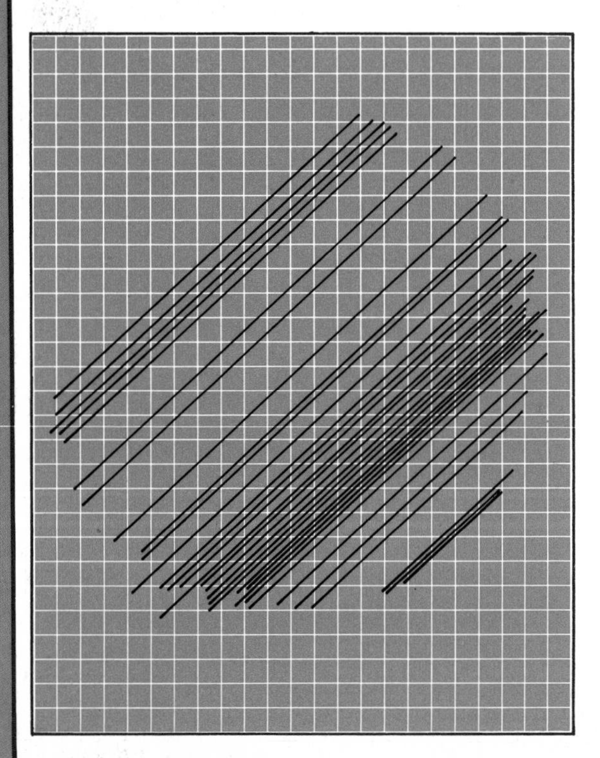

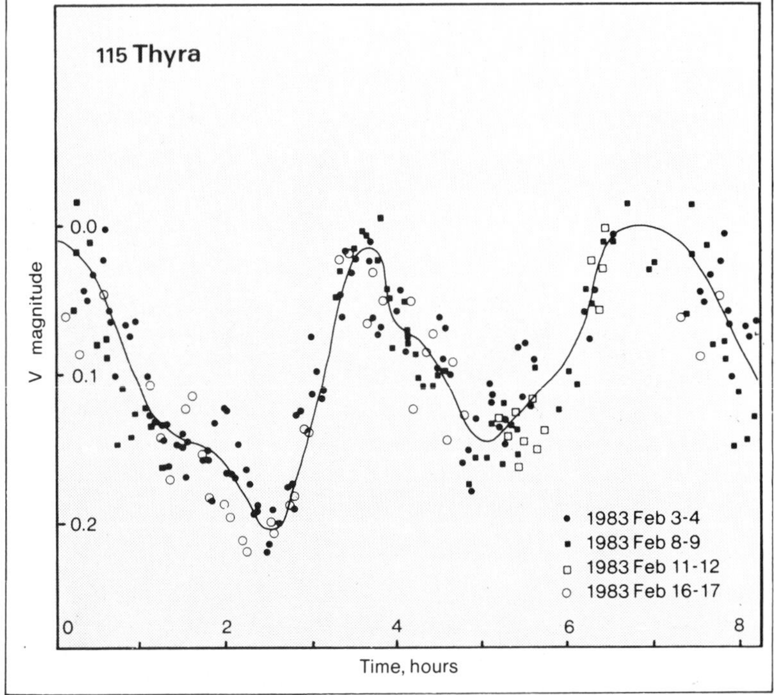

Some bright minor planets

No.	Name	Opposition Magnitude
1	Ceres	7·4
2	Pallas	8·0
3	Juno	8·7
4	Vesta	6·5
5	Astraea	9·9
6	Hebe	7·0
7	Iris	6·7
8	Flora	7·8
9	Metis	8·1
12	Victoria	8·1
15	Eunomia	7·4
18	Melpomene	7·7
20	Massalia	8·2
192	Nausicaa	7·5
324	Bamberga	7·3
387	Aquitania	8·2
433	Eros	7·2

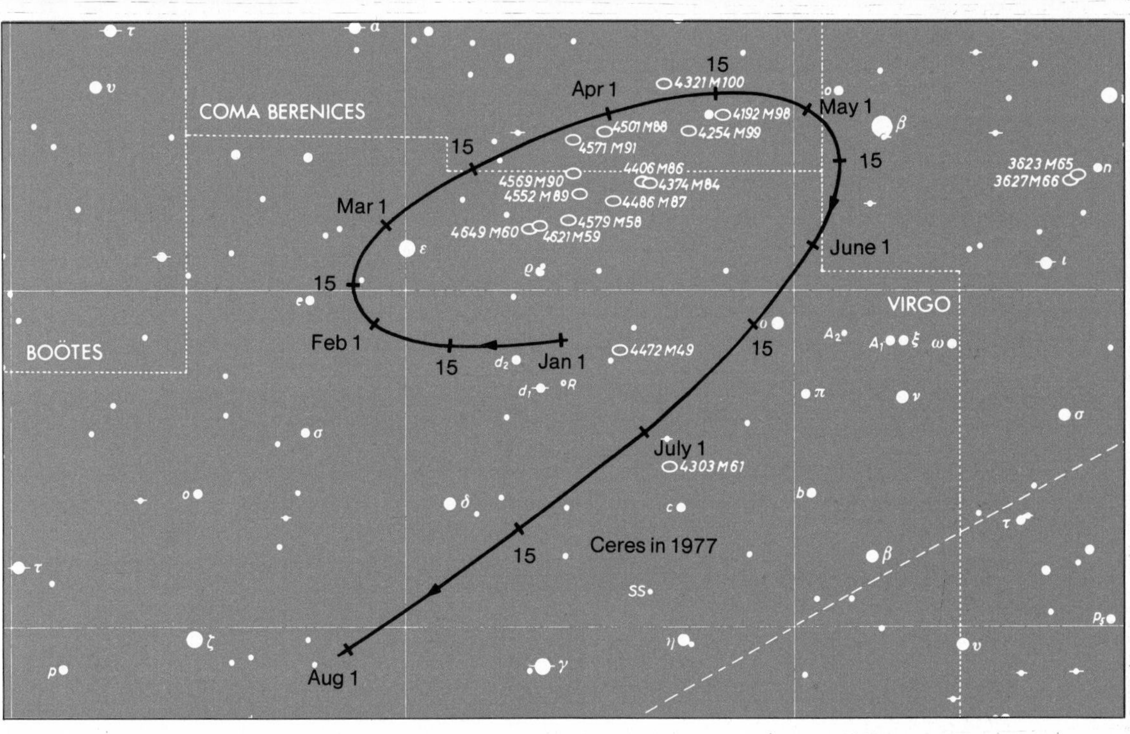

Comet Watching

The discovery of new comets has always been one sphere where amateurs have enjoyed great success; the names of Honda in Japan and Alcock in Great Britain come to mind, as well as the famous Peltier in the U.S.A. – not only for having discovered several comets, but for finding a number of novae as well. Naturally many comets are discovered by professional astronomers, but this is usually during work upon some other project rather than due to deliberate searching. However, there are not many observers who do undertake comet searches, as extreme dedication and an exceptional knowledge of the sky are required. Most searchers use large mounted binoculars or rich-field telescopes (see page 19), and many hundreds of hours of observation are needed before one can begin to remember the star patterns over the whole sky and down to the chosen limiting magnitude. Similarly, great dedication is required to search the sky on every possible occasion, most especially in those regions of the evening and morning skies close to the position of the Sun where comets may creep up and catch us unawares.

Although the orbits of most periodic comets are reasonably well known and predictions usually account for all the planetary perturbations, it is still important that the objects should be monitored at their successive returns. In this way the predictions and orbital elements can be checked and refined. Indeed any information on precise cometary positions is of use, and, needless to say, in new comet discoveries such positional measurements are all-important. Positions are best obtained by photography and the use of proper measuring engines – as mentioned in connection with minor planets (page 148).

Photography normally requires the use of wide-aperture optics. The field must be sufficiently wide to incorporate fairly bright stars having well-determined positions themselves, hence enabling accurate positional information to be derived from the comet. Similarly, a wide-aperture fast system is required to ensure that short exposures – minimizing the effects of the comet's motion – still record the comet. In some cases it may be necessary to guide on the comet itself during the exposure, and this procedure is used for detailed photographs aimed at recording the structure of the head and tail. As comets are diffuse objects this may pose considerable problems, although sometimes the presence of a star-like nucleus helps to lessen the difficulty. Depending upon the individual object it may be possible to arrange for the drive rates on the right ascension and declination axes to compensate for the cometary motion, but in extreme cases such as very fast-moving objects recourse may still have to be made to completely manual methods.

Comets are so varied in their brightness, size and features that visual observation may make use of any form of equipment. At times the full extent of long, faint tails may be best perceived by the naked eye, or at most with low-power binoculars. On the other hand, the perception of fine details such as the jets and shells – 'hoods' – of material being shed by the tiny nucleus may require a large telescope, a keen eye and a skilled observer. Photography just cannot record the finest details near the nucleus because of the usually limited resolution, the normal over-exposure in the central regions of the coma, and the short time-scale of the phenomenon. Drawings by experienced observers remain the major source of such information. Unfortunately, too few observers can gain much experience of comets because bright well-placed objects are so rare. With previously unknown comets the fact that no-one has any idea of the sort of features or activity which they may show, is one of the reasons – apart from the computation of the orbit – why there is such a flurry of activity amongst observers following the announcement.

Another aspect of cometary studies is that of magnitude estimates. Although these follow the general lines of variable star estimates (pages 56–57), there are many additional difficulties. As with minor planet work, one problem is that of obtaining accurate

A 15-minute exposure of Comet IRAS-Araki-Alcock obtained by H. B. Ridley on 1983 May 10, using a 500mm f/6.3 Ross lens.

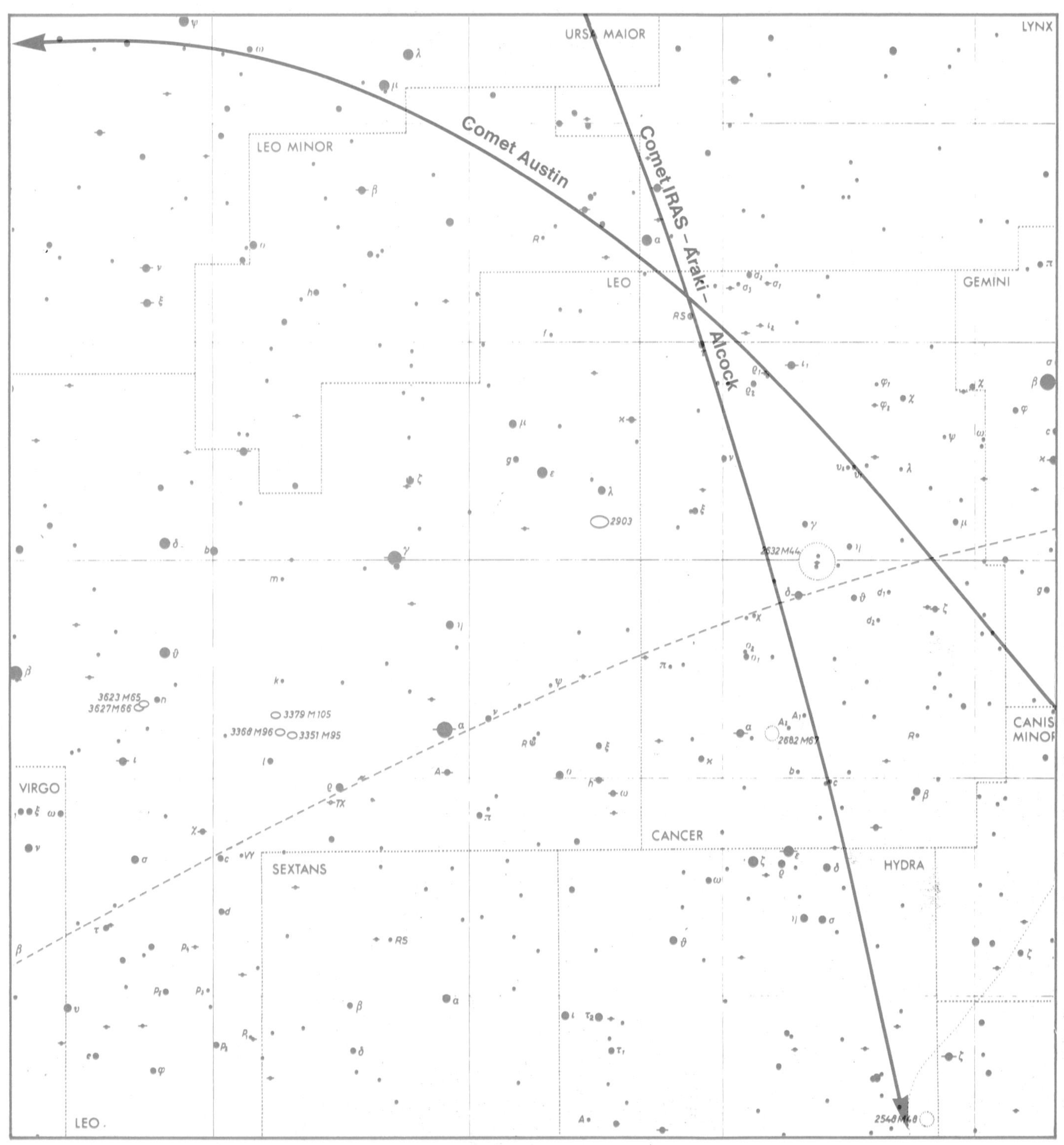

The paths of Comet Austin, 1982g, and the fast moving, close approach Comet IRAS-Araki-Alcock, 1983d.

magnitudes of comparison stars, but the real stumbling-block comes in trying to compare an extended object (the comet) with a point source (a star). Various methods involving the defocusing of the images have been devised, but although these are reasonably consistent within themselves, unfortunately they can never be as accurate as estimates between similar objects. Therefore, cometary magnitudes, especially early ones, remain uncertain overall.

Naturally estimates of a star-like nucleus can be more reliably undertaken, when one is visible. Under many circumstances photographs with ordinary cameras may well be as effective as visual estimates, as happened in the case of the large, close, and very fast-moving Comet IRAS Araki Alcock (1983d). Certain normally faint comets, particularly Comet Schwassmann-Wachmann 1, can undergo sudden outbursts of brightness, so observations are required on every possible occasion.

The general unpredictability of comets means that, as with other 'sudden' events such as novae and supernovae, information must be passed to the observers as rapidly as possible. The official channel through which this happens is the International Astronomical Union's Center for Astronomical Telegrams in Cambridge, Massachusetts, as well as the various individual countries' telephone and postal alert networks. When a cometary return is expected special watches may be mounted. The most significant example of this will come with the return of the Comet Halley in 1985/6, for which an International Halley Watch is being organized, involving professional and amateur astronomers around the world.

Pluto

The four inner planets have many common characteristics, as do the four gas giants, but Pluto, the outermost planet of the Solar System, seems to have little resemblance to any of them. Its mean distance from the Sun is so great (39·4 au) that it is very difficult to study and the few details which we have are given in Table 5·19. It has not been possible to determine the diameter directly, although from occultation results in 1965 it could not exceed 6 800 km. Indirectly it may be estimated from observations that the planet is covered in methane frost. Such a surface layer would have an albedo of 40–60 per cent, which, from the known brightness, leads to a diameter of 3 300–2 800 km, if it is completely frost-covered.

In the past, only very rough estimates of the density and mass have been possible. Despite the fact that Pluto was discovered in 1930 close to the predicted position calculated from the assumption that an unknown planet was perturbing Uranus, these calculations now seem to have been erroneous, and Pluto apparently has little effect on Uranus and Neptune. With the discovery in 1978 that it was accompanied by a large satellite, since named Charon, the situation has become much clearer. The total mass of the whole planet/satellite system now seems to be only about one-fifth of that of the Moon alone, and the Pluto:Earth mass ratio is certainly lower than 1:400. The diameter of Pluto appears to be close to 3 000 km, while Charon could be as large as 1 400 km, and probably orbits at a distance of approximately 20 000 km. The density of both bodies is very low, probably of the order of 1 000–2 000 kg per m³, and it is almost certain that both are composed nearly exclusively of volatile materials. Charon is not fully resolved, being so close to Pluto (photographs merely appearing elongated) but its period seems to be 6·39 days, probably synchronized with Pluto's own rotational period which had been inferred from a variation in brightness. A longer-term change in the planet's magnitude almost certainly related to the position in its orbit suggested that the inclination of the axis was greater than 50°. Charon has a highly inclined orbit, and could well be orbiting in the plane of the planet's equator.

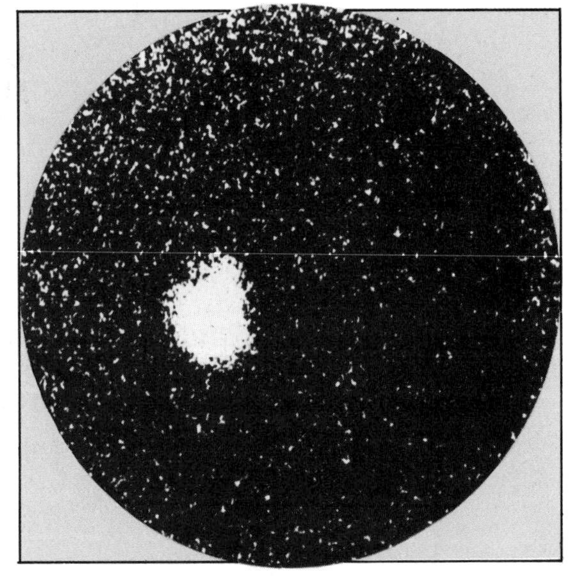

A photograph showing the combined images of Pluto and its satellite which orbits from north (top) to south. The orbital plane is highly inclined to the plane of the sky. If the satellite has an albedo similar to Pluto's, it may have a diameter as great as 1 200 km and be the largest satellite in the Solar System relative to its primary.

Table 5·19 **Pluto**

equatorial diameter (km)	3 300–2 800?
sidereal period of axial rotation	?
inclination to orbit	50°?
density (kg per m³)	2 000?
mass (Earth = 1)	0·0025?
surface gravity (Earth = 1)	?
escape velocity (km per s)	?
albedo	0·40–0·60?
mean Sun-Pluto distance	39·4 au

Although it had been long suspected that Pluto had no atmosphere, and that the surface temperature would lead to any gases being permanently frozen, there is now some evidence that gaseous methane does exist. Uncertainties arise because the mass of the planet is poorly established, and hence the surface gravity and surface pressure are unknown. It is quite possible that Pluto only has a gaseous atmosphere when it is close to perihelion, and that at other times it is frozen out on to the surface. With certain assumptions it can be shown that with a surface temperature of 60 K a gaseous atmosphere of methane could just exist.

Orbit

The planet's orbital inclination and eccentricity are far greater than those of any other planet, including Mercury, and for part of its orbit the distance from the Sun is less than that of Neptune. In order to explain both Triton's highly inclined retrograde orbit and Pluto's unusual characteristics, it has been suggested that Pluto was once a satellite of Neptune. Tidal interaction between the two satellites could have placed Triton in its present orbit and caused Pluto to escape from Neptune's gravitational field. However, objections to this theory are that the two orbits do not intersect, that the closest approach Pluto makes to Neptune today is still a distance of several astronomical units, and that the circularity of Triton's orbit remains difficult to explain.

This small wanderer seems to have practically nothing in common with the rest of the planets and, indeed, with the discovery of object 1977 UB (Chiron), which will be discussed with the minor planets, it has been pointed out that it and Pluto have many similarities to certain minor planets and comets.

With confirmation that Pluto is of low mass, the problem of the motions of Uranus and Neptune remains unresolved. There seems to be a good chance that there is some body, as yet undiscovered, with sufficient mass to cause the observed perturbations. The suggestions as to what this could be are: another planet, a dark stellar companion to the solar System, and inevitably, a black hole! A planet would have to be quite large and massive, and even if it were in a highly inclined orbit would have been unlikely to escape detection. The most probable candidate body is another star, perhaps a 'black dwarf'. Discovering such a body, which could be located at quite a distance from the Sun, is likely to be very difficult, although it may possibly be feasible to detect any gravitational influence on the two Pioneer spacecraft (in particular) which are now heading out into interstellar space in approximately opposite directions. For as long as they continue to operate it will be possible to determine from the Doppler shift of their radio signals whether they are being subjected to any additional gravitational influences.

The Minor Planets

The minor planets are probably the least generally known members of the Solar System. Only one body, Vesta, occasionally becomes bright enough to be seen by the naked eye, and nearly all the remainder are very small faint objects, which have been found photographically. However, the earlier members were discovered visually, the first being Ceres, noted on 1 January 1801 by Piazzi, who was compiling a star catalogue. Its orbit fitted in with the 'missing' planet of the Titius-Bode 'law' (see p. 28), but it was soon realized that it was very small and another three objects had been found by the end of 1807. The term 'asteroid' was introduced by William Herschel in 1802 to describe their star-like appearance, and this word has continued in use, although 'planetoid' and 'minor planet' describe them more correctly.

About 2 000 orbits are sufficiently well-known for the objects to have received permanent identification numbers and names, and there are some 1 000 others which have temporary designations. The latter consist of the year of discovery followed by two letters (and occasionally an additional number). Examples are 1976 AA (a planetoid which crosses the orbit of the Earth) and 1977 UB (Chiron, which will be discussed later). When positive identification is certain, they are given individual numbers and usually named by their discoverers. Examples are 1 Ceres, 4 Vesta and 1566 Icarus. Statistical calculations based upon the observed sizes suggest that the total number runs into several hundreds of thousands.

Orbits

The majority of the minor planets have orbits which lie between those of Mars and Jupiter (Fig. 5·20), in a belt from 2·2–3·3 au. Some significant groups lie outside this belt, but within it the distribution is uneven (Fig. 5·21), with distinct peaks and depressions at particular distances. The dips are known as Kirkwood gaps after their discoverer and are due to gravitational perturbations by Jupiter.

Under certain circumstances, groups of planetoids can be locked into simple orbital relationships with Jupiter. The most important of these is the group which has the same orbital period as Jupiter itself; its members are known as the Trojan planetoids. Their stable positions are 60° ahead of, and 60° behind Jupiter, positions which are known as Lagrangian points after the mathematician who predicted their existence. Since the minor planets do not have completely negligible masses, they are perturbed by the planets and one another, so that they actually oscillate about the theoretical positions (Fig. 5·20). For some unknown reason the Achilles group, which leads Jupiter, has about twice as many members as the Patroclus group, which follows the planet.

All the minor planets show direct orbital motion (unlike the comets, many of which have retrograde orbits), but they generally have greater eccentricities and orbital inclinations than the major planets. Table 5·20 gives data on some of their orbits. A few planetoids have very large orbits; 944 Hidalgo, for example. The object known as 1977 UB (Chiron) is truly remarkable as its orbit ranges from 8·5 au (inside that of Saturn) to 18·9 au (close to Uranus' mean distance of 19·2 au). The nature of this object is obscure, as its distance and orbit might suggest a cometary nature but its apparent size is rather too great for this. However, investigations of Chiron's orbit by means of computer modelling have shown

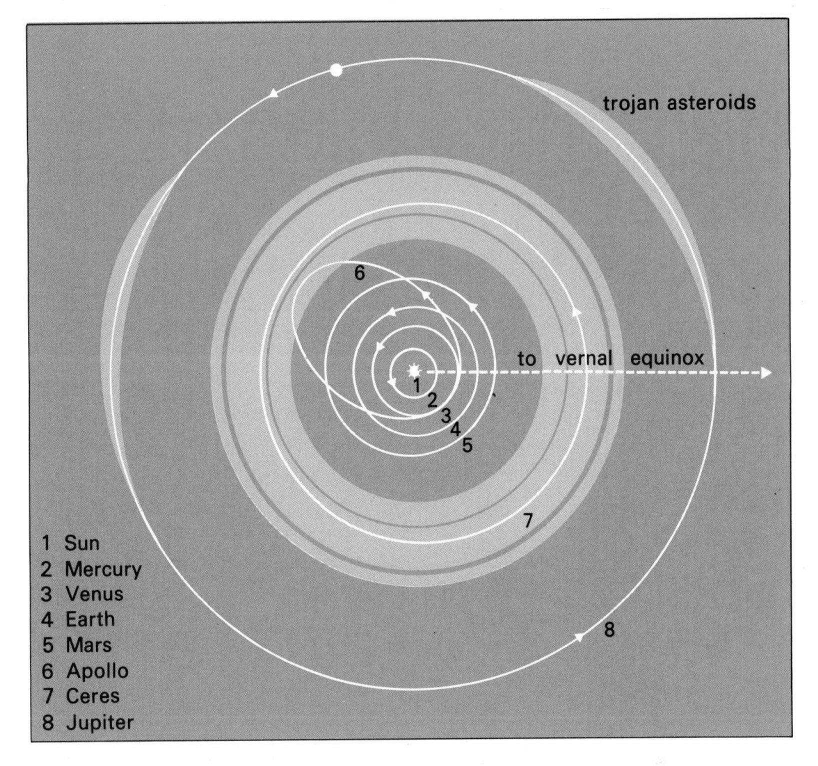

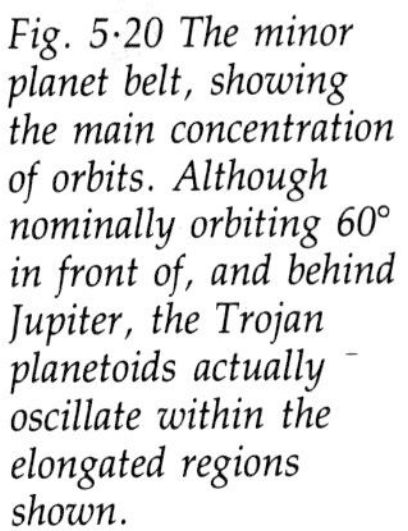
Fig. 5·20 The minor planet belt, showing the main concentration of orbits. Although nominally orbiting 60° in front of, and behind Jupiter, the Trojan planetoids actually oscillate within the elongated regions shown.

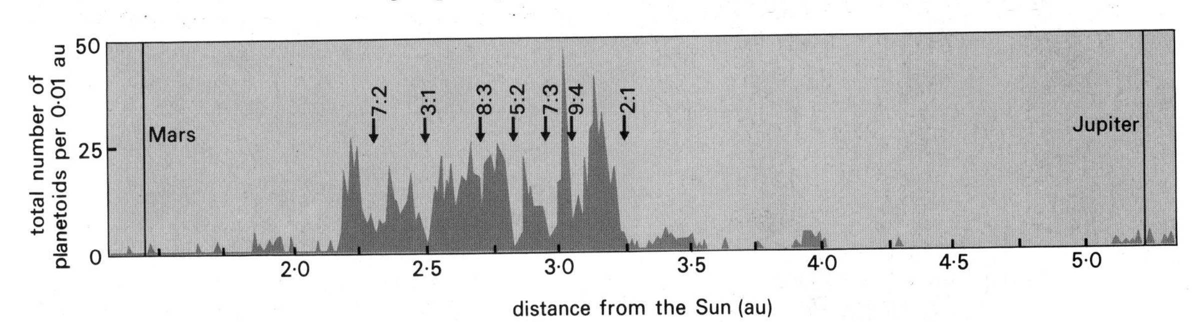

Fig. 5·21 The distribution of the minor planets. The arrows indicate distances at which bodies have orbital resonances with Jupiter, the number of orbits made by the planetoid being given first. The Trojan objects cluster about Jupiter's orbital distance.

Table 5·20 **Minor planet orbits**

number	name	distance (au) mean	distance (au) perihelion	period (years)	inclination	eccentricity	diameter (km)
1	Ceres	2·7663	2·5488	4·6012	10·604°	0·07863	1003
2	Pallas	2·7687	2·1136	4·6069	34·848°	0·23662	540
4	Vesta	2·3619	2·1528	3·6301	7·137°	0·08851	538
433	Eros	1·4581	1·1333	1·7607	10·828°	0·22286	23*
532	Herculina	2·7728	2·2878	4·6173	16·340°	0·17493	240*
588	Achilles	5·2112	4·4384	11·8964	10·316°	0·14829	53
944	Hidalgo	5·8201	1·9991	14·0413	42·494°	0·65651	16
1566	Icarus	1·0777	0·1868	1·1188	22·994°	0·82667	1
1862	Apollo	1·4697	0·6468	1·78	6·360°	0·55988	?
1973 NA		2·4470	0·8796	3·83	68·056°	0·64053	?
1975 YA		1·2901	0·9054	1·47	64·013°	0·29821	?
1976 AA		0·9664	0·7899	0·95	18·935°	0·18255	1
1976 UA		0·8440	0·4643	0·76	5·852°	0·44983	0·3?
1977 UB	(Chiron)	13·6991	8·5126	50·70	6·923°	0·37860	150–650?

*maximum (equatorial) diameter

that it is not in a stable orbital path, and that at some time in the future it is likely to be ejected from the Solar System. This will be because of interaction with Jupiter or Saturn, and it would also appear that it was originally captured by one or other of those two planets. This would support the idea of the body being more like a comet than a minor planet from the main belt.

Inside the main belt there are about twenty bodies which cross the orbit of the Earth, and which are called the Apollo group after the name of the first to be discovered. Two objects, 1976 AA and 1976 UA, have orbital periods less than that of the Earth and one body, 1566 Icarus, has a perihelion distance of only 0·186 au, inside the orbit of Mercury (0·308–0·467 au). Because they may closely approach the Earth, some very small, faint bodies are known in this group, including the minor planets 1973 NA and 1975 YA with the greatest orbital inclinations, approximately 68° and 64° respectively.

The discovery photograph of 1977 UB (Chiron), a 75 min. exposure with the 1·2 m Schmidt telescope at Mount Palomar. The short, bright trail of 1977 UB (arrowed) may be compared with that of a closer, more typical minor planet which is seen towards the upper right.

Sizes and composition

For a long time, approximate diameters were known for only a few of the largest minor planets. However, modern techniques allow albedos to be determined and, with measurements of apparent magnitude, it is possible to calculate sizes. By this means the diameters of some 200 objects have been established, ranging from about 1 000 km for Ceres down to 1 km for 1976 AA, while 1976 UA may be only a few hundred metres across. At least fourteen have diameters of 250 km or greater and most of these orbit in the outer part of the belt. Although the older theory that all of the minor planets originated from the break-up of a single object is now discounted, calculations suggest that there have been many collisions between the bodies and that probably only the three largest, Ceres, Pallas and Vesta may be substantially unaltered. Fragmentation of individual bodies is considered to account for the observed families of minor planets with similar orbits and characteristics.

The majority of the planetoids show variations in their magnitudes implying that they are irregular and reflect varying amounts of light as they rotate; periods of about 2·3–18·8 hours having been found. From various results, 433 Eros has been found to have diameters of about 10, 15 and 36 km, thus showing similarities to the Martian satellites in shape. Study of the results of an occultation in 1978 show that 532 Herculina has probable equatorial and polar diameters of approximately 240 and 210 km and is accompanied, at a distance of about 975 km, by a small satellite body of 45–50 km diameter.

The most important result of the determination of albedo is the fact that the majority of bodies fall into two main groups, with the larger (88 per cent) having very dark grey surfaces which reflect from 5 per cent to as little as 0·02 per cent of light. Spectral measurements show that the material is similar to the type of meteorite known as a carbonaceous chondrite (see p. 158). The other major group (1 per cent) is reddish, has a higher albedo of 10–20 per cent and corresponds to silicate material, like that of the stony-iron meteorites. The distribution of orbits shows that the silicate bodies are most numerous at the inner edge of the planetoid belt, and the larger carbonaceous type at the outer edge. Spectra of many planetoids do not resemble any meteorite type, suggesting that known meteoroids may be derived from only a few original bodies. However, some of the rarer meteorites can even be identified with individual minor planets from which they have presumably been fragmented.

Various studies, including the detection of radar echoes from some of the larger bodies, are gradually establishing the distribution of the different types of body and of their possible composition. It now seems likely that in the initial stages of the formation of the Solar System (page 160), many icy bodies formed in the outer regions, and that some of these have been captured to form the icy satellites of Jupiter and Saturn (and presumably of the other outer planets as well), while many were ejected completely from the Sun's gravitational influence. Some bodies were injected into very elongated orbits, and these are likely to be the comets which we see today.

Comets

The appearance of a great comet is undoubtedly one of the most striking celestial phenomena which can be seen by the naked eye. The fact that the occurrence of comets is generally unpredictable, and that they may suddenly appear as large and prominent objects, only serves to make them the more remarkable. However, it is now known that vast numbers are so faint that they can only be observed with the largest telescopes and that many others must be missed because, as viewed from the Earth, their orbits are too close to the Sun for them to be seen. Study of their frequency and their orbits suggests that comets in the Solar System number many millions.

Like the minor planets, so many comets are known that a special classification has been introduced. It is usual for the name of the discoverer (or discoverers, up to a maximum of three) to be given to a comet, although this is sometimes varied, as in the cases of Comets Halley, Encke and Crommelin, by using the name of the person who has made extensive orbital calculations. Further identification is given by the year and a letter awarded in order of discovery. If the comet is found to be periodic the letter P is used as a prefix. At a later date, a final designation is given in accordance with the date at which the various comets came to perihelion by giving the year in which this occurred and a Roman numeral. For example Comet Bennett 1969i (the ninth to be found in 1969) became Comet Bennett 1970 II (the second to pass perihelion in that year). Recoveries of periodic comets are included in this scheme and, occasionally, when a comet has been lost for a considerable time, the name of the rediscoverer is added to that of the original finder, as with Comet P/Perrine which was not recovered for five periods, and was then found by Mrkos. The object is now known as Comet P/Perrine-Mrkos.

Orbits and periods

The vast majority of comets which have been fully studied have closed (elliptical) orbits and are thus true members of the Solar System. Some orbits are so eccentric that, initially, for computational purposes, they may be treated as PARABOLIC. Calculation shows that those comets with open HYPERBOLIC paths which are escaping from the Sun's influence have been perturbed by the planets (especially Jupiter). Despite the fact that comets have been lost by the Solar System, none have been observed to enter it from interstellar space along hyperbolic paths. There are indications that some comets have aphelia at many thousands of astronomical units, perhaps even halfway to the nearer stars, but accurate determination of aphelion distances and periods in such cases is extremely difficult. It has been suggested that a large reservoir of comets exists at such great distances from the Sun, and that they are occasionally perturbed by the gravitational effects of nearby stars into orbits approaching the centre of the Solar System.

Comets may approach the Sun from any direction, but there are indications that the major axes of the orbits are concentrated towards the **galactic plane** (see p. 166) and that aphelion positions are clustered around the solar ANTAPEX. The orbital planes may have any inclination. However, the orientations in space are not completely random, as there are definite groups of objects which may be recognized from their similar orbital characteristics. This is particularly noticeable in the case of the so-called Sun-grazing comets, which have perihelion distances of 0·01–0·005 au, and are thus well within the Sun's corona (page 76). Since 1979 satellite-borne equipment has revealed three comets which have collided with the Sun, or at least have approached so close that they have been completely disrupted and have not appeared after perihelion. None of these objects was visible from the Earth, and the fact that three have been detected in such a short span of time suggests that similar comets are much more

Comet West, 1975n, photographed on 1976 March 13, showing the broad and relatively featureless dust trail above, and the narrow, blue, finely structured gas (plasma) tail below.

Comet Ikeya-Seki 1965 VIII, a typical Sun-grazing comet, photographed shortly after perihelion passage by Alan McClure on 1965 November 1, from Mount Pinos, California

numerous than previously thought, although there is the possibility that such close-approach comets tend to occur in groups. Both these 'colliding' comets and other Sun-grazers follow practically identical orbits and are thus probably fragments of individual objects.

A distinction may be drawn between cometary periods which are of thousands or even millions of years and those which are comparable with the planets'. The division is somewhat arbitrary, but is generally taken as being at about 200 years, and there are suggestions that planetary influences are responsible for a large number of the short-period orbits. Comets which have periods equal to, or shorter than Jupiter's have been captured from longer-period orbits by the planet, which has also caused a gap in the periods, similar to those in the planetoid belt and Saturn's rings.

Table 5·21 gives some details of typical cometary orbits.

Appearance and composition

At great distances from the Sun, comets are small, faint, indistinct objects, but as they approach, volatile materials begin to be vaporized to produce the main head or **coma**. At times this may become exceptionally large; in the Great Comet of 1811, for example, the diameter exceeded 2×10^6 km (nearly one-and-a-half times that of the Sun). The tail may begin to develop at a considerable distance, as happened with Comet Schuster 1976c, which has the greatest perihelion distance known (6·882 au, that is, beyond Jupiter) but which nevertheless had a moderate tail. Tails may not only be highly conspicuous (*left*), but also exceptionally long; that of the Great Comet of 1843 (a Sun-grazer) had a length of about $3{\cdot}2 \times 10^8$ km, considerably greater than the mean distance of Mars from the Sun. At times, a small star-like point, or **nucleus**, may be seen in the centre of the coma, and, occasionally, multiple nuclei develop. A further feature, which can only be observed from spacecraft, is a vast hydrogen halo surrounding the visible portions of the comet.

Despite their apparent size even the largest comets may have total masses of only 5×10^{16} kg (less than one-millionth of that of the Moon). Observations of recent comets confirm that the main parts are small, with diameters of the order of 10 km, and that they are concentrations of ice and dust particles – graphically described as 'dirty snowballs'. Variations in the rate at which the volatile materials are vaporized and gases released easily account for the observed changes in brightness of the head and in the structure of the tail.

Many molecules can be detected in cometary spectra, including water vapour and molecules such as hydrogen cyanide (HCN) and methyl cyanide (CH_3CN), which are also found in interstellar space (see p. 182). In the early stages of a comet's approach the light is reflected and scattered sunlight, but later the appearance of emission lines indicates that various elements have been vaporized by solar radiation. Within cometary tails there are usually two distinct components, one of which is ionized gas and its direction is controlled by the solar wind, while the other is usually comparatively featureless and consists of electrically neutral gas and dust.

The mass lost by a comet may amount to as much as 1 per cent per orbit, and short-period comets are seen to become noticeably and progressively fainter. Some even lose most of their volatile materials. This implies that even long-period comets cannot have followed their present orbits since the formation of the Solar System and that they are comparatively recent introductions.

Cometary particles are known to be responsible for many meteor showers, but the density of a cometary tail is so low that passage of the Earth through it (as happened with Comet Halley in 1910) is not likely to produce any observable effects. On very rare occasions, collision with the main body of a comet may be expected, and this is almost certainly the explanation for the brilliant fireball and immense explosion which occurred on 30 June 1908 in the Tunguska River area of Siberia. Trees were uprooted as far away

Table 5·21 **Cometary orbits**

name and designation*		perihelion distance (au)	eccentricity	period† (yr)	inclination
Gt. March Comet	1811 I	1.035412	0.995124	—	106.9397°
Gt. Comet (Flaugerges)	1843 I	0.005527	0.999914	513	144.3484°
P/Halley	1910 II	0.587211	0.967298	76.1	162.2160°
Stearns	1927 IV	3.683902	0.998179	—	87.6574°
P/Temple-Tuttle	1965 IV	0.981730	0.904396	32.9	162.7092°
P/Perrine-Mrkos	1968 VIII	1.272212	0.642630	6.72	17.7619°
Bennett	1970 II	0.537606	0.996193	—	90.0437°
P/Encke	1974 —	0.338125	0.847450	3.30	11.9820°
West	1975n	0.196630	0.999955	—	43.0710°
Schuster	1976c	6.882188	1.0‡	—	112.0176°

* Periodic comet designations are those of the last (or a recent) return. P/ before a comet's name denotes that this is a short-period comet, i.e. one whose period is less than 200 years. Apparitions of P/Halley can be traced back to 86 BC, and P/Halley and P/Encke have been well observed since 1682 and 1786 respectively.

† Periods are not given where they exceed 1000 years. ‡ Assumed parabolic orbit.

as 40 km, and the pressure waves recorded as far off as the British Isles, while vast quantities of meteoric dust remained suspended in the upper atmosphere for months. Yet no major fragments have ever been found, only microscopic iron and silicate particles having been recovered from the soil. These results are consistent with an encounter with the head of a small comet largely composed of ice and small solid particles.

Meteors and meteorites

At the end of the eighteenth century it was finally recognized that meteors (the so-called 'shooting stars') were produced by small bodies burning up in the Earth's atmosphere, and that meteorites were remnants which had survived to reach the Earth's surface. These terms have continued to be used and recently **meteoroid** has become widely employed for particles in space, while **fireballs** are meteors brighter than magnitude − 5.

Visual observations can give useful information about the number and direction of meteors, but increasing reliance is now placed upon photographic and radar methods of detection. Although the mass of a meteor is about 0·1–1g, in passing through the atmosphere it forms a trail of ionized gas which is an efficient radar reflector and may be detected during the day, giving valuable information about meteor rates which would otherwise be unobtainable. All three types of observation can give information about the height at which meteors occur (80–50 km), while radar and photographic techniques enable the velocity of some meteors to be obtained. From information about the direction and speed of meteoroids, their orbits may be established, and in every case these have proved to be elliptical, showing that they originate in the Solar System.

Meteor showers

Meteors are noticeably more frequent at certain times of the year, when they form meteor showers. These objects have tracks which appear to diverge from a single small area of the sky, known as the **radiant**, but this is an optical effect; they are actually travelling along parallel paths. The showers are usually named from stars close to the radiant position, or the constellation in which this is situated, as with the κ Cygnids and the Perseids. Examination of the orbits of showers shows that they are similar to those of short-period comets; the Bielids, for instance, following the path of the lost Comet P/Biela. The shower meteoroids are composed of cometary debris spread out along the parent cometary orbits and the showers occur when the Earth passes through the meteoroid streams. However, planetary perturbations can cause rapid changes in these orbits, causing sudden alterations in the number of meteors encountered by the Earth. Moreover, variations in meteor rates occur due to differences in density in various parts of the streams, while meteor storms may be observed if the Earth intercepts a compact cluster of particles recently derived from the parent comet. Such an event happened on 17 November 1966 when the Leonid rate became as high as 150 000 per hour, due to a dense cloud derived from Comet P/Tempel-Tuttle 1866 I about 100 years earlier.

Sporadic meteors and micrometeorites

Non-shower meteors occur at all times of the year with a rate of about eight per hour, and are known as **sporadics**. Very occasionally their analysis has revealed a hitherto unknown radiant and shower, but the majority are clearly the widely dispersed debris of extinct comets. A similar origin is responsible for the minute particles known as **micrometeorites**, which are so small that they do not burn up when they encounter the Earth's atmosphere. Their deposition rate over long periods of time may be studied by examination of cores from ocean sediments and polar ice caps.

Vast numbers of such tiny particles exist in interplanetary space, reflecting sunlight and producing the **zodiacal light**, which is a diffuse, weakly luminous area of sky, appearing ellipse-shaped and centred on the Sun. It is only visible under favourable conditions in the evening after the end of twilight, or in the morning before the dawn. As the ellipse lies along the ecliptic its visibility varies greatly with the seasons and the observer's geographical latitude. On occasions, a luminous area known as the **gegen-**

schein or 'counterglow' can be seen directly opposite the Sun's position and, under exceptionally good conditions, a faintly luminous band may be observed joining the brighter regions. The total mass of the zodiacal light particles has been calculated at 3×10^{16} kg (comparable with that of a single comet).

Meteorites

A large number of meteorite craters are now recognized on Earth (Table 5·22), although very big bodies causing explosion craters or events such as the Siberian impact of 1908 (pages 156–7) are rare, with perhaps no more than one or two bodies falling on the Earth's land area per century. However, about 100–150 smaller meteorites per year should fall on land, although only about ten of these are usually recovered. In recent years thousands of well-preserved meteorites have been collected in Antarctica, and these have very substantially increased the total quantity of objects available for study. Moreover, it has been possible to collect these meteorites under sterile conditions, in the knowledge that they are free from terrestrial contamination. This is particularly important when dealing with carbonaceous chondrites.

There are three main types of meteorite, described as irons, stony-irons and stones, and from objects which are seen to fall it is possible to state that their percentages are about 6, 1 and 93 respectively. However, the stones have a greater tendency to fragment, are less resistant to weathering and are more difficult to recognize, so that iron meteorites are more frequently recovered. Large numbers of irons are known, ranging in size up to the Hoba West meteorite, which is estimated at 60 000 kg. In contrast to this, only two stony-irons exceed 1 000 kg, the Bitburg, Germany (1 500 kg) and Huckitta, Australia (1 400 kg) meteorites. The largest stone is the Kirin, China 1976 meteorite, weighing 1 770 kg, and only two other falls are known to exceed 1 000 kg.

Composition and origin

The average composition of iron meteorites is about 98 per cent iron-nickel, where the nickel may range from about 4–30 per cent or more, and they have densities of up to 7 900 kg per m^3. The stones are rich in various forms of magnesium and iron silicates, frequently contain a distinctive form of iron sulphide and have densities ranging from 3 450–3 800 kg per m^3, while the comparatively rare stony-irons have intermediate compositions. Stony meteorites have two important sub-classes, **chondrites** and **achondrites**, based upon the presence or absence of tiny, approximately spherical features within the body, which are known as **chondrules**. A rarely recovered but very important type of meteorite is the class designated carbonaceous chondrites, which have densities of 2 200–2 500 kg per m^3. These contain a significant amount of organic (carbon-based) compounds but research has shown that these have not been formed by biological activity. As a class the carbonaceous chondrites have a composition close to that of the Sun, and radio-isotope dating and mineralogical techniques show that they are very ancient and have undergone comparatively little chemical alteration. It is now generally accepted that they are fairly representative of the less volatile material from which the planetary bodies were formed. However, they have been subject to certain chemical changes which suggest that water was present at some time to a significant degree, and that they once formed part of some small bodies (perhaps like minor planets or even comets) from which they were fragmented early in the lifetime of the Solar System. The presence of water (and presumably other volatile materials) at such an early stage supports the general theory of a 'cold accretion' from interstellar material, rather than condensation from a heated nebula.

As far as the other forms of meteorite are concerned, the various minerals have not been produced under very high pressure and temperatures, again indicating that the parent bodies were similar in size to the minor planets. Chondrites appear to have been subjected to temperatures of about 1 100 K, while in other bodies a higher temperature of about 1 700 K would have caused melting and chemical separation similar to those produced in planetary interiors. Later breakup of such bodies would give rise to the achondrites, the stony-irons and the irons. Mineralogical studies suggest that iron meteorites have originated in about ten different masses comparable in size to the larger present-day minor planets (250 km or more), and that there were a similar number of parent bodies for the chondrites, which had rather smaller diameters.

In only one case amongst all the known meteorites is it possible to suggest a body of origin with any degree of certainty. This particular Antarctic meteorite is a quite distinct form, being a breccia (page 107) containing anorthosite, a material highly characteristic of the Moon's highland regions. Theoretical studies show that in certain impacts it should be possible for a very tiny fraction of the ejecta to be given velocities greater than that of escape for the Moon, and that such particles could come from a reasonable depth below the lunar surface. The fact that the meteorite is essentially indistinguishable from any of the lunar samples strongly supports its lunar origin.

An even more surprising body of origin has been suggested for a rare class of meteorites known collectively as the shergottites. These quite closely resemble basaltic lavas, and have been found to have radiometric ages of no more than $1{\cdot}3 \times 10^9$ years – very young for meteorites, and certainly most unlikely to have come from the Moon or the minor planets. A possible body of origin is Mars, although it is difficult to see how an impact could have accelerated any fragments to a sufficiently high velocity for them to escape from the planet's gravitational field, which is much greater than that of the Moon. Their peculiarities, and their source of origin, remain highly problematic.

Some meteoroids which have resulted in very bright fireballs and meteorite finds, such as the Pribram (Czechoslovakia), Lost City (Oklahoma) and Innisfree (Alberta) objects, indicate a close connection with the minor planets (Fig. 5·22). However, the vast

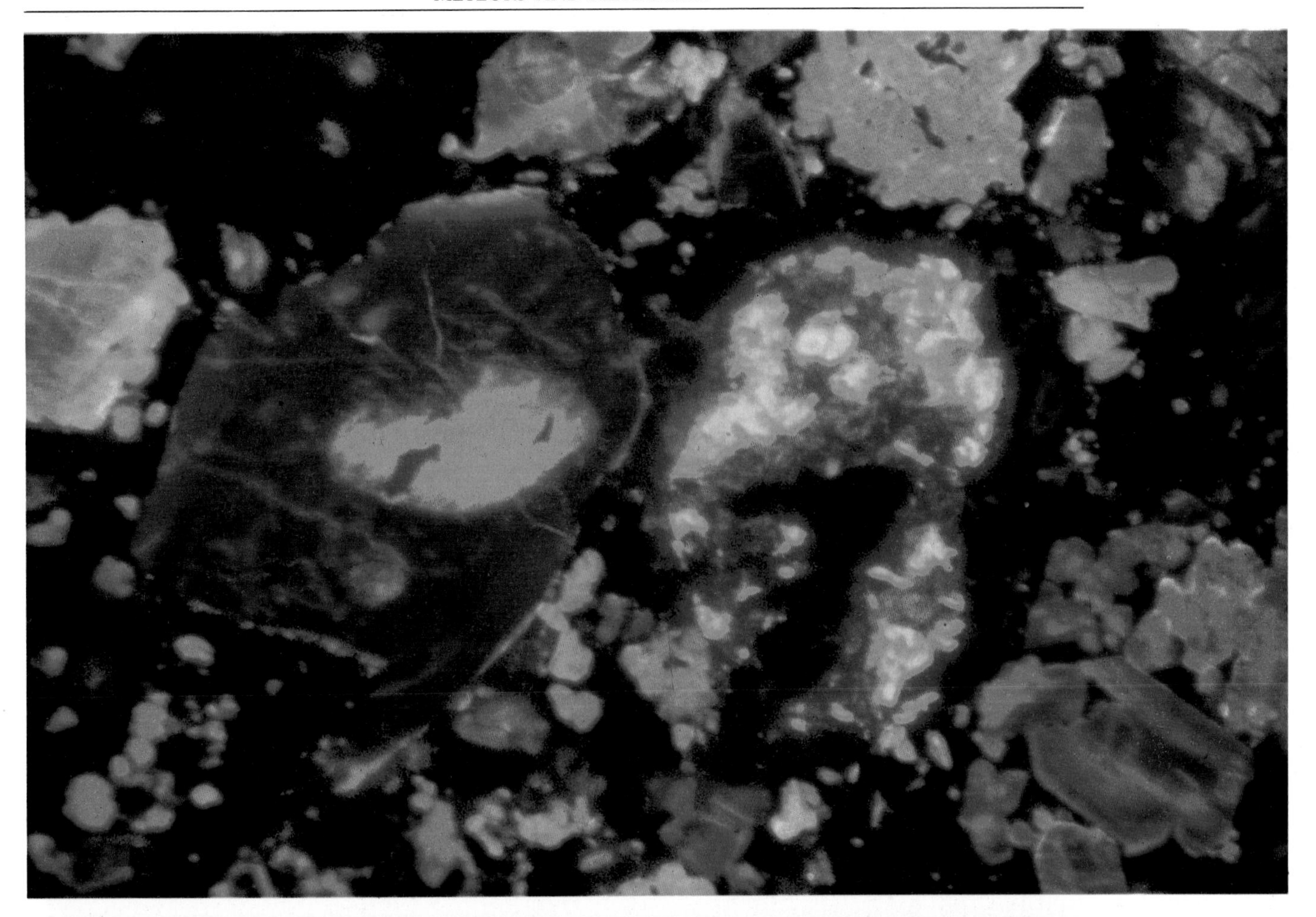

Above:
Mineralogical and chemical analysis of meteorites provides information about conditions both before and shortly after the formation of the Solar System. This is a section of the Indarch meteorite, a chondrite which fell in 1891 in Azerbaijan.

The Perseid meteor shower produces many bright meteors and fireballs (above magnitude –5), such as this one photographed by P. Parviainen, Finland.

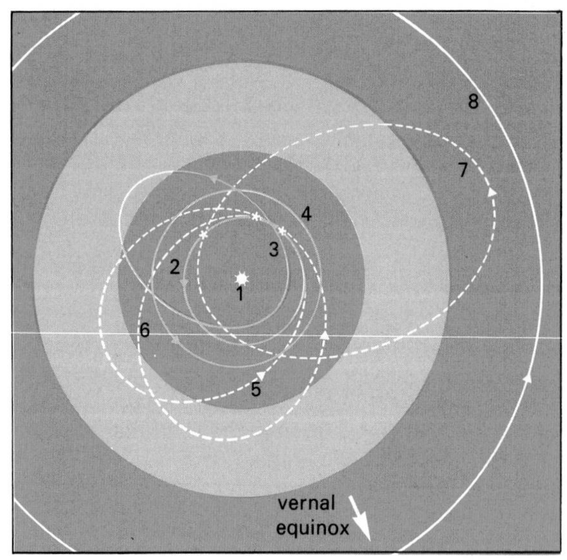

Fig. 5·22
The orbits of certain Apollo minor planets and three well-established meteorite bodies.

majority of fireballs do not result in meteorite falls, and observation of their break-up and slowing down in the atmosphere shows that they are fragile and probably of cometary origin.

Table 5·22 **Some terrestrial meteorite craters**

	diameter (km)
Manicougan (Quebec)	64
Richat (Mauretania)	50
Vredefort (South Africa)	40
Manson (Iowa)	31
Clearwater Lakes (Quebec)	26, 14
Ries Kessel (Germany)	23
Gosses Bluff (Australia)	19
Chassenon (France)	15
Ashanti (Ghana)	11
Lake Mien (Sweden)	6
Brent (Ontario)	4
Steinheim (Germany)	2·8
Meteor Crater (Arizona)	1·2
Wolf Creek (Australia)	0·8

Origin of the Solar System

The information which we now possess about the various bodies in the Solar System enables us to make reasonable assumptions about the way in which the System originated. Many different theories have been proposed in the past, but the one generally accepted today suggests that the Sun and planets were formed at about the same time from a concentration of gas and dust like that found in a galactic nebula or molecular cloud.

The processes of stellar formation, which have been discussed already (see p. 60), show that a protostar will form in the centre of the rotating primitive nebula of gas and dust. The major constituents of the protostar which formed the Sun and the nebula around it were hydrogen and helium, with the addition of a small amount (about 1·4 per cent) of the heavier chemical elements. The latter were themselves formed inside an earlier generation of stars and then dispersed as interstellar grains by supernova explosions (p. 63). The presence in meteorite samples of some products formed by very short-lived radio-active elements suggest that the Solar System originated about $4{\cdot}65 \times 10^9$ years ago, soon after a supernova exploded in the vicinity. Ices and the more volatile substances condensed on the surface of the grains; such particles accumulating into clumps with a maximum size of a few millimetres or centimetres. Within the primitive solar nebula, such clumps would rapidly settle towards the central plane, perhaps in as little as 100 years, so that the protostar would be surrounded by a disc of dust and gas. Objects have been observed elsewhere in space which show just such a structure, and where, it is believed, planetary systems are forming at the present time.

The actual mechanisms by which the planets condensed from the disc of particles are complex and somewhat uncertain. Because direct formation into large bodies could not occur, it seems likely that loose minor planet-sized bodies were produced. Collisions between clusters of such bodies would then result in the release of sufficient energy for more rigid objects to form, and by a repeated process planetary-sized bodies would develop. In the disc there would be a marked change of temperature outwards from the hot protostar, so that one would expect only the less volatile substances to be concentrated in the central regions. This accounts very satisfactorily for the differences in density of the inner planets; further out ices would not have been vaporized and, together with some denser materials, would collect together to form the cores of the giant planets. Moreover, these would be sufficiently large to capture a significant amount of hydrogen and helium from the nebula by gravitational attraction. In the cases of Jupiter and Saturn, subsidiary discs of gas and dust probably collected around the planets, eventually forming the satellite systems, by processes similar to those acting to produce the planets.

While the planets were forming, the central protostar continued to grow by collecting material until it eventually reached the point at which fusion of hydrogen began. At this time a considerable amount of material was lost because of an intense solar wind, this stream of atomic particles sweeping the Solar System clear of gas and dust, effectively putting an end to any more accumulation of material by the planetary bodies except by collision. Such mass loss is observed in T Tauri type variable stars (p. 60), which are young objects, just as the Sun would have been.

The energy acquired during collection of material is the main cause of the internal temperatures of Jupiter and Saturn, while the inner planets have been heated due to intense early bombardment by remaining interplanetary material and by RADIOACTIVE DECAY. Such heating caused the planets and the larger planetoids to separate into dense cores and lighter overlying layers.

The cratering produced by major impacts during an early period of bombardment can be seen on the Moon, Mercury and Mars, while as far as the Earth is concerned, it has been suggested that the impact of major planetoid-sized bodies broke up the original crust, producing a continental area and an ocean area.

Most of the larger-sized bodies had been swept up

by about 3×10^9 years ago, and since then impacts have occurred at a lower rate. There are also some indications of fluctuations in this lower rate, possibly due to the break-up of larger bodies to give showers of meteoroids, or else by planets pulling groups of bodies into collision orbits.

There are suggestions that the planetary atmospheres could have been formed by later accretion of volatile materials, and that this could account for some of the differences between Earth and Venus in particular. There are three methods by which later accretion of atmospheric materials could take place: condensation from a hot solar nebula; capture of a significant amount of the solar wind; and the impact of volatile-rich comets and planetesimals. Although there are certain points which favour the first two as subsidiary causes of the Venus:Earth difference, isotopic ratios do not suggest that the atmospheres of these two planets formed from solar-type material. Although the impact of 'cometary' bodies is certainly likely to have occurred, there is no real evidence that it was the sole source of atmospheric materials, so that the outgasing theory remains the most likely candidate for the time being.

The early atmospheres of Earth, Venus and Mars are likely to have been largely composed of carbon dioxide, carbon monoxide, nitrogen and some hydrogen, with some additional trace gases. It appears probable that liquid water has been present at some time on both Venus and Mars, but the former planet has evolved to its present atmospheric state largely because of the temperature conditions, while Mars has lost most of its early atmosphere to space and by chemical reactions with the surface materials. On Earth, plants have made a considerable contribution towards the amount of free oxygen which is now present in the atmosphere.

The origin of comets remains uncertain but there are two principal classes of theory which might explain their occurrence. In the first, comets formed from small cloud fragments in orbit about the nebula, doing so at about the same time as the planets were formed.

Perturbations by nearby stars are probably responsible for the injection of comets into the inner Solar System. The other suggestion is that comets were formed at a later date either by the gravitational pull of the Sun as it encountered a concentration of dust and gas, or by compression in front of an advancing galactic spiral arm. Either class of theory could account for the observed distribution of cometary orbits.

Other planetary systems

Although it has been possible to gain a reasonable idea of the way in which the Solar System was formed, it is difficult to obtain direct evidence for the existence of other planetary systems. However, there are various indications that such systems do exist and have been observed.

The nebular clouds from which stars form rotate slowly, but this rotation speeds up as they contract so that ANGULAR MOMENTUM is conserved; protostars will therefore be rotating much more rapidly. Spectroscopic observations show that there is a tendency for the earlier spectral classes 0 to about F4 (see page 48) to remain fast rotators. This had been taken to indicate that stars of later spectral classes (F5 to M) had transferred most of their angular momentum to companion bodies where most of the energy is expressed in orbital motion. However, with greater realization that even stars of late spectral classes may still emit intense stellar winds in the process of their formation, and thus could lose much of their initial angular momentum, this argument has lost much of its validity. It is quite possible that even in the case of the Solar System, where the massive planet Jupiter possesses about 98 per cent of the total angular momentum, the Sun, with its low rotational velocity of about 2 km per s at the surface, may have lost most of its angular momentum through an intense stellar wind.

The minimum mass which can give rise to thermonuclear fusion and produce a true star is about $0{\cdot}06\ M_\odot$. Between this and about $0{\cdot}01\ M_\odot$, objects will be luminous for about 10^9 years due to the release of gravitational energy as they contract, but after this time they will become non-luminous bodies or dark stars (sometimes called black dwarfs). Any mass smaller than about $0{\cdot}01\ M_\odot$ may fairly be described as a planet, as it will only radiate away heat at infrared wavelengths – as in the cases of Jupiter and Saturn.

From observations of Doppler shifts caused by the orbital motion of some stars it has proved possible to estimate roughly the number of stars which have companions, and to assess how many of these companions may themselves be stars, black dwarfs or true planets. Apparently the vast majority of stars have one or more secondaries, and, although the calculations are rather uncertain, possibly about 20 per cent of stars of classes F5 to M have planetary companions.

By careful photographic techniques it is possible to chart the proper motion of nearby stars. In the absence of any companions the proper motion track should appear as a straight line, but the presence of other bodies will cause a 'wobble'; from this it is possible to estimate the mass and distance of its companions. The work is naturally very difficult, but at least two stars (and as many as seven) have been suspected of having planet-sized companions. Recently however, considerable doubt has been expressed about these determinations, which have been largely the work of a single observatory. Even the cases of ε Eridani, which had been thought to have a massive planet about 6 times the mass of Jupiter, and Barnard's star (a faint red dwarf and one of the closest stars), suspected of having two planets of about 0·9 and 0·4 Jupiter masses, must now be considered suspect. It will probably be necessary to await more advanced methods which will allow planets to be detected directly (such as the proposed Project Orion system or, more probably, the Space Telescope) for this matter to be resolved – or at least for some progress to be made. It is likely to be a long time before Earth-sized planets become detectable by any means.

Meteor Watching

The visual observation of meteors is exciting work and provides an excellent introduction to observational astronomy because little equipment is required, other than a suitable atlas or set of star charts. Under good conditions a keen-eyed observer can expect to see about eight sporadic meteors per hour on average. On nights when showers are active the rate may be very much higher, although none of the major showers can be expected to rise much over 100 per hour under normal conditions. The spectacular displays such as those of the Leonids in 1799, 1833, 1866 and 1966, with many thousands of meteors per hour, are very much the exception. The rate which one can normally expect from the Leonids is not much greater than that for sporadics. The clumps of meteoroids which give rise to high rates are frequently very compact, so that even with the regular showers the peak frequency may only be seen over a small area of the Earth. Various other factors will affect the number of meteors seen, the most important being interference from the Moon, the altitude of the radiant, the clarity of the atmosphere and the time of night – more meteors are seen after midnight when the Earth's rotation is carrying the observer 'into' the stream. It is also best for the observer to watch an area of the sky about 45° away from the radiant.

A single observer can obtain very interesting and useful records, but obviously it is not possible to watch the whole sky at one time, so small groups of persons frequently observe from one site, each observer covering a portion of the sky, with perhaps one member recording details called out by the observers. This form of meteor watch is frequently used when a shower is close to maximum. The methods adopted and the information recorded vary, depending upon whether the observer is alone or in a group, and whether the meteor rate is low or high. Basically, however, records are made of the time at which meteors are seen, their magnitudes, paths – particularly start and end points – whether they seem to belong to a shower or appear to be sporadic, and any special features such as persistent trails. (The latter can persist for some considerable time after a meteor, and recording its position, either visually or photographically, can give important information about motion in the upper atmosphere.)

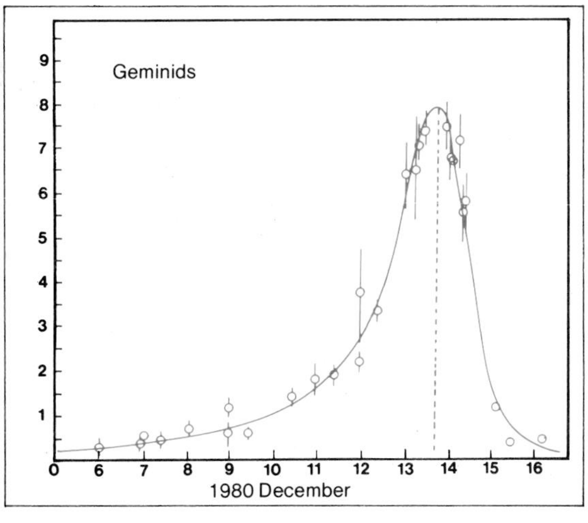

It is best for watches to last for a set period – say 30 or 60 minutes – but the beginning and end times must in any case be recorded fully.

Although it is of advantage to know the position of the radiants, these do alter slowly from day to day due to the Earth's motion relative to the orbit of the meteoroids. However, the identity of individual meteors can usually be determined by plotting their paths on star charts. (Only one special map projection – known as the gnomonic – shows paths as straight lines, but provided that the start and end points are correctly identified, accurate tracks may be derived from any chart using nothing more complicated than a scientific calculator.) The detailed determination of radiants from visual observations is difficult and has now largely been superseded by other means. However, during showers the recording of meteor magnitudes and of the numbers seen allows very important information about the distribution and sizes of the meteoroids within the stream to be deduced.

Many meteor observers supplement their visual observations with photographic records and even simple undriven camera mounts can be used for time exposures. However, photography will only ever record the brightest meteors. As with visual observing it is most effective if a single camera is pointed about 45° from the radiant. During a time exposure, the starting and finishing times of which must be recorded, the observer should, if possible, also watch the same area of the sky, recording the details of any meteors as usual. Any meteors recorded on the film will then be fully documented. More advanced amateurs use a battery of cameras, or special mirrors or fish-eye lenses to record the whole sky. Rotating shutters may be fitted to either single or multiple-camera arrangements, in order to break the trails at fixed intervals, thus providing a means of determining the velocity of any object.

Co-operative photographic projects are well worthwhile, the simplest technique being to place two observers some distance apart – perhaps 50 km or more – to photograph the same volume of sky. The paths of any meteors photographed by the two cameras can then be determined exactly, and the heights derived by triangulation processes. National organizations frequently arrange for special coverage on particular nights, asking all observers to direct one camera towards one of a number of points, usually 70km above specific positions on the ground. (In a small country, such as the United Kingdom, only three such points may be needed.)

Similar techniques are used by the fireball patrols which regularly operate cameras, sometimes quite automatically, on every clear night. Unlike ordinary meteor photography, however, this is only suitable for very dedicated observers who are prepared not only to expose the films, but also to develop and check them as quickly as possible. This is especially important when there has been a bright fireball which may have resulted in a meteorite reaching the surface of the Earth. Here photographs are needed to determine the most probable impact point so that a search may be mounted as soon as possible.

A daytime fireball is a very important event and should be recorded as fully as possible, giving the altitude and azimuth of the ends of the track – or those parts which were seen – as well as the usual

Geminids

1980 December

The plot of meteor numbers during the course of the Gemenid meteor shower, observed over several nights in 1980 December (B.A.A. data).

sort of information about magnitude, colour and other details. Large meteoroids can give rise to sound effects which are, of course, only heard after the event, and may take several minutes to travel to the observer after the fireball has been seen. The time at which such sounds are heard must be noted as it can be used to derive the distance from the observer to the closest part of the track. National organizations will frequently send someone to check upon details reported by members of the general public, such is the importance of these major events. Such reports are useful material to add to the more detailed information provided by amateur astronomers, and indeed, may sometimes be the only information available.

Some amateurs are working in a very difficult field in trying to obtain spectra of meteors. As only the very brightest will give sufficient light to expose the film adequately after being dispersed by a prism or grating, very few meteor spectra have ever been obtained, but a few amateurs are achieving quite considerable success. Similarly, one or two advanced amateurs have constructed equipment which will record meteors and fireballs photoelectrically. With the general growth in expertise in electronics, doubtless more observers will be making such observations in future.

One fairly simple, but strangely and regrettably neglected, aspect of meteor observing concerns telescopic meteors – i.e. those seen when using any form of instrument, even binoculars. Some observers undertake special telescopic meteor watches, but many events are seen when other objects, especially variable stars, nebulae and galaxies, are being observed. As they form part of faint streams, the study of these meteors is difficult and requires a large number of observations. They should always be recorded and reported.

Finally a word about what could be regarded as the visual observer's most important piece of equipment – a piece of string! If this is held up along the meteor's path it makes the task of determining the position relative to the stars much easier and more precise. A stick, suitably graduated in degrees, is a more sophisticated version, rather better and easier to use, but less portable.

On this type of chart the paths of meteors should actually appear curved, but if the start and end points are accurately marked the true paths may be calculated. Meteors are taken to belong to a particular stream if they appear to come from anywhere within a small circular area around the theoretical radiant position.

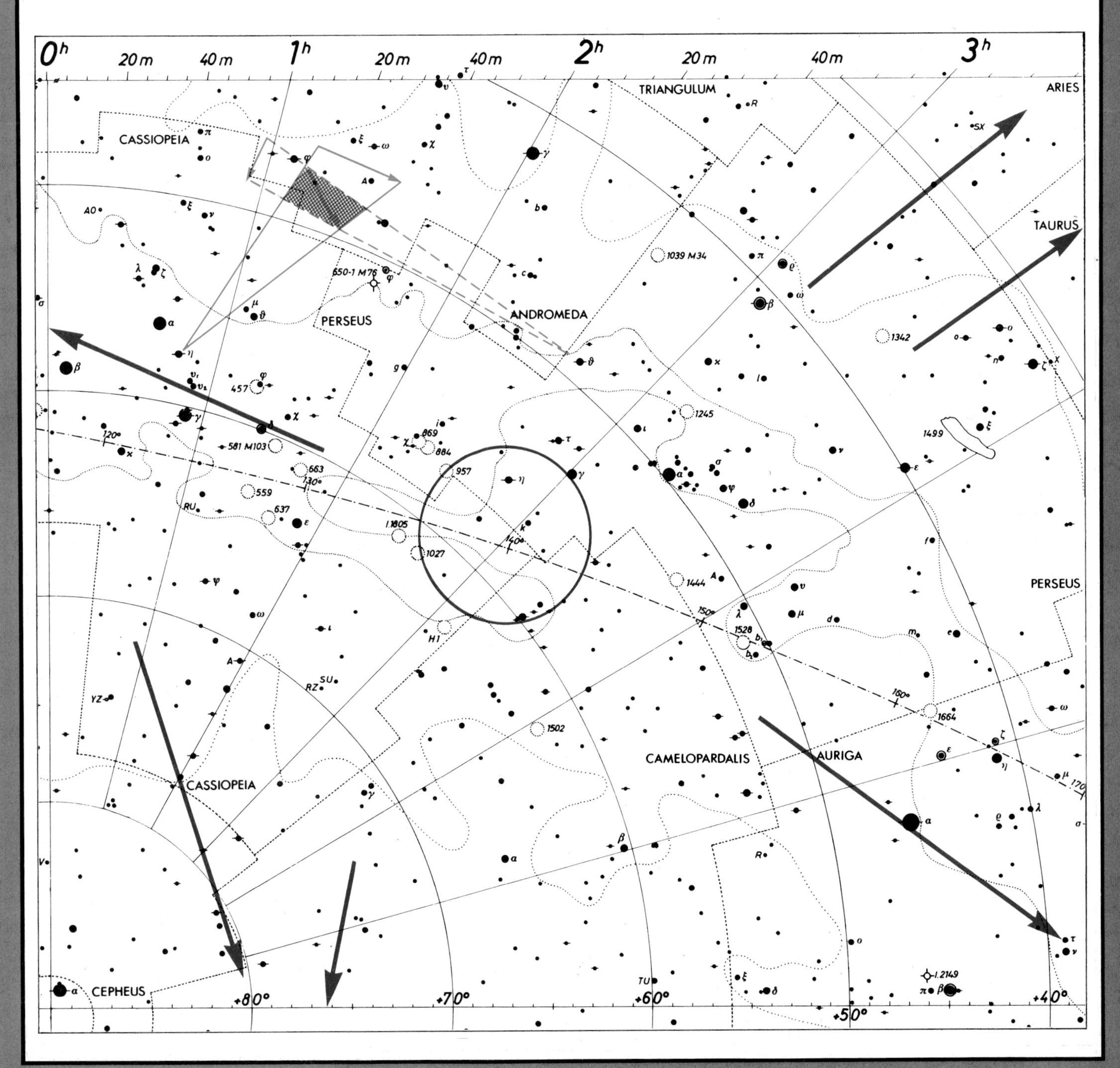

CHAPTER SIX

The Galaxy

On a dark, moonless night, away from the glare of city lights, the heavens themselves seem ablaze. Even familiar constellations, like the Plough or Orion, are difficult to trace on the crowded vault, and countless faint stars cover every fragment of the inky black sky. Yet the sky is not completely dark. Threading its way among the stars is a narrow, misty band of light which spans the sky from horizon to horizon. In some places the band becomes so faint as to be almost invisible; in others so dense and glowing that it looks like a nearby cloud. To the astronomers of old, unhampered by the dubious blessing of artificial lights, it was an important and prominent feature. Its shape suggested to them a path, a road, or a river, and they devised legends and stories to describe and explain it. One such legend ascribed its origin to milk spilt from the breast of the goddess Juno when nursing the thirsty infant Hercules; hence the name *Via Lactea*, the Milky Way.

However, not until the early years of the seventeenth century did a picture of the real nature of the Milky Way begin to emerge. It was then that Galileo directed the newly-invented telescope towards the luminous band, and found that its light derived from countless thousands of stars, too faint to be seen individually without optical aid. Little more progress than this was made in the following century and a half, because successive generations of telescopic observers busied themselves with the nearby Moon and planets, leaving philosophers to grapple with the problems of the remote stars. Among them, Thomas Wright of Durham (1711–86) and Immanuel Kant (1724–1804) made the far-sighted suggestion that the Milky Way might represent the extremities of a vast, flattened star system in which we live.

Sir William Herschel (1738–1822) – widely regarded as the greatest observational astronomer of all time– was not satisfied with such a qualitative assessment. In 1784, with the aid of his sister, Caroline, he undertook to count all the stars visible through his great telescope, in order to determine just how they were distributed over the sky. A survey covering the whole sky would have taken a prohibitively long time, as his telescope had a field of view only 8 arc minutes across – a quarter of the size of the Full Moon. So Herschel decided to make sample counts in some 700 regions, widely scattered over the sky, which he believed would give a representative picture.

Herschel found that the stars increased in number towards the plane of the Milky Way, reaching a maximum density in the plane itself. A great distance above and below the Milky Way, the stars were spread thinly. Perhaps, Herschel reasoned, the stars are arranged in a kind of grindstone- or lens-shaped system, with the Sun somewhere towards the centre. Then, by looking along the diameter of the disc, we would see nearby stars scattered all over the sky, while ever more remote stars became blurred with distance into a misty band. As only relatively close stars inhabited the regions above and below this diameter, we should see them spread thinly over the sky.

Although Herschel himself came to doubt this model in his later years, and nineteenth-century astronomers gave it little support, we now know that it is essentially correct. One of its greatest stumbling-blocks was in accounting for the patchiness in the Milky Way itself. Herschel believed the gaps in the Milky Way to be true voids in space where there were no stars; places where we could see through our star-system to the greater universe beyond. But his successors considered it to be too great a coincidence that so many long, starless tunnels were centred on the Sun. Instead, they envisaged the Milky Way as a thin, rather broken ring of stars which crossed a smaller belt of stars containing our Sun.

As we shall see later, the star-voids turned out to be almost the reverse: regions of dense, obscuring material which appeared black because they blotted out the light from distant stars. So there is no need to explain coincidences; Herschel's 'lens' was right, and the nineteenth-century conception wrong.

Herschel had no way of ascertaining the dimensions of his star-lens, or Galaxy, as it was becoming known (from the Greek *galaxias kyklos*, meaning milky circle). The first star distance was not measured until 1838, sixteen years after Herschel's death; and so he could only make a rough estimate. The way in which stars were distributed suggested to him that the long diameter of the lens was some 800 times the average separation between stars, while the short diameter was only 150 times this distance. Taking as he did, the Sun-Sirius separation to be an average distance, Herschel arrived at a star-system measuring 8 000 by 1 500 light-years – over ten times smaller than the Galaxy as we conceive of it today.

The story of how our Galaxy increased its bounds over the years is a long and fascinating one, and we will touch on fragments of it as we explore our star system's contents and structure. However, now it is time to make the step to the present: to examine the modern astronomer's conception of our Galaxy and see how it fits into the universe. Although many astronomers have contributed to the complex picture, our debt must rest ultimately with Herschel, the perceptive and diligent observer who wove the original canvas.

Structure of the Galaxy

Our Milky Way Galaxy is just one of the millions of galaxies in the universe, but it is the one galaxy we can probe in great detail, and so it becomes a standard by which we can judge others. There is, however, one unexpected drawback about our location deep inside the Galaxy; it means that we are limited to studying the sociology of its contents, rather than its large scale geography. For the latter, we must resort to observations of galaxies beyond our own, where the structure is clearly visible. Both types of observation complement each other well. Studies of external galaxies, when combined with a knowledge of the contents of our own, lead to improved understanding of the make-up of galaxies in general; and by comparing the contents of our own Galaxy with other galaxies located close by, we can build up a detailed picture of its structure.

Seen from outside, our Galaxy would look like a vast, slowly-rotating Catherine wheel or pin wheel made up of stars (Fig. 6·1). The central hub, or 'nuclear bulge', is marked by an ELLIPSOID of densely packed stars, and from it spring spiral arms, like curving spokes, threading their way through the Galaxy's disc. To an astronomer, our Galaxy is an intermediate-type normal spiral galaxy, notable only for its size. Measuring 30 kiloparsecs in diameter, and containing some 10^{11} stars, our Galaxy is bigger than most other spirals, and is classified as a giant system.

Contrary to earlier ideas, the Sun and Solar System are by no means centrally placed within the Galaxy. They lie some 10 kpc from the centre – about two-thirds of the way towards the edge – and so our view of our surroundings is rather lopsided. When we look at the Milky Way in the constellation of Sagittarius, we are facing towards the galactic centre. Here the Milky Way is at its brightest, showing how the star density increases towards the central regions. In the opposite direction (the galactic anticentre), the Milky Way in the constellation of Auriga is only faintly traceable, because we are looking straight out through the rim of our Galaxy.

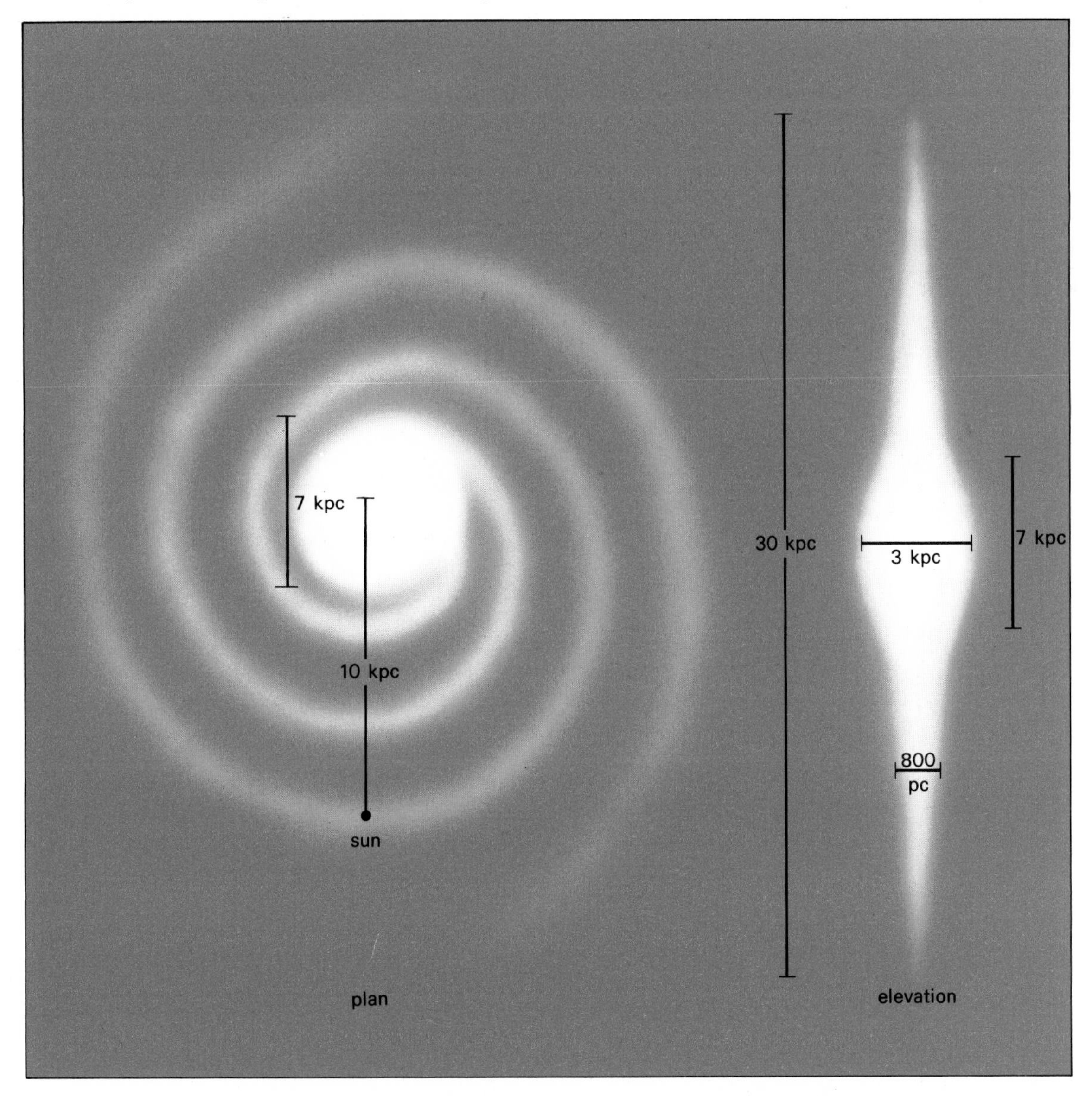

Fig. 6·1: Our current ideas on the shape and size of our Galaxy, as seen from above (left) and edge-on (right).

Opposite page: *The Trifid nebula* (top) *and the Lagoon* (bottom) *photographed by 1·2 m UK Schmidt telescope.*

The difference in brightness between the two regions of the Milky Way is far less marked than might be imagined, as the glow of the Sagittarian star-clouds is considerably dimmed by intervening obscuring matter. The effects of this material are strikingly obvious in the 'Great Rift' in the constellation of Cygnus – where the Milky Way appears to divide into two – and in the aptly-named Coalsack in Crux. These are some of Herschel's so-called voids in space. They lead us into considering two further important constituents of our Galaxy: the gas and dust of the interstellar medium.

Mixed in with the stars of the disc is a considerable quantity of cold gas. It is mainly very rarefied hydrogen, and makes up only 10 per cent of the mass of the Galaxy, but its consequences are far-reaching, for it is the future material of new stars. In some regions – near glowing gas clouds like the great Orion nebula – we see the gas under compression and collapse, as the star-formation process sets to work. Hot, glowing gas clouds are strung out along the arms of other galaxies, pointing up the spiral structure, just as they undoubtedly do in our own.

Only 0·1 per cent of the mass of our Galaxy is in the form of interstellar dust, but it is this component which produces the most severe observational consequences. Choking the plane of the Galaxy, the tiny dust grains – comparable in size with particles of cigarette smoke – act just like city smog. At best, they scatter and redden the light of distant stars; at worst, they absorb it completely. A leading American astronomer, Bart J. Bok, has likened the exploration of our Galaxy to 'trying to map a large city from a suburb on a misty day'!

All the contents of the disc rotate about the centre of the Galaxy, the inner regions more rapidly than those towards the edge. This **differential rotation** has an important application in measuring the way in which the Galaxy's mass is distributed, as we shall see later. At the Sun's distance from the galactic centre, the rotational velocity is 250 km per s (about 900 000 km per hour!), and yet it still takes us some 250 million years to do one circuit of the Galaxy. The Sun and its family of planets have travelled round the Galaxy only twenty times since their formation.

In addition to their rotation about the galactic centre, the objects in the disc share another important property: they are all relatively young. Gas, dust, open clusters of stars and bright blue supergiants are all indicators of youth, and are found only in the galactic disc. Historically, they are referred to as 'Population I objects', following on from a decisive

Star clouds in one of the densest regions of the Milky Way, crossed by starless voids where clumps of interstellar dust obscure our view.

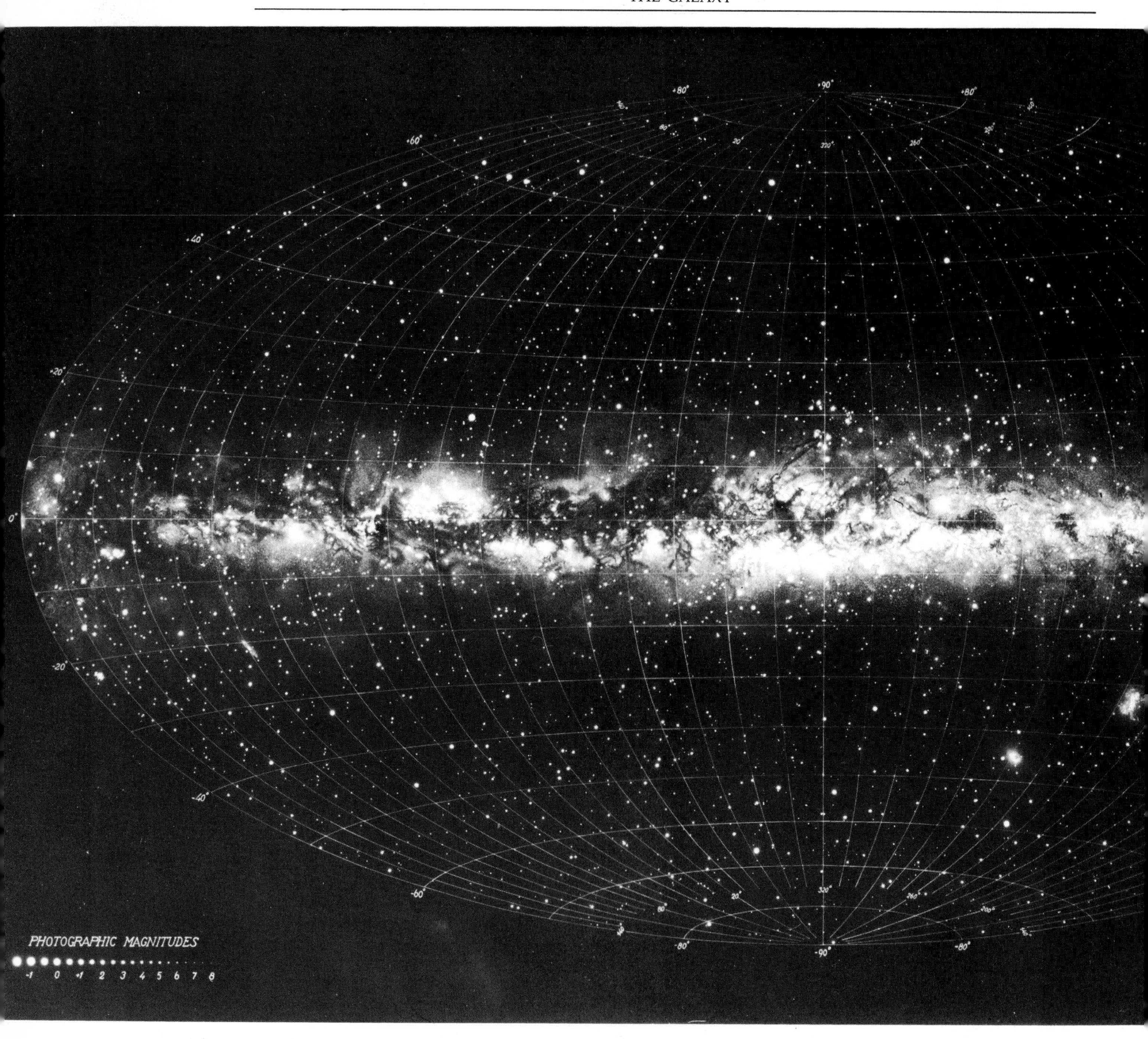

The luminous band of our Milky Way is seen stretching across the whole sky on this map. The detached portions, lower right, are our Galaxy's closest companions, the Magellanic Clouds.

piece of research by Walter Baade. Observing with the 100-inch (2·5 metre) telescope at Mount Wilson during the blackout of World War II, Baade was partly able to resolve the Andromeda galaxy – the nearest spiral to our own – into stars. He noticed that the disc, and spiral arms embedded therein, contained only young objects. On the other hand, the nuclear bulge, and the vast SPHEROIDAL region surrounding the entire galaxy – the 'halo' – comprised only evolved red stars, with no dust and gas mixed in to build future generations. Baade termed these objects, and the globular clusters which demarcated the galaxy's halo, 'Population II'.

The two populations – old and young – were quickly identified in our own Galaxy. Although these definitions were useful as guidelines at the outset, the classification of objects slowly became more unwieldy. Terms such as 'extreme Population I', or 'intermediate Population II' crept into the literature, and now it is recognized that there is simply an AGE GRADIENT between the outermost regions of the spheroidal halo, and the innermost plane of the disc. Clearly, this gradient is telling us how and when our Galaxy was created out of the gas in intergalactic space. Let us now trace the history of our Galaxy by examining the various objects which formed at different epochs in its past.

Formation of the Galaxy

Fifteen thousand million years ago ($1{\cdot}5 \times 10^{10}$ years), our Galaxy, as we know it today, did not exist. It is believed that it was a huge, roughly spherical volume of gas, so tenuous that its density was only 10^{-22} kg per m^3 – equivalent to one atom in a 3 cm cube! For

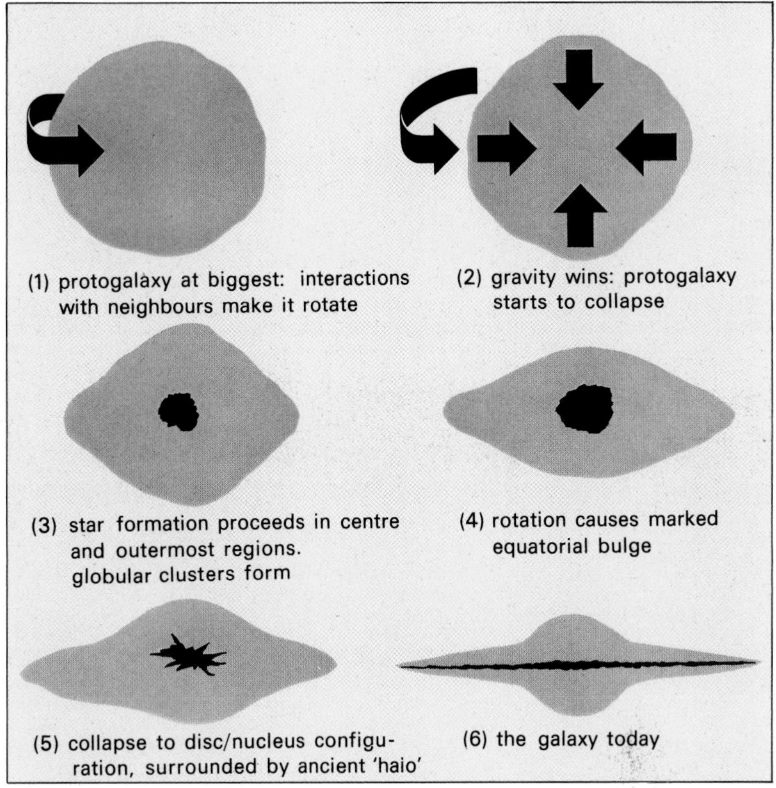

Fig. 6·2 above: *Stages in the formation of a galaxy like our own. In (1), the protogalaxy begins to rotate under the gravitational influence of its neighbours. At stage (2), it starts to collapse under its own gravity, spinning faster as it does so. By (3), the first stars have formed in globular clusters, left stranded by the shrinking protogalaxy, while a great burst of star formation takes place in the dense central regions (4). Rotation flattens the galaxy into a disc (5). Our Galaxy currently appears as in (6), with star formation still taking place in the outer regions.*

some time, this lump of gas, or **protogalaxy**, had been taking part in the expansion of the young universe, growing steadily larger until it was able to exert influence on, and be influenced by, its neighbours – other protogalaxies lying close by. The nature of these interactions is speculative, but it is possible that the young cloud may have been set into slow rotation by the interplay of forces, and its gas stirred up into turbulent motion.

Our protogalaxy probably reached a maximum diameter of 100 kpc before its own gravity slowed and then halted its expansion. The expansion of the cloud changed to a slow collapse, which proceeded increasingly quickly as gravity took over. Towards the centre, where collapse was most rapid and the turbulence greatest, the gas density became high enough for the first stars to be born, forming the nuclear bulge of our Galaxy (Fig.6·2).

The first stars were quite unlike the stars we are used to dealing with today. They were formed only from hydrogen and helium (some 75 per cent and 25 per cent by mass, respectively), the two elements present in the protogalaxy gas. Some of these stars were extremely massive, perhaps more than 100 $M_\odot$, and so they went through their life cycles very rapidly, ultimately exploding as supernovae. Heavy elements – some processed inside the stars, others produced in the supernova explosions themselves – were spewed out into the surrounding medium, enriching the gas which was to form the next generation of stars. Today, this enrichment of the interstellar medium still continues. It gives valuable clues as to the age of stars.

Not all the stars which formed were as massive as these. In the outer regions of the protogalaxy, gas clouds of 10^5 or even 10^6 $M_\odot$ began to fragment into individual, normal-sized stars. Possibly as a result of the slow rotation and smaller turbulence in these outer regions, or because there were insufficient stars forming nearby to disrupt these huge gas clouds by their ultraviolet radiation, enormous clumps of stars formed together. Today we still see these globular clusters marking the farthest outposts of our Galaxy.

As the collapse of the protogalaxy continued, ever more rapidly, the gas drew away from the outermost regions, leaving the globular clusters stranded. To conserve angular momentum, the young Galaxy spun faster and faster as it shrank, gradually developing a distinct bulge around its equator. Had star formation been rapid and efficient at this stage, the Galaxy might have been frozen into this ellipsoidal shape forever; for, dynamically speaking, stars interact little with their surroundings once they have been formed. However, sufficient gas remained to shape our Galaxy into the way it is today.

Opposite, top: *Globular cluster M13 in Hercules, dimly visible to the unaided eye, is revealed on this photograph as a dense ball of almost a million old stars.*

Further collapse meant faster rotation, a more pronounced equatorial bulge, and an increased gas density. The atoms in the gas, once so dispersed, now became close enough to interact and collide with one another, a process which most efficiently removed energy from the gas. No longer able to support itself, the gas collapsed quickly into a very thin disc, rotating rapidly about the starry nuclear bulge of the Galaxy. The entire collapse process, from protogalaxy to disc system, probably took no longer than a few hundred million years – a mere one-hundredth of the Galaxy's present age.

Since then, the disc of our Galaxy has been in a perpetual state of change. Gas clouds continue to collapse, stars still form, while old stars die, returning their debt with interest to the ever-evolving interstellar medium. But all around, in the remote halo, are the faded relics of a bygone age. By studying the halo population, we learn of our Galaxy's past history; and we also unearth vital clues as to the formation of galaxies in general.

The galactic halo

Of all the regions in our Galaxy, the halo is the most poorly understood. Its extent, mass and density are not known with any certainty, and there are even indications that our ideas of its contents are in need of a review. However, there is no dispute over its most prominent members: the 130 or so globular clusters which can be observed to distances approaching 100 kpc (Fig. 6·3).

Early in this century, the globular clusters played an important role in finally dethroning the Sun from its assumed position at the centre of the Galaxy. Using the newly discovered period-luminosity relationship (page 66) of Cepheid variable stars to measure the distances to these clusters, and thereby gauge the extent of the Galaxy, Harlow Shapley found that they were strongly concentrated towards the Milky Way in Sagittarius. He reasoned that this grouping reflected the underlying distribution of matter in the Galaxy and, consequently, that the massive central regions lay some 16 kpc away from the Sun. More recent investigations have modified this last figure to between 8–10 kpc, but Shapley's position has held remarkabiy well.

Despite their distance, globular clusters can still appear bright in our skies. Two clusters in the southern hemisphere – ω Centauri and 47 Tucanae – were originally catalogued as stars by mistake; and M13, the brightest globular cluster in the northern hemisphere, is visible as a small, misty patch in the constellation of Hercules. In long-exposure photographs they are revealed as tight balls of 10^4–10^6 stars packed into a region averaging only 30 pc across. Despite appearances, the stars are far from touching one another – it is just that star images are spread out on photographs – but the stars in the central region of a globular cluster must still be extremely high; perhaps 1 000 times higher than in the neighbourhood of the Sun. Here the skies must be truly spectacular, with the closest stars outshining the planet Venus, and 1 000 other stars brighter than even Sirius.

Opposite, bottom: *Clouds of dusty gas surround members of the Pleiades (or 'Seven Sisters'), revealed here as a cluster of over two hundred stars. The presence of gas and dust betrays the cluster's youth: it is thought to have been formed a mere sixty million years ago.*

The skies of a globular cluster would have none of the variety which we are accustomed to. As the first in our Galaxy to form, the stars in a globular cluster are all now in an evolved stage – red giants, or red and white dwarfs. This is strikingly revealed on globular cluster H–R diagrams where the luminous stars form a prominent **giant branch** in the red giant region.These H-R diagrams tell us how long ago the globular clusters came into being: around $1{\cdot}3 \times 10^{10}$ years ago.

Although most globular clusters superficially resemble one another, there are some significant differences. There is a factor of 100 between brightest and faintest, and a similar range in their star numbers. The most distant globular clusters are noticeably larger, because their outer regions are not stripped away by galactic tides to the same extent as those clusters closer in. Some astronomers have called these outermost systems 'Tramp' globular clusters, believing them to be moving freely in intergalactic space. It seems more likely that these clusters formed in the outermost halo and are now leaking away from the Galaxy, having perhaps suffered slight perturbations from passing galaxies far off in space.

Striking differences are found in the proportions of heavier elements ('metals') making up globular cluster stars. As would be expected from their great age, the stars in globular clusters are relatively metal-poor, but the range in this deficiency is surprisingly large. Stars in the cluster 47 Tucanae have metal abundances of about 25 per cent that of the Sun (about 0·6 per cent). This figure is down to 5 per cent in the case of M 5, and the stars in M 92 and M 15 have only 0·2 per cent of the solar metal abundance. A general trend in metallicity is observed, those globular clusters near the galactic disc having the highest abundances, while the clusters in the outer halo are very metal-poor. Differences such as these are believed to reflect the times and places at which globular clusters formed, and can tell us much about the collapse of our Galaxy. Information about the brightest globular clusters is summarized in Table 6·1.

Globular clusters travel about the Galaxy on long, highly-inclined orbits, another relic from the early days of the protogalaxy. Some clusters are observed to be slowly rotating, and all of them – although it may not be at all obvious – are slowly losing stars to the halo. Interactions between the closely-packed stars in globular clusters can speed up or slow down their motions; and some interactions are sufficiently energetic to eject completely a star from the cluster. It is unlikely that a globular cluster could be totally disrupted in this way, although these processes can speed the break-up of looser star clusters. Over their total lifetimes of 10^{10} years, it appears that the globular clusters have remained virtually intact.

The stars which do leak away become members of the general halo population, which appears to make up the greatest part of the halo's mass. What proportion of this population started off their lives as members of clusters is uncertain; it is currently estimated to be a few per cent. Individual halo stars can be identified readily by their high velocities (greater than 63 km per s) and steeply inclined orbits, which sometimes intersect the galactic plane. Although they

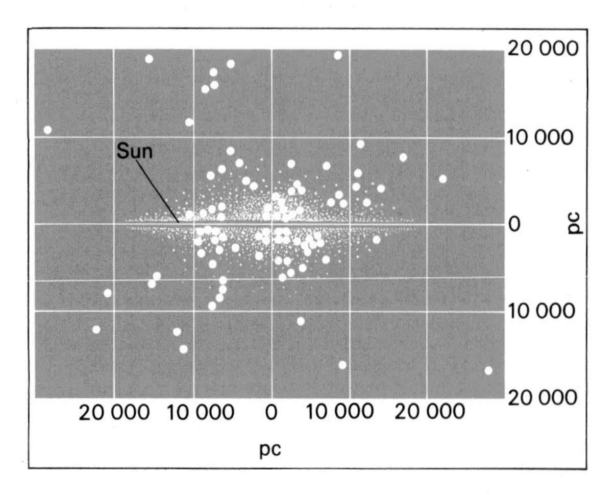

Fig. 6·3: The distribution of globular clusters around our Galaxy. These objects are the most prominent tracers of the galactic halo.

are often called 'high velocity stars', this is really a consequence of the Sun's high velocity relative to them, and because the orbits are in different planes.

Many halo stars are still more readily identifiable, for they are RR Lyrae variable stars. Their brightness (M_v = 0·5) qualifies them as excellent halo tracers, and their distances can be immediately ascertained from their apparent magnitudes. All the halo stars and RR Lyrae field stars share essentially the same distribution as the globular clusters, with a gradient of increasing metallicity towards the disc. There still remains a slight mystery concerning the metal-rich RR Lyrae stars, which are found in the field population, but do not appear to be present in even the most metal-rich globular clusters.

By combining local counts of halo stars with an overall picture of star densities in the halo, we can derive a figure for its mass. Current estimates put this at about 10 per cent of the mass of the disc, but this figure could be uncertain by a factor of at least ten, because of our lack of knowledge as to what constitutes the halo. There are indications that the halo may contain streams of hydrogen gas, connected in some way with the streamers in our Local Group of galaxies (see page 198): how much mass might they contain? Astronomers have also to consider the unseen white dwarf stars which must populate the halo, for they too could contribute significant mass. In fact, there are several theorists who believe that the disc of a galaxy like our own would be unstable unless surrounded by a massive halo with perhaps ten times the mass of the disc, but it is difficult to envisage what form this mass might take. Suggestions ranging from old red dwarfs to mini-stars – stars insufficiently massive to have ever commenced nuclear reactions – have been put forward; but all of these are undetectable with present techniques.

There is now increasing evidence that the total mass of the Galaxy may be very much larger than previously supposed. Determinations of the velocity of some very distant stars (particularly of one of the RR Lyrae type) for which the absolute magnitude may be accurately determined, suggest that total mass may amount to about twice the mass of the galactic disc, i.e. somewhere around $1{\cdot}5 \times 10^{12}$ solar masses overall. This is in accordance with the discovery that many galaxies are surrounded by 'coronae' of gas which also make a contribution to their total mass.

For some time there has also been the question of whether the halo – or other parts of the galaxy – could contain any of the so-called 'Population III' stars. It was suggested that these were stars which had formed from the primordial hydrogen and helium remaining after the Big Bang, and which could themselves have provided some of the heavier elements present – albeit at low amounts – in the old Population II stars. There is now some slight evidence that such stars may still exist in very low numbers, but that the vast majority are long extinct. However, the contribution of such stars to the early evolution of the universe could have been very considerable as they may have been responsible for some of the microwave background radiation (see page 206).

The disc of our Galaxy

The disc was the last part of our Galaxy to be formed, and, in a sense, it is still forming. Young stars are created even now in the central plane of the disc; although the majority of the disc stars have by now reached sedate middle age. Like the halo, the disc shows an age gradient from its outermost regions inwards, with younger objects occupying successively smaller distances above the galactic plane. The oldest disc stars, presumably the first to form after the Galaxy's sudden collapse, have a total spread above and below the plane of some 700 pc. By way of comparison, those O stars which have recently formed have a thickness in the plane of only 80 pc. The distribution of all the other stars lies in between these two extremes, although our Sun, very much a middle-aged member of the Galaxy, has an orbit which strays by only 80 pc above and below the plane.

The vast majority of disc stars are only 0·1 $M_\odot$, and spend almost their entire lives as faint red dwarf stars. With absolute magnitudes of about $M_v = +15$, they are extremely difficult to detect, and we can only pick out those which lie within 100 pc of the Sun. On the other hand, bright young O and B stars shine out like beacons all around the Sun's neighbourhood, and to great distances beyond. They give an impression of being very numerous, when in fact they are rarities among stars; it is their great intrinsic brilliance ($M_v = -8$ to -10) which makes them stand out.

Unlike the halo, where stars appear to be spread out uniformly, the disc has an uneven distribution of stars. Clumping and clustering are the rule here. Young stars are found in groups called 'galactic' or 'open' clusters (to distinguish them from the more compact and populous globular clusters), or looser aggregates called **stellar associations**. Some of these open clusters are easily visible to the unaided eye. The Pleiades, or Seven Sisters, shine like a tiny jewel-box of stars in Taurus, and were deemed worthy of mention in the Chinese records of the twenty-fourth century BC. Also in Taurus are the V-shaped Hyades, the nearest star cluster to the Sun, whose distance is the foundation-stone of the current extragalactic distance scale.

In all, over 700 open clusters have been listed, the

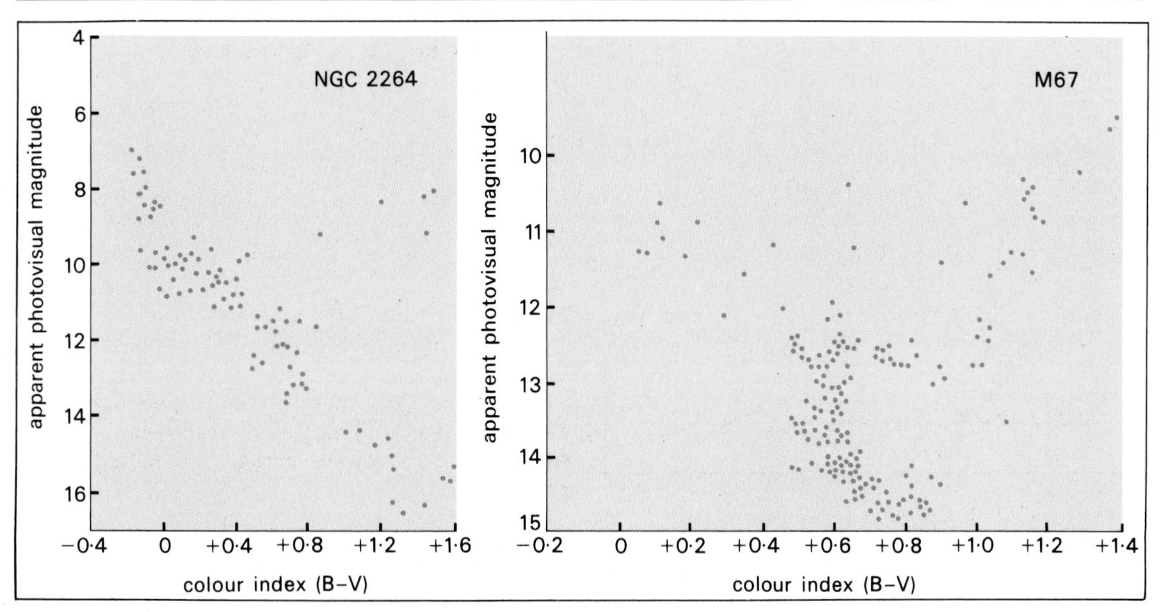

Fig. 6·4 Hertzsprung-Russell diagrams for two open clusters of different ages. NGC 2264 is a young cluster: very few stars have evolved off the Main Sequence. On the other hand, M 67 contains several stars which have by now reached their red giant phase.

majority of them lying within 2 kpc of the Sun. At distances greater than this, they blend into the stellar background, but it has been estimated that the Galaxy may contain in excess of 10^4 such clusters. Most are small, with diameters of about 2 pc and a membership of about 100 stars, but the larger clusters (such as Praesepe, or h and χ Persei) may contain several hundred stars in a region 10–15 pc across. These clusters play a major role in furthering our knowledge of stellar evolution. The stars in each cluster represent a sample which formed simultaneously in the same region of space, and whose present range in brightness and temperature can be ascribed to differences in their initial masses. Studying the H–R diagrams of open clusters can tell much of how stars with different masses change with age, as well as giving an overall age for the cluster (Fig. 6·4). These ages range from a few million years for the very youngest clusters (the Pleiades are some 6×10^7 years old) to several thousand million years (as in the case of M 67, whose age is estimated at 4×10^9 years). (See Pleiades photo, page 171.)

Few open clusters are much older than this, which probably reflects the fact that they disperse on this sort of timescale. As in a globular cluster, all the stars in an open cluster move around and undergo gravitational interactions with one another. The 'leaking-away' of stars as a result of these interactions, while not fatal to the populous globular clusters, does appear to be capable of disrupting the sparser open clusters in a period of about 10^9 years. Details of some of the brightest open clusters are given in Table 6·2.

Measurements of the motion of stars in open clus-

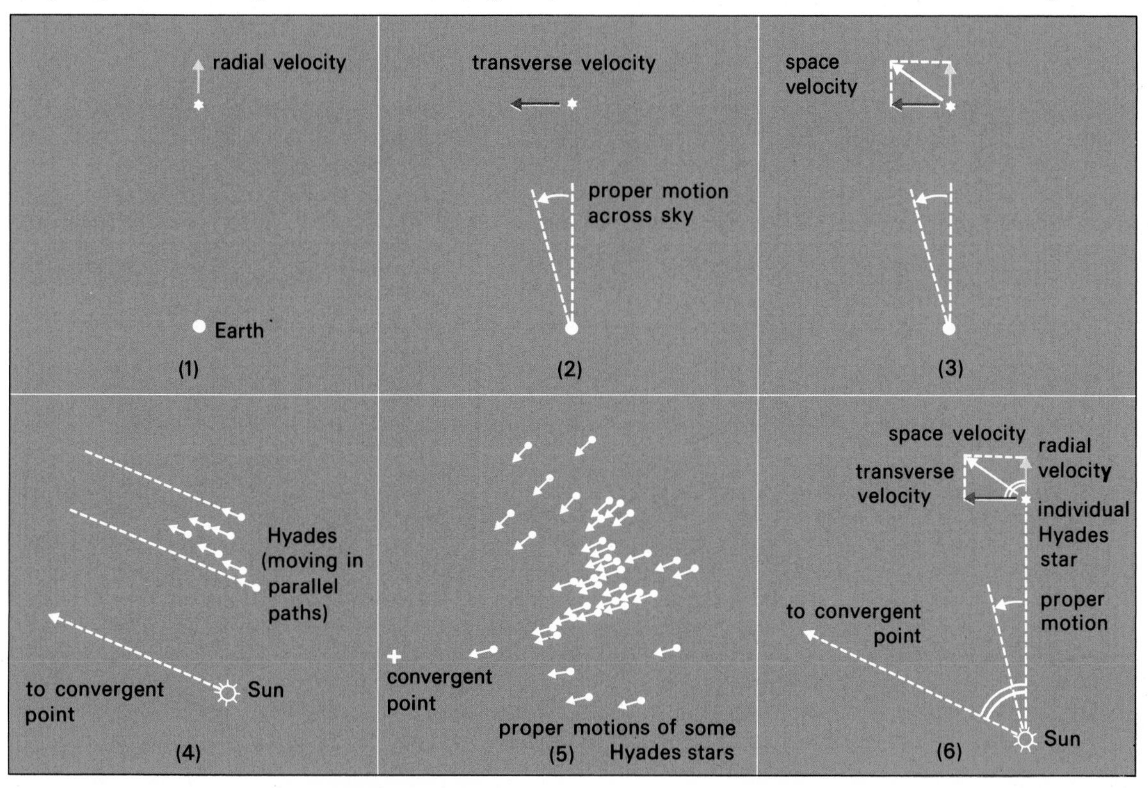

Fig. 6·5 below left: Two 'components' of a star's motion can be detected from Earth. The radial velocity directly towards or away from us is measured from the Doppler shift in the star's spectrum (1), while the proper motion (2) – the star's real motion across the sky – reveals the transverse velocity if its distance is known. Combining these two velocities yields the star's true speed and direction of motion through space, the space velocity (3). On the other hand, if a star's direction of travel and radial velocity are known, then its distance can be estimated by measuring the proper motion. This is the basis of the Moving Cluster method, which relies on the fact that all the stars in an open cluster move in parallel paths through space which appear, by perspective, to converge. (4), (5). A measure of each cluster member's radial velocity, when combined with a knowledge of the direction of this convergent point, gives the transverse velocity (6). When this is compared with the magnitude of the proper motion, the star's distance can be found. The accuracy of this method rests in the large number of stars considered.

The double cluster in Perseus, h and χ Persei. These twin open clusters each contain between 300–400 very young stars.

ters form the basis of a very accurate method of distance determination. The **moving cluster method** is, apart from trigonometrical parallax, the only direct means of measuring distances to normal stars, and it can be used out to far greater distances (Fig. 6·5). The method relies on the fact that all stars in a cluster move in parallel paths in the same direction through space; but, because of perspective, these paths appear to meet at a convergent point. Plotting the proper motions of cluster stars reveals the position of this convergent point. A measure of each cluster member's radial velocity, when combined with a knowledge of the direction of the convergent point, yields the transverse velocity; and from comparison with the proper motion the star's distance can be found. The accuracy inherent in this method lies in the large number of stars considered. By using over 150 member stars, the distance to the Hyades, the nearest open cluster, can be measured to an accuracy of a few per cent. The most recently derived distance of some 45 pc forms the basis of all successive distance measurements, right out to the most remote galaxies.

In measuring the distance to a far-flung star cluster whose proper motion is not detectable, the essential step is to compare its average stars with the members of the Hyades. In practice, H–R diagrams are drawn for both clusters, using absolute magnitudes for the Hyades stars, and apparent magnitudes for the distant cluster. Considering ordinary main sequence stars in each case, the degree to which the cluster stars are fainter than those of the Hyades is due to their greater distance. By exactly matching the cluster stars to the Hyades – a process called **main sequence fitting** – we can see how much fainter they appear to be, and consequently how much more distant they are. In the case of very distant clusters, the obscuring dust also creates considerable dimming and reddening, and this too must be taken into account.

Such measurements, along with parallax determinations of distances to nearby stars, enable very detailed H–R diagrams to be built up, and absolute magnitudes derived for almost all types of star. In fact, the distance to any star – whether a member of a cluster or not – can be estimated by comparing its spectral type and apparent magnitude with the stars on such an absolute magnitude H–R diagram. This technique of distance measurement is known as **spectroscopic parallax**.

Spectroscopic parallax can be used as a tool for mapping the Galaxy. In this case it involves measuring the distances to 'stellar associations', stars of the same spectral type which form loose, unbound groups in space (the five inner stars of the Plough, for example). Such associations disperse very quickly; since they cannot exist for long, those which are recognizable must comprise very young stars – O and B supergiants – which have strayed little from their birthplaces.

Mapping the distribution of these associations shows that they are not spread at all uniformly. As well as occupying the thinnest region in the galactic plane, they are concentrated into clumpy bands – a distribution shared by other youthful objects such as nebulae and young open clusters. All these objects are in fact 'spiral tracers', because the bands they delineate are the nearest spiral arms of our Galaxy. There are three local arms, or portions of arms, historically named the Sagittarius, Orion and Perseus arms, after the general directions in which they lie. Today it is recognized that the arms picked out by the spiral tracers are more extensive than their names would suggest. The Sun is associated with the stars in the Orion arm, whose identity as a *bona fide* arm is currently in dispute: there are indications that it may be merely a bridge between the Sagittarius and the Perseus features.

It is no coincidence that the spiral arms contain extremely young objects. Astronomers believe that arms are zones of compression which travel around the disc of a galaxy and trigger the gas into forming stars. Glowing nebulae and dark, clumpy clouds trace the line of this compression wave. We will now turn our attention to these regions.

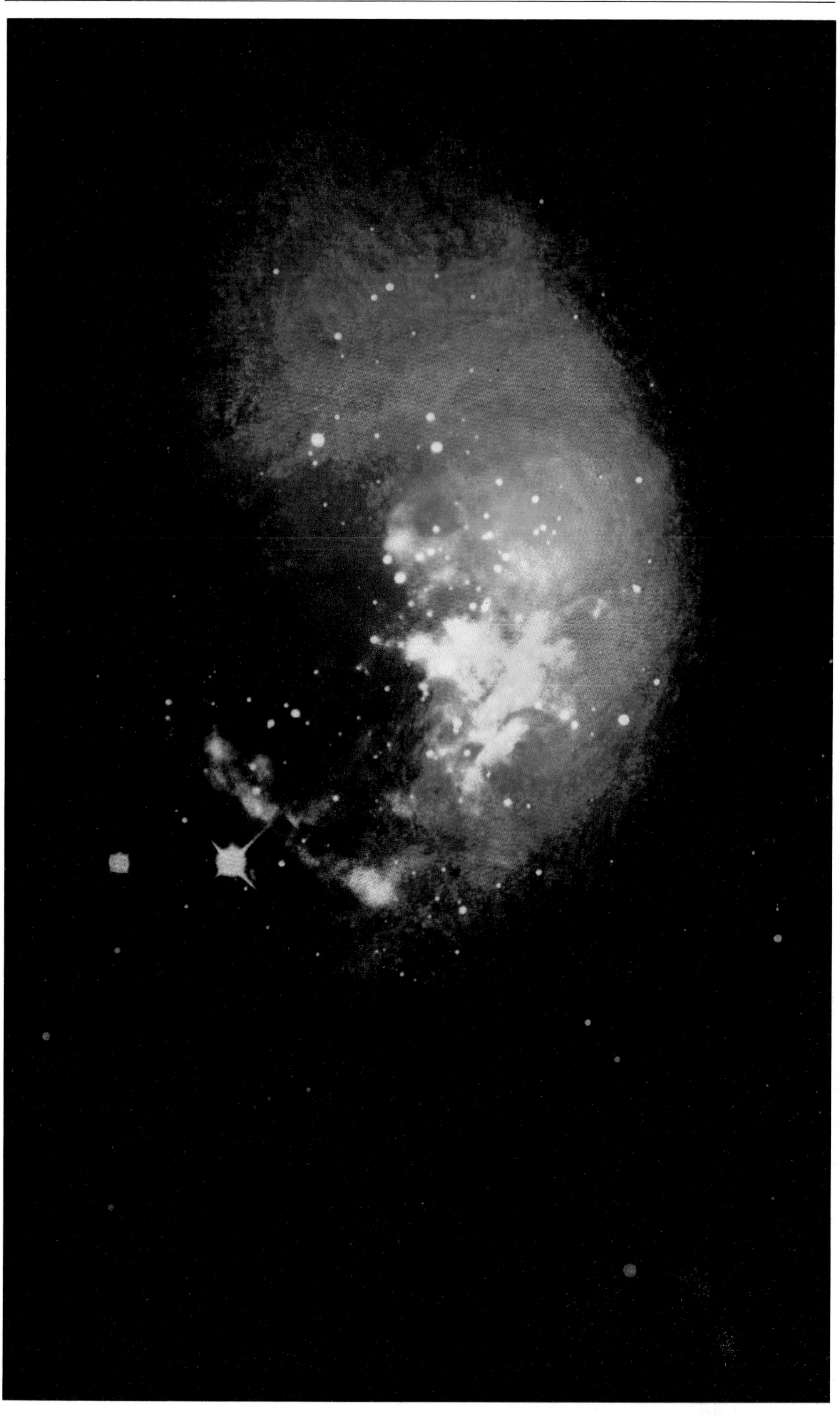

An infrared image of the Orion nebula and the Trapezium star cluster (centre). *This may be compared with the optical pictures on pages 59 and 179, and with Fig. 6.6 on page 180.*

Observing the Galaxy

The Galaxy offers an amazing range of different types of objects to be observed, and if we exclude double and variable stars (dealt with elsewhere), allows all sorts of observations to be undertaken for sheer enjoyment, rather than for serious 'scientific' purposes. The naked eye is more than adequate – indeed it is admirably suited – to take in the overall splendour of the Milky Way itself in its course across the sky, provided dark-sky conditions can be found, away from the bane of street lighting. Those persons who are fortunate enough to live or to be paying a visit south of the Equator, where the magnificent star clouds of the Southern Milky Way stretch from Sagittarius all the way round to Carina, are indeed to be envied. The bright northern section, best seen in the summer months, extends from Sagittarius through to Cygnus and is only a little less spectacular, with the brilliant Scutum star cloud and the Great Rift in Cygnus. On the side away from the galactic centre the Milky Way is naturally not so striking, but it can still be followed all the way round the sky, as it takes in constellations such as Cassiopeia, Auriga, Monoceros and Puppis.

The naked eye and low-power binoculars are ideal for tracing the intertwining star clouds and dark lanes, and there is quite a large number of observers who derive great pleasure from trying to distinguish all the dark nebulae described by Herschel and others. As some of these are of very low contrast, the finest seeing conditions and the very keenest eyesight are required for the details to be perceived, so that even this apparently simple pastime may require considerable skill and perseverance.

Some of the Galaxy's most spectacular objects can be located by the naked eye alone, e.g. the globular clusters M13 (in Hercules), 47 Tucanae and Omega Centauri, as well as the Orion Nebula itself and the various open clusters like the Pleiades, Praesepe and Kappa Crucis (the Jewel Box). However, binoculars or a wide-field telescope are needed fully to appreciate their beauty, which so captivates many observers that they take great pains to render them in drawings, checking their impressions and representations against those of the old-time observers. There are many other tasks which observers set themselves purely for the sake of interest, such as seeing how many of the closest stars they can locate, finding the faintest planetary nebulae, or even on a rather simpler level, finding the star with the greatest Declination South (or North) which they can see from their usual observing site. In the process of 'sight-seeing' many observers learn to find their way around the sky with ease, a skill sometimes envied by their colleagues who follow a more specialized programme of observation.

However, it is in photography that the Galaxy both offers the greatest challenges to the skill and patience of the observer, and also affords the opportunity for the simplest equipment to produce useful and striking pictures. With only an undriven camera and moderately fast film a short exposure will show the major constellation patterns at least as well as they can be seen by the naked eye.

The length of exposure which can be given on an unguided mount will depend upon the focal length of the lens being used and on the amount of trailing which can be allowed. Although long undriven exposures may be quite spectacular, they are confusing, and unsuitable for individual objects. However, even the simplest form of camera mount will give very worthwhile results, especially with wide field lenses. When a telescope is used, with its long focal length and high magnification (compared with normal camera lenses), accuracy of mounting and of drive become all-important. In most cases, and particularly when long exposures are in prospect, it is essential for the observer to guide the telescope all the time during the exposure, to compensate for the various inevitable inaccuracies which occur. This can become very tiring, and one can only admire the early photographers (and a few modern ones) who spent hours obtaining a single exposure. Modern fast black and white and colour films have naturally helped a great deal, although grain size may be a distinct problem with some films under certain circumstances.

Nevertheless quite long exposures are still required for very many faint objects, particularly nebulae (and galaxies), and here all films – with the exception of those expensive emulsions especially developed for astronomical photography – suffer from the limitation known as 'reciprocity failure'. This has the effect of failing to produce an increase in the density of the image corresponding to the increase of exposure time: when dealing with exposures of a fraction of second, as in most normal photography, exposure time and density are directly linked, but when exposures run to many minutes the relationship

The North American and Pelican nebulae in Cygnus, photographed by C. R. Martys, Bakewell, Derbyshire, using a 350mm × 500mm Schmidt telescope, in a 20-minute exposure on N2/H2 baked hypersensitized Technical Pan film, with high-contrast copying.

breaks down. The overall result is to reduce the speed of the film, which in the case of colour films may give rise to strange colour casts and effects due to the different emulsion layers having differing colour responses.

There are several ways in which this problem may be overcome by advanced amateurs. The first and simplest is to use the expensive special astronomical emulsions. Because they are so expensive, these tend to be employed only by those few observers who have obtained complete mastery of all aspects of astronomical photography, and who can be confident of wasting little film. Next in order of simplicity in that it requires no modification to the actual camera, is the technique of hypersensitization. In this the film may be baked, chemically treated, or subjected to both processes, sometimes in several stages, with the general aim of increasing the film speed without increasing the grain size, and of delaying the onset of reciprocity failure. The overall process can be quite complicated, even though each individual step may be simple, and usually requires some knowledge of laboratory techniques and safety measures, the latter being very important in those procedures where hydrogen gas is used for hypersensitizing the film. The film is treated before exposure, but there is also a similar technique, known as latensification, which can be applied after the exposure has been made. This latter process, however, is little used by amateurs.

A further technique is that of cooled-emulsion photography, in which the film is maintained at a very low temperature during the exposure. The most frequently used substance for this purpose is carbon dioxide, either as a solid ('dry ice' or 'snow') or as a gas, and at a temperature of 194.5 K (−78.5°C). Because of this low temperature a number of problems have to be overcome, such as brittleness of the film, and most especially the prevention of condensation, which in some cases has been solved by the use of optical windows and evacuation of the air from the film chamber. As will be obvious, special cameras have to be built for use with this technique, so it is tending to become less frequently employed than hypersensitization. However, it gives excellent results with colour films, with very little alteration of the colour balance.

Another type of approach to obtaining colour photographs is that used by some advanced amateurs who make a set of exposures through different filters, and then combine the images, with appropriate colour filtration, to give a final print in 'natural' colours. There are also various darkroom techniques which may be employed to enhance the visibility of faint nebulosity or detail. These include the superimposition of more than one negative in a technique sometimes used for planetary photography, the 'unsharp masking' method used on many professional photographs, recopying to give higher contrast, or by some combination of all these methods.

Despite all these apparent complexities, however, a lot can be done by a patient observer prepared to spend some time at the eyepiece guiding a telescope which is coupled to a camera in a perfectly straightforward manner.

Certainly there are enough objects in the Galaxy to provide anyone with a lifetime's programme of photography.

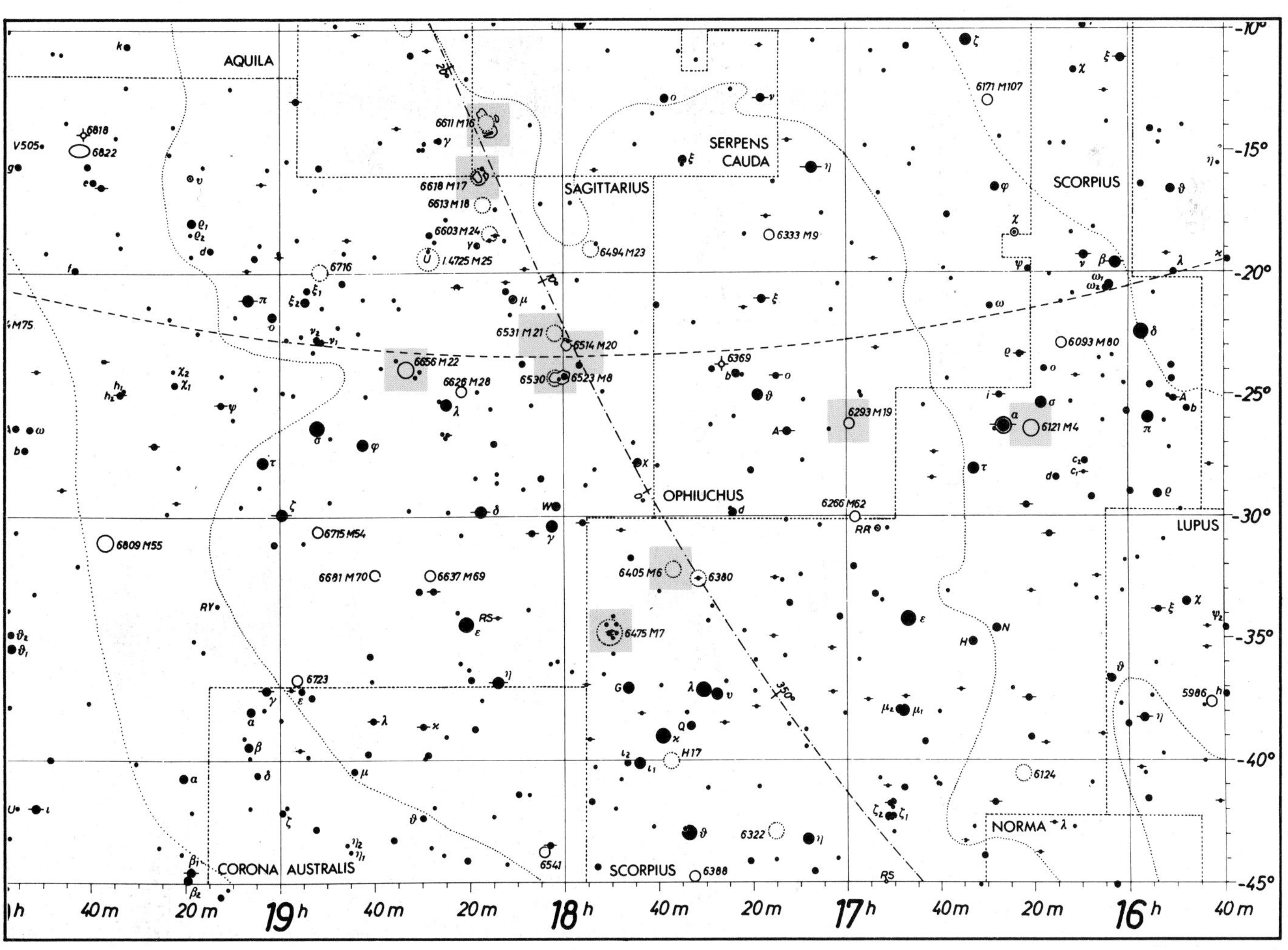

The region of the galactic centre of Sagittarius is very rich in objects to observe, particularly globular and open clusters.

Right:
The Lagoon Nebula in Sagittarius, also shown in colour on page 167.

Opposite page: *The great Orion Nebula, photographed by the 1.2m UK Schmidt telescope in Australia. The vivid colours demonstrate the different processes at work: red from the bright Hα hydrogen emission line, and blue from reflection off dust grains in the nebula.*

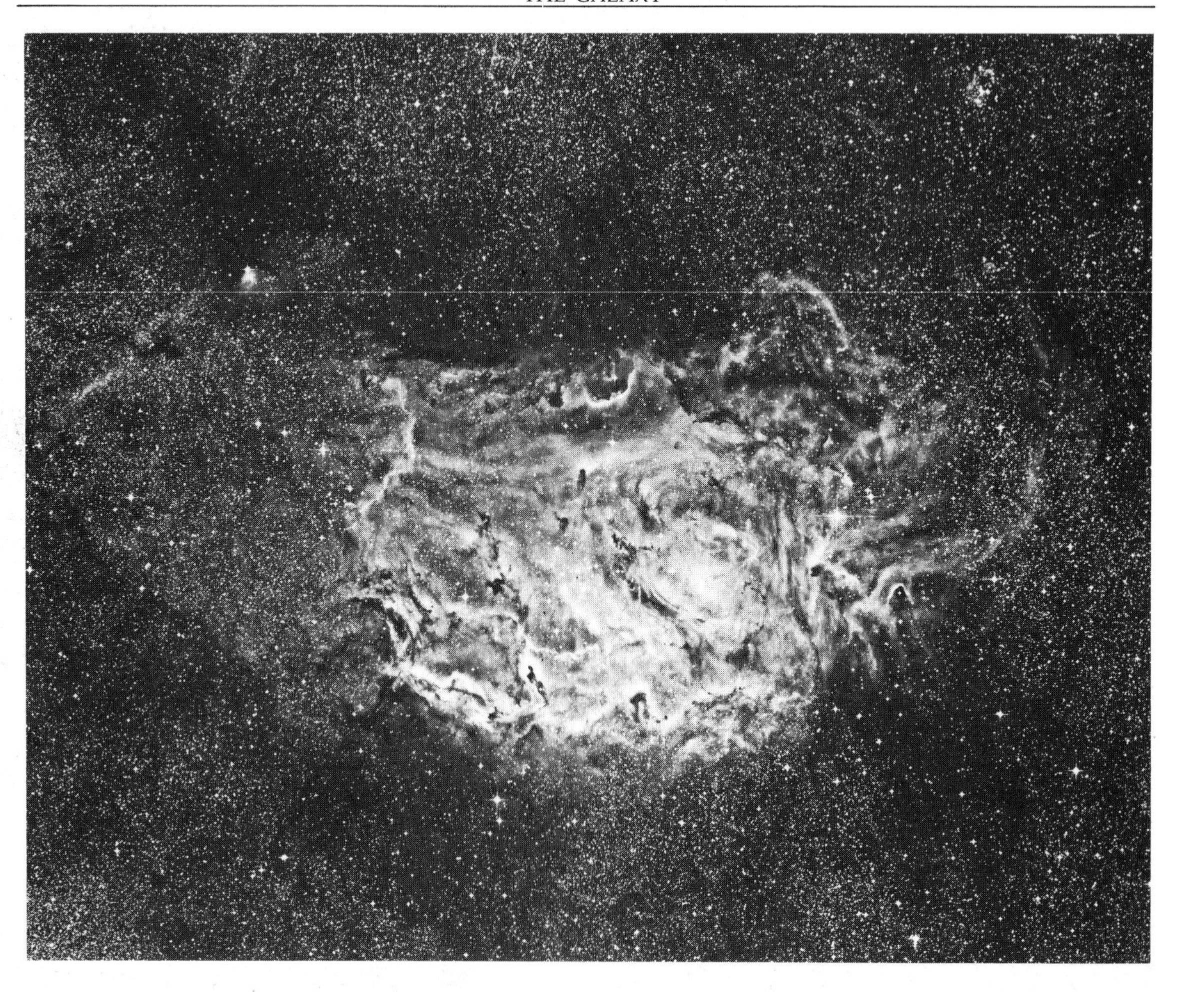

The clouds of the Milky Way

Up until now, we have been concerned in the main with the stars of our Galaxy. However, in pursuing a policy of working from the oldest regions inwards, we have now reached a point where there are no stars: only gas, the precursor of stars. This gas is spread thinly throughout the Galaxy in the very innermost plane of the disc, averaging a density of only 1 atom per cm^3, but the distribution of the gas is by no means smooth. In places, it is clumped together in spectacular cloud complexes, where hot, glowing nebulae lit by young stars within contrast starkly with the dramatic, looming outlines of dark, obscuring material. In this section, we shall discuss these clouds and their structure before going on to consider the general nature of the interstellar medium.

Most obvious of all the cosmic clouds are the glowing nebulae. Our Galaxy contains hundreds, many known by fanciful names after the objects they resemble – the Rosette nebula, the Lagoon nebula, the North America nebula. The brightest were listed in Messier's 1784 catalogue of 'objects which could be confused with comets'; a list which also includes such other objects of fuzzy appearance as star clusters and other galaxies. The great Orion nebula, for example, is listed as object number 42 in Messier's catalogue: hence its designation M 42.

A view of the Milky Way in the direction of the galactic centre in Sagittarius. Although the centre of our Galaxy is hidden from our view by interstellar dust, our line of sight traverses some of its densest regions when we look in this direction, giving the appearance of vast clouds of stars.

These bright clouds are huge regions of heated, ionized gas – gas at a temperature of 10^4 K, whose atoms have been split up into their constituent nuclei and electrons. Because most of the gas in these clouds is hydrogen, they are called **H II regions**, to denote that the hydrogen is in a singly ionized state: that is, with its one electron removed. H II regions are intimately associated with sites of star formation in the Galaxy. The source of their heating and ionization is the ultraviolet radiation from the hot, young OB stars embedded deep inside them. The small Trapezium star cluster, for example, supplies all the energy to ionize the great Orion Nebula, some 5 pc across (Fig. 6·6).

The size of an H II region depends on the density of the gas and the number of energetic young stars inside; the largest H II regions of all measure some 200 pc across. Most are roughly circular, and sharply defined against the sky by an abrupt boundary around their outer edge. This **ionization front** clearly indicates the limited ionizing range of the stars inside the nebula.

As would be expected, the spectral lines emitted by H II regions are those characteristic of a hot, low-density gas. Hydrogen lines of the **Balmer series** are strong, particularly the vivid pink Hα at 656·3 nm, which gives nebulae their distinctive coloration. The greenish tinge noticeable (especially optically) in some H II regions comes from a couple of oxygen [O III] lines at 495·9 and 500·7 nm – at first mis-identified as a new element and called 'nebulium'. These lines arise in what are termed 'forbidden transitions': transitions which would be impossible in the high densities of a terrestrial laboratory, but which occur freely in the near-vacuum of space. The observation of

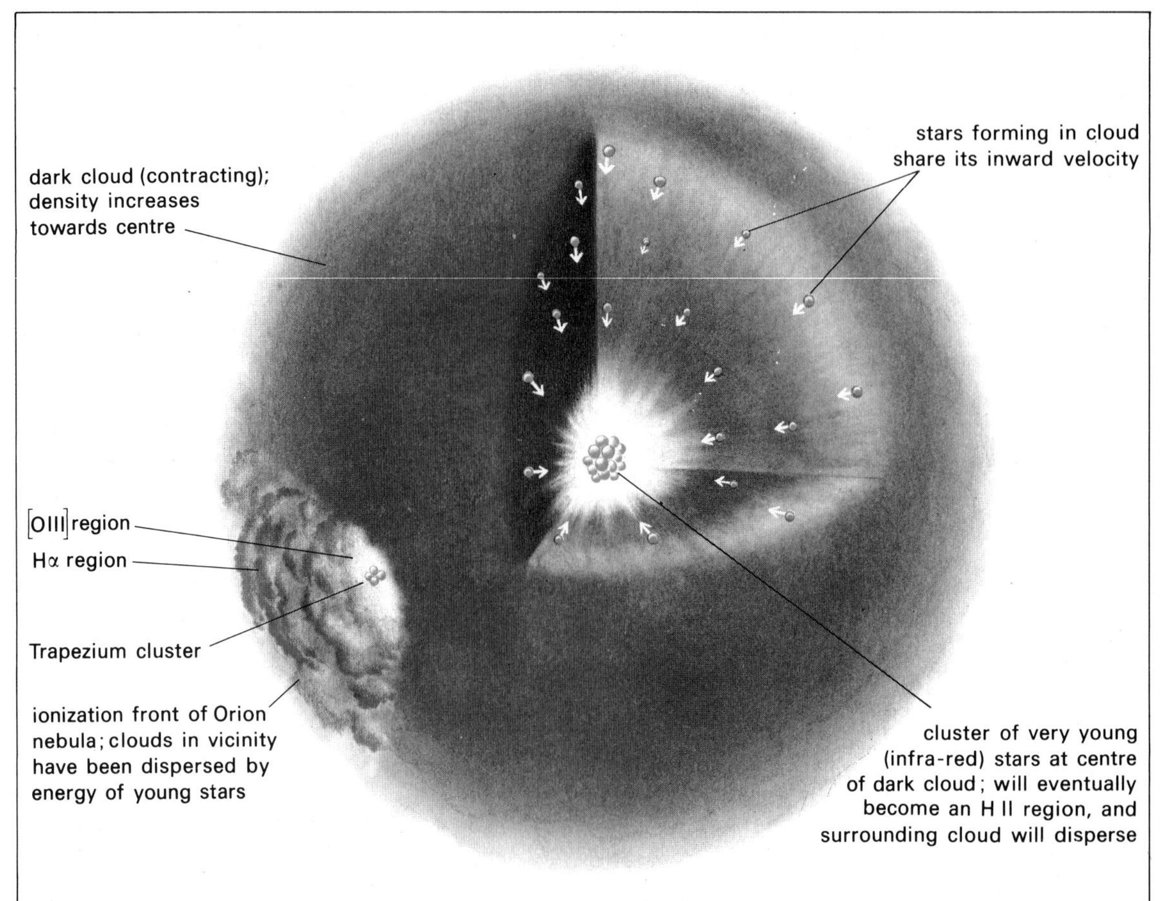

Fig. 6·6
The Orion Nebula is just a small part of a gigantic, non-luminous cloud complex. In this cutaway diagram, we see the Orion Nebula (left) as a region in the cloud where newly-formed stars have excited the gas surrounding them, causing it to glow and expand. The rest of the cloud is collapsing, and many stars are being born in the heavily obscured central regions. Although we cannot yet see these young stars, their radiation will eventually excite the whole cloud complex. (The Sun is situated to the left of the diagram.)

forbidden lines such as these yields much information on the temperatures and pressures in H II regions.

Mixed in with the gas of an H II region are large quantities of obscuring dust particles. Radio waves can penetrate the dust, allowing us to examine the structure of the nebula behind; and, of course, the dust can also be studied for its own sake. Heated by the young stars, the dust grains reach temperatures of up to 300 K, and give out tremendous quantities of energy at infrared wavelengths (from 3–3 000μm). These dust zones in H II regions are among the most powerful infrared sources in the Galaxy, with luminosities up to 10^7 times greater than the Sun.

As we have already noted, there is a strong association between H II regions and the occurrence of star formation. But an H II region does not itself contract and fragment to form stars; in fact, H II regions are observed to be expanding outwards at an average rate of 1 pc every 10^5 years. To find the actual sites of star formation, we must investigate the dark, cool clouds found in association with H II regions: Herschel's 'holes in space'.

Clouds such as these – the Horsehead nebula and the Coalsack are well-known examples – appear to be similar in size and composition to H II regions, but their hydrogen is almost entirely in molecular form (H_2). This is a reflection on the high densities (10^9–10^{10} atoms per m^3) and low temperatures ($T \simeq 10$ K) which prevail: ideal conditions for collapse and fragmentation into stars. Protected from the hurly-burly of ionizing radiations by a cocoon of dust, a cloud may begin to contract undisturbed.

Less common than the dark clouds, but believed by many astronomers to be the sites of active star formation, are the warmer, denser **molecular clouds**. These have densities as high as 10^{10}–10^{12} atoms per m^3 and temperatures between 30–100 K. Many are found in association with giant H II regions and, because they contain large quantities of heated dust, they are strong infrared sources as well. However, their novelty and uniqueness stems from the large and bewildering variety of surprisingly complex molecules which they harbour, most of them unsuspected and unpredicted until their discovery.

Interstellar molecules

A large number of interstellar molecules are now known (Table 6·3), the first few of these being discovered by their absorption lines in the spectra of distant stars. Discoveries by radio methods began in 1963, when the hydroxyl radical (OH) molecule was found at a wavelength of 18 cm. Although molecules also emit energy at ultraviolet and infrared wavelengths, their most intense lines are produced in the millimetre or microwave region of the spectrum, arising out of a change in their rotation rate (ROTATIONAL TRANSITIONS). Many molecules have no net spin unless they collide with another molecule (usually hydrogen, the commonest molecule in space), which sets them into rotation. After a few hours, a spinning molecule will lose all its energy by emitting a photon of microwave radiation, and then return to its GROUND STATE.

It is no coincidence, then, that the sudden rash of

Table 6·1 **Bright globular clusters**

cluster	NGC	right ascension h m	declination ° ′	apparent magnitude	distance (kpc)	mass ($\times 10^4$ $M_\odot$)	spectral type	number of variables
47 Tucanae	104	00 21·9	−72 21	4·0	5·0		G3	11
ω Centauri	5 139	13 23·8	−47 03	3·6	5·2		F7	165
M 3	5 272	13 39·9	28 38	6·4	10·6	21	F7	189
M 5	5 904	15 16·0	02 16	5·9	8·1	6	F6	97
M 4	6 121	16 20·6	−26 24	5·9	4·3	6	G0	43
M 13	6 205	16 39·9	36 33	5·9	6·3	30	F6	10
M 92	6 341	17 15·6	43 12	6·1	7·9		F1	16
M 22	6 656	18 33·3	−23 58	5·1	3·0	700	F7	24
Δ 295	6 752	19 06·4	−60 04	6·2	5·3		F6	1
M 15	7 078	21 27·6	11 57	6·4	10·5	600	F2	103
M 2	7 089	21 30·9	−1 03	6·3	12·3		F4	17

Table 6·2 **Bright galactic clusters**

cluster	NGC	right ascension h m	declination ° ′	apparent magnitude	distance (pc)	spectral type	estimated age ($\times 10^6$ yr)
h & χ Persei	869,884	02 15·5	56 55	4·2	2 360	B0·5	12
M 34	1039	02 38·8	42 34	5·6	440	B8	160
Pleiades (M 45)	—	03 44·1	23 57	1·3	126	B7	63
Hyades	—	04 17	15 30	0·6	45	A2	400
M 38	1 912	05 25·3	35 48	7·0	1 320	B8	50
M 36	1 960	05 32·8	34 06	6·3	1 260	—	32
M 37	2 099	05 49·1	32 32	6·1	1 280	B8	200
M 35	2 168	06 05·8	24 21	5·3	870	B5	—
Praesepe (M 44)	2 632	08 37·2	20 10	3·7	158	A5	252
M 67	2 682	08 47·8	12 00	6·5	830	F2	4 000
ϰ Crucis (Jewelbox)	4 755	12 50·7	−60 04	5·0	830	B3	16
M 21	6 531	18 01·6	−22 30	6·8	1 250	O9·5	6·3
M 16	6 611	18 16·0	−13 48	6·6	2 500	O5	1·3
M 11	6 705	18 48·4	−06 20	6·3	1 740	B8	160
M 39	7 092	21 30·4	48 13	5·1	250	B9	252

discoveries of new molecules in the late 1960s took place at the same time as dramatic improvements in microwave receiver design. However, the range of molecular species found in the cloud complexes was a great surprise. A few of these molecules were inorganic (lacking carbon); but the vast majority were organic compounds, ranging from simple diatomic varieties such as methylidyne (CH), cyanogen (CN), and carbon monoxide (CO), up to nine-atom molecules like ethyl alcohol – ethanol – (C_2H_5OH). Quantities of the latter in the huge cloud complex Sgr B2 have been estimated as sufficient to make up more than 10^{27} litres of whisky! Even larger molecules have now been detected including cyanooctatetrayne (HC_9N) and the 13-atom cyanopentaacetylene ($HC_{11}N$), although the latter is probably only associated with a circumstellar shell, rather than a true interstellar molecular cloud.

A number of molecules, which we would expect to be relatively common in space, such as oxygen (O_2), nitrogen (N_2), carbon dioxide (CO_2) and, most especially, hydrogen (H_2) are not detectable by their rotational transitions on account of their symmetry. Although we do not observe them directly, we can infer their presence. Hydrogen is the most common molecule in space, followed by carbon monoxide (with 10^{-4} of the hydrogen abundance), and then by hydroxyl (OH) and ammonia (NH_3) (with only 1 per cent of the CO abundance). Most of the 50 or so molecular varieties so far discovered have been found in fewer than a dozen locations in the Galaxy, where the density is greater than 10^{10} atoms per m^3 and where there is enough dust to shield the molecules from the dissociating effects of ultraviolet radiation from young stars.

The fourth molecule to be discovered, hydroxyl, still poses problems for astronomers. Although it is usually seen as a single, broad line at 18-cm wavelength, there are regions of the sky where the OH line is observed to be split into several bright, narrow emission lines with strengths a million times greater than normal. These OH sources are very tiny: condensations in them are less than 10 au across, whereas a normal molecular cloud is several pc in

The 'Snake' Nebula, Barnard 68 and 72. It is easy to see on this photograph why such patches of obscuring material were once regarded as being starless voids in space.

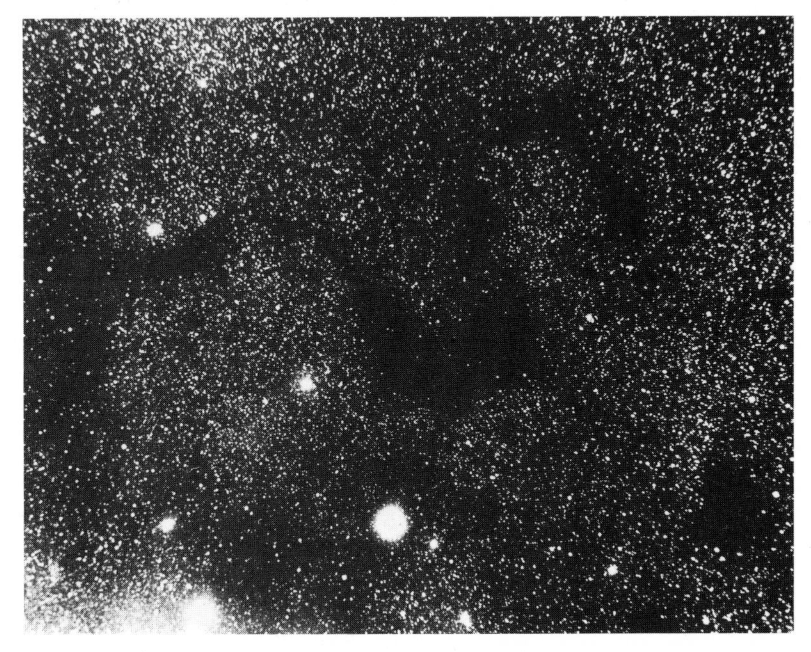

Table 6·3 **Interstellar molecules**

year	name	formula	wavelength
1937	methylidyne	CH	4 300 Å (0·43 μm)
1940	cyanogen	CN	3 875 Å (0·3875 μm)
1941	methylidyne ion	CH^+	3 745–4 233 Å (0·3745–0·4233 μm)
1963	*hydroxyl	OH	2·2, 5·0, 6·3, 18 cm
1968	ammonia	NH_3	1·20–1·26 cm
1968	*water	H_2O	1·35 cm
1969	*formaldehyde	H_2CO	{ 1, 2·2, 6·2 cm 2·0, 4·0 mm
1970	*carbon monoxide	CO	2·60 mm
1970	*cyanogen	CN	2·64 mm
1970	hydrogen	H_2	1 100 Å (0·11 μm)
1970	*hydrogen cyanide	HCN	3·38 mm
1970	*formyl ion	HCO^+	3·36 mm
1970	*cyanoacetylene	HC_2CN	3·30 cm
1970	methanol	CH_3OH	0·3, 1·2, 36 cm
1970	formic acid	HCO_2H	18·3 cm
1971	carbon monosulphide	CS	2·04 mm
1971	formamide	$HCONH_2$	6·49 cm
1971	silicon monoxide	SiO	2·30 mm
1971	carbonyl sulphide	OCS	2·74 mm
1971	methyl cyanide	CH_3CN	2·72 mm
1971	isocyanic acid	$HNCO$	0·34, 1·36 cm
1971	*hydrogen isocyanide	HNC	3·31 mm
1971	propyne	CH_3C_2H	3·51 mm
1971	acetaldehyde	CH_3CHO	28·1 cm
1971	thioformaldehyde	H_2CS	9·5 cm
1972	hydrogen sulphide	H_2S	1·78 mm
1972	methyleneimine	CH_2NH	5·67 cm
1973	sulphur monoxide	SO	3·49 mm
1974	ethynyl	C_2H	3·43 mm
1974	dimethyl ether	CH_3OCH_3	3·47, 9·6 mm
1974	*methylamine	CH_3NH_2	3·48, 4·1 mm
1974	*hydrodinitrogenyl ion	N_2H^+	3·22 mm
1975	cyanamide	NH_2CN	2·98, 3·73 mm
1975	silicon sulphide	SiS	2·75, 3·30 mm
1975	ethanol	CH_3CH_2OH	2·8, 3·3, 3·5 mm
1975	sulphur nitride	SN	2·60 mm
1975	sulphur dioxide	SO_2	3·46, 3·58 mm
1975	acrylonitrile	CH_2CHCN	21·86 cm
1975	methyl formate	HCO_2CH_3	18·6 cm
1976	*formyl	HCO	3·46 mm
1976	cyanodiacetylene	HC_4CN	2·80, 11·28 cm
1976	methyl cyanoacetylene	CH_3C_2CN	3·46 mm
1977	cyanoethynyl	C_2CN	3·03, 3·37 mm
1977	ketene	CH_2CO	2·94, 3·00, 3·67 mm
1977	nitrosyl hydride	HNO	3·68 mm
1977	ethyl cyanide	CH_3CH_2CN	2·58–3·06 mm
1977	cyanohexatri-yne	$HC_2C_2C_2CN$	2·95 cm
1978	nitric oxide	NO	1·99 mm

*Several isotopic forms are known.

diameter, that is more than 600 000 times greater. The strength of the OH lines varies markedly over periods of a few months.

All these factors point to some sort of amplification occurring in the production of the OH lines, and such sources are called MASERS (an acronym for *M*icrowave *A*mplification by *S*timulated *E*mission of *R*adiation), the microwave equivalent of LASERS. The maser process is, as yet, not well understood. It is clear that the OH molecules absorb energy by some process and convert it into radiation; but the amplification can only occur under extremely critical conditions of density, temperature and magnetic field. Some masers appear to be identified with stars which are in the process of formation, while others are associated with very late-type red stars, such as Mira variables and NML Cyg. Recently, other masers – water vapour (H_2O), silicon monoxide (SiO), methyl alcohol (CH_3OH) and methylidyne (CH) – have been discovered, and it is hoped that they will provide more clues to the working of the mechanism.

All molecules, whatever process they are undergoing, are basically unstable in the hostile environment of space. Those which have been discovered must, therefore, be relatively young; and astrochemists are currently working hard on theories of their formation. It appears that the simpler diatomic molecules can be built up on the surfaces of dust grains; but the complex varieties need to be assembled by collisions between atoms in dense gas clouds and there is no general agreement as to how this happens.

Conditions in space are vastly different from those in terrestrial laboratories, which makes prediction of the end result that much more difficult. A few outspoken scientists have even suggested that the complex molecules which make up living creatures were originally built up in space.

Matter between the stars

The space between the stars is far from empty. Although it is a far more perfect vacuum than any we could create on Earth, the minute amounts of matter present – less than 1 atom per cm^3 on average – begin to add up across the vast gulfs which separate the stars. Optical and ultraviolet astronomers see these effects in the spectra of distant stars as thin absorption lines: lines caused by atoms of calcium (Ca), potassium (K), sodium (Na) and iron (Fe) in space. Broader absorption lines of unknown origin, the **diffuse interstellar bands**, also cross the spectra of remote stars. Radioastronomers can tune into a cacophony of broadcasts from the matter between the stars, each telling a different part of the story. Those near 21-cm wavelength tell of the distribution of cool hydrogen gas (H I) in space; while a radio astronomer listening in between 300 m and 3 mm will learn of the convoluted paths travelled by cosmic ray electrons. Also, X-ray, γ-ray and cosmic ray researches all have their part to play in unravelling the structure of the interstellar medium. Surveys by future spacecraft similar to COS-B (gamma-rays) and Einstein (X-rays) should be of the greatest value in this particular field.

This all-sky photograph of the Milky Way reveals how the dust grains clump along the plane of our Galaxy.

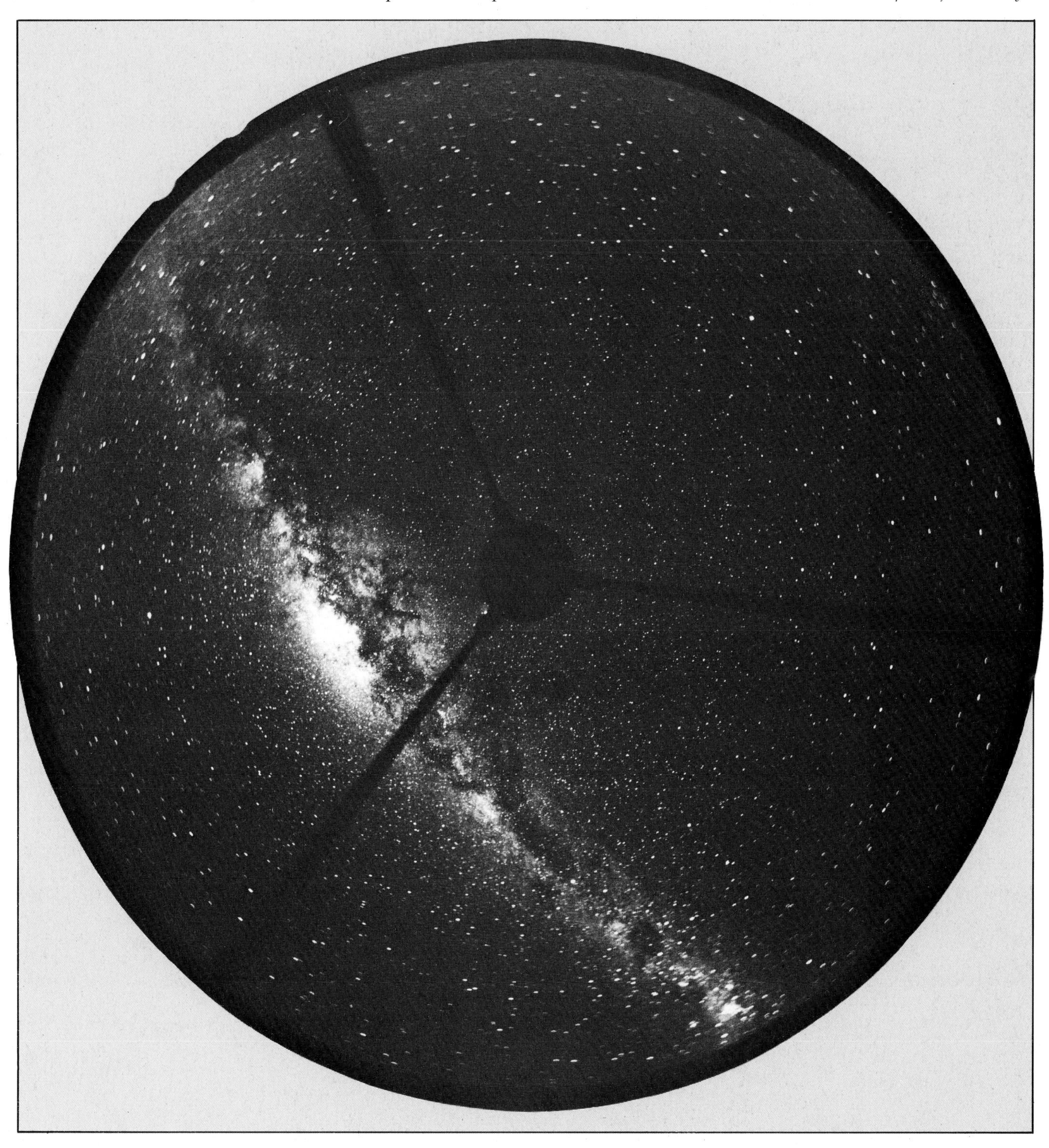

The Rosette Nebula in Monoceros, a cloud of gas made to glow by radiation from newly-formed stars.

Opposite page, bottom: *Dust-laden clouds in the young Pleiades star cluster reflect starlight with a bluish glow. This nebula surrounds the star Merope.*

Opposite page, top: *A 'radio photograph' of the supernova remnant Cassiopaeia A, the remains of an exploded star. Such energetic sources may be responsible for some of the cosmic rays which fill the disc of our Galaxy.*

Ironically, it is the least abundant component of the interstellar medium which is the most obvious. Dust in space only accounts for 0·1 per cent of the mass of our Galaxy, and although it is spread thinly inside the disc, its cumulative effect is to scatter and absorb the light from distant stars to such an extent that our view of the Galaxy becomes somewhat parochial. The amount of this absorption is uneven; it averages 1–2 magnitudes per kpc at visual wavelengths, but it can become tens of times higher in the dark, dusty clouds.

Absorption by dust is sometimes called **interstellar reddening**, because blue stars are dimmed far more by the dust than red stars of the same brightness. If we now compare the dimming at a wider range of wavelengths, we find that the dust is practically transparent to infrared radiation longer than 10 μm, while the amount of absorption at ultraviolet wavelengths ($\sim$0·1 μm) is far greater than in the optical. The resulting **extinction curve**, showing the absorption by dust as a function of wavelength, yields a great deal of useful information about the nature of the dust. It tells us at once that we can avoid the problems of dust obscuration altogether if we probe the Galaxy at long wavelengths (with infrared and radio waves). Even the size of the dust grains can be inferred from the extinction curve. The large dimming of optical and ultraviolet light indicates that the particles responsible must have sizes comparable with these wavelengths, that is, around 0·1 μm. Some astronomers believe that the three 'bumps' on the extinction curve (at 0·2 μm, 3 μm and 10 μm) reveal the actual composition of the grains. These are wavelengths at which graphite, ice and silicate particles absorb radiation, and it is plausible – although by no means certain – that the dust grains are made up of these substances. The model which best fits the data is that of a silicate core coated in ice, the whole grain weighing 10^{-18} kg!

In addition to being spread out through space, dust grains are frequently found in large clumps around sites where stars are forming. It appears that grains may play a key role in star formation by protecting collapsing gas clouds from the disrupting effects of ultraviolet-bright stars nearby; and it is likely that the dust has a further role to play in the formation of planetary systems around young stars. Many very young stars are cocooned in shells of heated dust; and even some of the stars in the 6×10^7-year-old Pleiades cluster are seen to be wreathed in dust. These particles reflect starlight, shining in the sky as a glowing **reflection nebula**.

Although grains are found in association with the birth of stars, astronomers believe that they originate in the atmospheres of extremely old stars. It is possible that the condensation of gas in the outer atmospheres of cool red giants, such as μ Cephei and IRC + 10216, produces dust grains, ultimately blown away by a stellar wind into space. In this way, stars of varying compositions could create different types of grain.

Magnetic field and cosmic rays

Another facet of the interstellar dust is revealed in the light from distant stars. Their starlight is **polarized**, that is, the light vibrations tend to occur in one plane. The amount of POLARIZATION increases with the extinction due to dust, showing that some alignment of the dust grains is responsible. This is one of the many indications that a large-scale magnetic field, with a strength of about 3×10^{-10} TESLA (T), permeates the disc of our Galaxy (Fig. 6·7).

Radioastronomers can detect the field more directly, for it 'splits' the 21-cm hydrogen line into two closely-spaced lines (by the ZEEMAN EFFECT); and hot interstellar gas in the general magnetic field affects the polarization of radio waves from pulsars and extragalactic radio sources. The magnetic field also has an important effect on the very high speed particles called **cosmic rays**. These are electrically charged particles, moving through space with almost the velocity of light, and without the restraint of a magnetic field they would escape from the Galaxy in a few thousand years. Charged particles become 'tied' to magnetic fields, however, whirling along them in extended helixes, and consequently the disc magnetic field can bottle up the cosmic rays for millions of years.

Nine-tenths of cosmic ray particles are protons (hydrogen nuclei), and almost one-tenth helium nuclei, with a small admixture of the nuclei of the heavier elements; in addition there is one electron to every hundred protons. Energies of cosmic rays are customarily measured in ELECTRON VOLTS (eV), the energy gained by an electron moving through a POTENTIAL DIFFERENCE of 1 volt (V). Cosmic rays with energies higher than 10^8 eV can penetrate into the Solar System, and are detected by balloon-borne detectors, or more recently by satellites and deep-space probes. The number of cosmic rays drops off sharply with increasing energy, falling to an expected detection rate of less than 1 per m^2 per year for particles with energies higher than 10^{16} eV. Clearly, not many such particles can be picked up by the relatively small instruments carried by balloons, satellites and probes, but they are detectable indirectly from the ground. These very high energy cosmic rays produce a large 'shower' of elementary particles when they hit the top of the Earth's atmosphere, and this air shower can be picked up by particle detectors spread over several square kilometres on the Earth's surface. The most energetic cosmic rays known have energies of around 10^{20} eV – the energy of a tennis ball travelling at 100 km per h.

Although Solar-System-based experiments can only sample cosmic rays in a tiny portion of the Galaxy, the new science of γ-ray astronomy tells of their wider distribution. Cosmic rays colliding with hydrogen atoms in interstellar space can produce a shower of neutral PIONS, which quickly decay to γ-rays. The gamma ray satellites SAS-II and COS-B show enhanced emission from the central regions of the Galaxy, just where a higher density of gas and cosmic rays is expected. Unfortunately the generally low sensitivity and poor directional qualities of satellite-borne detectors means that very little detail can be distinguished in these all-sky surveys, and this is likely to remain the case until better detectors can be developed.

All except the highest energy cosmic rays (which may well be extragalactic) must be accelerated to their high speeds by energetic processes in our Galaxy. Supernova explosions are the most likely cause. Particles can either be accelerated by the neutron star relic (like those energized by the Crab pulsar, which make the Crab nebula shine), or by the shock wave of the exploded gas shell expanding out into the interstellar gas. These **supernova remnants** are generally faint optically because of their high temperatures (around 10^6 K), but this very hot gas makes them prominent X-ray sources. They also 'shine' at radio wavelengths by the synchrotron process, as particles accelerated at the SHOCK FRONT whirl around the lines of magnetic field in the expanding shell.

The interaction of the general cosmic ray background with the magnetic field of interstellar space also causes the whole of the Galaxy's disc to emit radio waves by the synchrotron process. The important particles here are the electrons, for although they are a rare component of cosmic rays, their small mass makes them very efficient synchrotron broadcasters. This radio background predominates at 1 cm wavelengths upwards, because of the preponderance of low energy cosmic rays. Since it arises in the interstellar region of the disc, the distribution roughly follows the path of the visible Milky Way across the sky, but without troublesome obscuration by the radio-transparent dust grains.

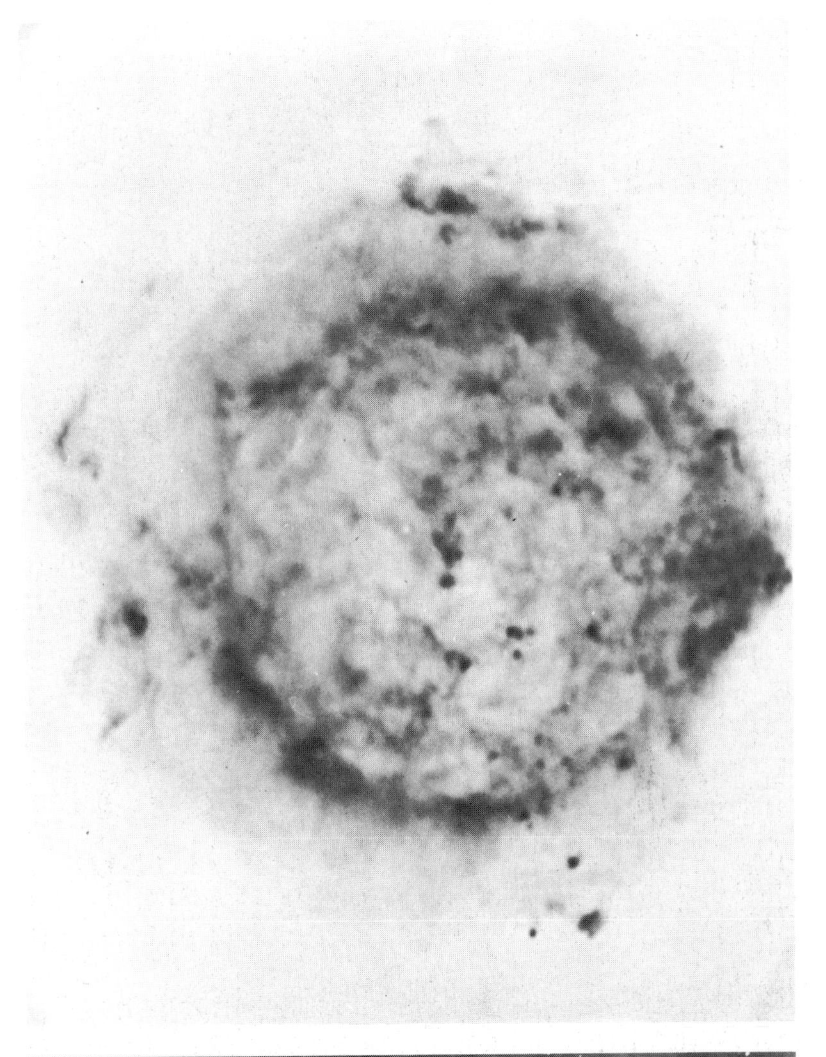

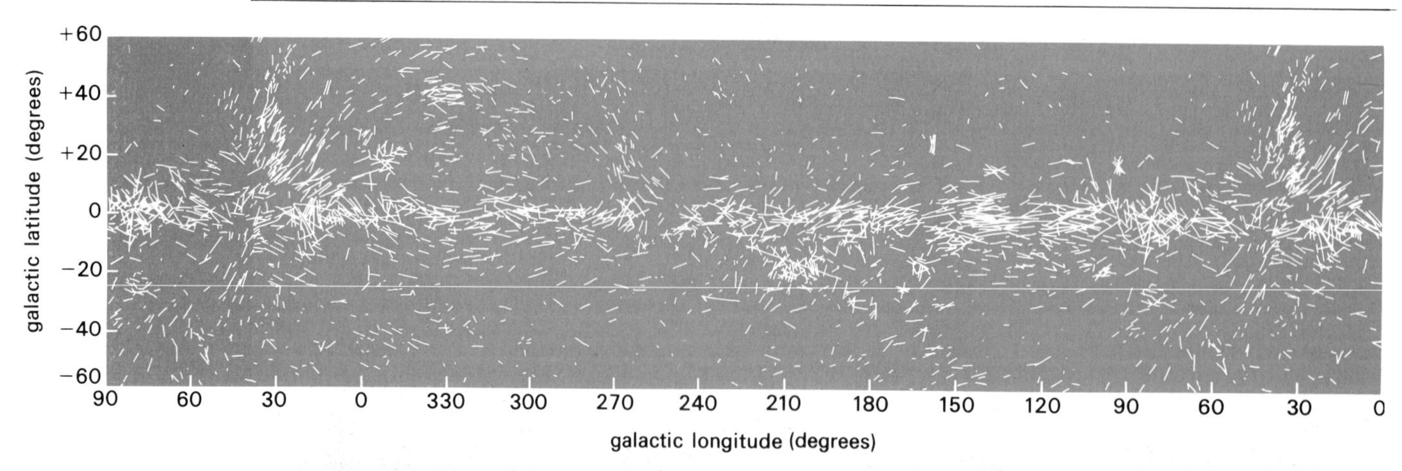

Fig. 6·7 above: *Polarization observed in the light from distant stars reveals that there is a magnetic field in our Galaxy which aligns the dust grains. Measurements such as this tell of both the strength and direction of this magnetic field, and the composition of the grains.*

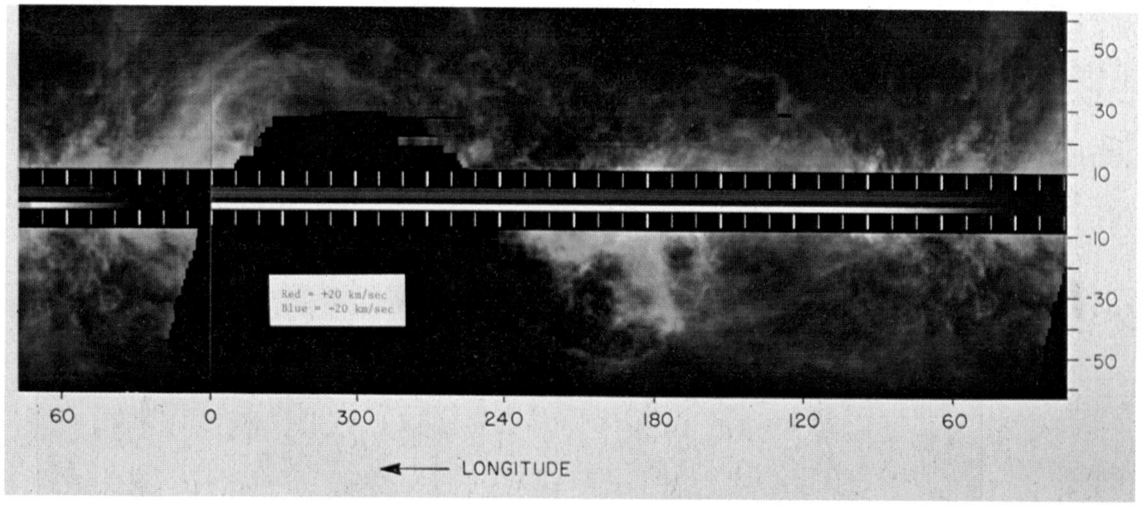

Above right: *Half a million observations with radio telescopes were combined in a computer to construct this map of the distribution of neutral hydrogen (HI) in our Galaxy. The plane of the Galaxy (where the hydrogen is densest) is blanked out in order to reveal the higher-latitude gas. Colours indicate motion: blue towards us, red away.*

Radio studies and hydrogen distribution

The transparency of the interstellar medium to radio waves, and the sophistication of modern radio-telescopes make these frequencies ideal for studying the large-scale structure of the Galaxy. Most useful of all, neutral hydrogen (H I) atoms emit a spectral line in the radio region, the famous 21-cm line. Hydrogen makes up 70 per cent (by mass) of the interstellar gas, and so it is ideal for mapping the interstellar medium; moreover, slight deviations from this frequency due to the Doppler effect reveal motions of the gas along the line of sight (its radial velocity).

A hydrogen atom consists of a single electron orbiting a proton, and each of these particles is spinning about its own axis. The laws of quantum mechanics insist that the spin axes of proton and electron be parallel, and so the hydrogen atom can exist in only two states: either with both particles spinning the same way (higher energy state), or spinning in opposite directions (lower energy state). In an interstellar hydrogen cloud, atoms are continually colliding, and once in a few hundred years the electron in any given atom will flip over during the collision, and hence change the atom's spin-state. There is a very small chance (about one in 10^5) that the electron will flip spontaneously from having the same spin as the proton to the opposed spin state, and in this case it emits the excess energy as a radio wave at 21-cm wavelength. Despite the very low emission rate – any given atom emits this radiation once in 11×10^6 years, on average – the vast amount of hydrogen in space makes the radiation readily detectable.

Although other lines, for example those of OH, also occur in the interstellar medium, the hydrogen line is the most generally useful. Careful study of the H I emission from a cloud reveals its extent, its velocity, whether it is rotating, its density and its mass. If it happens to lie in front of a bright background source giving a continuous radio spectrum, the cloud's H I line can be see in absorption, and its temperature can then be determined.

From these observations, the interstellar medium seems to be extremely non-uniform. The simplest interpretation is that the hydrogen is clumped together in clouds about 10 pc across, with a density of some 2×10^7 per m^3 and a temperature of around 100 K. These clouds occupy only a few per cent of the disc's volume, and they are separated by a much hotter, more tenuous **intercloud medium** with a density 100 times lower and a temperature of about 3 000 K. These parameters are only rough, for individual clouds and patches of intercloud medium vary considerably. The clumping into clouds may well have been caused by the expansion of old supernova remnants, which eventually sweep up thin dense shells of cool gas. The centres of old supernova remnants are even hotter and more tenuous than the intercloud gas.

It is now thought that extreme heating by supernovae has given rise to the exceptionally hot gas observed at considerable distances from the galactic plane. The so-called high-velocity clouds have been

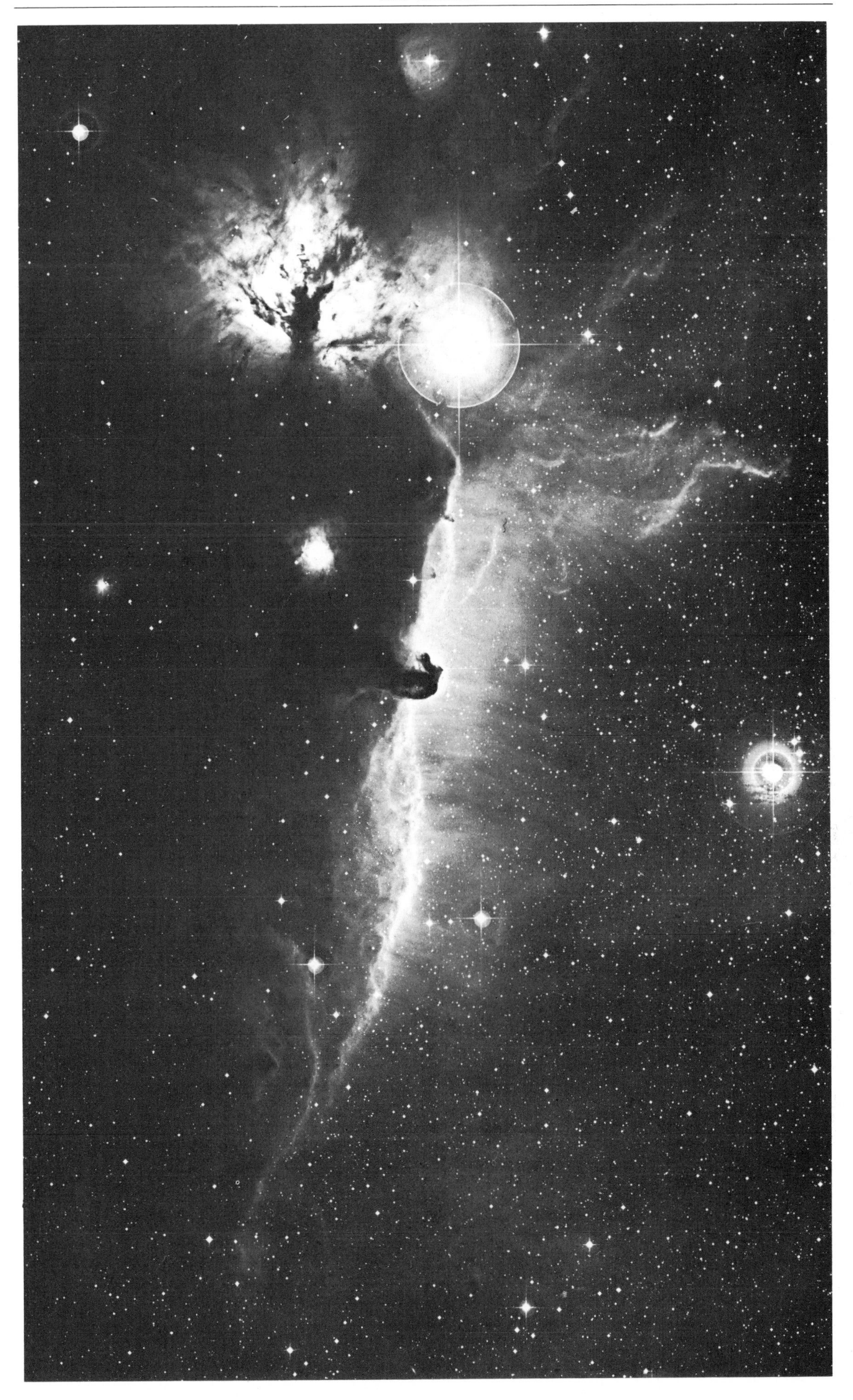

Like a celestial chess-piece, the Horsehead Nebula stands out starkly against clumps of glowing gas clouds. Such dark nebulae were once believed to be starless voids in space, but are now known to be concentrations of dense, obscuring material.

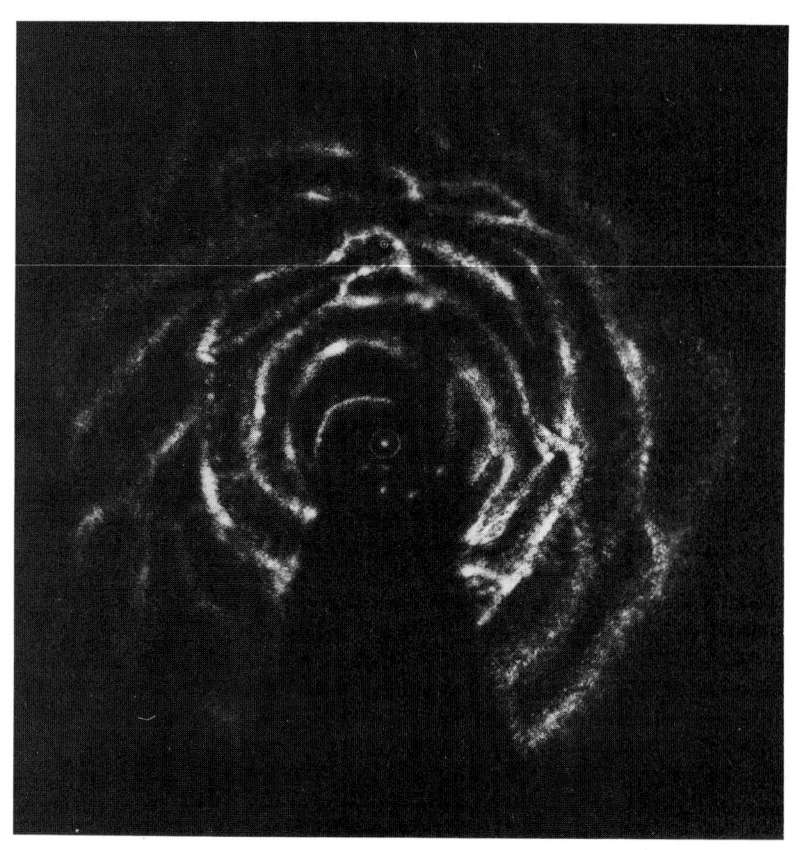

Radio surveys at a wavelength of 21 cm reveal our Galaxy to be spiral in form. The sun lies in the position circled; the region opposite it is blank because the hydrogen there does not show a Doppler shift (due to the Galaxy's uniform circular rotation) and so no information about its distribution can be obtained.

known for some time, but understanding of them has only come through research with the highly successful IUE (International Ultraviolet Explorer) satellite. This has shown the presence of very hot gas clouds with temperatures as high as 10^5 – 10^6 K at distances of at least 5 000 pc from the galactic plane; there are indications that they may extend out to about 8 000 pc on each side. It now appears that intense heating caused by supernova explosions causes 'fountains' of hot gas to rise into the outer regions of the Galaxy, and that cooler gas is flowing back towards the galactic plane. The considerable mass of gas at such distances forms the **galactic corona,** and similar coronae have been shown to exist around other galaxies.

The structure and distribution of the Galaxy's disc is revealed by a study of the hydrogen distribution. A radiotelescope pointing in a direction within 90° of the galactic centre is receiving radiation from all along a long line stretching right across the Galaxy. All the hydrogen clouds in this line of sight are orbiting the centre of the Galaxy at different radii from it, and simple geometry shows that the highest-velocity cloud is that situated where our line of sight passes closest to the galactic centre. By looking for the highest velocity in the hydrogen spectrum, we can thus find the rotational velocity of a cloud at that particular distance from the galactic centre. A different line of sight lets us measure the velocity of a cloud at a different distance from the centre; and by building up observations in this way it is possible to construct the **rotation curve** of our Galaxy. This curve simply shows how the orbital velocity of a gas cloud (or a star) in a circular orbit around the Galaxy depends on its distance from the centre.

One immediate use of the rotation curve is in determining the total mass of our Galaxy. Using Newton's law of gravitation, it is relatively easy to calculate the mass of the Sun, for example, by knowing the Earth's orbital velocity and the Earth–Sun distance. Unfortunately, the situation is rather more complicated with the Galaxy, for the Sun is not orbiting just a central compact body, it is moving in the gravitational field of all the stars in the Galaxy. The rotation curve does allow us to make a reasonable estimate, although the result depends on exactly how the mass of the Galaxy is distributed. Within the orbit of the Sun (radius 10 kpc) the Galaxy contains about 10^{11} $M_\odot$; and although it is not possible to measure directly the rotation curve outside the Sun's orbit, the matter further out is estimated to raise the Galaxy's total mass by another 50 per cent.

Armed with the Galaxy's rotation curve, the distance to any H I cloud can be calculated from its radial velocity and its angular distance from the galactic centre. The distribution of clouds so obtained clearly show the spiral arms of the Galaxy far beyond the region mapped by optical spiral arm tracers.

Present theories of spiral structure indicate that the material making up a spiral arm is just passing through it: there is a spiral gravitational pattern, or **density wave**, which bunches up stars and gas as they travel around the galactic centre. Since this gravitational pattern must affect the gas velocity by some 10 km per s, speeding it up as it enters the arm, and slowing it down on leaving, distances derived simply from radial velocities are not exact. The positions of distant spiral arms are thus not known as accurately as was once hoped, but the spiral nature of our Galaxy is established beyond doubt by the H I observations.

The galactic centre

The picture of our Galaxy which we have built up is a calm and orderly one, and until recently there was little reason to question this view. The first indications that matters might be different came to light in the late 1950s, when radioastronomers first mapped the H I distribution in the Galaxy. All the spiral arms so far discovered had been found to be smoothly rotating about the galactic centre. In particular, all hydrogen clouds along the line of sight from the Sun to the centre of the Galaxy had zero radial velocity, just as would be expected on a uniformly rotating model. But in 1957 Jan Oort, working in Holland, found two clouds of H I lying in this direction which did have significant velocities along the line of sight. In addition to their normal rotation, these clouds actually appeared to be moving outwards from the galactic centre, one towards and the other away from us. Further work established that the closer feature was an armlike extension of gas (called the **3 kpc arm**, after its estimated distance from the centre), and that the cloud on the far side of the centre was expanding outwards at 135 km per s (hence its name of the **+ 135 km per s feature**).

The following years saw great improvements in radioastronomy methods, and the consequent discovery of several smaller 'expanding' features in the direction of the galactic centre. Some even appeared

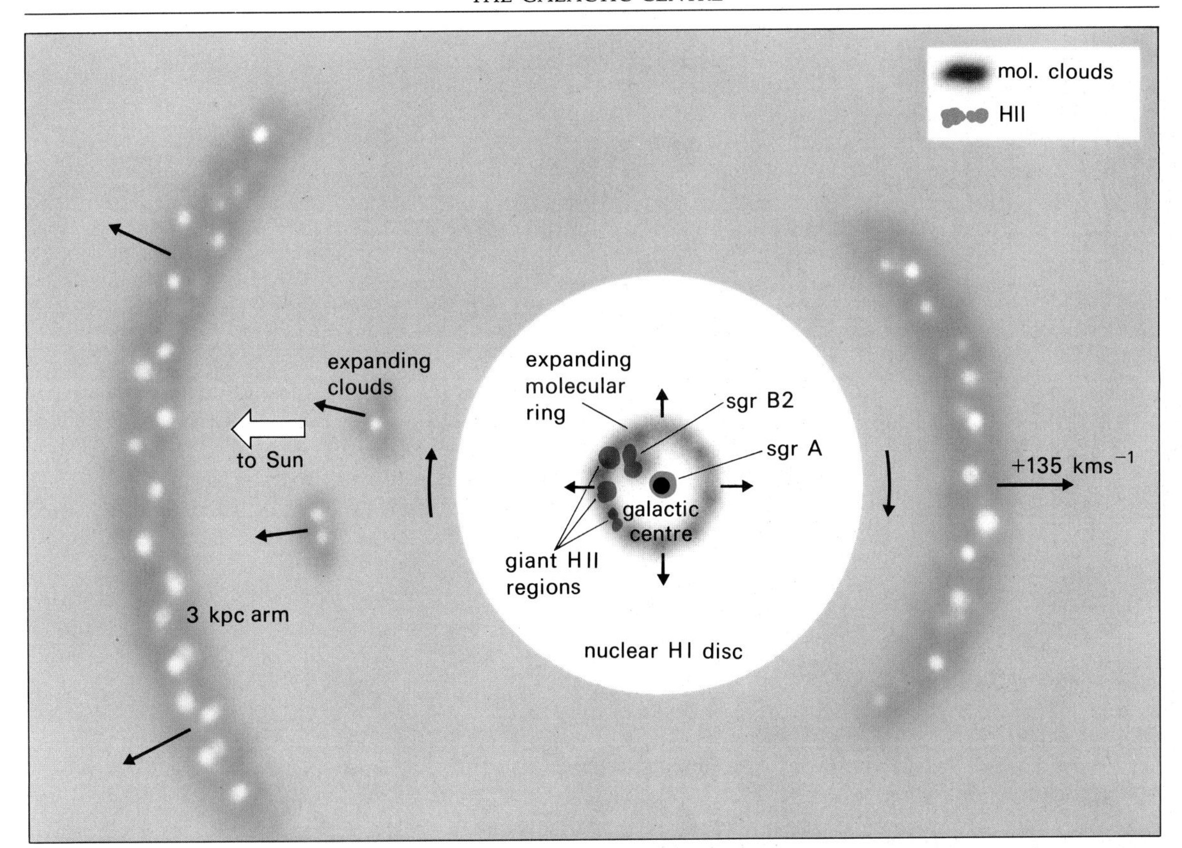

Fig. 6·8
Features surrounding the centre of our Galaxy, showing the expanding ring of molecular clouds embedded in the rotating disc of neutral hydrogen. Outside are several expanding features, including the 3 kpc arm (left).

to be thin jets of material moving at steep angles to the galactic plane. The evidence strongly suggested that some disturbance or explosion had occurred at the galactic centre. However, astronomers in the early 1960s were unused to such concepts and preferred to explain the observations in more conventional terms. It is just possible, for example, that some of the 'expanding' features arise from the ever-changing orientation of a bar-like distribution of mass at the galactic centre, such as we see in the so-called barred spiral galaxies. Other astronomers believed that Oort's features could be RESONANCE effects in the Galaxy's gravitational field. But later work has turned up many more anomalous features, and these have generally lent further support to the explosive hypothesis.

Because of the 28–30 magnitudes of visual absorption between ourselves and the galactic centre, our picture of its structure has been built up almost entirely from observations at radio and infrared wavelengths and, more recently, from X- and γ-ray observations (Fig. 6·8). Such observations show that, after an initial drop inside the 3 kpc arm, the H I density rises as we approach the centre, and the gas forms a uniformly rotating disc, some 1 500 pc across. Embedded in this disc, with a diameter of 380 pc, is an expanding ring of cool, dense molecular clouds jostling with supergiant H II regions. There is an extraordinarily high concentration of molecules in this region, including the most enormous complex of molecules in the entire Galaxy: the 10^6-$M_\odot$ cloud Sagittarius B2. Some 30 pc across, and containing at least seven H II regions each as bright as the Orion Nebula, Sgr B2 sits just inside the expanding ring. Every type of molecule so far discovered in the Galaxy has also been found in the Sgr B2 complex.

Moving inwards from the molecular ring, we reach the radio source Sagittarius A. This, in fact, comprises two regions: Sgr A East, a bright supernova remnant, and Sgr A West, which surrounds the galactic centre.

Between them is a giant molecular cloud, rushing away from the centre at a velocity of 40 km per s. Sgr A West is another unique feature in the Galaxy, a very evenly lit H II region whose gas begins to form a thin uniformly rotating disc towards the centre. This ionized disc is only a couple of parsecs across, 1 000 times smaller than the H I disc we discussed earlier. Infrared observations show it to be in turbulent motion, studded with hot clouds of gas, possibly stripped off stars lying within a few parsecs of the galactic centre (Fig. 6·9).

However, their infrared emission gives an estimate of their space density, which reaches the alarming figure of 10^8 times the solar neighbourhood value as we approach the central star cluster of our Galaxy. Even at these densities, the stars are not touching, although near-neighbours would be separated by a mere 4 light-days. Astronomers have estimated that the central star cluster – made up largely of cool, Population II stars – would have a mass of around $3 \times 10^6 M_\odot$. There is, however, a puzzling discrepancy, for the rotation speed of the 1·2 pc ionized gas disc indicates a central mass twice as great as this. Where is this hidden mass? Current thinking indicates that the innermost regions of our Galaxy may contain several million solar masses which are not in the form of stars.

Right at the centre of the ionized gas disc lies a tiny, possibly variable, radio source. Using very long baseline interferometry (see page 242), radioastronomers have tracked down the emission to a region 150

X-ray observations such as this one from the Einstein satellite are beginning to reveal some of the structure of the very centre of the Galaxy.

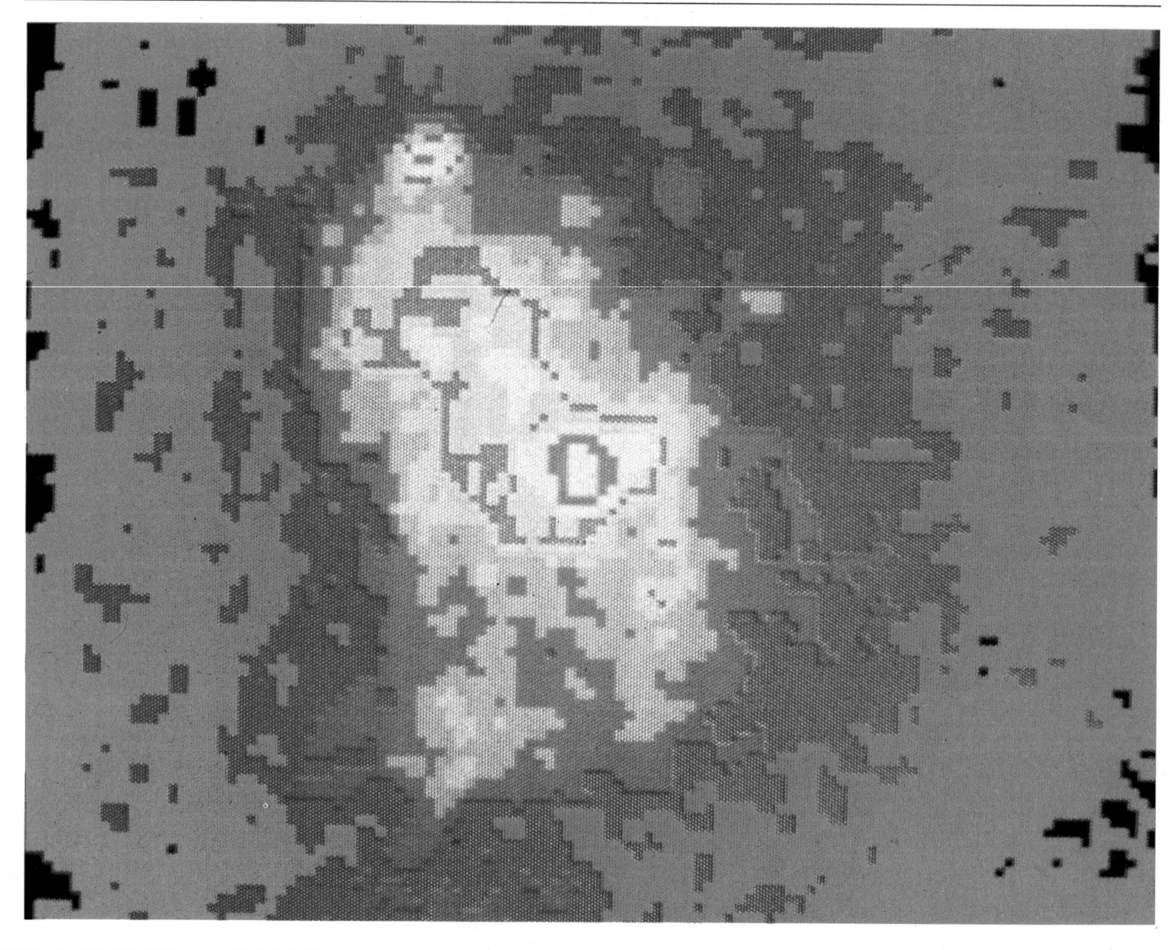

Fig. 6·9 below: *Surrounding the compact radio source which marks the exact centre of our Galaxy is a rapidly-rotating disc of ionized hydrogen, part of the radio source Sagittarius A.*

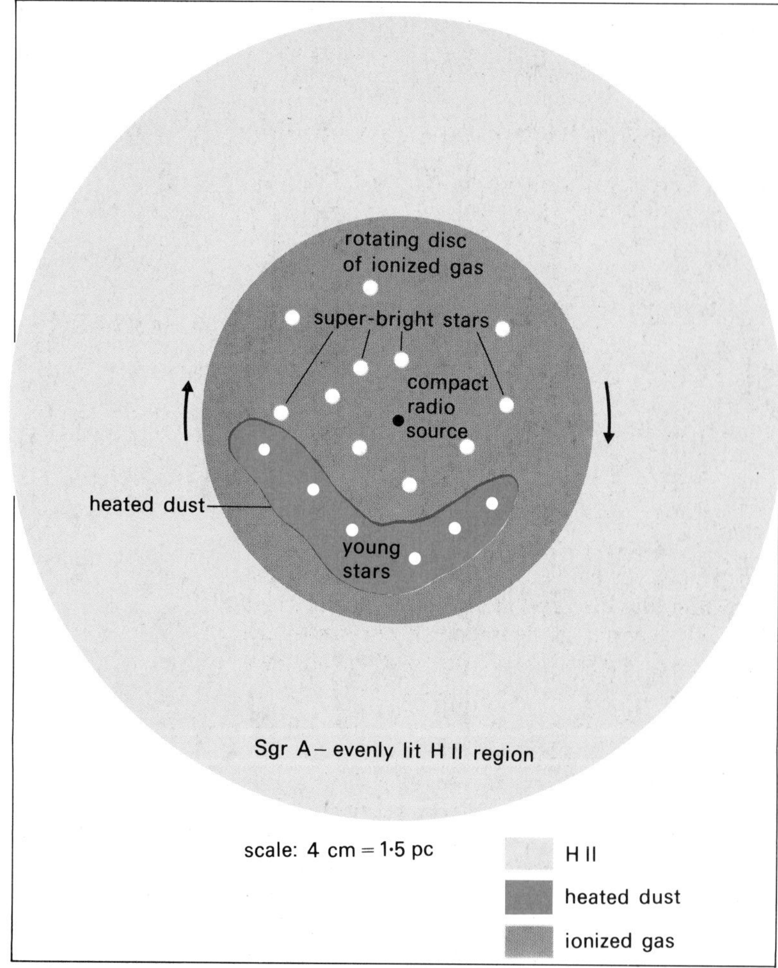

au across, with 25 per cent of the power coming from an ultracompact source only 10 au in diameter. Although the output from this source is some tens of millions of times weaker than the fantastically powerful nuclei of exploding galaxies, its emission, volume for volume, is comparable. And it is staggering to consider that our galactic centre radio source would comfortably fit inside Jupiter's orbit, were it in the Solar System.

Present-day physics knows of only one mechanism which can produce such concentrated energy: the acceleration of charged particles in an accretion disc around a black hole. Current ideas of galaxy evolution indicate that a central massive black hole may form early on in a galaxy's life, following on from the first rapid infall of matter. The black hole need not always be 'active', with an energetic accretion disc, but when there is sufficient gas present, the nucleus will be able to glow, and there will be a source of energy capable of pushing away clouds of gas at high speeds.

At present, our Galaxy is certainly not active. But there are other galaxies – Seyfert and radio galaxies, for example – which show evidence for intermittent explosions.

If this picture is correct, when did our Galaxy's nucleus last flare up? To account for the 3 kpc arm, there must have been an explosion some 12×10^6 years ago; while the expanding molecular ring points to an outburst 10^6 years in the past. In addition, there may be less violent activity on a timescale of about 10^4 years, maintaining the turbulent gas motions in the small ionized disc.

CHAPTER SEVEN

Extragalactic astronomy

Complete acceptance of the fact that some astronomical objects are outside the Milky Way system came only in the mid-1920s. During the previous seventy years it had been established that many nebulae have a spiral shape and the spectra of such objects suggested that they are collections of stars, but their distances were still unknown. Novae were recognized and their apparent brightness suggested that the distances are great, but just how great was uncertain since the distinction between novae and supernovae was not yet clear. Only with the advent in 1918 of the 100-inch (2·5-m) reflector on Mount Wilson was it possible to detect individual stars and for approximate distance measurements to become practical. The conclusive step came in 1924 when Edwin Hubble used observations of Cepheid variables (page 66) in several nearby nebulae to determine distances accurately. By this time, too, it was coming to be realized that the distribution of nebulae in the sky, avoiding the plane of the Milky Way, is an observational effect caused by interstellar obscuration and is not due to an uneven distribution in space, which would have required them to be attached to the Milky Way system. These results were related to the earlier concept of **island universes**, isolated separate star systems which later came to be called galaxies.

In spite of these discoveries, astronomers still refer to bright nebulae by their reference numbers in the 1784 catalogue of 103 objects drawn up by Charles Messier (M), or in Johann Dreyer's *New General Catalogue* (NGC) of 1888 which, with its later supplements, the *Index Catalogue* (IC), contains more than 13 000. These all list objects of extended appearance lying beyond the Solar System, objects which are of very different natures and distances so that, for instance, it turns out that only 34 of the items in Messier's catalogue are extragalactic.

The nearest external galaxies are the two Magellanic Clouds, satellites of our own Milky Way system, although they have never been regarded as nebulae and do not appear in the M or NGC catalogues. They are seen prominently in the southern sky and look like detached pieces of the Milky Way.

The word nebulae is now restricted to interstellar clouds of gas and dust, except sometimes colloquially in phrases such as extragalactic nebulae or the Andromeda nebula. The term island universes is not now used; the **universe** is defined as the totality of observable things and necessarily is unique.

It is now customary to write Galaxy, with capital G, for the Milky Way system and to use small g for other galaxies. Generally, although not quite always, use of the adjective **galactic** is restricted to our own Galaxy, with increasing use of the new word **galaxian** for another galaxy or galaxies in general. This convenient usage is followed here. Consistent with it, **extragalactic** means anything outside the Galaxy, including other galaxies, and, if not consistent, at least unambiguous, is the use of **intergalactic** to mean what is between the galaxies.

In the past thirty years, radioastronomy has become as important as optical astronomy in extragalactic studies; indeed the two are largely complementary. In general, optical observations show more detail, while spectral lines give information about motions, distances and chemical abundances. Radio signals, on the other hand, can be detected often from galaxies or parts of galaxies, which are too faint to be studied optically. Newer areas of observation in the infrared, ultraviolet and X-ray regions of the spectrum also contribute important information.

Classification of normal galaxies

Most galaxies are **normal**, in contrast to **active** galaxies, discussed later, which involve violent non-thermal processes in the central region or nucleus. Normal galaxies can be classified in a small number of basic types; the outer regions of galaxies with active nuclei can be classified in the same scheme.

Hubble's classification

The first classification was made by Edwin Hubble in the 1920s and, with modifications, is still in use. It is morphological, that is, based on the appearance of a galaxy in the sky, and contains three major classes: galaxies with no evident internal structure, which appear as ellipses (**elliptical galaxies**); those with a thin, flat disc containing a structure of spiral arms and a central condensation or nucleus (**spiral galaxies**); and those which are neither of these (**irregular galaxies**). These classes are subdivided, and there is also a sequence of **barred spirals**, SB, parallel to the spiral galaxies, S.

Elliptical galaxies are classified by the shape of the ellipse seen in the sky. Mathematically, if we take a and b as the axes of an ellipse, the ellipticity or flattening is given by $(a-b)/a$, which runs from zero for a circle towards 1 for a very flat system. The galaxies are classified by writing ten times this value, taken to the nearest whole number, after the letter

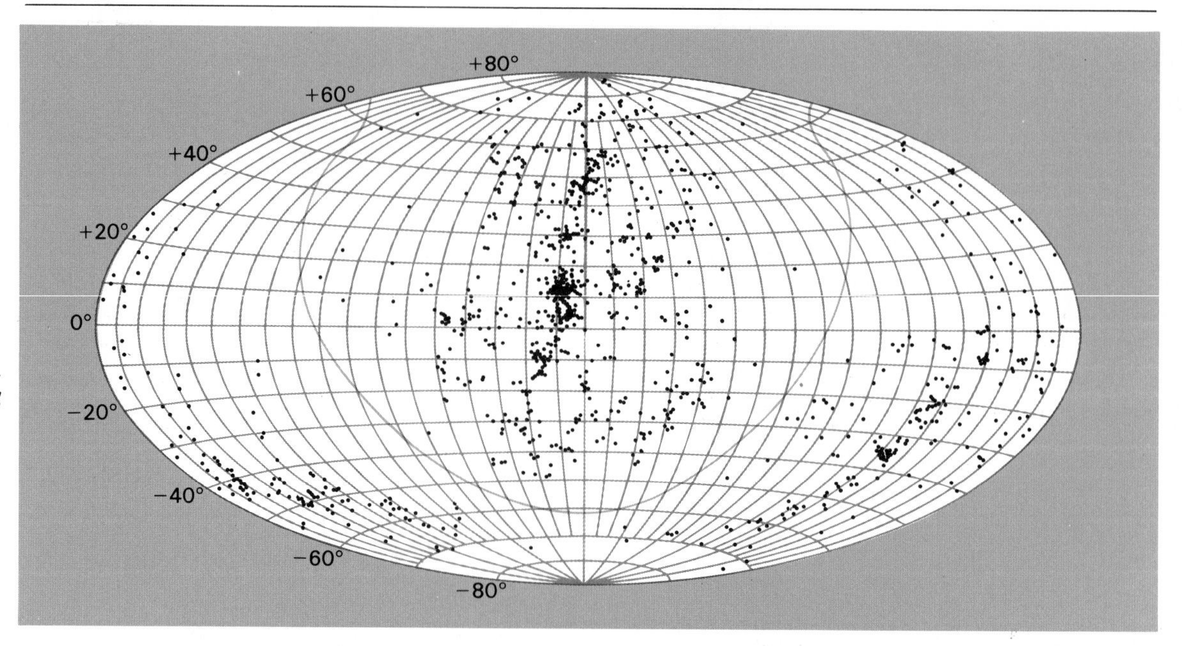

Fig. 7·1
Positions in the sky of the thousand brightest galaxies, plotted using an equal-area projection of the entire sky using right ascension and declination as co-ordinates. The curved line represents the galactic equator. The distribution of galaxies avoids directions near the plane of the Galaxy, and also shows a strong tendency towards clustering (page 203).

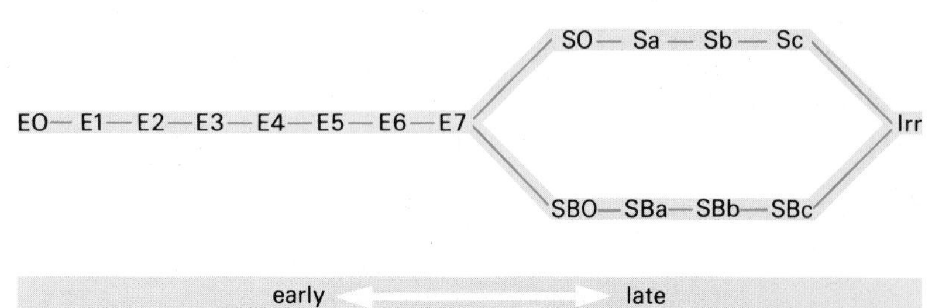

The Hubble diagram of galaxy types.

E. Classes observed run from E0 to E7 only, systems flatter than E7 no longer being ellipses but displaying a central condensation and a disc instead. Next after E7 come the lens-shaped galaxies S0 and SB0, which have discs but no spiral arms. They lie in intermediate positions in the continuous Hubble sequence from E to S and SB types.

The classification Sa, Sb, Sc for spiral galaxies is related to the size of the central condensation and the tightness with which the spiral arms are wound; they vary together. For Sa, the arms are tightly wound and the central condensation is large; Sb have more open arms and a smaller central condensation; Sc have very open arms and a very small nucleus. The amounts of interstellar gas and dust increase in sequence from Sa to Sc. For barred spirals, SBa and so on, the structures are the same except that the arms emerge not from the nucleus but from the ends of a prominent bar passing through the nucleus. Unlike an elliptical, the classification of a spiral galaxy is absolute and does not depend on its orientation in the sky.

Two types of irregular galaxy are recognized. The more common type Irr I (or Ir I or Im), which includes both the Magellanic Clouds, clearly follows in sequence after Sc and SBc, with the break-up of arms into a confused structure and the presence of still more dust and gas. The second type Irr II (or Ir II or IO) is rarer; they are very different from Irr I galaxies, being peculiar, chaotic objects, highly obscured and reddened by internal dust. They lie outside the general sequence of galaxies, and often display strong activity.

Significance of the classification

The Hubble sequence is based purely on appearance. There are, however, systematic physical variations – the relative number of bright blue stars and the content of gas and dust increase steadily from E and S0 galaxies through the sequence of S or SB to Irr I. So too does the galaxian angular momentum, except that it is small in irregular galaxies.

The sequence is continuous, but there are strong reasons to believe that galaxies do not evolve along it, in either direction. One reason is that galaxies of all classes contain highly-evolved stars (red supergiants), so they must all be around the same age, at least 10^{10} years old. Secondly, the giant ellipticals are more than ten times more massive than the largest spirals, and it would not be easy for the difference in mass to be gained or lost. Further, the rotation could not easily be changed to give the different values observed for angular momentum.

Despite these results, galaxies to the left in the sequence (E and S0) are called **early** and those to the right (S and Irr) **late**. This is similar to a terminology used for stars in the main sequence of an H–R diagram, but it has even less justification: late-type galaxies contain more early-type stars.

Since the galaxies described in the Hubble sequence are all relatively nearby, the times taken for their light to reach us are much shorter than their ages. There is no question, then, of their appearance being affected by changes occurring within the long ages of a cosmological time scale.

Extensions to the classification scheme

Over the years, more detail has been introduced into Hubble's original classification by various people, including particularly Allan Sandage and Gérard de Vaucouleurs (Fig. 7·2, page 196).

In the sequence between E7 and Sa, the S0 class may be conveniently divided into three (and similarly for SB0); de Vaucouleurs uses the labels $S0^-$, $S0^\circ$, $S0^+$, and precedes these by a class E^+. Considering the relative dimensions of disc and nucleus, the S0 sequence can be regarded as running parallel to the

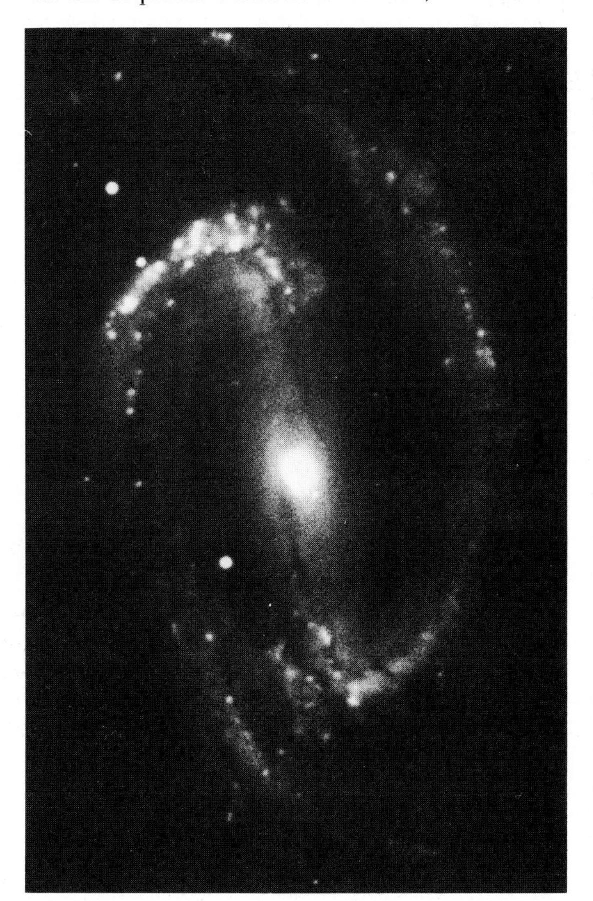

Four typical galaxies. This page, top: *the elliptical galaxy NGC 205, Class E5, a companion to the Andromeda Galaxy.* Opposite page: *the Sb galaxy NGC 4569.* Above right: *the Sc galaxy NGC 4565, which we see edge on. Notice the strong obscuration by dust, in a narrow belt.* Above left: *the barred spiral SBb, NGC 1300.*

The Large Magellanic Cloud. Note the strong bar structure and the slight suggestion of weak spiral arms.

S sequence, but of course without any spiral arms and containing much less gas. In addition, Sidney van den Bergh has described a sequence of gas-poor **anaemic spirals**, intermediate between S and S0, which are common in clusters of galaxies.

The original classes Sc and Irr I cover a wide range in appearance and each has been split into two, Sc and Sd, Sm and Im. Sm contains irregular galaxies with slight but definite traces of spiral structure: in this system, the Large Magellanic Cloud is SBm. Im contains irregular galaxies without such structure: the Small Magellanic Cloud is IBm. Both clouds have distinct bar structures.

Combination symbols, such as Sab, are used for intermediate types; (s) may be added if the spiral arms start in the nucleus and (r) if they start in a ring around the nucleus, as Sbc(s), SBa(r), with (rs) as intermediate type. De Vaucouleurs calls the ordinary spirals SA, to match the notation SB for barred spirals, with SAB intermediate between them. There is, thus, a continuous range covering three aspects, 0-a-b-c-d-m, A-B, (r)-(s). Further, spiral arms of similar structure can differ in strength; some galaxies have thick, massive arms while in others, of the same Hubble type, they are thin and filamentary. They are sometimes distinguished by a subscript m or f.

About 1958, William Morgan recognized a special class of giant ellipticals which he called cD galaxies. These are very large and bright elliptical galaxies with extended outer envelopes, and frequently the largest galaxy in a rich cluster is of class cD; most of them are also strong radio sources.

Luminosity classification

About 1960, van den Bergh found that the appearance of an Sb, Sc or Irr I galaxy is related to its luminosity; for example, the most luminous galaxies have the longest and most fully developed spiral arms. He, therefore, introduced luminosity classes numbered I to V in decreasing order of brightness; for Sb galaxies, however, only classes I, II and III are used, for it seems that all the intrinsically faint spiral galaxies are of class Sc. Also, there are no class I irregular galaxies.

Luminosity classification gives a simple method for obtaining the relative distances of large numbers of spiral galaxies.

Relative numbers of galaxies

Among the brightest galaxies observed, spirals amount to about 75 per cent, ellipticals and S0 20 per cent and irregulars 5 per cent. The relative numbers vary with limiting magnitude, however, for there are very many dwarf elliptical galaxies (sometimes counted as a distinct class dE) and also more irregular galaxies of low luminosity. For galaxies as a whole, therefore, the numbers of ellipticals probably exceeds 60 per cent and numbers of spirals and irregulars are approximately in the ratio 3 : 1.

The three groups, the ordinary spirals S (or SA), the barred spirals SB, and the intermediate group SAB, are present in about equal numbers.

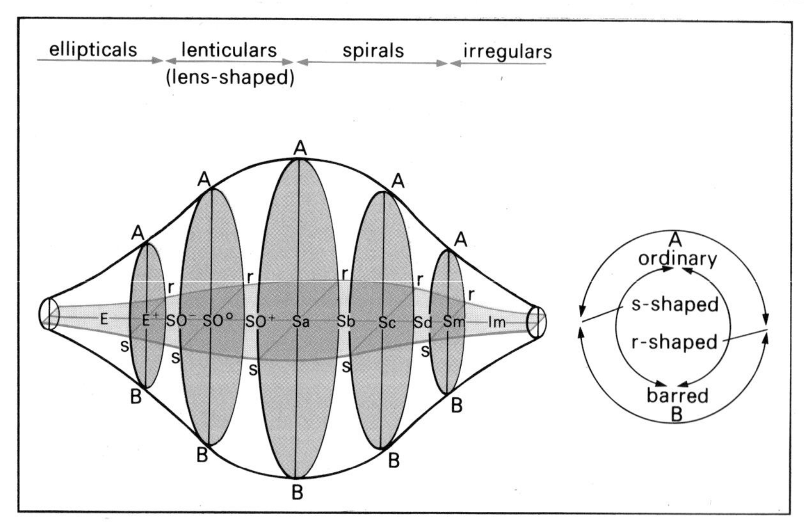

Fig. 7·2
Above: *A three-dimensional representation of de Vaucouleurs' classification scheme.*

Below right: *A cross-section through the above figure, near the region of the Sb spirals, showing transition cases between the ordinary (SA) and barred (SB) spirals and between those with (r) and without (s) inner rings.*

The redshift

It is found that for all galaxies, apart from a few of the very nearest ones, the spectral lines are shifted to longer wavelength; for optical lines, this means a shift towards the red. The more distant galaxies have larger redshifts, and the exact relationship between redshift (z) and distance – **Hubble's law** – was established by Hubble in 1929. It states that the redshift of an extragalactic object is equal to its distance multiplied by a constant. Mathematically it can be written in the form

$$cz = H_0D$$

where c is the velocity of light and d is the distance corresponding to redshift z. The constant, H_0, is now called Hubble's constant. (If λ_0 is the wavelength at the source and λ the wavelength observed, $z = (\lambda - \lambda_0)/\lambda_0$. The value of z is the same for all lines in the spectrum.)

The implications of the redshift, z, are discussed fully in Chapter 8, but for the moment it suffices to note that if the redshift is a Doppler shift caused by motion away from us, it is evidence for the expansion of the universe.

The rate of expansion may have changed as the universe has evolved; H_0 is the value at the present epoch, corresponding to all but a very few of the

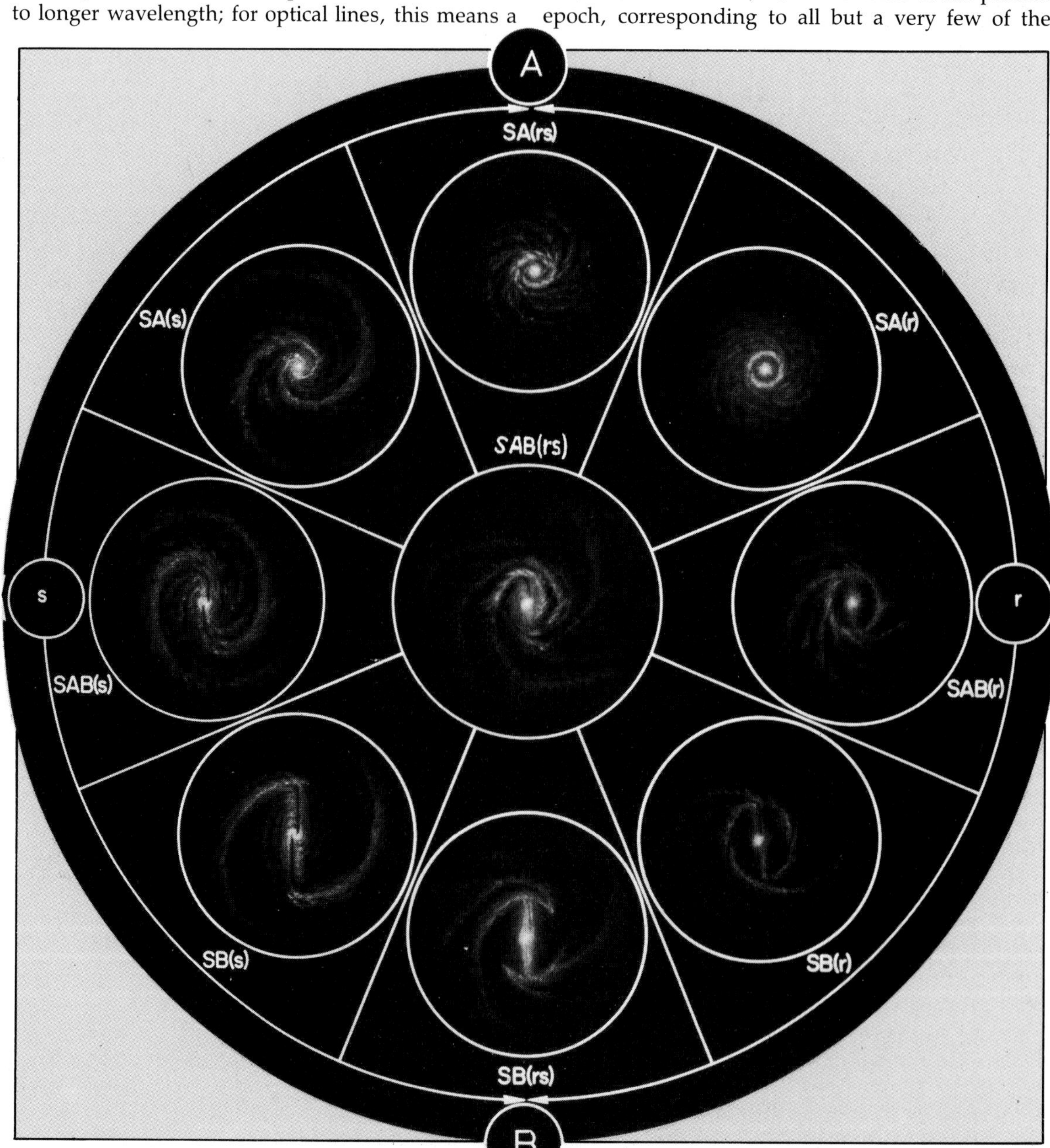

Fig. 7·4 Opposite page, bottom:
The range of distance over which various extragalactic distance indicators can be used. The step-by-step nature of distance determination is clearly seen.

most distant observable galaxies. Velocities are measured in kilometres per second and distances in megaparsecs, so it is convenient to express H_0 in the units km per s per Mpc. Hubble's own best estimate of H_0, given in 1935, was about 530, but we now accept that H_0 is much smaller than this, because, since 1935, estimates of the distances d of galaxies have increased. This revision of the distance scale has happened in many small steps over the intervening years, although the largest single change came in 1952 when Walter Baade realized that the Cepheid and RR Lyrae variable stars do not have the same period-luminosity relation. As a result of Baade's revision, the extragalactic distance scale was doubled, and, whereas before 1952 the Andromeda galaxy (M 31) was thought to be distinctly smaller than our Galaxy, since 1952 it has been known to be distinctly larger.

The use of the most modern instrumentation, especially CCD equipment (see page 235), has enabled us to observe even more distant galaxies and obtain their spectra. At the time of writing the most distant known galaxy (as distinct from a quasar) is one associated with the radio source 3C 324, where z is 1·21, but the vast majority of galaxies which can be studied easily have redshifts lower than 0·1.

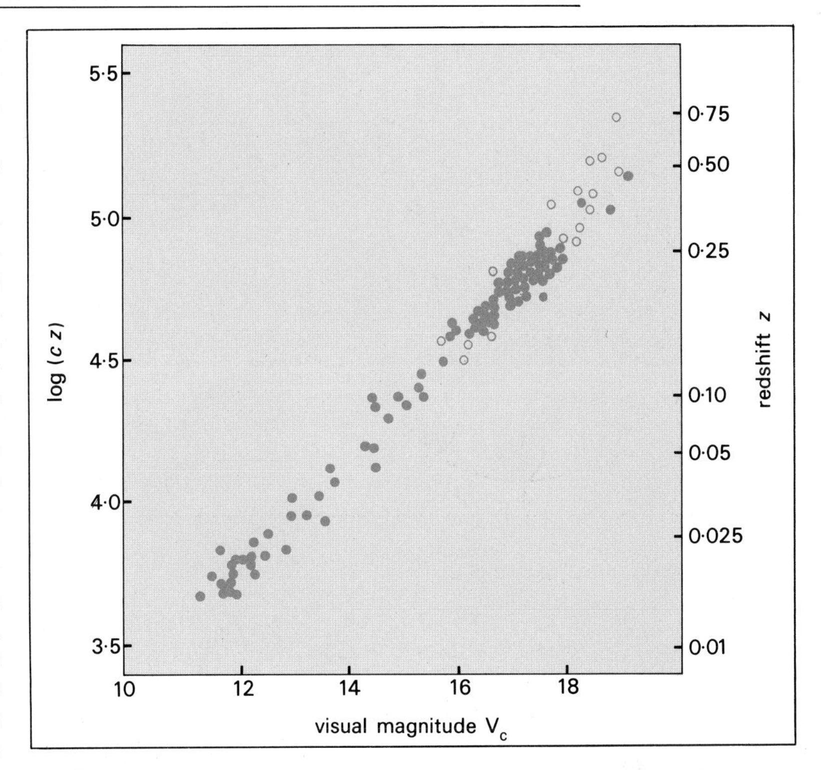

Fig. 7·3 above: The Hubble relation for the brightest galaxies in large clusters. z is the redshift and V_c is a corrected apparent visual magnitude. If it is assumed that all these galaxies have the same absolute magnitude, V_c is related to distance. The open circles come from more recent observations and suggest that for larger z the plot departs slightly from a straight line.

The galaxian distance scale

Hubble's law relating distance to the observed redshift means that the scale of galaxian distances can be expressed by stating the numerical value of Hubble's constant H_0. However, to determine H_0 means determining the distances of a suitable sample of galaxies, and this is not easy. In fact, such astronomical distances must be determined step-by-step, starting with radar ranging of planets which gives the scale of the Solar System. Distances to nearby open clusters of stars may then be found using the cluster method (page 174), which leads to absolute magnitudes for main-sequence stars. We are then in a position to find distances for more distant open clusters which contain Cepheid variables, and this allows the Cepheid period-luminosity relation to be calibrated. This is important because Cepheids can be observed and used to establish distances to galaxies within about 4 Mpc. (See Fig. 7·4)

A recent systematic study to determine H_0 has been made by Allan Sandage and Gustav Tammann. They use Cepheid distances to determine the absolute diameters of the largest H II regions in the galaxies concerned, then, assuming that the diameters of large H II regions are much the same in different galaxies, distances to others can be found. It is a useful technique because H II region diameters can be measured out to about 25 Mpc. The next step is to use these H II region distances to find the absolute magnitude for Sc I galaxies; for, being of the same luminosity class, all Sc I galaxies have the same absolute magnitude. Finally, samples of more distant Sc I galaxies, whose distances are known from this absolute magnitude, are used to establish the relation between distance and redshift and so provide the value of H_0. According to Sandage and Tammann H_0 is 55 km per s per Mpc, with a probable error of ± 5 km per s per Mpc. (In this context, probable error is a measure of the internal consistency in a determination, and does not give the full range of possible inaccuracy of the result.)

This is the best value available at present, in the sense of having the smallest probable error, and it is quite widely accepted, but some other recent determinations disagree with it. Certain details of the work have been questioned, and several other workers deduce that H_0 is about 77 km per s per Mpc, using the same observational data.

Alternative methods can be used at different stages in the determination. Cepheid calibrations can be confirmed and extended using novae, for which the absolute magnitude at maximum and the subsequent rate of decline in intensity are related, and bright

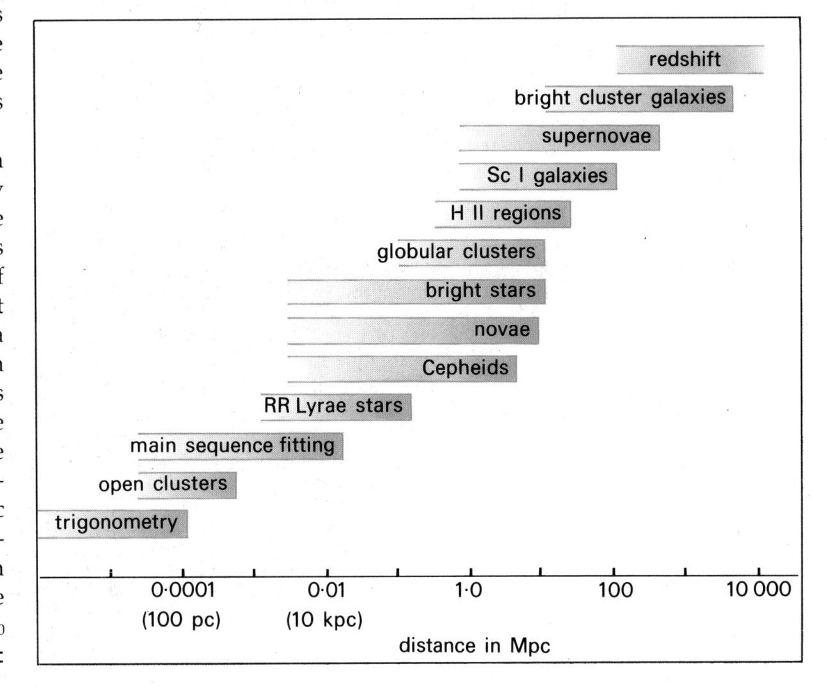

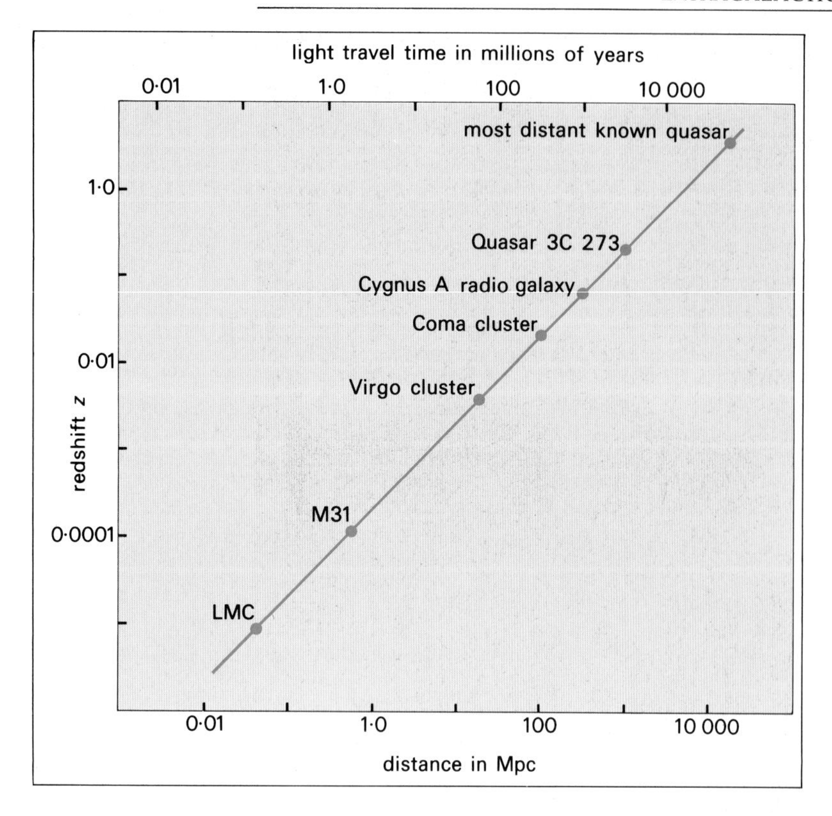

Fig. 7·5
The relation between redshift, distance and light travel time, using a very simple theory. A value 55 has been adopted for Hubble's constant H_0.

main sequence stars, for which absolute magnitudes are well established. Again, supernovae can be observed at large distances, up to about 400 Mpc, and all reach about the same maximum luminosity, but it is difficult to calibrate them accurately. A recent study using supernovae gives H_0 as 60 km per s per Mpc but with a large probable error, ± 15 km per s per Mpc. Another method uses the third brightest galaxies in small clusters of galaxies, presuming that they all have about the same luminosity; for some reason, this is truer for third brightest than for brightest galaxies in clusters. In 1974, van den Bergh analysed results from these and several other methods and found H_0 = 93 km per s per Mpc, with probable error ± 7 km per s per Mpc. An only slightly smaller value, 86 km per s per Mpc, has been suggested by recent work of de Vaucouleurs, based on the brightest globular clusters associated with different galaxies; and the same author has also obtained a similar value from various other methods.

A method using radioastronomy has been proposed recently by Brent Tully and Richard Fisher, who find a close relationship between the total width of the hydrogen 21-cm line emitted by a galaxy and its absolute magnitude in blue light. With this method, they obtain H_0 = 75 km per s per Mpc, but Sandage and Tammann obtain H_0 = 50 km per s per Mpc, consistent with their own other value. The point of disagreement is different from the one which concerns the optical determination by Sandage and Tammann, and so, if they are wrong on both counts, this method would give a value of about 100 km per s per Mpc.

At present there remains a considerable divergence of view between those who favour a small value of H_0, and those who believe that the true value is nearer to 85–100 km per s per Mpc. We retain the value of 55 km per s per Mpc in this discussion, but must bear in mind that a revision to a higher value would require all the distances and masses mentioned to be adjusted in the ratio of 55:85, or 55:100, or in accordance with whatever value is finally adopted. Any such adjustment would of course affect the calculations which can be made concerning the mean density of the universe, for example, as will be discussed later (page 205).

Variation of H_0

In 1973, Vera Rubin, Kent Ford and Judith Rubin, using a sample of distant Sc I galaxies, presented evidence for a variation of H_0 with direction in the sky, but their interpretation was questioned. More recently, Vera Rubin and Ford, with other colleagues, have studied a larger sample of galaxies and confirmed their earlier result.

Different values of H_0 in different directions could be produced by different rates of expansion of the universe in different directions, by large random velocities of groups and clusters of galaxies, or by a large random velocity of the Galaxy and the Local Group. In the last case, the effect of our motion would be that the apparent velocity of recession of galaxies towards which we are moving is less, producing the appearance of a smaller H_0 there than in other directions. Rubin and her colleagues favour this third interpretation, and conclude that the Sun is moving at 600 ± 125 km per s relative to the distant galaxies. After allowing for the Sun's motion around the centre of the Galaxy and the motion of the Galaxy itself, the velocity of the Local Group is found to be 454 km per s.

This result is not yet accepted without question, for the observations could be interpreted in other ways. For example, an effect similar to what is observed might be produced by unexpected fluctuations of interstellar absorption in different directions within the Galaxy. It is also possible that there is still something special about the sample of galaxies being used.

There is some evidence for systematic differences in the motions of relatively nearby galaxies seen in different directions, an effect which can be interpreted as due to systematic motions within what is called the local Supercluster.

The Local Group

The Milky Way Galaxy forms an association with the nearest other galaxies, called the Local Group, which has nearly thirty known members. Most of their distances are known accurately from measurements of Cepheid variables. The Group contains a reasonably mixed sample of galaxy types, except that there are no conspicuous giant ellipticals or barred spirals.

It is hard to establish the precise Hubble class for our Galaxy because we are inside it and so do not easily see its large-scale structure. The Galaxy has several small satellite companions, of which the Large and Small Magellanic Clouds are the most prominent, being clearly visible to the naked eye in the southern sky. Satellites of our Galaxy also include

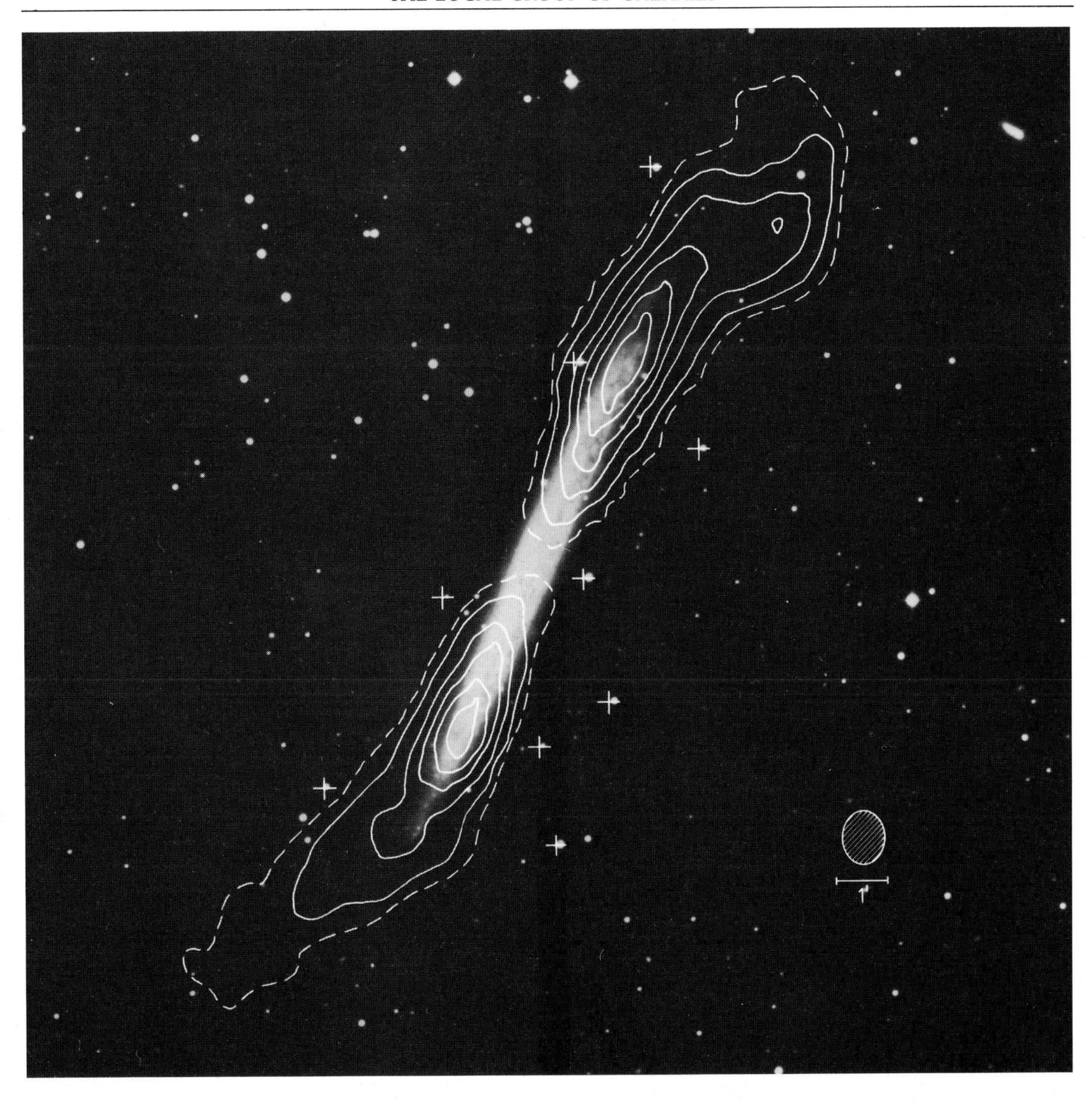

The hydrogen distribution in the spiral galaxy NGC 5907 which is seen edge-on, showing clear evidence for a warping of the galaxian plane farther out than the limit seen in the optical photograph.

half a dozen dwarf ellipticals; those in the constellations Sculptor and Fornax are the most conspicuous. The Andromeda galaxy (M 31) also has several companions. There could be many more dE galaxies in the Local Group, more distant from us and very faint.

Proper motions of galaxies are undetectable; our only information about their motion through space comes from their line-of-sight or radial velocity v, and from this it is evident that the Local Group galaxies are all in orbit about a common centre of mass, although observation suggests that the system is not gravitationally stable and will break up. For the more distant of the galaxies concerned, distances are not so accurately known and membership of the Local Group is uncertain. It is generally assumed that the Group includes all galaxies out to a distance of 1 Mpc.

It has already been mentioned that galaxies near the galactic plane suffer obscuration. An intrinsically bright nearby galaxy could therefore be heavily reddened and appear faint optically, but it would be brighter in the infrared. Several heavily reddened galaxies are known, some of which are probably in the Local Group. In 1968, two galaxies, called Maffei 1 and 2, were discovered by infrared observation. Maffei 1 is a large E, or S0 galaxy and could well be a Local Group member about 1 Mpc away; Maffei 2 is a spiral and is almost certainly outside the Local Group.

Local intergalactic matter

Radio observations of hydrogen at 21 cm wavelength (page 186) show that there are high velocity clouds of gas at high galactic latitudes, with radial velocities up to ± 250 km per s. Since it is not possible to determine their distances, there is some doubt about their true nature. They could be nearby interstellar clouds, or be part of the outer structure of the Galaxy, or even be intergalactic, either as satellite clouds of the Galaxy at tens of kpc or as separate member clouds of the Local Group at hundreds of kpc from the Sun. It is possible that the clouds include objects of all these different types.

In 1974, D. S. Mathewson and colleagues, in Australia, detected an extended, continuous belt of high-velocity hydrogen gas, passing near the south pole of the Galaxy and enveloping the two Magellanic Clouds. Evidently this material, called the **Magel-**

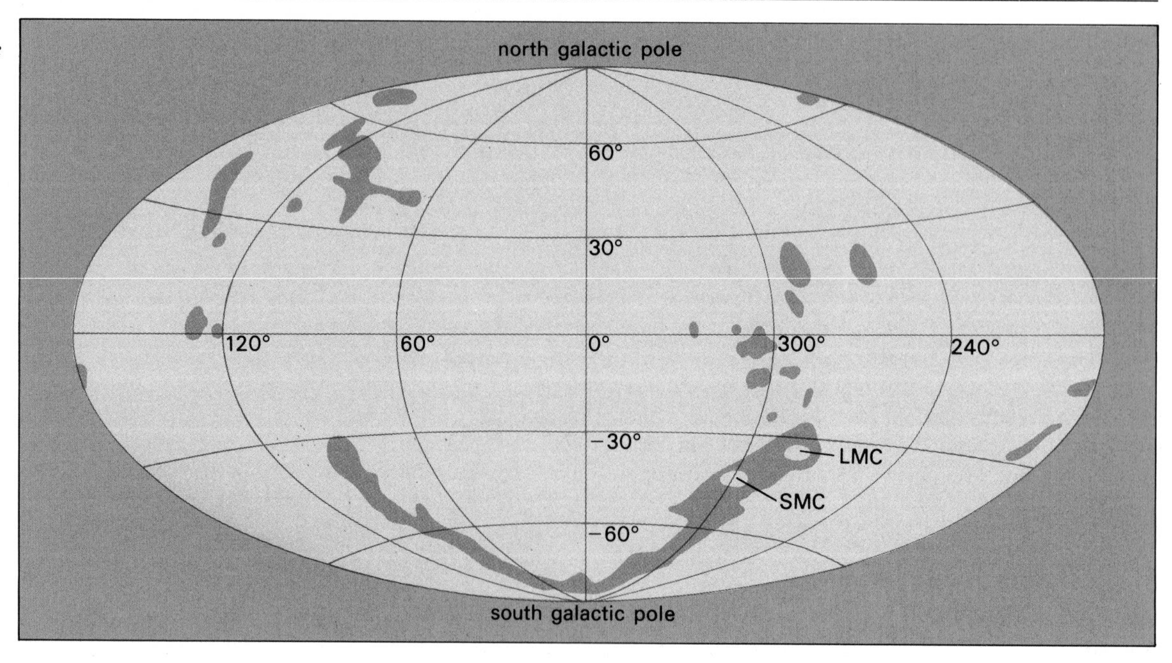

Fig. 7·6
Outline positions in the sky of the Magellanic Stream and other high velocity clouds, plotted in galactic co-ordinates with the centre of the Galaxy at the centre of the plot. The positions of the two Magellanic Clouds are indicated.

lanic Stream (Fig. 7·6), lies at the distance of the Clouds. There are two theories about it, one that the material was drawn out of the Clouds during a supposed recent close approach to the Galaxy, the other that it originated as an intergalactic gas cloud and never formed part of any galaxy. The existence of the Magellanic Stream makes it reasonable to think that at least some of the other high-velocity material also lies outside the Galaxy.

In addition, two clouds have been found, in directions near M 31 and M 33, which probably lie at the same distances as these galaxies. In a similar way, certain other intergalactic hydrogen clouds may be associated with galaxies in the Sculptor group. However, the possibility cannot be excluded that some or all of these are smaller clouds lying much nearer to us.

Stars in galaxies

In 1944, soon after the introduction to astronomy of a red-sensitive photographic emulsion, Walter Baade noticed that the bright stars in the spiral arms of the Andromeda galaxy are blue, while those in the nucleus are red; with only blue-sensitive plates this had not been observable. It led to the concept of population types. Population I contains interstellar gas and dust, blue stars and other essentially young objects, and is found in spiral arms. Population II contains evolved objects, in particular red supergiants, and is found in elliptical galaxies, globular clusters and the nuclei of spiral galaxies.

Even in the nearest galaxies only the very brightest individual stars can be studied, so the stellar content of a galaxy is investigated with the spectrum of the total light, **integrated starlight**. Galaxian colours change continuously through the sequence from E to Sm, reflecting an increase in numbers of Population I stars. Colour also varies (by a smaller amount) with radius in a galaxy and with galaxian size, indicating a higher abundance of heavy elements in the central regions of galaxies and in larger galaxies. These results are confirmed by the intensities of spectral lines in the integrated starlight. In regions of higher density, star formation is more rapid and so the total rate of processing of interstellar material through stars is faster there.

Recently, a few nearby elliptical galaxies have been found to contain small populations of blue giant stars. These observations disturb the conventional idea that ellipticals contain only old material.

Interstellar matter in galaxies

Spiral and irregular galaxies

Interstellar matter is observed in galaxies of later type in four main ways: optical obscuration by dust, optical spectral lines due to emission from H II regions, a radio background or **radio continuum** emission from ionized gas and spectral line emission at specific radio wavelengths from neutral gas. As in the Galaxy, dust and gas appear to be well mixed together. The dust is seen by the dark lanes which it produces in spiral arms and is conspicuous in some galaxies which are seen edge on. It can also produce significant reddening of starlight.

H II regions are brighter than individual stars and can be studied in more distant galaxies; but few spectral emission lines can be detected in the integrated light from a galaxy and HII regions are best studied individually. From theoretical considerations rather precise abundances may be derived for the atoms observed, atoms which, besides hydrogen and helium, include oxygen, nitrogen and sulphur. Heavy element abundances tend to decrease with distance from the galaxian centre, and this is consistent with the general picture obtained from the spectra in integrated starlight.

Continuum radio emission from normal galaxies comes from the electrons in ionized regions – H II regions, supernova remnants and more extended regions of lower density. As a free electron passes near a proton or other positive ion, it is accelerated and emits radiation. It is a type of thermal radiation (from the German often called **thermal bremsstrah-**

lung) and in this case it indicates that the gas has a temperature of about 10^4 K. Galaxies are mapped at various radio continuum wavelengths, although because of the longer wavelengths the power to see detail or ANGULAR RESOLUTION is less good than at optical wavelengths.

By observation of spectral lines at specific radio wavelengths, a few of the most abundant galactic interstellar molecules – carbon monoxide (CO), hydroxyl (OH), water (H_2O) and formaldehyde (H_2CO) – have been detected, but only in a handful of nearby galaxies. Similarly, radio spectral lines from atoms in H II regions have been detected in a few galaxies. Only one radio spectral line, the 21-cm line of atomic hydrogen, is easily observed in other galaxies. Therefore, it is important for mapping galaxies, just as it is for mapping our own Galaxy (page 186). Of course, viewing a galaxy from outside, we can see more clearly the spiral arm pattern and the general distribution of hydrogen gas. Several hundred galaxies have been studied using the 21–cm line, the objects being to determine the amount of hydrogen present (or to establish an upper limit), to map its detailed distribution and, using Doppler shifts, to determine galaxian rotation. Rotation can lead to a value for the mass of the galaxy.

The value deduced for the mass of hydrogen as a fraction of the mass of the galaxy is proportional to the value assumed for the distance and so depends on the value of Hubble's constant. This introduces uncertainty, but at least the data show clearly a very strong increase of hydrogen content from early to late type. In contrast, the average density of hydrogen gas is found to depend weakly if at all on galaxy type; thus in the later-type galaxies the gas is distributed within a larger volume.

The contour lines showing the strength of 21-cm radiation follow the optical spiral arms quite closely. In most cases, the hydrogen is detected in the plane of the galaxy much further out than any optical emission; indeed, often it extends twice as far out, as in our Galaxy. In many cases, the outer parts of the plane of this hydrogen distribution are warped. There is notably little hydrogen gas in the centre of a spiral galaxy, where it is brightest optically; this is consistent with the distribution of other Population I material, and is also found in our Galaxy.

Elliptical galaxies

There is little optical evidence for interstellar matter in normal elliptical galaxies, although emission lines from ionized gas are observed in the nuclei of some giant ellipticals. Most ellipticals are not radio sources either, but a few contain very strong radio sources and are classified among the **radio galaxies**. These rare elliptical galaxies are a type of active galaxy; their radio emission is not thermal.

The first detection of 21-cm emission from an elliptical galaxy came only in 1977 when two groups, in the United States and in France, found approximately 0·1 per cent hydrogen gas in the active galaxy NGC 4278. Since then, a few other detections have been made, but for most galaxies studied an upper limit in the range 0·1–0·01 per cent is all that can be established. Similar upper limits have been set for galactic globular clusters.

These upper limits are unexpectedly low, because Population II galaxies and globular clusters contain many highly-evolved stars which must have lost substantial mass by ejecting material into interstellar space within the galaxy. To account for the upper limits observed for both neutral and ionized gas we must suppose that these galaxies themselves are losing material, in what could be called a **galaxian wind**.

Structures and dimensions of galaxies

Elliptical galaxies appear as ellipses in the sky. Their true three-dimensional shapes are presumed to be oblate spheroids with flattening produced by rotation, although recent measurements of rotational velocities suggest that this could be an oversimplified picture. Since we see galaxies orientated at random, the relative numbers with different shapes can be deduced statistically. It is found that there are not many truly spherical galaxies. Also, there are no elliptical galaxies with true ellipticity flatter than 0·7: flatter galaxies have discs and nuclei.

Studies of the distribution of luminosity in elliptical galaxies show an increase of brightness towards the centre, indicating an increase in the density of stars there. Moreover, as the intensity falls off faster near the outer boundary, the size of ellipticals can be determined relatively accurately.

For a spiral galaxy, the basic structure is a nucleus plus a disc containing spiral arms. The spiral arms contain Population I material, while Population II material is found in the nucleus and distributed evenly over the disc; the relative size of the nucleus and the strength and degree of openness of the arms are, as we have seen, classification parameters. Within the nucleus there is a distribution of luminosity similar to that in an elliptical galaxy. The disc, averaged round in angle to smooth out the spiral structure, is bright beside the nucleus and becomes fainter quite rapidly with radius, but there is no sharp boundary. A longer photographic exposure reveals material farther out, while in the 21-cm line, hydrogen gas is often seen at twice the radius detectable optically. It is not easy, therefore, to determine, the full dimensions of spiral galaxies. Further, it has been suggested that the dynamical stability of a rotating disc system requires a massive outer halo, so far unobserved, although recent observations do indicate that in our own case this may exist. The discovery that the Galaxy (and others) have coronae (page 172) is further evidence of material occurring outside main discs of galaxies.

Although in barred spirals the bar is prominent optically, it does not, in fact, represent a large distortion from circular symmetry.

Galaxian diameters range from 1 kpc or less for dwarf ellipticals up to 50 kpc or more for the brightest cD galaxies seen in rich clusters. The largest galaxies have absolute magnitude about −23 (if Hubble's constant H_0 = 55 km per s per Mpc) corresponding to a luminosity of about 10^{11} Suns.

Mass values

Elliptical galaxies have masses which range from about 10^5 $M_\odot$ for the smallest dwarf ellipticals up to about 10^{13} $M_\odot$ for the giant cD galaxies. Spirals have a smaller range, between about 10^9 and 10^{12} $M_\odot$. Irregular galaxies all have masses below about 10^{11} $M_\odot$.

Interacting galaxies

Systems where a large galaxy has several small companions, as with our Galaxy and M 31, are common, and, in addition, there are some other cases where two or more galaxies of more nearly equal mass are closely associated. Galaxies of different classes are found together, supporting the idea that they have essentially similar ages.

When the galaxies are close enough, optical filaments, bridges and other structures are observed optically and at the 21-cm radio wavelength. Elegant computer models based on gravitational interaction have been constructed by Alar and Juri Toomre, and lead to some particular forms observed, thus indicating that many of the strange shapes we find are indeed produced by mutual tidal distortion of the two galaxies (see diagram on facing page). It is possible, though, that some of the other distorted shapes could be due to activity in the nucleus.

One explanation of the warping commonly found in the discs of spiral galaxies is that it is a tidal distortion. For our Galaxy, the warping we observe could have been produced by the Large Magellanic Cloud, but, acording to Alar Toomre, it would need to have passed close to the Galaxy (about 20 kpc from the centre) some 5×10^8 years ago and also to have a mass rather larger than is otherwise supposed. Another problem in using a hypothesis of tidal interaction is that warping is observed in some galaxies which have no known companions. Other possibilities are that unseen intergalactic material is responsible for the warping, or that it is due to purely internal dynamical processes in the galaxy, although neither of these explanations is fully satisfactory.

Computer processing can reveal many faint features which are otherwise quite invisible. In this case a dark 'jet' of uncertain origin crosses a field of galaxies in Centaurus.

Clusters of galaxies

The distribution of galaxies in the sky is not random. Over large areas of sky the numbers are affected by interstellar obscuration within our Galaxy; within small areas, there is a strong tendency towards clustering. A cluster is a real physical clumping of galaxies, with a higher number density of galaxies inside than outside; the diameter may be as large as 10 Mpc. Clusters range from groups with ten or twenty known members, similar to the Local Group, up to rich clusters containing several thousand. Tens of thousands of discrete groups and clusters can be identified in survey photographs; the most conspicuous ones are named after the constellations in which they lie.

Following George Abell, clusters are classified as **regular** or **irregular**. All regular clusters are large, with thousands of members, and have a spherical shape with a high concentration of galaxies towards the centre, where there is often a cD galaxy; a well-studied example is the Coma cluster. Irregular clusters extend in a continuous range from small groups to large, rich clusters. They have little symmetry and little concentration towards a centre; examples are the Local Group and the Virgo cluster, the nearest large one, which has about 2 500 known members. The centre of the Virgo cluster is about 20 Mpc from the Sun (if Hubble's constant H_0 = 55 km per s per Mpc).

Compared with the statistics for galaxies in general, irregular clusters contain slightly fewer spiral galaxies, and regular clusters very few, particularly in the central regions. Some years ago, it seemed that the probable explanation was that in the denser clusters direct-hit collisions between galaxies would be more frequent, with the result that gas and dust would be swept from the colliding galaxies, causing spirals to become ellipticals. However, using the increased distance scale of recent years, such collisions would be too infrequent. It now seems probable that the gas and dust are pushed out of a galaxy as it passes through relatively dense intergalactic material in the central region of a cluster. There is, however, another possibility – that, for some unknown reason, fewer spirals formed in these higher-density regions in the first place.

About half of the nearby regular clusters and a quarter of the irregular clusters contain radio galaxies; indeed, about a fifth of all radio galaxies lie inside rich clusters, and it may be that many of the more distant radio galaxies which appear separate belong to clusters in which the other galaxies are too dim to be seen. In any case, the figures we have are consistent with giant elliptical galaxies having about the same chance of being strong radio sources whether they are in rich clusters or not.

About a third of the nearby regular clusters, but many fewer irregular ones, have been identified as sources of X-ray emission. They include most of the optically identified, extragalactic X-ray sources. Generally, the X-rays are emitted from an extended region in the cluster centre, larger than a galaxy, and in some cases there is also emission from a central active galaxy. The most probable source of X-ray

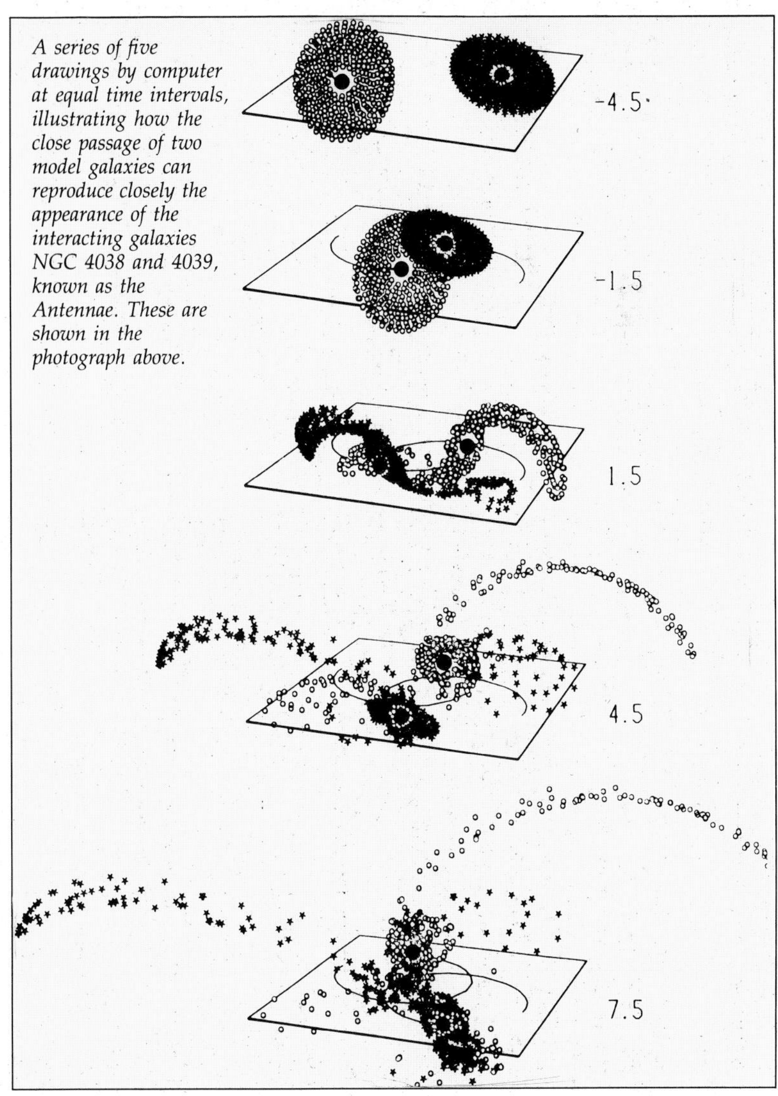

A series of five drawings by computer at equal time intervals, illustrating how the close passage of two model galaxies can reproduce closely the appearance of the interacting galaxies NGC 4038 and 4039, known as the Antennae. These are shown in the photograph above.

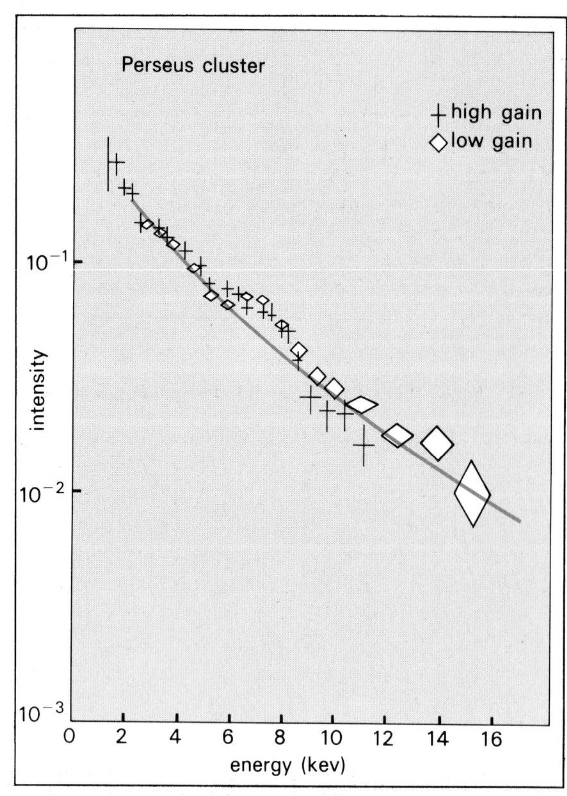

Fig. 7·7
X-ray spectrum of the Perseus cluster obtained by the Ariel V satellite.

emission is a very hot ionized gas or plasma lying in between the galaxies of the cluster, an intergalactic and intracluster gas, with a temperature of about 10^8 K and density about 100 electrons per m^3, for such a plasma would emit X-rays by the process of thermal bremsstrahlung (pages 200–201). An alternative suggestion, that there is an interaction between the observed background microwave radiation in the universe (page 206) and supposed very high speed electrons, seem much less probable.

In 1976, astronomers from University College London detected an X-ray spectral line feature at a wavelength of about 0·2 nm – just the emission to be expected from a hot plasma containing very highly-ionized iron atoms, atoms where twenty-four or twenty-five of the normal twenty-six electrons have been lost (Fig. 7·7). Not only does this observation support the idea of a hot intergalactic plasma, but the presence of iron suggests that the gas has been processed through nucleosynthesis in stars, since primaeval gas would contain only hydrogen and helium. It seems that the gas must have been ejected from galaxies within the cluster.

Superclusters

Clusters do not appear to be distributed at random, and this is generally interpreted as good evidence for a clustering of clusters, or a superclustering (Fig. 7·8). The Local Group is believed to be part of a supercluster of diameter about 75 Mpc centred on the Virgo cluster, and known as the **Local Supercluster**. Going further, there have even been suggestions of a clustering of superclusters and possibly even higher-order aggregations, but there is no clear evidence for these.

The problem of the 'missing mass'

The virial theorem which gives a method for determining the mass of a spherical distribution of stars can be used also to determine the mass of a cluster of galaxies. We have to know how velocity is distributed, and for stars in a galaxy this is derived using integrated starlight, but the galaxies in a cluster are studied individually. The mass deduced in this way can be compared with the total mass of all the galaxies in the cluster, but in every case studied the results disagree. The mass given by the virial theorem is larger than the sum of the masses of the galaxies, by a factor of about 8 for large regular clusters dominated by elliptical galaxies (such as the Coma cluster) and by more for some smaller clusters containing more spirals.

To examine this discrepancy it is necessary first to consider possible errors in the virial theorem mass determination. The value could be too large: if there were a significant number of pairs of galaxies much closer together than the average in the cluster, which

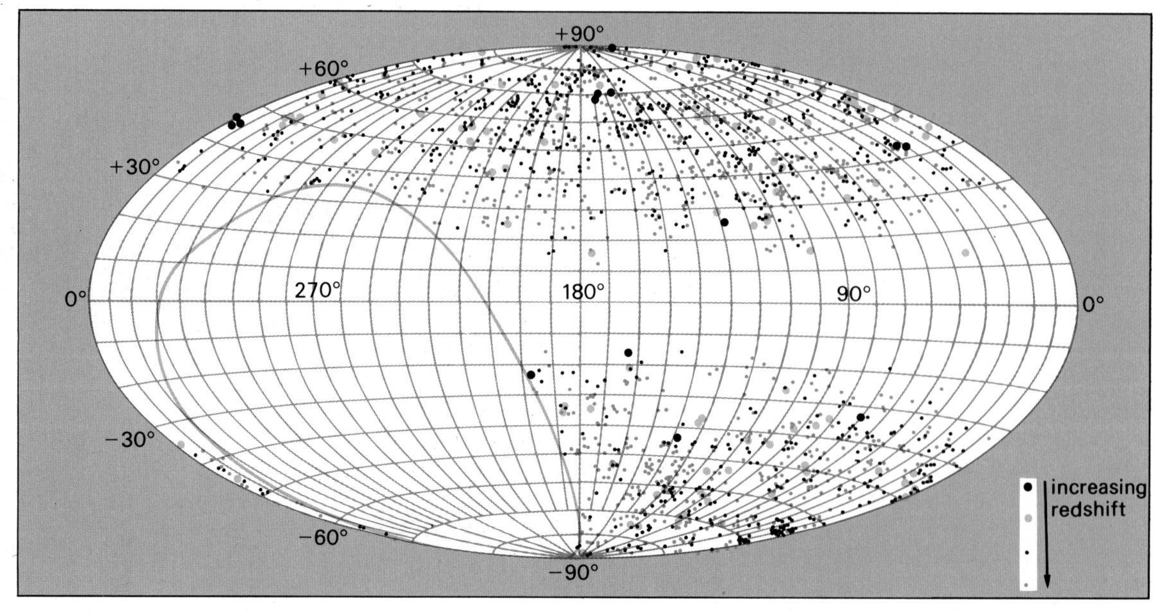

Fig. 7·8
The distribution of clusters of galaxies in the sky, plotted in galactic co-ordinates with the galactic anticentre near the centre of the plot. The large empty oval was not covered by this survey. Clusters show a clear tendency themselves to cluster together. Notice also the zone of avoidance on either side of the galactic plane (compare with Fig. 7·1 on page 192).

would affect the potential energy V; or if the galaxies were not, as assumed, all in the same cluster but in neighbouring clusters which overlap in the line of sight, which would affect the kinetic energy T. It is just possible that these and other observational errors could account for all of the 'missing mass'.

There are, however, four other possibilities:

a. that the clusters are indeed unstable and are dispersing;
b. that the masses of galaxies are currently underestimated by large factors;
c. that intergalactic matter in the cluster contributes the required mass;
d. that the mass is present in some other form, as yet undetected.

The first of these runs into the severe difficulty of explaining the large numbers of galaxies still in clusters, since clusters would disperse completely within some 10^9 years, short in comparison with their ages. The first point to be made about b is that masses have been determined for galaxies in clusters as well as for galaxies outside, and there are no significant differences. Now, radio observations of spiral and irregular galaxies have led to mass estimates about three times larger than the earlier ones derived from optical studies alone, so it is possible that even these revised values are underestimates. Arguments that disc systems require substantial outer halos to retain stability also suggest that there might still be quite a large underestimate of their masses but, on the other hand, studies of binary galaxies suggest that there is not.

The third possibility concerns the intergalactic matter in clusters. It is possible to set very good upper limits on the amount of neutral intergalactic gas, by looking for absorption lines (optical or radio) in the spectra of galaxies seen through the gas. If it is evenly distributed in the cluster, such gas certainly contributes insignificantly to the 'missing mass', and although more gas could be present without being detected if it were in separate high density clouds, recent observations suggest that there is unlikely to be enough mass present in such clouds either.

As discussed earlier, the detection of X-ray emission from clusters indicates the presence of an ionized intergalactic gas. Support for this comes from the distorted shapes of several extended radio sources found in clusters, which could be due to the expanding material of the source interacting with intergalatic gas. Further support comes from a recent report by radioastronomers at Cambridge University, England, that the cosmic background radio radiation (page 206) has a lower intensity when seen through clusters, presumably because of absorption by the intracluster gas. Moreover, ionized gas at the temperature of 10^8 K, mentioned previously, would be distributed in the cluster in the same general way as the galaxies, and a dynamical analysis for the Coma cluster suggests that the 'missing mass' must have such a distribution.

From the total X-ray intensity the mass of ionized intergalactic gas can be deduced, but it is only about 10 per cent of the virial theorem mass. However, combined with the known mass in galaxies, this reduces the discrepancy to a factor of about 4. It is reasonable to suppose that the balance could be made up of a combination of errors in galaxian masses and in virial theorem masses. Because of this, it may well not be necessary to invoke d, the presence of other, less normal sorts of intergalactic objects. The possibility of a supermassive black hole or other unseen condensed object at the centre of a cluster would in any case be excluded by the extended distribution required for the Coma cluster.

The mean density of the universe

The rate of expansion of the universe is becoming slower because of gravitational forces. If its average density is larger than a certain critical value, gravity will take over and expansion will eventually stop, to be followed by contraction; the universe is said to be **closed**. If the density is less than this critical value, expansion will continue for ever; the universe is **open**. Using the value of 55 km per s per Mpc for Hubble's constant, this critical density is equivalent to 3·4 hydrogen atoms per m^3. A reasonable estimate for the matter present in galaxies gives, when aver-

The radio structure of the galaxy NGC 1265 in the Perseus cluster, superimposed on an optical print. The swept-back structure is evidence for intergalactic gas in the cluster, sweeping past the galaxy.

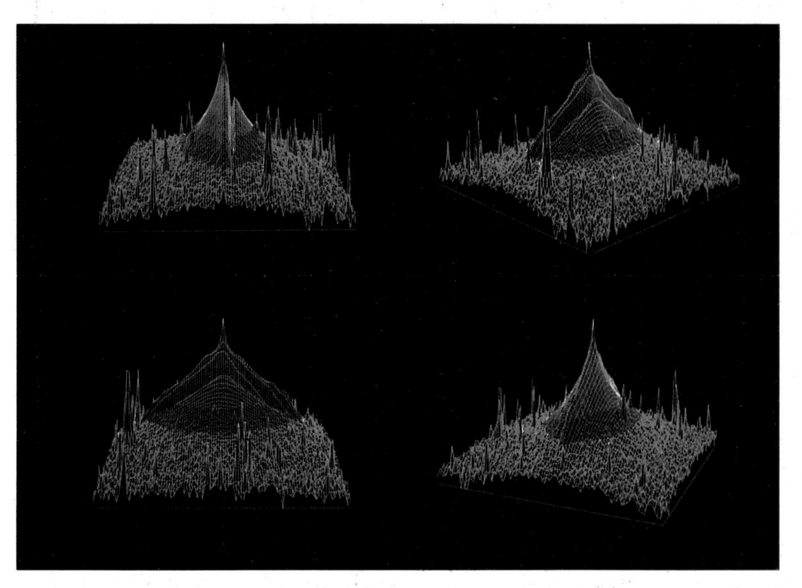

The use of computer graphics such as these displays of optical emission from part of M104 make it possible to 'view' galaxies from various angles.

aged through space, only one-fiftieth of this value. Supposing, though, that the mass discrepancy in clusters were caused entirely by underestimating the masses of galaxies (which is possible but unlikely), then the averaged mass would be brought up to about one-third of the critical value. It seems reasonably certain that galaxies and whatever it is that binds clusters together would not be sufficient to close the universe. Some other material, in between the clusters, is required.

Intergalactic material between clusters

It is possible that gas exists in the space between clusters. Such gas could be electrically neutral or ionized, and, because of the expansion of the universe, it should display a redshift. Observations show that if there is neutral gas spread throughout space, it can be there only in tiny amounts, and that even if it were concentrated into dense clouds, there could be no more than about one-third of the critical density. One instance of a fairly massive cloud of neutral hydrogen is known at a distance of about 10 Mpc. The indications are that this is at least 100 kpc across, and although the average density is low it appears to contain about as much mass as a galaxy. Whether other examples exist remains unknown at present.

We do observe an X-ray background coming apparently from every direction over the whole sky, and this can reasonably be interpreted as emission by an intergalactic ionized gas at about 10^8 K. There is an alternative explanation for this radiation, which is that it comes from a host of faint discrete sources. If this were so it should be possible to detect fluctuations in the amount of radiation over small areas of sky. At present observations are not precise enough for us to establish whether such fluctuations exist. In any case, it seems likely that the density of intergalactic gas required to produce the observed radiation is less than the critical density, although it is just possible that it could equal it. Moreover, there are several problems concerned with maintaining the high temperature required for the radiation and with the effects of the hot gas on clusters and galaxies embedded in it, which make it appear necessary that the density of a 10^8 K intergalactic gas must be significantly below the critical value.

Another possibility is the presence of intergalactic dust. This would redden the light from galaxies, but no such reddening is observed. The amount of any dust present must be far too small to contribute significantly to the total density. Indeed, this is to be expected, because intergalactic material probably contains scarcely any atoms heavier than hydrogen and helium, since heavy atoms are created by nuclear processes inside stars. Moreover, this expected lack of heavy elements also rules out the presence of very large numbers of larger solid objects, of sizes between a few centimetres and planetary sizes, even though they are not ruled out from an observational point of view.

Intergalactic populations of subluminous objects – very cool main sequence stars, white dwarfs, neutron stars and black holes – which could in total contribute more than the critical density are not ruled out by observation. However, it appears unlikely that a large population of low mass main sequence stars could exist outside galaxies, while the other objects form as end products of stellar evolution, so they are also to be expected in substantial numbers only inside galaxies. A final possibility of meeting the mass deficiency is that there might be significant quantities of neutrinos or **gravitons** (gravitational wave packets), but there is no possibility of detecting them if they are there.

Because of all the many uncertainties, the question of whether the mean density is greater or less than the critical value is still unresolved. Nevertheless, since all forms of matter which can be studied with accuracy contribute substantially less than the critical density, and most other forms of material appear implausible, it is fairly widely felt that the total density probably is below the critical value, in which case the universe will expand for ever.

The microwave background radiation

In 1965, Arno Penzias and Robert W. Wilson discovered an unexpectedly strong radiation from space at the short wavelength of 7 cm, in the microwave region of the radio spectrum. It was some 100 times stronger than the radiation to be expected at this wavelength from known sources, and it was found to come equally from all parts of the sky. Studies of it, extended to other wavelengths, have concentrated on two questions: what is the form of the spectrum of the radiation, and how accurately is it independent of direction?

Early observations showed that the spectrum has the form to be expected from a black body (page 47) radiating at a temperature of about 2·7–2·9 K. Such black-body radiation should, at this temperature, have a maximum intensity at a wavelength of about 2 mm, so one important question is whether the intensity actually reaches a maximum at about 2 mm or continues to increase beyond there to shorter wavelengths. However, the answer can not be obtained from ground-based observations, since wavelengths of 2 mm and shorter can not penetrate the Earth's atmosphere. Nevertheless, observations made at wavelengths throughout the range detectable from the ground (from a few mm to nearly 1 m) are consistent with a uniform-temperature, black-body radiation.

At each wavelength observable from Earth, the radiation is found to be independent of direction within about 0·1 per cent. This has been done both for small areas up to about 1° across and for widely separated regions of the sky, and the results clearly exclude the possibility that the radiation is produced by a very large number of discrete sources, not otherwise detected: to produce what is observed, such sources would need to be far more numerous than galaxies.

Recently, a very small difference of intensity has been measured between opposite directions in the sky. It is believed to be caused by motion of the Sun relative to the material which last scattered the radiation, in the very earliest stages of the history of the universe.

Measurement at shorter wavelength

First attempts to obtain measurements at wavelengths which cannot penetrate the Earth's atmosphere were indirect, and were based on optical observations of the interstellar molecules CH, CH^+ and CN, first detected in 1937–41. This was possible because in low density interstellar clouds, excited molecular states are energized by radiation from the microwave background. The relative strength of optical absorption by excited CN molecules gave a measure of the intensity of the background radiation at wavelength 2·64 mm, while failure to detect any absorption from other excited molecules gave upper limits to the intensity at several particular shorter wavelengths. Unfortunately, the results were not accurate enough to make certain that the curve reached a maximum.

However, direct observations have now been made, from both rockets and balloons. The first observations, carried out around 1969, disagreed, but this was later found to be caused by instrumental errors; more recent results, obtained by astronomers from London, England, in 1974 and Berkeley, U.S.A., in 1975, show very clearly the maximum and the descending part of the curve. The results are quite accurately consistent with a black body curve at about 2·8 K.

It has now been established that the background radiation is not perfectly isotropic a (i.e. it does not come perfectly evenly from all parts of the sky, being slightly 'hotter' in one direction and 'cooler' in the opposite region). This is taken to indicate the motion of the Local Group, at a velocity of about 600 km per second with respect to the 'primordial' radiation. This was not entirely unexpected, and indeed would have caused some concern if it had not been discovered. The motion of 600 km per second can only be reconciled with the peculiar motion of about 450 km per second found by Rubin and Ford (page 198) if it is assumed that the galaxies which they used themselves share a common motion of some 800 km per second in a different direction. This could be taken to indicate large-scale turbulent motions in the universe, but in view of the reservations expressed about the Rubin-Ford value (as discussed earlier), such conclusions must await further research.

An optical photograph of M33, a nearby spiral galaxy, image-processed to accentuate the difference between the young blue stars in the arms and the older, red stars in the nucleus.

The background radiation has very important consequences for cosmology, because its presence has to be explained for any theory to be acceptable. Its significance in this respect is discussed in the next chapter.

Active galaxies

In a normal galaxy, everything seems to be stable and in equilibrium. By contrast, an active galaxy is one where there is strong non-thermal emission, and where rapid variations of intensity are observed.

Active galaxies have been found in two different ways, by being strong radio sources or by having a peculiar optical appearance. These two groups overlap but do not coincide: strong radio sources may be optically normal or optically peculiar; and some peculiar galaxies are not strong radio sources.

Seyfert galaxies

In 1943, Carl Seyfert published a list of six unusual galaxies, with a small nucleus which was bright compared with the rest of the galaxy, and now about 100 galaxies are classified as **Seyfert galaxies**. The optical spectrum of the nucleus contains emission lines not seen at all strongly in normal galaxies, but commonly observed in galactic gaseous nebulae. The lines are particularly broad, implying expansion out from the nucleus at rather high velocities, around 500 km per s.

Daniel Weedman has identified two classes of Seyferts. In Class 1, the hydrogen lines are broader than the forbidden lines (page 178), so emission is from different regions; in Class 2 the lines have the same width. In Class 1, the emission lines are weak compared with the continuum emission from the nucleus, in Class 2 they are strong. The nucleus appears smaller in Class 1. In many ways the nuclei of Class 1 Seyferts resemble quasars, which we shall

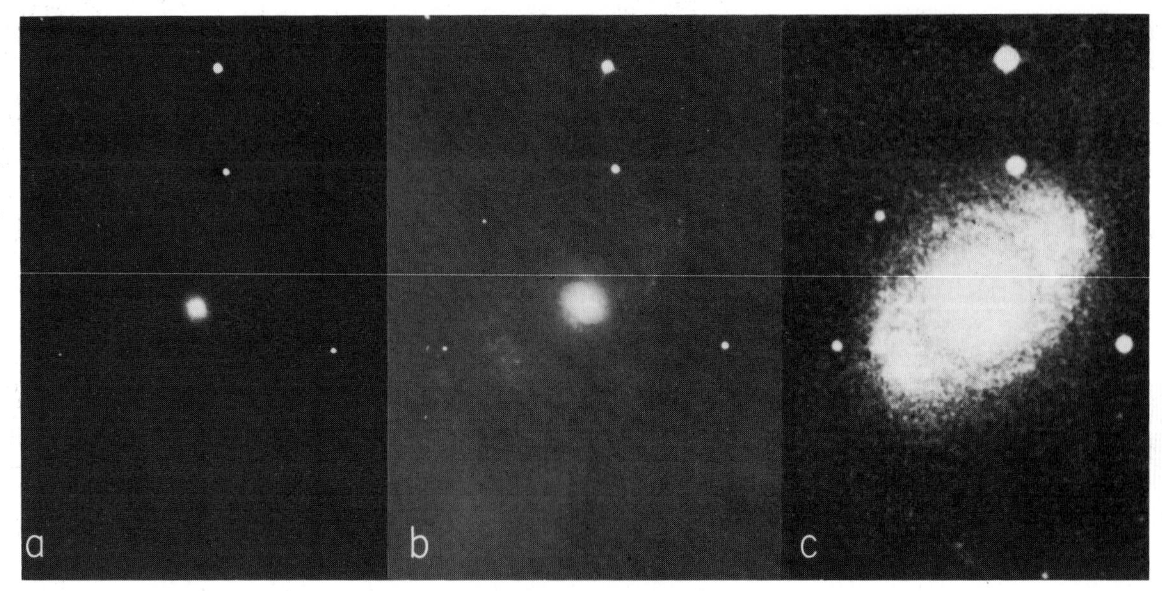

The Seyfert galaxy NGC 4151 in three photographs on the same scale but with different exposures. In (a) only the star-like nucleus is seen (making it look like a quasar); outer spiral structure is seen faintly in (b) and strongly in (c).

discuss shortly. Most Seyfert galaxies radiate strongly in both the infrared and the ultraviolet, but although all emit radio waves more strongly than ordinary galaxies only a few are very strong radio sources. Several Class 1 Seyferts have been identified as very strong X-ray sources.

It is just possible that Class 2, the less extreme type, could generate their radiation using only normal thermal processes, if the nucleus contains a very large number of hot stars and a large amount of interstellar dust, which could produce the excessive quantities of ultraviolet and infrared radiation observed. It is certain, however, that Class 1 Seyferts could not be explained in this way.

Almost all Seyferts appear to be S or SB galaxies with active nuclei, and it is estimated that about 1 per cent of all spiral galaxies are Seyferts.

N-type galaxies

The N galaxies are basically similar to Class 1 Seyferts but the nuclei are relatively brighter and more compact and the surrounding galaxy is less clear; both have redshifts rather larger than those for ordinary galaxies. N galaxies were identified as a separate optical class by William Morgan in 1958, and came to be studied particularly in the 1960s when optical surveys were made of radio sources. This clearly influences the fact that very many of those known are strong radio sources, by a process of observational selection.

Markarian galaxies

An extensive survey of galaxies which are bright in the ultraviolet has been carried out since 1967 by B. E. Markarian at the Byurakan Observatory in Armenia. His lists so far contain about 700 objects, of which about 10 per cent are Seyfert galaxies; indeed, it is through his work that most known Seyferts have been discovered.

Markarian galaxies are of two major types, one where the ultraviolet emission is from a bright nucleus, and the other where it comes from the whole galaxy. The galaxies with bright nuclei include the Seyferts but most are of a new type; they appear to have an excess of hot stars in the nucleus, which causes the emission of narrow spectral lines. There is also a variety of other objects, ranging in size down to small dwarf galaxies which, unlike dwarf ellipticals, have strong ultraviolet radiation, and so must contain very young, hot stars and ionized gas. They are essentially what could be called intergalactic H II regions.

Compact galaxies

Starting in the 1930s, Fritz Zwicky prepared lists of objects appearing in the Palomar Sky Survey photographs which are only just distinguishable from stars there, and called them **compact galaxies**. They include objects now known to be at very different distances: a few of them are probably nearby 'intergalactic H II region' dwarf galaxies; many, however, appear to be ordinary galaxies seen very far away; while some are N- and Seyfert-type objects, resembling quasars.

Quasars

A quasar or QSO (quasi-stellar object) is defined as a starlike object whose optical spectrum contains bright emission lines with large redshift, although marginal cases, such as 3C 48 which has weak nebulosity or 3C 273 with a jet feature, are not excluded. (Both of these objects are now known to be surrounded by considerable nebulosity in the form of an associated galaxy.) Usually the optical light varies irregularly over a time scale of months, there is strong ultraviolet radiation, the emission lines are broad, and often there are also narrower absorption lines. Many quasars are identified with strong radio sources, often with radio components of very small angular size. A radio-strong quasar is denoted by QSS (quasi-stellar source). The intense ultraviolet radiation or the strong emission lines are used in searching for new quasars, particularly those which are radio-quiet.

Although the first identification of a radio source with a starlike object was in 1960, it was three years before the emission lines could be identified; only in 1963 was Maarten Schmidt able to recognize them as familiar lines but with unprecedentedly large redshifts. Within the next two years Allan Sandage recognized the existence of radio-quiet quasars. It has been estimated that around a million QSOs are detectable and that the QSS are outnumbered about 100 to 1 by radio-quiet objects.

Emission line redshifts have been determined for some hundreds of quasars. The distribution in redshift is smooth, and values range from $z = 0{\cdot}04$ up to 3·78, although almost all are below 2·4. If the redshifts are related to the quasar distances by Hubble's law, most quasars are very much more distant than ordinary galaxies or even active galaxies. Most astronomers now accept that there is a continuous progression through Class 1 Seyferts, N galaxies and quasars and that their distances are indeed great. From Seyferts to quasars we progress towards relatively more intense nuclei, a quasar appearing starlike because its distance makes the rest of the galaxy undetectably faint. As for Class 1 Seyferts, the emission lines in the quasar spectrum suggest the presence of a low-density plasma similar to a gaseous nebula.

BL Lacertae objects

The object BL Lacertae, once classified as a variable star, was identified with a radio source in 1968, and now about 40 objects like it are known. They are essentially similar to quasars and the nuclei of Seyfert and N galaxies, except that the optical spectrum of the compact object is a continuum; no emission lines are seen. Some appear in the nuclei of elliptical galaxies, some are associated with nebulosity, but most appear stellar on the best available plates.

In many of their properties, they are like more extreme versions of quasars and N galaxies; for example, in the form of the spectrum, the rapid variations of intensity at all wavelengths (within a few weeks), and the strong polarization of their radiation, which may reach 30 per cent. Almost all of those known are radio sources, but none is associated with a large extended source as many quasars are. None is yet identified as an X-ray source. They radiate most strongly in the infrared.

It has been suggested that the BL Lacertae objects are young quasars which have not yet ejected much plasma or gas. This would explain the lack of association with extended radio sources and also the absence of emission lines, believed in quasars to come from ejected gas outside the central region. On the other hand, the fact that some BL Lacertae objects are seen in elliptical galaxies has led to the suggestion that they are old active objects.

Extragalactic radio sources

The first radioastronomy surveys of the sky, made at various wavelengths in the late 1940s and 1950s, showed hundreds of discrete sources, but it required optical identification to establish how many are extragalactic, for from radio observation alone we have no certain knowledge of the distance. It turns out, however, that fewer than 10 per cent of the identified sources are galactic; these are mostly H II regions and supernova remnants, and are concentrated towards the plane of the Milky Way. Identified extragalactic sources are more evenly distributed over the sky and, although nearly a third of the brightest sources are not identified, it is clear from their distribution that almost all the unidentified sources are extragalactic too. Some very strong sources are associated with faint galaxies, and similar objects farther away will still be reasonably strong radio sources but invisible optically.

Extragalactic optical identifications which have been made are, in rough order of increasing absolute radio luminosity:

a. normal spiral galaxies. These are relatively nearby and have thermal radio emission similar to galactic sources. They are not classed as 'radio galaxies';
b. Seyfert galaxies;
c. certain bright early type (E and SO) galaxies. Most such galaxies are not detected as radio sources at all, but some of the largest ones, particularly cD galaxies, are very strong sources. There is generally nothing special about the optical appearance. Often, the brightest galaxy in a cluster is a strong radio source;
d. N-type galaxies;
e. radio quasars, QSS, if their redshifts are interpreted as indicating their distances.

Structures of radio sources

Some sources are compact, with a single component centred on the nucleus of the optical object, but most have an extended structure, typically in two similar lobes roughly symmetrical about the optical nucleus (Fig. 7·9). Both structures are found associated with all the different types of optical object, although most compact radio sources are found to be associated with quasars or galaxies with bright nuclei. It should be noted that a particular radio source can be iden-

Fig. 7·9: The relative linear sizes of a sample of typical extended radio sources. Sources of such different extent have essentially similar structures. Notice the small central component of 3C 236, with structure aligned with the outer lobes. (See page 210).

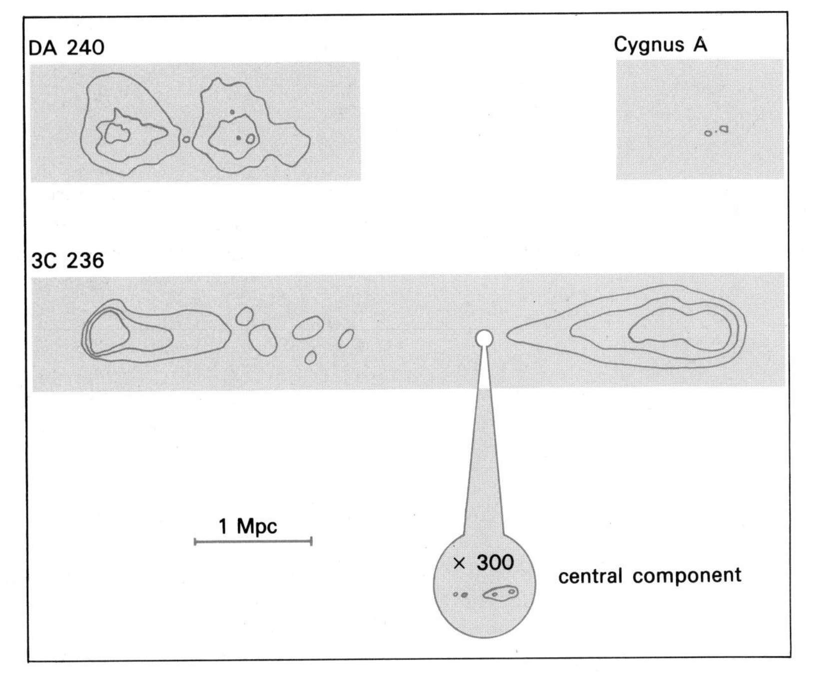

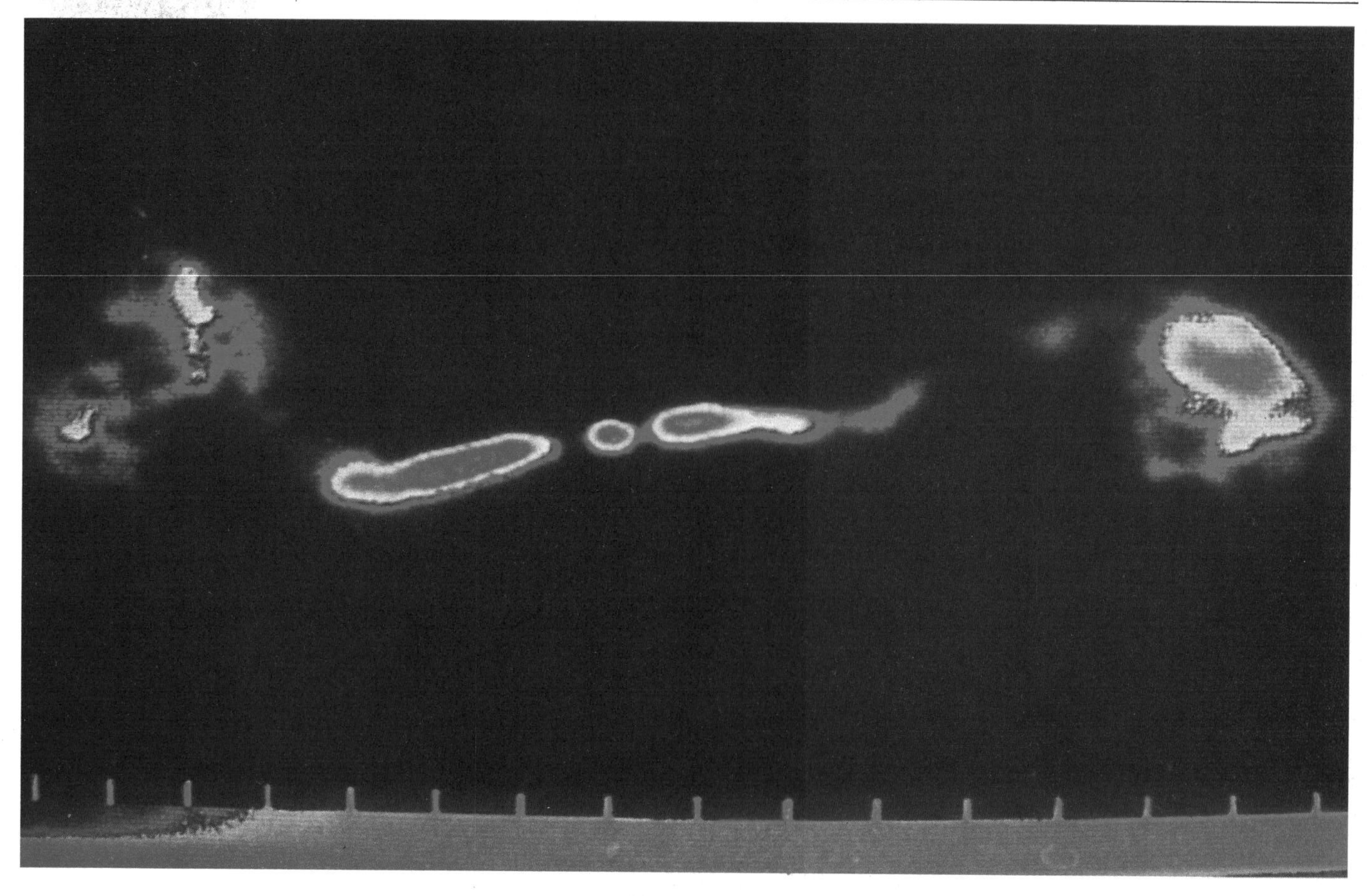

A 'photograph' made from radio observations of 3C 499, showing the extended lobes of radio emission on either side of the parent galaxy.

tified as a quasar only if it is related to a starlike optical object.

Some two-lobed sources are very large, extending over several Mpc. The normal structure within each component is a strong head with a tail extending back towards the centre, and quite often there are inner maxima in the intensity, much closer in than the outer lobes but aligned along the same axis.

In most cases, the radio waves show polarization, in amounts up to 20 per cent. Polarization is related to the strength and direction of a magnetic field, and in some sources the field is uniform over remarkably large volumes of space. The most powerful radio sources radiate about 10^7 times more energy at radio wavelengths than the Galaxy does, while irregular variations within a few months are observed in many compact sources.

Opposite page, top:
Image-processed optical CCD picture of the active galaxy M87, showing the various bright knots along its jet.

Opposite page, bottom:
The apparent superluminal velocity (see page 214) of the quasar 3C 273. Over a period of 3 years it has appeared to expand by 25 light-years.

The active galaxy M 87

M 87 is a giant E0 galaxy, about 10^{13} solar masses, and the third brightest member of the Virgo cluster. Optically it displays a strong jet, similar to the jet seen in the quasar 3C 273. As the third brightest radio source in the sky, it is also known as Virgo A; it appears bright because it is very close to us, only about 15 Mpc away. At radio wavelengths it is seen to have two jets, one coincident with the optical jet and one opposite it, as well as a fainter outer halo. The core is very compact, emitting about 1 per cent of the radio energy from a region no more than about 0·1 pc across. M 87 is also an X-ray source.

Early in 1978, evidence from optical astronomy helped towards an understanding of the nucleus; it is found to be very bright in relation to the rest of the galaxy, compared with ordinary ellipticals, while the spectral lines reveal a sharp increase in the velocity range in the nucleus region. These optical observations, together with the observed radio structure, indicate not only an unusual concentration of mass at the galaxian centre but also an energy source there, and the most plausible explanation is considered to be the presence of a black hole of about $5 \times 10^9\ M_\odot$. Although not at all conclusive, this is the strongest evidence available for the existence of a black hole in an extragalactic object.

Energy sources

The first energy source proposed for radio galaxies was the collision of two galaxies in a cluster, partly because the very strong source Cygnus A looks like two galaxies colliding. However, many radio sources appear optically normal, and in any case collision would not give either synchrotron radiation or a two-lobed structure. It is now believed that radio sources originate in violent events within single galaxies, which are related to the energy sources in quasars and active galaxian nuclei.

Various possible sources for such violent releases of energy have been suggested. The energy release from the gravitational collapse of an object of some 10^6–$10^8\ M_\odot$ was one, but it would tend to happen too quickly and again not produce two lobes or the fast particles which give synchrotron radiation, and the collapsing body would probably be unstable anyway. Another was a multiple outburst of supernovae where, in conditions of high stellar density in

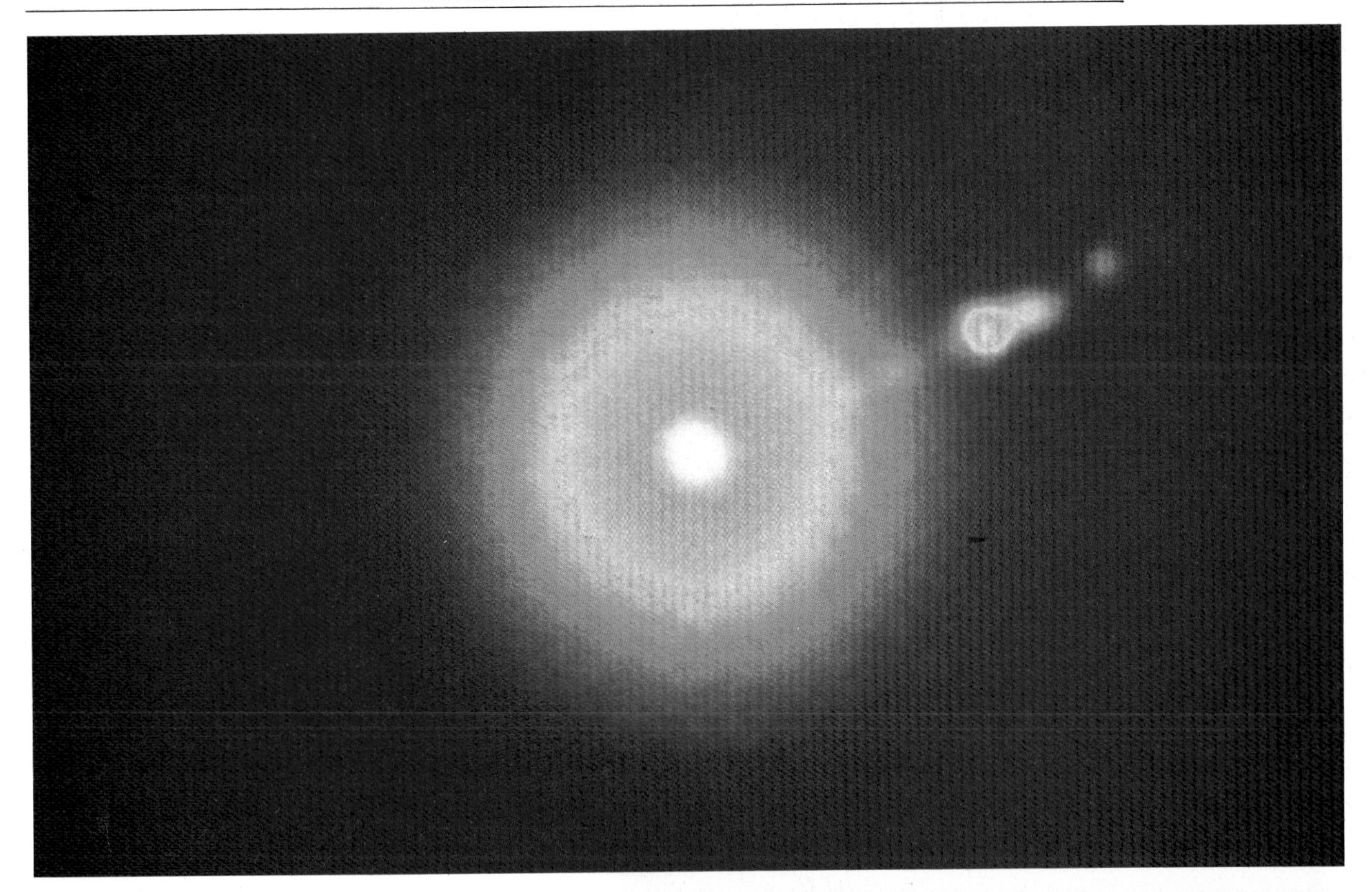

a galaxian nucleus, one supernova might trigger off another in a chain-reaction sequence; but again energy release would be in all directions and too fast. A massive black hole in the galaxian nucleus, as discussed above for M 87 and with less confidence for normal galaxies, has been proposed, but as the source of energy for an extended radio source, it presents similar problems. An opposite idea is that there could be a **white hole** in the nucleus, and that material for the entire galaxy could be emerging from it. This has been considered particularly by Viktor Ambartsumian, who further suggests that entire clusters of galaxies could have emerged from such holes and still be dispersing, an idea which gives a possible resolution of the 'missing mass' problem.

Another class of theories, introduced by L. M. Ozernoy and developed by Philip Morrison and others, considers a rapidly rotating region in the centre of the galaxy, of 10^8–10^{10} $M_\odot$ and called a **spinar**. Magnetic interactions at its outer edges would release energy. In a version due to Martin Rees, energy is released as a highly energetic plasma along the two ends of the rotation axis; thus a double structure is produced along a defined axis, as is observed. Such 'jet processes' are coming to be accepted as important in many unusual objects, both in accounting for the double structure of many radio sources, and also on a smaller scale for individual objects within galaxies. (An example of the latter is the object known as SS 433, which can be explained by a compact object accreting matter from a companion in a binary system and giving rise to two jets of energetic particles which are ejected along its rotational axis.) With a spinar, plasma and energy are supplied

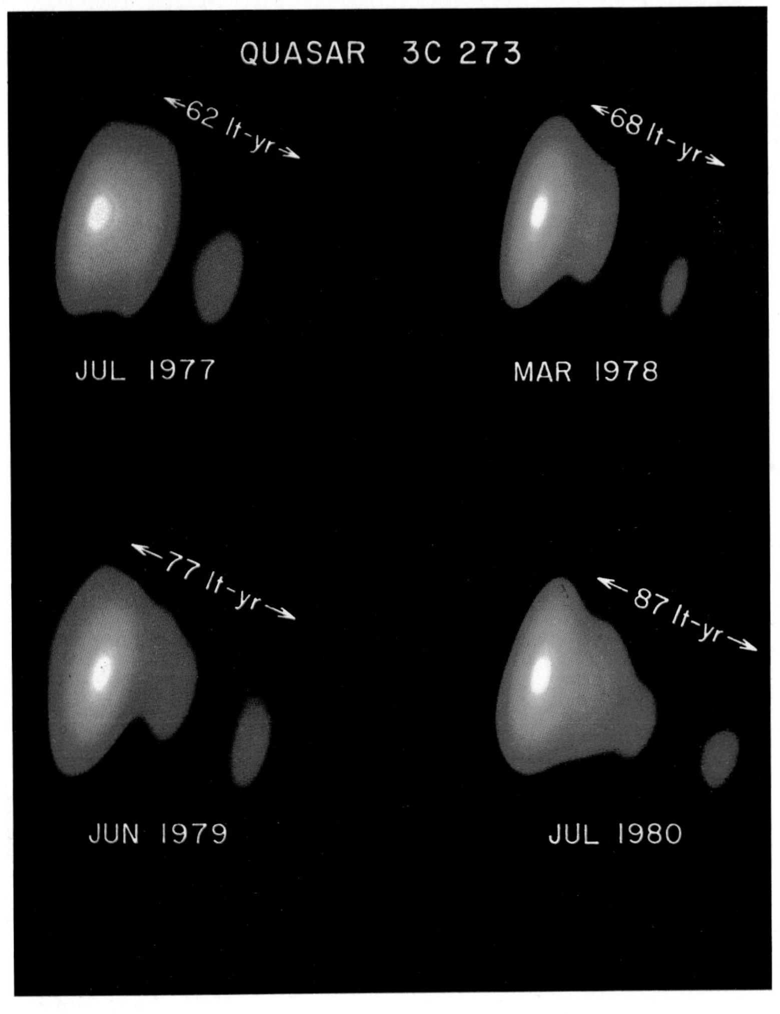

Extragalactic Astronomy for Amateurs

It is the undoubted fascination of identifying extragalactic objects which drives many amateurs to spend long hours searching for them and looking at them through their telescopes. Although no one could ever expect to see them as they are shown in the photographs obtained with some of the world's largest telescopes, there still remains the pleasure in comparing one's own impression or drawing with those photographs, and in being able to identify certain features, even if they are only faintly visible. There are also people who like making comparisons between their own drawings and those of some of the famous earlier observers. In particular, they wish to identify with their own eyes all the objects in Messier's catalogue, and as many of the NGC and IC nebulae and galaxies as possible. Some observers even limit themselves to equipment similar to that used by Messier himself – for most objects roughly the equivalent of a modern 90mm refractor. However, rather larger equipment is needed for most of the galaxies to be visible in detail.

Three extragalactic systems (the Large and Small Magellanic Clouds and the Andromeda Galaxy, M31) are certainly visible to the naked eye, and one other spiral galaxy (M33) is reputedly so. Of the fainter objects on Messier's list, some are naturally of very considerable astronomical and cosmological interest in their own right, e.g. M77, one of the Seyfert galaxies (page 207), M82, a galaxy apparently showing violent activity, and M87, the very active galaxy in Virgo (page 210).

Both of the Magellanic Clouds are so close to our own galaxy that they can be partially resolved into individual stars with only moderate optical aid. They are both well worth examining with binoculars, particularly the Large Cloud, as this contains the vast region of nebulosity known as 30 Doradus (or the Tarantula Nebula) which is also visible to the naked eye alone. Close to the Small Magellanic Cloud is the large bright globular cluster NGC 104, 47 Tucanae, which belongs to our own Milky Way system. The Andromeda Galaxy (M31) is not particularly striking in binoculars, but becomes worthy of close examination with larger instruments. It would be pointless to attempt here to deal in any detail with all the other galaxies which become visible with even a moderate telescope.

For anyone with rather larger equipment – e.g. a reflector of 200mm aperture or more – some even more exotic objects are available for study. These include BL Lacertae itself (page 209) and the quasar 3C-273 (page 208), which has a nominal magnitude of 12.3 but is, of course, variable. Some amateurs follow the brightness changes of these, and a few other similar objects, using the methods of variable star observing described on pages 56–57.

It is perhaps necessary to offer a word of warning about the magnitudes of many extragalactic objects given in catalogues. Very frequently a magnitude is actually an integrated magnitude for the whole of an extended object, including the faint outlying regions, and it may well have been derived from a photograph. In most cases the observer's eye will not perceive these faint outer regions and the apparent magnitude will be considerably less than the catalogue value. A failure to realize this has caused observers to spend a lot of time fruitlessly searching for objects which they believed to be above the limit of their telescopes, but which in actual fact become visible only in much larger instruments.

As previously noted, it is essential that the 'unexpected' eruptive variables, such as the novae and supernovae, be discovered as early as possible. Searching for novae has already been discussed (page 56), but the methods employed in attempting to catch supernovae are rather different. There are one or two professional patrols, but most discoveries come from workers in other fields who happen to obtain a photograph which shows a supernova. However, more amateurs are now turning their attention to this subject, and are beginning to achieve some success in discoveries of supernovae.

As with the novae, searching may be carried out either visually or photographically. Each method has its advantages – speed and low cost in the case of the visual search, permanence of records and generally fainter limiting magnitudes in the photographic method. In both cases suitable charts of candidate galaxies are needed. These must not only identify suitable comparison stars for use if a supernova outburst occurs, but must also plot any foreground stars, condensations or H II regions (page 178) in the galaxy which might be confused with a true supernova. Obviously discoveries can and do take place in galaxies for which no charts are available, and at present this occurs in the majority of cases. In these instances checking has to be even more thorough, as there may be considerable

The Andromeda Galaxy, photographed with a 350mm × 500 mm Schmidt telescope on N2/H2 baked, hypersensitized Technical Pan film, with high-contrast copying.

differences between the visual and photographic appearances of any galaxy; and even with a photographic method, different lengths of exposure can considerably change the apparent structure. The use of photographs taken with large professional telescopes can prove especially confusing: long exposures are usually employed to capture the faint detail in the outer regions, but this results in over-exposure of the central regions and consequent loss of information. With the relatively small apertures available to most amateurs it is usually only those central regions which are recorded adequately, so that the problem of comparison is a very real one. This can be partly overcome by the use of 'master' photographs taken with the same telescope as will be used for the searching. In the case of a supernova in a previously uncharted galaxy, photography is used in order to establish a suitable series of comparison stars for subsequent visual or photographic monitoring. This is usually the major problem when a supernova is discovered, and sometimes the sequence of comparison stars may only be satisfactorily decided at a very much later date.

Because of the problems involved in the correct identification of true supernovae, great care has to be taken to ensure that all suspect objects are properly checked before any major alert is issued. Both for this reason and because there are so many galaxies which could be examined, some discretion has to be used in selecting the objects which are to be kept under surveillance. Spiral galaxies are the most favourable candidates, and in those which are face-on to us it is less likely that supernovae will be obscured by gas and dust in the galactic plane.

Photography of extragalactic objects is naturally of interest to many people. As with all the various galactic objects, a whole range of techniques may be employed to obtain good images keeping exposures as short as possible (page 177). Hypersensitization or cooled-emulsion methods can be exploited to produce pictures of the faintest objects.

Just as professional astronomers are turning more and more to highly sophisticated electronic equipment, so some amateurs are also exploring this field. Some are investigating the use of image intensifiers, for example, partly for the sake of interest, and partly for more serious photography and the detection of supernovae. Others are applying video and image-enhancement techniques (using microcomputers) to photographs of galaxies. Only time will tell whether these experimental projects will have lasting, positive results for the observation of distant galaxies.

Extragalactic objects

Large Magellanic Cloud (LMC)	Nearest system, contains Tarantula Nebula ***
Small Magellanic Cloud (SMC)	**
M31 And Andromeda Galaxy	Nearest major system **
M32 And	Companion to M31
M33 Tri	Large nearby spiral **
M49 Vir	Elliptical galaxy
M51 CVn Whirlpool Galaxy	Striking spiral galaxy **
M58 Vir	Spiral galaxy
M59 Vir	Elliptical galaxy
M60 Vir	Elliptical galaxy
M61 Vir	Spiral galaxy
M64 Com Black-eye Galaxy	Spiral galaxy **
M65 Leo	Spiral galaxy
M66 Leo	Spiral galaxy
M74 Psc	Spiral galaxy
M81 UMa	Spiral galaxy **
M82 UMa	Spiral galaxy *
M83 Hya	Spiral galaxy **
M84 Vir	Elliptical galaxy
M85 Com	Elliptical galaxy
M86 Vir	Giant elliptical galaxy
M87 Vir	Elliptical galaxy, powerful radio source
M88 Com	Spiral galaxy
M89 Vir	Elliptical galaxy
M90 Vir	Spiral galaxy
M91 Com	Spiral galaxy
M94 CVn	Spiral galaxy **
M98 Com	Spiral galaxy
M99 Com	Spiral galaxy
M100 Com	Spiral galaxy
M101 UMa	Spiral galaxy *
M104 Vir Sombrero Galaxy	Edge-on spiral galaxy
M106 CVn	Spiral galaxy
M110 And	Elliptical galaxy, companion to M31

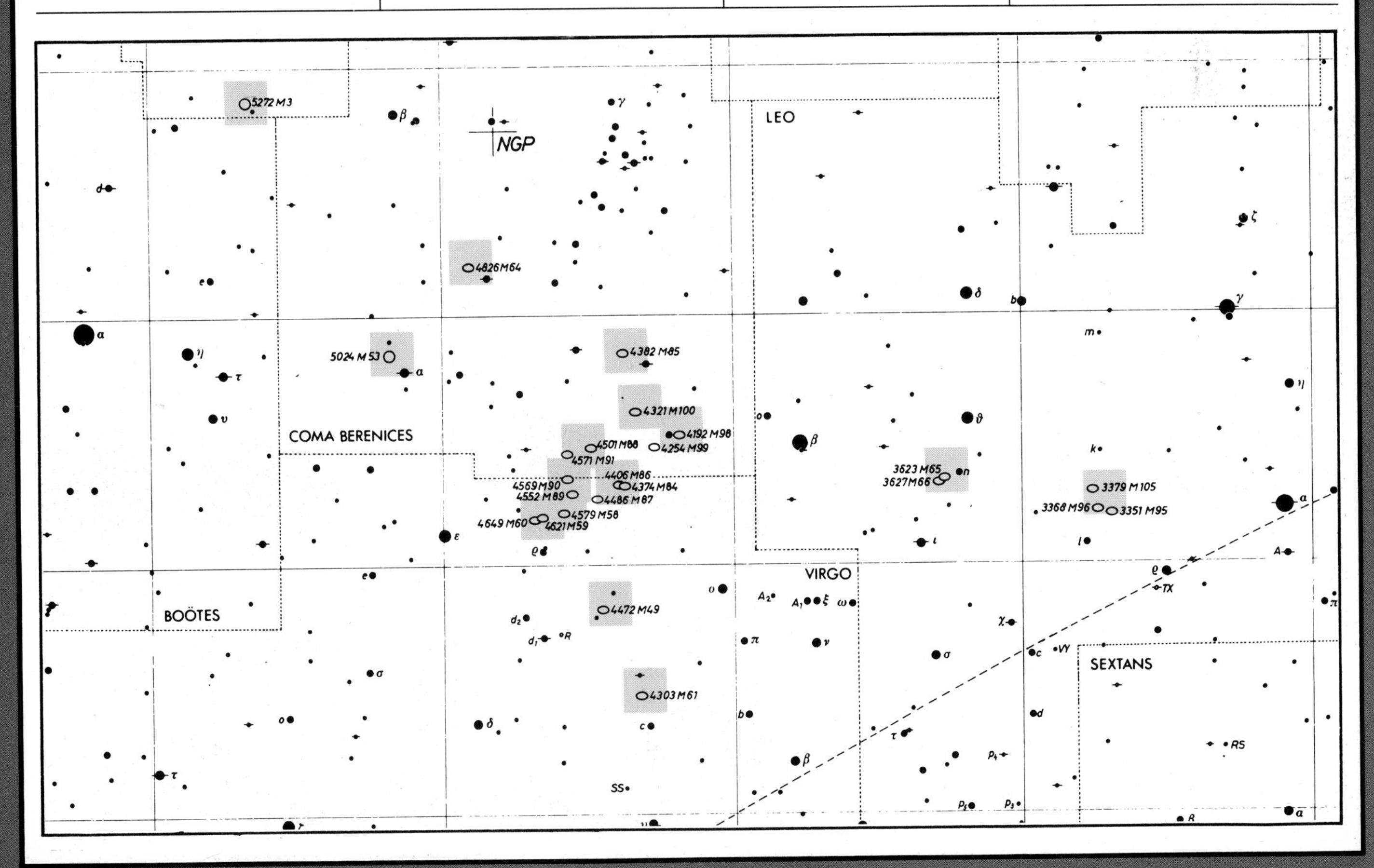

Bottom:
The regions of Coma Berenices, Virgo and Leo, where we are looking away from the crowded plane of the Galaxy, contain many interesting extra-galactic objects.

in a continuous outflow from the nucleus. The origin of the hot plasma is another problem; again, it could involve a black hole.

Apparent superluminal velocities

Since about 1972 it has been possible to observe the structure of some of the compact radio sources, using very long base-line interferometry (page 242), and they are found to be double-lobed, similar to extended sources. The angular sizes of less than 0·01 arc sec. correspond to actual dimensions of only a few parsecs. Now, in at least four cases – three quasars and one galaxy – structural changes which have taken place over several years give an apparent velocity of expansion much greater than the speed of light. This is based upon the assumption that the distances are given by interpreting the redshifts in accordance with Hubble's law. As one of the sources is a galaxy (3C 120) this can not easily be questioned.

It is a fundamental result of modern physics, well verified experimentally, that nothing physical can move faster than the velocity of light. This result is, indeed, an essential part of the theory of relativity. However an *appearance* can move faster than light, so long as no direct *physical* motion is involved. This is now thought to be the explanation of the phenomena observed, especially after examination in detail of one such source, the nearby quasar 3C 273, well-known for its visible jet as well as for its association with a galaxy. It is believed that jet processes are producing a beam of highly energetic electrons travelling at close to the speed of light. Under such circumstances, despite what might be thought at first, if the beam is directed towards us rather than across the line of sight, the material excited by the electrons can appear to move at greater than the speed of light. In the case of 3C 273 the beam is probably within about 12° of our line of sight. The other instances almost certainly have the same cause, and statistical studies are likely to show that because we may expect the beams to be randomly orientated in space, only a few such examples will be visible to us.

As far as the gravitational redshift hypothesis is concerned, it requires a large mass in a small volume. We then have to face problems of stability, while, in addition, the nature of the emission lines in the spectrum implies a large region of relatively low density gas through which the gravitational field is uniform. All in all, an object of the order of 10^{11} $M_\odot$ only 10 kpc away seems to be required, the observational consequences of which would be enormous.

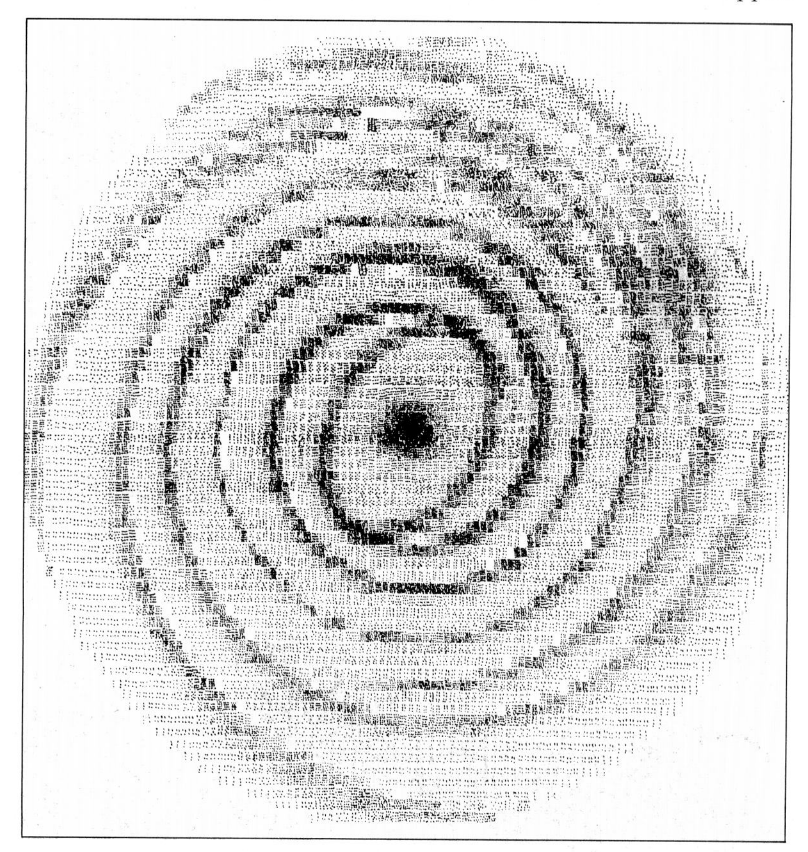

Fig. 7·10 A computer representation of the possible optical appearance of our Galaxy seen face on from outside, based on an interpretation of the observations using the density-wave theory. The structure is essentially a four-fold spiral. The Sun is located about two-thirds of the way towards the top edge of the picture.

The formation and evolution of galaxies

Galaxies formed about 10^{10} years ago, at an earlier epoch in the development of the universe, so galaxy formation must be related to cosmological processes. Our ideas of how galaxies formed are much less coherent even than our ideas about star formation. There are various competing theoretical models, all of them primitive.

It is generally accepted that the galaxies formed by condensation from a primaeval 'intergalactic' material, and that this must have contained irregularities from the earliest stages of the expansion of the universe: the growth of chance fluctuations in an initially uniform medium would not be efficient enough. In one class of theories, unevenness in the distribution of the material led to instabilities and so to the collapse under gravity of parts of the material to form galaxies. Rotation would then come about by interaction between neighbouring galaxies or clusters of galaxies. In another class of theories, the early irregularities took the form of vortexes in turbulent motion, which naturally led to the formation of rotating systems. Presumably galaxies formed in clusters, as stars do.

There is, however, a third, quite different idea: that galaxies emerge from white holes at their centres. A white hole is the exact reverse of a black hole; nothing can enter it and there is a steady outflow of matter. Such white holes (if there are such things) must have been in existence from the start of the universe, and each would cease to exist when it had thrown out all the mass which it contained.

Accepting the idea that the galaxies all formed from the same material at about the same time, the differences between them must have arisen from different evolutionary developments after they formed. Two major properties of a galaxy are its mass and its angular momentum, but since all the Hubble types have a wide range of mass, it seems that mass is not decisive. On the other hand, angular momentum appears to be linked with Hubble type and can in outline explain the different developments. The flattening of an elliptical galaxy can be related to its rotation, while in a disc galaxy, faster rotation means a slower collapse, so that up to the present time less of the interstellar gas has been used forming stars.

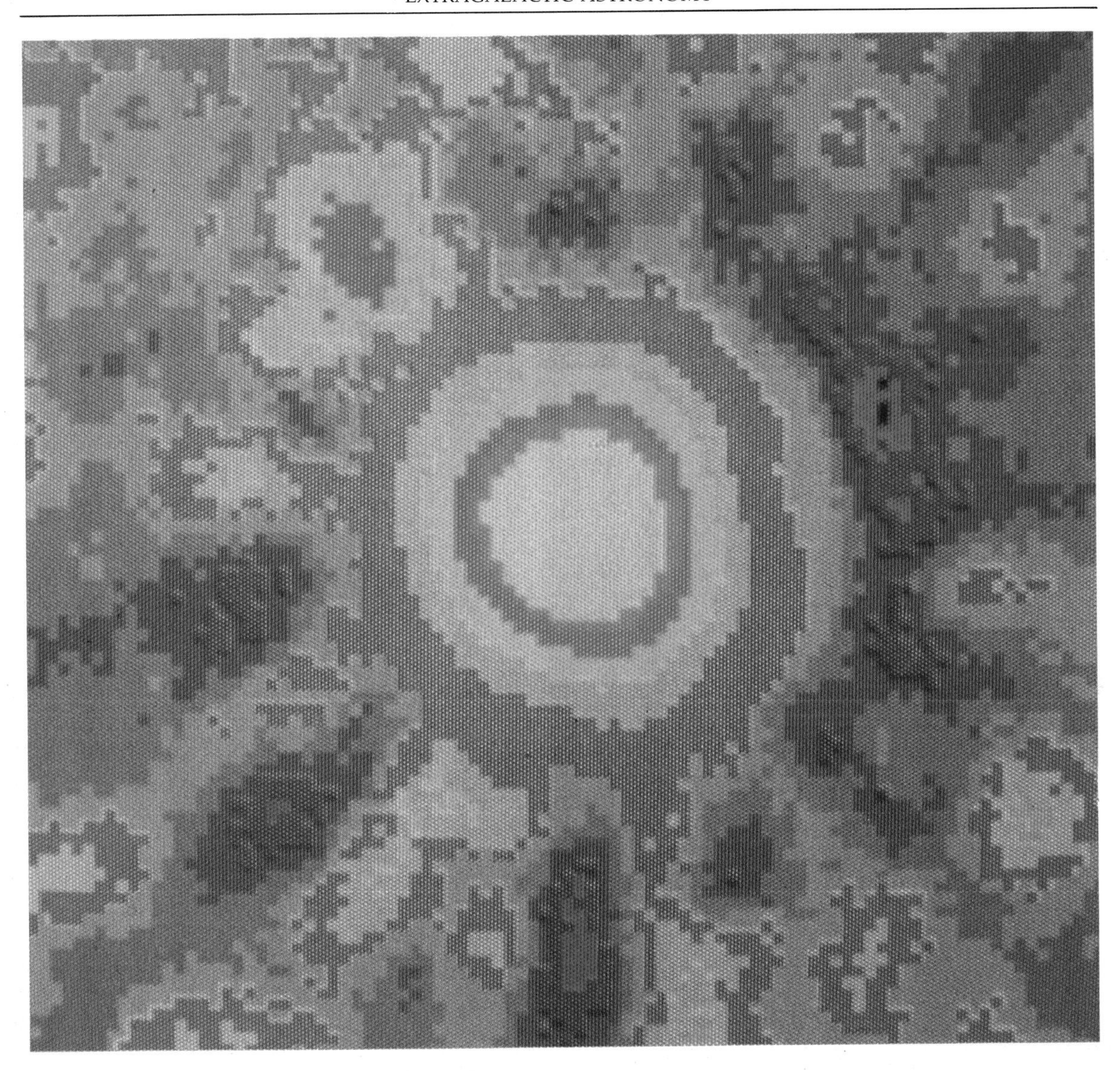

This X-ray picture from the Einstein satellite shows some of the structure of the quasar 3C 273 and its jet (at the 5 o'clock position).

Some, however, consider that the difference between elliptical and disc galaxies requires further explanation. It could be that outside influences have in some way affected the manner in which star formation proceeded, or possibly it has had something to do with activity in the nucleus. S0 galaxies could well be spirals which have been swept clear of gas, either through the pressure of the intergalactic gas in clusters or, less frequently, by direct collision between galaxies.

After forming, a rotating gas disc becomes thinner with time: gas passing through the plane of the disc collides with other gas and so tends to lose its motion perpendicular to the plane. Stars, however, can pass freely through the disc, and their distribution does not become thinner. This scenario is consistent with the distribution in our Galaxy of stars of different population types where the younger populations (stars which formed later) have thinner disc distributions.

Our general theories of galaxy formation and evolution have a long way to go before they can properly be related to real galaxies. It is not yet clear, for instance, whether all galaxies go through a phase of being active galaxies, perhaps early in their lives, although this is a question of fundamental importance.

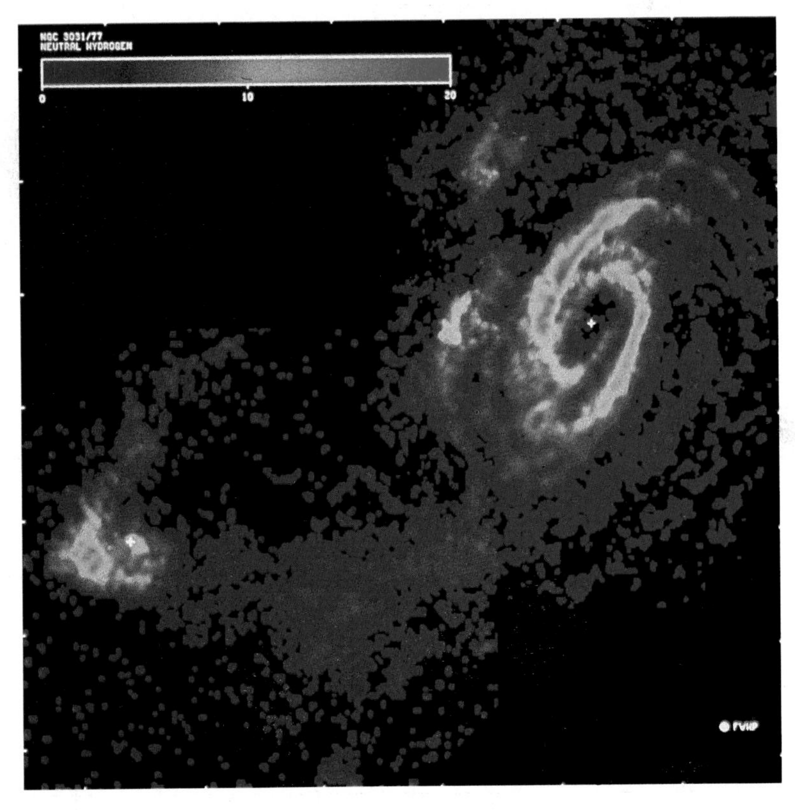

A map of the spiral galaxy M81 and its companion NGC 3077 at the neutral hydrogen wavelength of 21 cm.

Theories of the universe

Cosmology is the study of the universe as a whole. It is an overview which allows us to formulate a general picture of the cosmos, and to construct theories about its earliest stages and its ultimate fate. At the basis of this study are two physical theories: the quantum theory – the theory which states that energy is absorbed, emitted or transformed in discrete units or quanta; and the theory of relativity. Quantum theory, coupled with the contemporary picture of atoms and sub-atomic particles, affects all phenomena, as does relativity theory, but the latter has particular cosmological implications, and deserves further explanation.

Relativity theory

Relativity theory falls into two parts – the special theory and the general theory, the first being published by Albert Einstein in 1905, the latter by him in 1916. The special theory is concerned primarily with electromagnetic phenomena and the way electromagnetic waves travel in time and space, while the general theory was developed mainly to deal with gravitation; both are obviously important from a cosmological point of view and it will be convenient to deal with them in turn.

Relative and absolute motion

Before special relativity was formulated, Newton's theory of universal gravitation and his laws of motion were accepted. According to Newtonian theory there must be a basic 'frame of reference': for instance, Newton's first law of motion states that a body must be in either a state of rest or a state of uniform motion in a straight line, unless the body is being acted upon by some outside force. The basic frame of reference is necessary because if we talk of a body being at rest, it must be at rest with reference to something – its surroundings if you like. Again, if it is in motion then its motion must be relative to a frame of reference, which, again, we may specify as its surroundings. We should note that such a frame of reference involves time as well as space, for velocity is a concept which incorporates both space (distance travelled) and time (the time taken). By the end of the nineteenth century it had become clear that there was a problem in choosing an absolute frame of reference: should one choose the Earth, or the Sun (about which the Earth orbited), or the whole gigantic system of stars through which the Sun was moving?

Other research had led to the discovery that light is part of the electromagnetic spectrum, all wavelengths of which travel at the same speed, 3×10^5 km per s. It was clear too that light is a wavelike disturbance, and since one could not have a disembodied wave, there was a new problem – to discover the nature of the substance or medium in which the waves travelled. The medium, whatever it might be, was called the 'aether', and in 1887 Albert Michelson and Edward Morley set up an experiment to determine the velocity of light with respect to this aether. Because of its result, this experiment proved to be one of the most important in the whole history of physics.

The Michelson-Morley experiment

In order to avoid using clocks of any kind to time the motion of a beam of light, Michelson and Morley used two simultaneous beams of light from the same source, sending one in a direction at right angles to the other, then reflecting them back so that they interfered (Fig. 8·1). Because of the movement of the Earth through the aether there should be a difference in the time taken by the two light beams, the light beam perpendicular to the Earth's motion taking a shorter time than the light beam out and back in the direction of the Earth's motion. The difference is very small, but by rotating the whole interferometer, a shift in the interference lines due to this difference

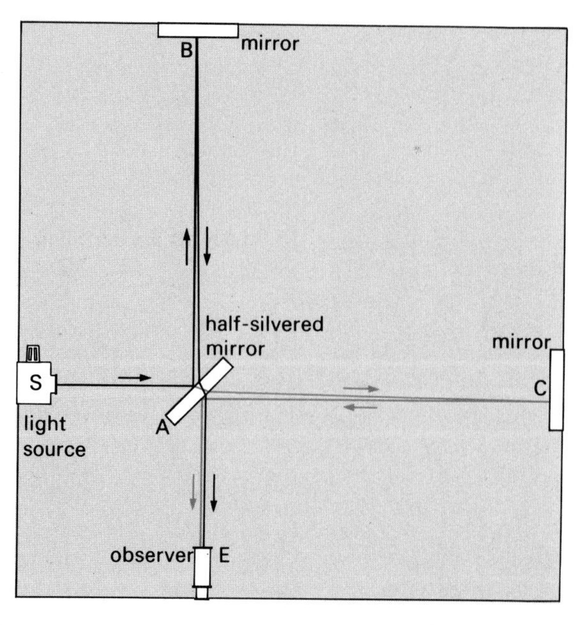

Fig. 8·1 far right: *The optical layout of the Michelson-Morley experiment. The light beam from the source S moves to a half-silvered mirror A, where it is split into two components. One component (black) is reflected to the mirror B and back again to A, when it passes through to the observer at E. The second component passes through the half-silvered mirror A and on to the mirror C; here it is reflected back to A and down to the observer at E. The two sets of waves interfere at E and produce a series of bright and dark fringes.*

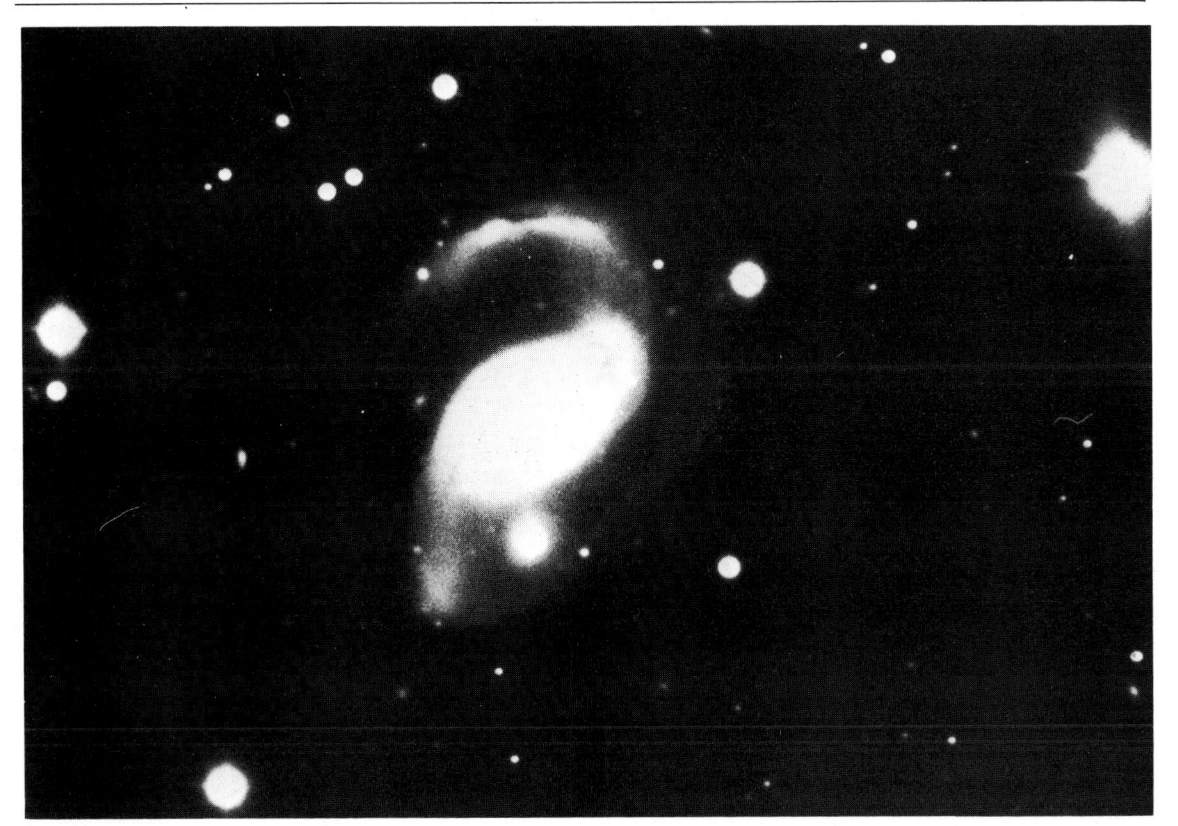

Quasars pose many problems for cosmology, particularly those cases where an apparently nearby galaxy is linked with a seemingly distinct quasar, as here where quasar Markarian 205 (the circular object) is close to the galaxy NGC 4319.

would nevertheless show up quite clearly. Yet Michelson and Morley could detect no shift. They thought this might be because the Earth was moving with the aether, and to check this they repeated the experiment six months later when the Earth would be moving in the opposite direction. Further repeats of the experiment also gave a nil result, and although various explanations were offered, none was satisfactory.

The aether did not seem to exist. Yet even more significant was that the result showed that the velocity of light was invariable – it did not matter how one moved relative to the source emitting it, the value was always the same.

This invariance of the velocity of light – indeed of all electromagnetic radiation – goes against our usual physical experience. For instance, consider two high speed trains moving towards each other (on separate tracks!). If both, say, are travelling at 240 km per hour, the velocity at which they are approaching each other is 480 km per hour. Supposing someone throws a ball out of the window of a carriage towards the other train at a speed of 20 km per hour, this will meet the other train at a speed of 500 km per hour (Fig. 8·2). This is straightforward enough – we just add the velocities to come to our final result. But if instead of throwing a ball someone leans out of the window and shines a torch, light will escape from it at a velocity of 300 000 km per s, and it will reach the approaching train at 300 000 km per s *not* at

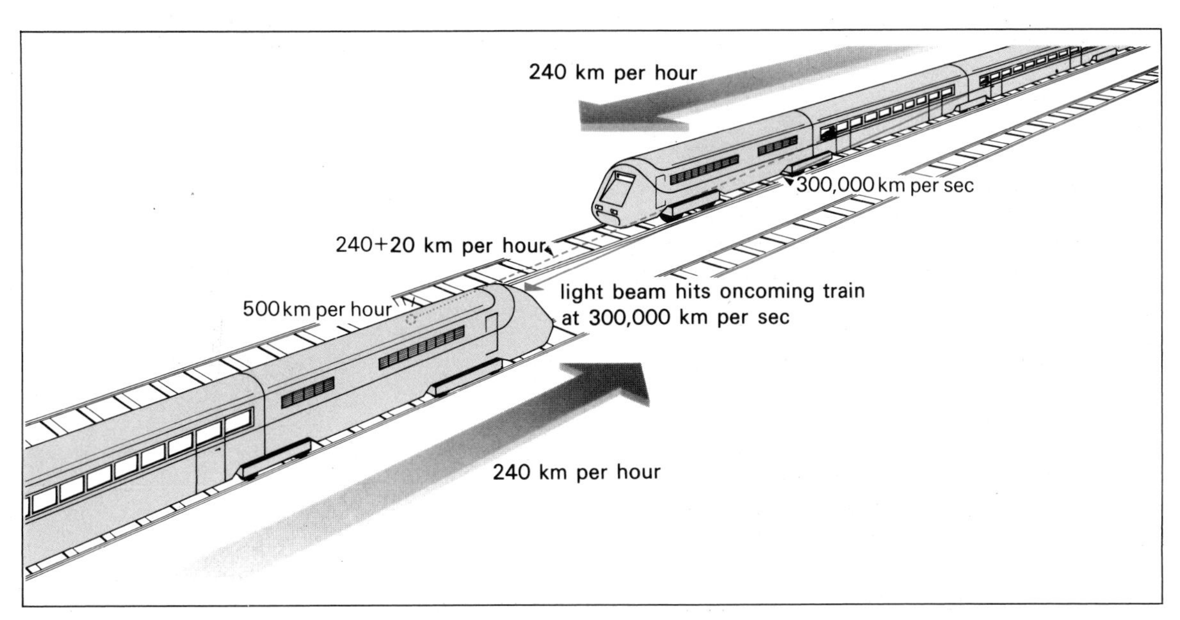

Fig. 8·2
A ball thrown at a speed of 20 km per hour from one high speed train travelling at 240 km per hour to another high speed train travelling in the opposite direction at 240 km per hour, will meet the second train at 240 + 240 + 20 = 500 km per hour. But the beam of light from a torch emitted at the speed of light (approx. 300 000 km per sec) will still meet the oncoming train at speed of light.

Opposite page; Fig. 8·3 and below: Astronauts under zero gravity conditions. The photograph shows Ed White during the Gemini IV mission in June 1965

300 000·13 km per s. Clearly there is something special about the way the universe is constructed for this to happen.

Special relativity

In considering a theory in which all motion is relative, in which Newton's concept of absolute motion did not exist, Einstein took into account the results of the Michelson-Morley experiment, paying special attention to the interpretation of the Dutch and Irish physicists, Hendrik Lorentz and George Fitzgerald. They had claimed that the nil result could be squared with the physics of the time – 'classical physics' – if one considered that length, mass and time changed in a moving body compared with a stationary one. Mathematically, Lorentz showed that the length of a moving body should be diminished by an amount which depended on the square root of the quantity $(1 - v^2/c^2)$, where v represents the velocity of the moving body and c the velocity of light. Then if a body moved with the speed of light, v would equal c, and v^2/c^2 would be equal to 1; but if this were so $(1 - v^2/c^2)$ would become zero, and the body would have no length at all – clearly, a nonsensical answer. Lorentz also derived relationships for the mass of a body (which increased with velocity) and for time (which also increased); these also contained the square root of $(1 - v^2/c^2)$, and gave nonsensical answers for travel at the speed of light.

Einstein saw the deep implications of these results and realized that they demanded a full scale revision of the laws of physics. The increase of mass with velocity, for instance, involved the whole question of kinetic energy (the energy of motion) and made him realize that there was an equivalence between mass m, and energy E, leading to the famous equation:

$$E = mc^2.$$

It was also clear to him that the velocity of light was not only an invariant, but was also a limiting velocity, so that it became one of the tenets of relativity that nothing can travel faster than light. Again, there could be no absolute motion because there was no fixed frame of reference in the universe. Also, since time was involved and was relative in bodies moving relative to one another, Einstein realized that the simultaneity of events was also relative – that is, because one observer might see two events happening at the same instant, this apparent simultaneity would not be seen by other observers moving relative to the first observer. For instance, suppose someone on our Galaxy observes two events – the explosions of two widely separated supernovae, say – in another galaxy, and sees these happen together. We cannot tell whether these explosions were truly simultaneous without knowing their distances. But consider another observer on another galaxy moving relative to our Galaxy. The other observer's values for length and time will be different to those made in our Galaxy (how different will be proportional to their relative velocity v), so they will not see the explosions simultaneously. Whether or not one observes simultaneity depends, then, on one's frame of reference. In other words, if there is no absolute reference standard of motion in the universe, there is no absolute reference standard of simultaneity either.

Since time as well as place are relative according to the theory, it would lead to errors to consider questions of space and time separately. Therefore, relativity is concerned with 'events', which incorporate the three 'dimensions' of space (length, breadth, height) and the 'dimension' time – a **space-time universe**. Although time intervals will differ from one frame of reference to another, and space intervals will vary too, by placing them in a space-time combination, relativity does give an unchanging or **invariant** interval between events. Of course if we are dealing only with events measured in a single frame of reference, then time and space differences are the same for all observers; this is what we are accustomed to in our everyday experience for, of course, our measurements are all made on our own frame of reference, the Earth.

To make it as simple as possible, in special relativity Einstein considered only frames of reference in uniform (that is, non-accelerated) relative motion. Appreciation of this explains the so-called twin paradox, which is sometimes claimed to show that the principle of relativity – the absence of any absolute frame of reference in the universe – is untrue. As usually stated the paradox is about twins, one of whom stays on Earth while the other goes off in a space vehicle and then returns. According to the Lorentz equation, time for the space traveller has gone more slowly than that for the observer on Earth; therefore, when he returns he will be younger than his stationary twin. In this way we can tell which frame of reference is the standard or absolute one. Yet the paradox is not real. The age difference does not invalidate relativity, and the paradox is false because we do not have two uniformly moving frames of reference; there is only one. Any second frame associated with the space traveller would have to move in uniform not accelerated motion: however, the space traveller undergoes acceleration as he leaves the Earth, slows down prior to his return, and so on.

General relativity

In considering the motion of bodies, Newton devised the concept of universal gravitation to explain why a body falls to Earth, or to the centre of any massive body, with accelerated motion. On Earth this acceleration is 9·81 m per s^2. Naturally, Newton thought of gravitation as acting instantaneously, but Einstein's relativity theory did not accept instantaneous action: no interactions can take place at a speed greater than that of light. Gravity may, therefore, be transmitted with the velocity of light, but no faster. To take this into account, Einstein had to develop general relativity so that it could incorporate not only the non-accelerated frames of reference of special relativity – **inertial** frames as they are called – but also frames of reference in relative accelerated motion. The incorporation of gravity raised many difficult problems, not least due to the relative nature of mass and length. In Newtonian gravity theory, the force of gravitational attraction is proportional to the masses

of the bodies and the distance between them, so in general relativity there was bound to be a problem since masses change when in relative motion; Einstein was able to remove this by what he called the **principle of equivalence**.

To appreciate this principle we must consider the mass or quantity of matter in a body, and there are two ways of defining this. On the one hand we can consider defining a body's mass as its tendency to resist change, either to the change from a state of rest to a state of motion (or *vice versa*), or from uniform motion in a straight line to motion in some other way. This is called the **inertial mass**, and can be seen in operation when a vehicle stops suddenly, propelling the driver towards the windscreen (Fig.

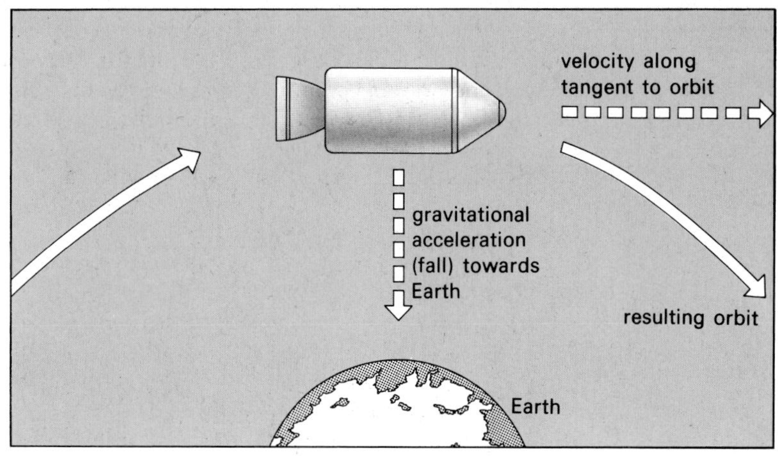

Fig. 8·4
An example of inertial mass – the movement of a driver towards the windscreen when his vehicle stops suddenly.

8·4). It is inertia that carries the driver forward. On the other hand we can consider the mass of a body as a measure of the force that acts on it in a gravitational field. This is its **gravitational mass** and when we weigh something we are measuring the force, which gives a gravitational mass its weight. In general relativity, Einstein realized that it must be a basic principle that inertial mass is exactly equal to gravitational mass, and experimental comparisons between them have shown that they appear to be equal, at least to one part in 10^{11}.

Because inertial and gravitational mass are equivalent, it is impossible (relativity states), to make any measurement which will distinguish between the presence of a uniform gravitational field and the uniform acceleration of a moving reference frame. In other words, a gravitational field of force is equivalent to an artificial field of force (due to an accelerating reference frame). There is also no way of detecting any difference between a non-accelerating (inertial) reference frame and freely falling frames of reference. Force – gravitational or otherwise – is relative. Two examples will illustrate this. First, consider an observer in a box, isolated so that he can not get any clues from the rest of the universe (Fig. 8·5). If the box is in a uniform gravitational field and its occupant drops a coin, he will see it fall to the floor. If there is no gravitational field and, instead, the box moves upwards at the correct acceleration, the coin 'dropped' by the observer will appear to fall just the same because the floor of the box will move upwards to meet it. Since any measurements of the gravitational mass (first case) and the inertial mass of the coin will be the same, the observer will be unable to determine whether the field is there or not.

For the second example, consider astronauts aboard an orbiting spacecraft. Here the spacecraft is moving in two directions – one straight ahead and, the other, falling towards the Earth – these two motions resulting in an orbit round the Earth (Fig. 8·3, page 219). In this case, the motion towards the Earth is called 'free fall', and is equivalent to being in non-accelerated motion – that is, in an inertial reference frame. Therefore, although subject to the Earth's gravity (otherwise the craft would not continue to orbit), the astronauts will experience 'free fall', that is they and everything else around them will experience a floating sensation, popularly referred to as **zero gravity**. From their isolated experience in the spacecraft they have no means of telling whether they are in free fall in a gravitational field, or whether there is no gravitational field at all.

The curvature of space-time

The principle of equivalence covers uniform gravitational fields, yet over anything but a very small volume of space, a gravitational field is not uniform. This is particularly noticeable when we deal with astronomical and cosmological distances; then gravitational attraction is different in different parts of so large a field. Einstein saw that this meant that space-time and gravitation must be intimately linked. Indeed a mathematical analysis showed that while, for a uniform field over a very small volume, space-time was 'flat' or not detectably curved (that is, its geometry approximated to that devised by Euclid and met with in small areas on Earth), for larger volumes, where gravitational attraction is not uniform, space was 'curved'. The curvature was best expressed in

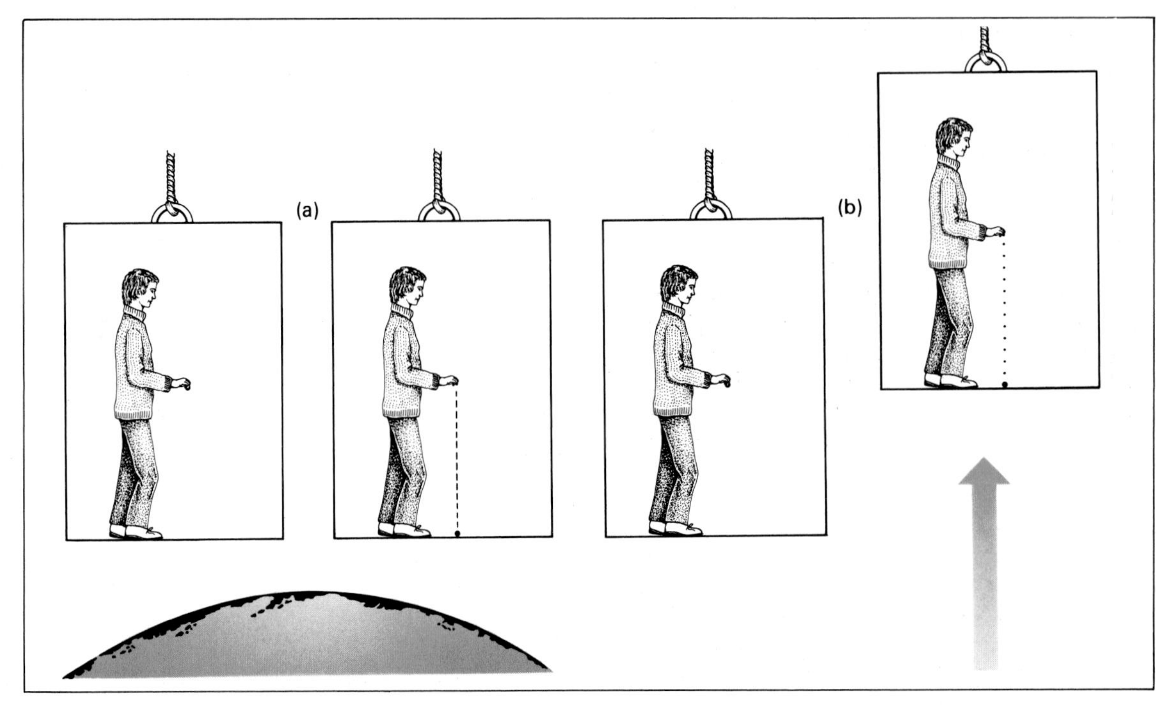

Fig. 8·5
An observer in an isolated box, (a) in a gravitational field, and (b) in an upward accelerating box with no gravitational field.

the curved-space or spherical geometry which the German mathematician Bernhard Riemann had devised in 1854 – a non-Euclidean geometry in which straight lines do not exist, and where the shortest distance between two points is a curve called a geodesic. Light and all electromagnetic radiation in curved space-time must therefore travel in geodesics.

The idea of curved space-time where nothing can move in a straight line may seem a little exotic, but one only has to consider navigating on the Earth's surface to see that it is no way peculiar. When navigating across an ocean no mariner can travel in a straight line because the Earth's surface is curved (Fig. 8·6). the shortest route is not a straight line (he would have to dive through water and ocean bottom for that!) – it is a GREAT CIRCLE, the equivalent on a three-dimensional sphere of a geodesic in curved four-dimensional space-time. It is worth noting too, that on Earth a very small area of its surface can be considered flat, just as a very small volume of curved space-time has a flat geometry.

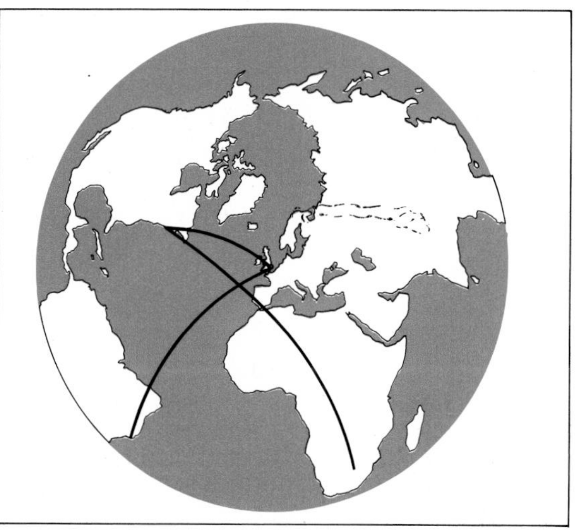

Fig. 8·6 On the 'spherical' surface of the Earth, the shortest paths between places are curved lines: these are in fact 'great circles' of which the equator is the best known example.

Observational proofs of relativity

No theory, however ingenious or attractive, is acceptable unless it can make valid predictions, that is, unless it can forecast some consequences which can be checked by observation. Over astronomical volumes of space there are three clear tests for relativity; one is the motion of the perihelion of Mercury, another the apparent displacement of stars near the Sun, and the third a redshift on the spectrum of a massive radiating body. The movement of the perihelion of Mercury's orbit is due to the perturbations of the other planets, and is observed to be 574 arc sec. per century, but Newton's gravitation theory predicts a motion which is too small by 43 arc sec. per century. Using relativity calculations the error virtually vanishes; Einstein's gravitation theory gives the correct result.

The apparent displacement of starlight passing close to the Sun (Fig. 8·7) is due to the Sun's gravitational field; if we compare photographs of a star field at night and the star field close to the Sun at the time of a total solar eclipse, this displacement can be measured. Newton in the eighteenth century had conjectured that light had mass, and according to his gravitation theory the deflection of starlight by the Sun should amount to 0·87 arc sec. According to relativity the curvature of space-time caused by the Sun should give a deflection twice this amount, 1·75 arc sec., and in 1919 the first measurements to test the theories were made. Observations in Sobral, Brazil by Andrew Crommelin and on the Island of Principe in the Gulf of Guinea by Arthur Eddington, were made at the particularly favourable eclipse of that year when the Sun was close to the Hyades open cluster, and the results obtained were 1·98 and 1·61 arc sec. Later eclipse observations by William Campbell and Robert Trumpler in 1922 gave a value of 1·72 arc sec. Although there was some spread in values of these delicate observations, Newtonian theory was obviously inadequate; the observations clearly confirming relativistic gravitation.

Another consequence of the deviation of light in a gravitational field, predicted earlier but only recently discovered, is the existence of **gravitational lenses**. If a sufficiently massive body occurs between us and another celestial object, not only will the light from the latter be displaced, but double images may be produced. This was first confirmed from radio observations of a distant quasar, and several cases are now known, including purely optical ones. In each case the intervening object is a galaxy, but their detection has sometimes been very difficult due to their great distance and near-invisibility, even though most of them seem to be massive elliptical galaxies.

Another prediction from relativity was that spectral lines emitted from a massive body will be redshifted. Thus, if one compares the spectrum of an element in the laboratory and a similar element on the Sun, the lines from the Sun should suffer a redshift. This test is again difficult to make, the gravitational redshift being only some two parts per 10^6, but it has been found to be present, while spectra from the dense companion of Sirius, where the expected shift is some thirty times greater, also show the predicted shift.

There is no doubt that these three independent confirmations of relativity put the theory in a very sound position, and in studying the cosmos, the universe as a whole, the validity of relativity is accepted.

Fig. 8·7 Space is distorted close to a massive body. Thus a beam of starlight is deflected and, in the case of the Sun, a star close in the line of sight appears in a different position.

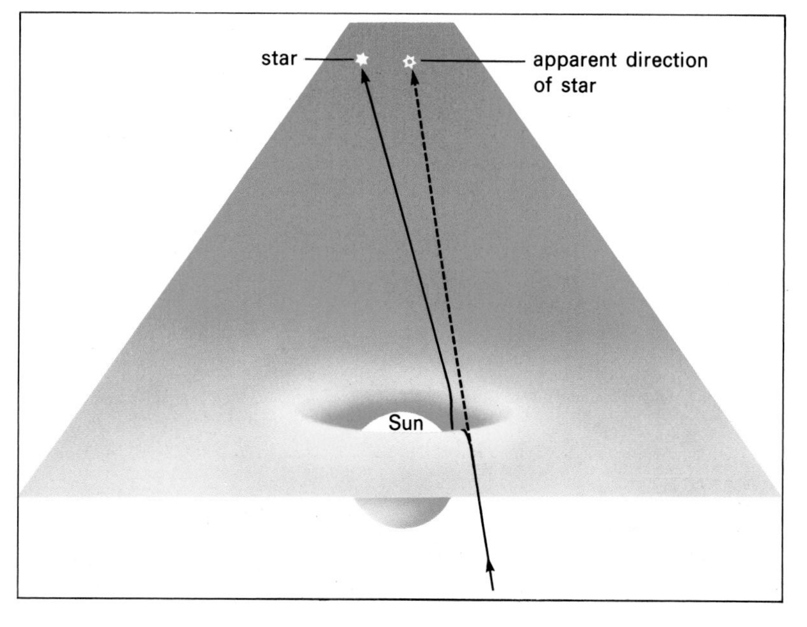

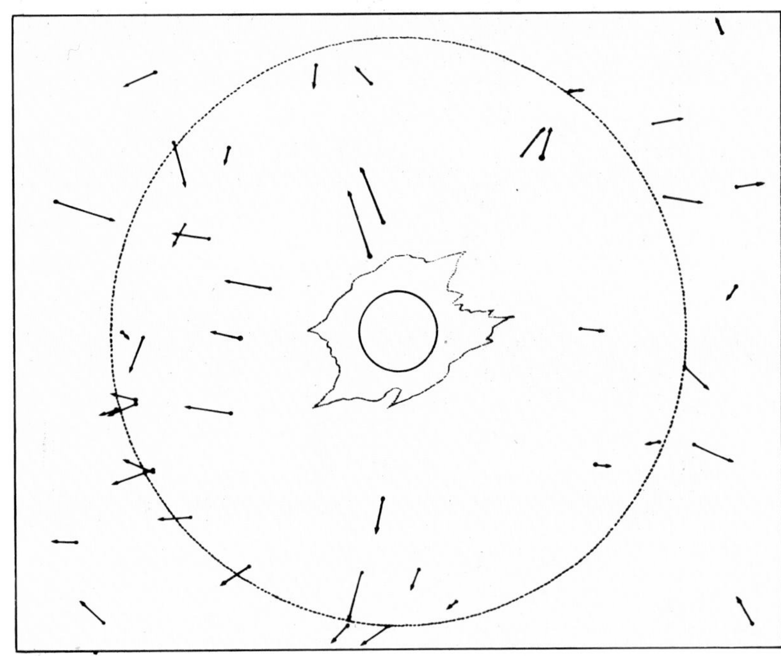

The 1922 September 21 total solar eclipse photographed from Wallal Downs, Western Australia. Similar plates were used to obtain the apparent shift of stars close to the Sun, and these are shown on the chart. Note that the scale of the chart (A) is very much smaller than that of the stellar displacements (B), with overall lengths representing 1 degree and 1 second of arc respectively.

Cosmological consequences of relativity

In determining redshifts and making measurements of the brightnesses and apparent diameters of galaxies, the curved space of general relativity results in different values to those which would be obtained if space were flat (Euclidean). For instance, in flat space the redshift is a straightforward quantity (page 196), such that if its value ever reaches 1, it shows that the body concerned is moving away with the velocity of light. In special relativity the formula for the redshift is more complex:

$$1 + z = \sqrt{(c + v)/(c - v)}$$

where z is the redshift. Here the value for the redshift can be much more than 1, reaching infinity for a velocity equal to the velocity of light. Relativistic redshift calculations have to be used for large redshifts, and in consequence we find quasars with redshifts of 2 and 3, or even nearly 4, a redshift of 2 indicating a velocity of recession at 80 per cent the speed of light. Not only does the redshift formula have to be changed in a relativistic universe, there is also the gravitational redshift to take into account.

The redshift plays its part in measurements of the apparent brightness of galaxies, since when present it reduces the apparent luminosity. Measurements of galaxy brightness will, therefore, all be underestimates unless a redshift correction is made, and a relativistic one is needed for all but the shortest distances. Perhaps one of the more unusual aspects of the effect on measurement of the relativistic universe is in the appearance of the diameter of distant galaxies. On Earth and in the nearer reaches of space, we are used to seeing the angular diameter of a source becoming less as it moves further away: the apparent angular diameters of the Sun, Moon and planets are smaller the further off they are. For example, as the relative positions of Venus and the Earth alter, the apparent diameter of Venus goes from almost 60 arc sec. when nearest, to less than 10 arc sec. when at its furthest. In curved space-time, this diminution of apparent angular diameter also occurs, but only up to a certain distance; after this the bending of light gives a similar effect to a lens and apparent angular diameters increase with distance. The distance at which this change occurs depends on the precise kind of relativistic universe we live in – a question to be discussed in a moment – but as an example, in the 'model' worked out by Einstein and Willem de Sitter the change happens at a distance of almost 5 000 Mpc.

Another very strange effect is the time dilation close to a black hole. A black hole, as we have seen (pages 73–75), emits no radiation and is called a hole because its excessive gravitational pull distorts space-time so much that it bends over on itself leaving the huge mass of material in what is virtually a hole in space-time. The material itself falls together so that it occupies no space and forms what is known in space-time as a singularity. Around the singularity there is a small spherical volume of space which acts as a boundary; inside this spherical boundary everything is crushed out of existence. As mentioned in Chapter 3, the gravitational field inside is so strong that an astronaut falling in feet first would be stretched out like a piece of spaghetti. We have already seen (page 74) that the radius of the sphere which forms this boundary, the Schwarzschild radius, depends on the amount of material inside. The condition under which a black hole exists is that the material of which it is composed must be highly concentrated: it has, in fact, an incredible density, so great that all the material in the Sun, which amounts to something of the order of 2×10^{27} tonnes, would have to be nearly 4×10^{16} times more concentrated. So great a concentration of mass causes the strange relativistic effect that we, outside the Schwarzschild radius, observe as time dilation. The space-time surrounding the region of the hole is so distorted, that at the Schwarzschild radius time is infinitely stretched out. We could, therefore, never see the astronaut fall into the black hole; he would appear to hover there for ever. It is for this reason that the sphere with the Schwarzschild radius is usually called the event horizon.

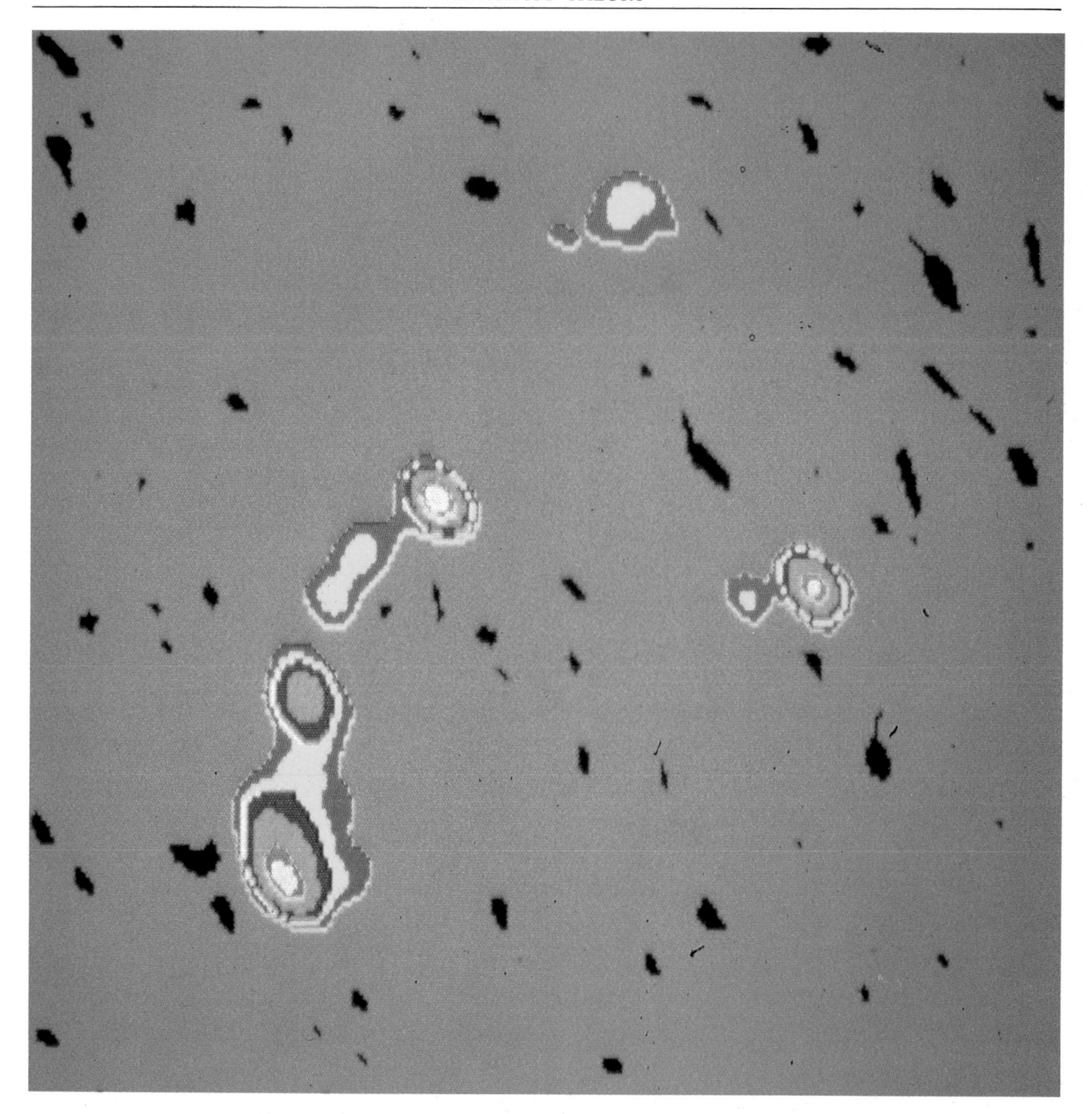

The twin images of the quasar 0957 +561 shown in this radio 'picture' are the two ellipses in the centre. The galaxy which produces the gravitational lens effect (page 221) is invisible here, but is just above the lower twin. Calculation shows that the second images of the radio lobes (left and right) *are too faint to be detected.*

The Schwarzschild model of the black hole is a stationary object, yet it is more probable that it rotates (Fig. 8·8). The mathematics of a rotating black hole are somewhat different from that of a stationary one and, in particular, the event horizon rotates and the hole is surrounded by a 'stationary limit', a spherical surface on which a body must travel at the speed of light if it is to appear to remain stationary. Although in theory nothing can escape from a black hole, recent research has shown that this may not be strictly true. Research by Stephen Hawking of Cambridge University, England, has shown that quantum theory leads to the idea that the whole of space is filled with virtual particles and their antiparticles, virtual particles being so-called because they can only be observed by their effects. Ordinarily, they combine with their antiparticles (that is, particles of the same kind but with an opposite electric charge) and both are annihilated, but Hawking has shown that in the presence of a black hole one of the pair may fall into the hole, leaving the other without a companion so it can not be annihilated. This companion may itself fall into the black hole, or it may escape; if the latter happens energy is removed from the black hole (Fig. 8·9). Calculation shows that few particles can be expected to escape from a very large black hole, because in such a hole the stationary limit is very close to the event horizon, but in a tiny black hole – one about the size of a proton and with a mass of 10^9 tonnes – particles would stream out at a terrific rate, emitting energy equivalent to three nuclear power stations.

Hawking has found that a black hole will lose mass as it radiates energy, so that in due course it will evaporate away. Small black holes will vanish after a life of only 10^{10} years, but large ones may last as long as 10^{66} years. Since the universe, as we shall soon see, appears to be a few 10^{10} years old, and since the tiny black holes – if they do exist – would have been formed in the early stages of the universe, they should now be evaporating, emitting vast amounts of high energy γ-radiation as they do so. Such γ-radiation could be detected by orbiting spacecraft and from Earth because the γ-rays hitting our atmosphere will create a shower of electron-positron pairs in a kind of electromagnetic sonic boom.

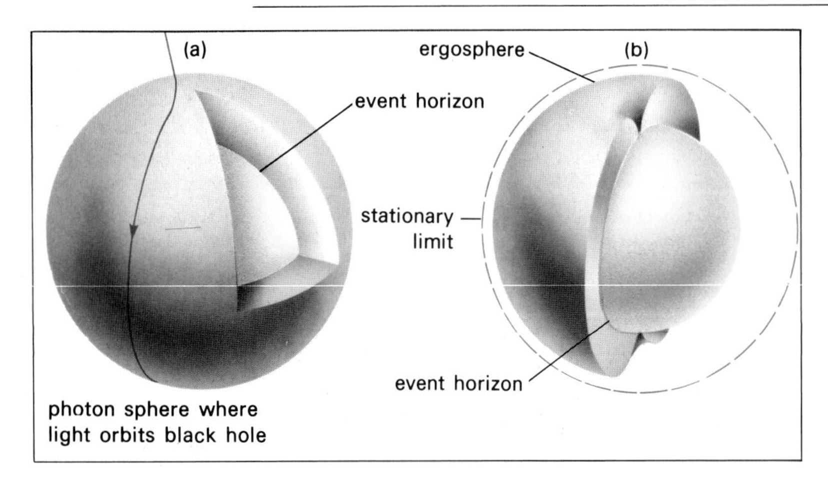

Fig. 8·8 above: *These drawings depict the limits of a non-rotating black hole (a), and a rotating black hole (b) as calculated by Roy Kerr. In the latter anything such as photons and particles lying below the 'stationary limit' must keep in motion on the 'ergosphere' or 'work space' otherwise they will fall into the black hole. In theory a beam of light or even an astronaut could escape from the ergosphere.*

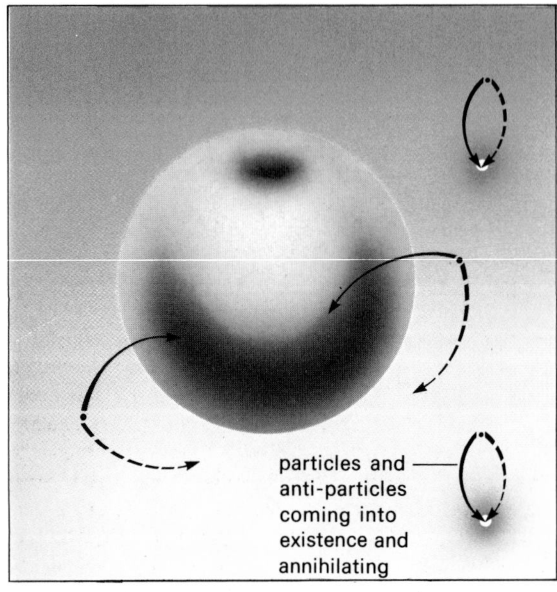

Fig. 8·9 top right: *Stephen Hawkings' idea that in the neighbourhood of a black hole some nuclear particles may escape.*

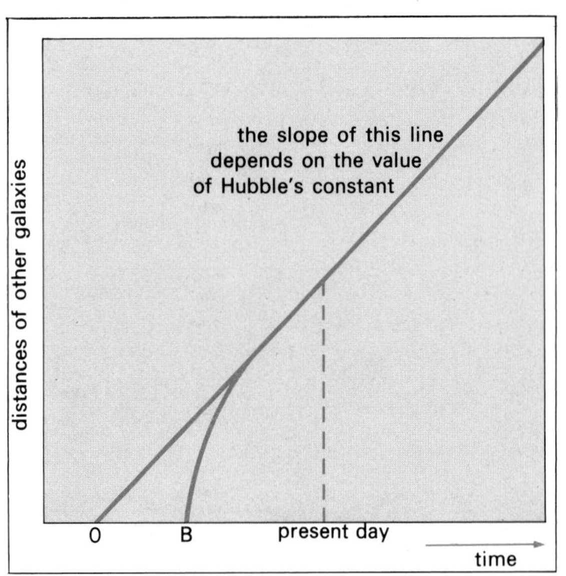

Fig. 8·10 far right: *The age of the universe can be calculated: supposing we have a certain value for the Hubble constant, we obtain a beginning at 0, but this is only valid if gravity did not have a greater effect when the galaxies were close together. Since gravity will have such an effect, the 'origin' will lie elsewhere. B is just such a point; its position will depend on the model universe we choose.*

Cosmological models

The cosmologist, applying the theory of relativity, can make various theoretical 'models' of the universe: these can then be compared with the results from observation, hopefully, to determine which model is the correct one. As we shall see, however, it is not possible to come down firmly in favour of one model only, although some models can be eliminated, and a general idea about the nature of the universe can be reached.

Relativistic models of the universe are based on what is sometimes called the **cosmological principle**, which states that on the largest scale the appearance of the universe at any given time is the same for all observers located in, and moving with, galaxies; and, for simplicity's sake, all matter in the universe is 'smoothed out', that is, it is assumed that the universe is one vast uniform 'fluid'.

In 1917, in his first relativistic model of the universe, Einstein assumed the universe was static and, to prevent it collapsing under the influence of gravity, he introduced a cosmological term which showed a repulsion. Once the expansion of the universe had been discovered, new models were devised in the 1920s and 1930s by Einstein and Willem de Sitter, by Edward Milne, the Abbé Lemaître, and the Russian, Alexander Friedmann. All were what we should today call big-bang universes in that they all considered the universe to have started in a concentrated form (Fig. 9·10).

The Milne model, sometimes known as the kinematic relativity model, uses Euclidean space and special relativity as far as possible. It takes the total mass in the universe to be small and the movement of the galaxies as uniform; thus the 'age of the universe' could be determined quite simply from galaxian distances and velocities. On this model, devised in the 1930s, the age of the universe came out at 2×10^9 years, although using more modern figures this would work out as $1{\cdot}5 \times 10^{10}$ years. The theory involved the use of two time scales, one basic in the universe, the other local to a moving observer. This gave rise to a kinematical redshift, greater for more distant galaxies, but very much smaller than the total redshift observed in any particular galaxy.

The Friedmann models were straightforward, giving an expansion which, after a time, continued at an ever decreasing rate. However, the age of the universe they produced seemed too short considering the evidence there was then for the age of the Earth. The Einstein-de Sitter model, which was also of the same kind, nevertheless gave what seemed to be a less unsatisfactory age. Age, however, was no problem for the Lemaître model, for here the cosmological term of Einstein's original model was involved once more, giving a repulsion which, after a certain point, began to take over from gravitation and so caused an increasing rate of expansion. Later, in the 1950s, the revision of the cosmic distance scale for galaxies by Walter Baade removed the age of the universe problems presented by the Friedmann and Einstein-de Sitter models, while the Lemaître model was modified by George Gamow and his colleagues Ralph Alpher and Robert Herman, to incorporate a hot **big bang** in which the universe was not only originally very dense but also very hot.

A model which avoided problems of the age of the universe entirely yet did not invoke the cosmological term was the **steady state** model, devised in 1948 by Hermann Bondi, Thomas Gold and Fred Hoyle (Fig. 8·11). It was based on what they called the 'perfect cosmological principle', namely that on the largest scale the universe is the same to all observers *all the time*. In other words, if one took a movie film of the universe now, and compared it with films taken, say, 10^9 years ago and 10^9 years in the future, it would be impossible to pick out one from another: they would all look the same. One of the consequences of this is that in any given volume of space-time the amount of material is always the same, yet as the galaxies are moving apart, the density of material is not always static. To resolve this dilemma one can either say that galaxies are not moving outwards, that the increase of redshift with distance is due to something other than motion in the line of sight, or that new matter is being formed to replace that which is being removed by the expansion of the universe. In this theory the universe never had a beginning and will never have an end; galaxies condense and form all the time, move outwards, and ultimately reach velocities taking them beyond the boundary or horizon of the visible universe.

The Friedmann- and Lemaître-type models not only have a beginning, they also present us with two possible future states of the universe: the universe may be open or closed (see page 205). Also, in the Lemaître model cosmic repulsion could come in to play to cause a 'bounce' when the universe is again very dense so that it starts expanding once more. In that case the universe would be an oscillating one, having always existed (as in the case of the steady state model), going through alternate cycles of expansion and contraction (Fig. 8·12). although, there are strong reasons for believing that conditions could not be exactly reproduced with each cycle, and that the universe would 'grow' with each oscillation.

Observational evidence

The obvious hope is that observations will allow us to decide which of the models of the universe is the correct one; for instance, the model chosen will give a specific relationship between redshift, velocity and distance. At the moment there is not precise enough information to allow us to make a definite choice, but there is enough to allow some models to be rejected. First of all we know that the universe is expanding, that is that the distances between all the galaxies are increasing, and that this is giving rise to their observed redshifts and apparent velocities which increase with distance. Attempts have been made to find other explanations, but none have proved satisfactory. Yet, as pointed out on page 209, there is doubt among a few astronomers about whether the redshifts of quasars should be interpreted in this way.

Secondly, there is the evidence of the microwave background radiation, mentioned on page 206. This appears evenly spread in every direction and is hard if not impossible to explain on the basis of the steady state theory, at least as originally envisaged. Indeed, it presents such a stumbling block that it has led to the abandonment of the theory for the present. Almost all astronomers favour the theory of the big bang, the start of the universe from a concentrated superatom, and in particular accept the idea put

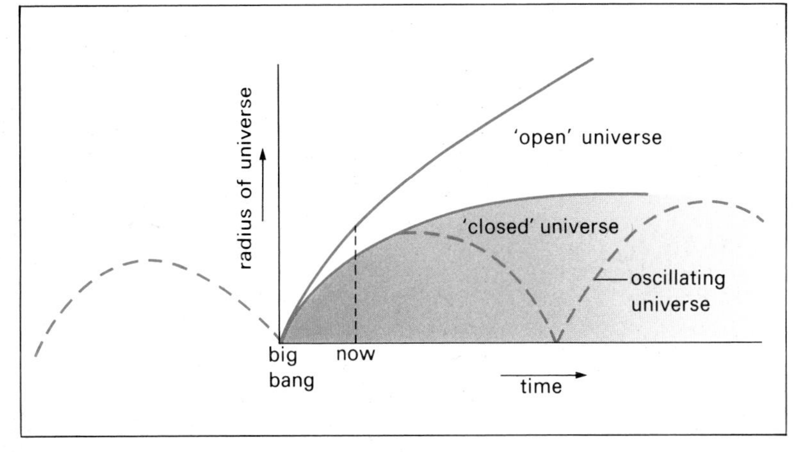

Fig. 8·12: According to the amount of matter in the universe, it can be 'open' (expanding for ever) or 'closed'. In the latter case there may be only one expansion and subsequent contraction, or an endless series of them, giving rise to an oscillating universe.

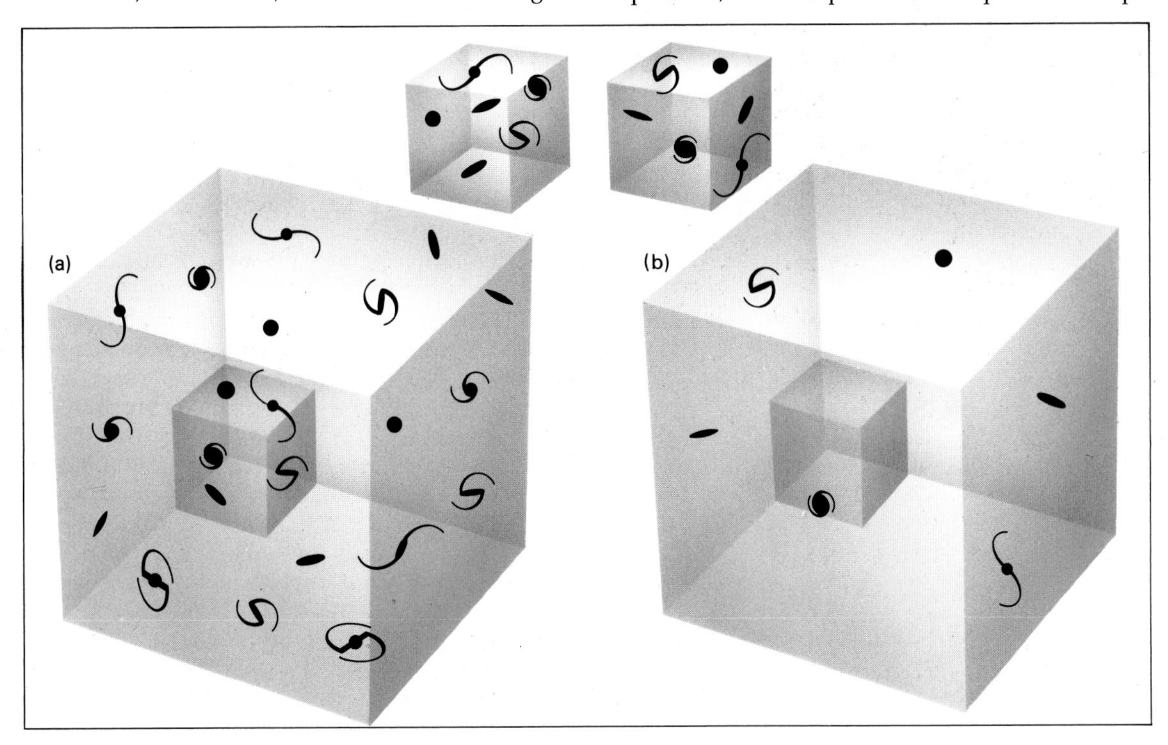

Fig. 8·11 Differences over a long time between the steady state (a) and big bang (b) universes. In the steady state, although material (galaxies) spread out with time, new ones are created to take their space, and the general appearance of a volume of space remains the same. This is not so with a big bang universe; the material becomes increasingly separated with time.

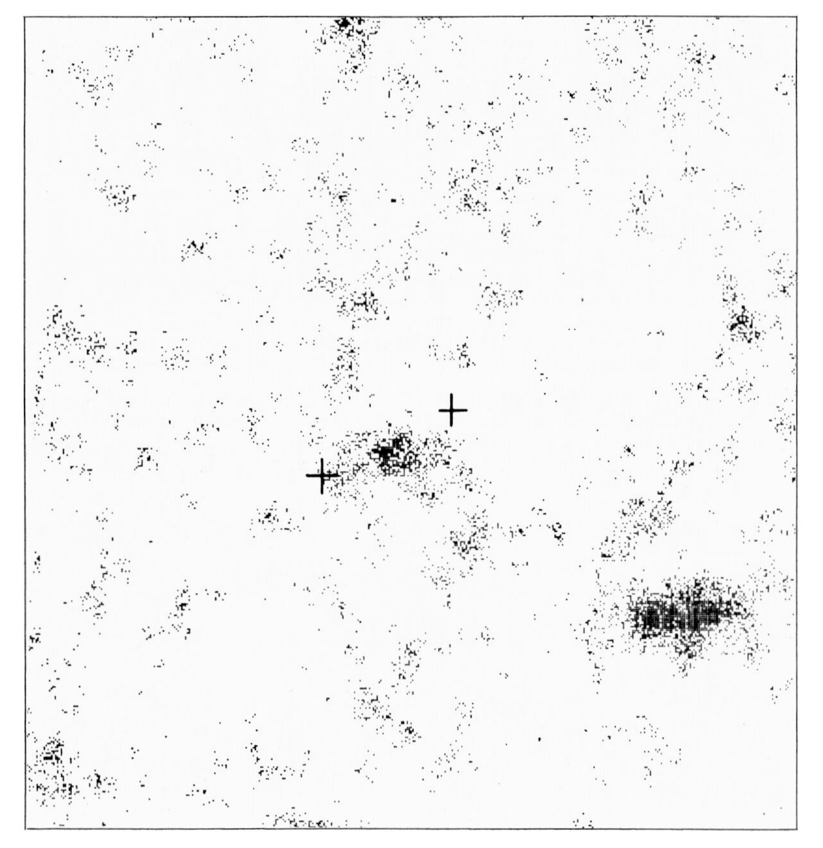

The radio galaxy 3C 184. The small chart (p. 227) *shows the observation which detected the galaxy, made with the 5 km aperture synthesis radio telescope at the Mullard Radio Astronomy Observatory at Cambridge University, and published originally in a paper in the* Memoirs of the Royal Astronomical Society *by C. J. Jenkins, G. G. Pooley and J. M. Riley. Compare this with the processed image on the facing page.*

forward by the American cosmologist George Gamow, that it was a **hot big bang**. It explains the origin of the microwave background and has, in fact, proved a very fruitful idea; in the light of modern nuclear physics it is possible to work out theoretically what occurred – or perhaps more correctly what could have occurred – in the very early stages of a big-bang universe, the so called **fireball stage**.

Early stages of the big bang

Some theoreticians have been extending their investigations back in time to the unbelievably short period of only 10^{-39} second after the initial explosion, when events may have had a profound influence upon the nature of the whole universe. However, we will begin our account at a much later time – 0·0002 s after the big bang! At 2×10^{-4} s, the universe entered what is called the **lepton era**, with a temperature of 10^{12} K and with many leptons (electrons and MUONS), followed by a reduction in temperature and the presence of protons and neutrons, some of which combine at a temperature of 4×10^{9} K to form the isotopes of hydrogen, deuterium and tritium.

The next stage was the **radiation era**, when the immense pressure of the radiation generated controlled the universe. This occurred some 1 000 s or rather more than 16·5 minutes after the initial explosion, when it was still very hot although the temperature was down to about 5×10^{8} K, and hydrogen began to be converted into helium. This state of affairs continued for some time, the temperature gradually dropping, until something like 3 years after the beginning not only were isotopes of hydrogen and helium formed, but some heavier elements too: the temperature was 10^{6} K. This is a very significant period because we find too much helium in the universe now to be accounted for by supposing it to have been formed inside stars. Its formation during the fireball stage seems the only satisfactory explanation, and the quantity makes it look as though it is more likely that the universe will expand forever, not alternately expand and contract. At all events the radiation era continued for a very long time, until 6×10^{5} years had elapsed and the temperature had dropped to 3 000 K.

Before this, the radiation had heated the hydrogen so that it was ionized, resulting in a vast number of free electrons. Such electrons are efficient scatterers of radiation, so radiation did not stream away; its pressure was a controlling factor and the universe was opaque. When the crucial temperature of 3 000 K was reached the universe became transparent, for the hydrogen was no longer ionized and radiation could pass outwards; its domination ended and we enter the phase we now experience where matter, and hence gravitation, takes over. The changeover point is often referred to as the point at which matter and radiation are 'decoupled'. From then on condensations occur, protogalaxies and then galaxies form, followed by condensation into stars, and so, after $1{\cdot}5 \times 10^{10}$ years we reach the present time. The temperature has dropped until it is no more than 2·76 K, the temperature of the microwave background.

Choosing the correct model universe

The hot big bang theory presents what seems, on present day evidence, to be a very satisfactory description of the early stages of the universe and the presence of the microwave background, and leads to figures for the abundance of hydrogen and helium similar to that found in our own Galaxy. But what of the subsequent expansion? Does it go on eternally – as the helium formation indicates – or will it cease, to be followed by a period of contraction? And if this does occur, will the contraction phase mean the universe ends up as a giant black hole, or will there be a bounce back to another period of expansion? There is the purely physical question of how much matter there is in the universe, and particularly whether there is sufficient to lead to a contraction. In Chapter 7 it was argued that the question is undecided, even though from present evidence it looks as if there is probably too little for this.

How, then, does all this affect our choice of a correct model of the universe? If an analysis is made of the various models other than the steady state, which has no satisfactory explanation for the microwave background, additional points emerge. In its early stages the universe looks rather like the Einstein-de Sitter model, but what of the later periods? Does the Hubble distance-velocity relation for galaxies hold? In fact the big-bang models do give slight differences for velocities at very large distances, and observations tend to favour the Einstein-de Sitter and, more particularly, the Eddington-Lemaître type of model. One can only say 'tend to favour' because the optical observations of very distant galaxies which can be made are not yet precise enough to lead to a definite decision.

It is now clear that radio observations can extend the detection of galaxies and the measurement of radiation received from them out further still and can, therefore, provide significant information about types of model. To this end **source counts** have been made, the number of sources with radiation at various intensities having been plotted: these show a significant departure from what should be expected for a steady-state universe (and so are yet another reason for discarding it), and make it clear that the universe does undergo evolutionary change (Fig. 8·13 and 8·14). But they do not lead to any unequivocal result over whether or not the universe will contract or continue expanding. Counts of quasars also tend to confirm this, but, of course, there is the problem of whether the redshifts of quasars are really due to recession or to shells of gas ejected from them, or even to the quasars themselves being objects which have been ejected from surrounding and not too distant galaxies. The American astronomer Halton Arp, for instance, is strongly of the opinion that there are too many photographs of quasars lying almost in the line of sight or close to galaxies, as well as having redshifts of related kinds, to allow of their being all – or mostly – very distant, but most astronomers remain unconvinced at the moment. Indeed, they feel not only that quasars are truly at cosmological distances but that they may well represent galaxies in an early stage of development when the nucleus could have been very active, for the more distant the objects are which we observe, the longer light has taken to reach us and so the further back in time we see.

We can not, then, be certain at the present time which model of the universe is correct. Evidence from the perihelion of Mercury, the deflection of starlight by the Sun, laboratory observations of the increase of the mass of atomic particles when travelling at velocities approaching light, the equivalence of gravitational and inertial mass, together with the evidence for neutron stars and black holes, all make it seem as if it must be a relativistic universe; if gravity waves are detected then this will be yet another confirmation. But astronomical observation does not allow us yet to choose a specific relativistic model; we can not at present say whether the universe will expand for ever, or whether it will one day contract, and if it does contract whether it will bounce back to a subsequent stage of expansion.

General comments

All through this section on cosmology we have assumed, for instance, that gravity is a constant unchanging with time, but there have been some suggestions that this may not be so. Originally proposed by Paul Dirac and Pascual Jordan, this theory has been investigated in recent years by many cosmologists, including Robert Dicke and Fred Hoyle, and it has interesting consequences. For instance, if, as is suggested, gravity decreases with time, then continental drift (page 93) could have been caused by the Earth's expansion as gravity decreased.

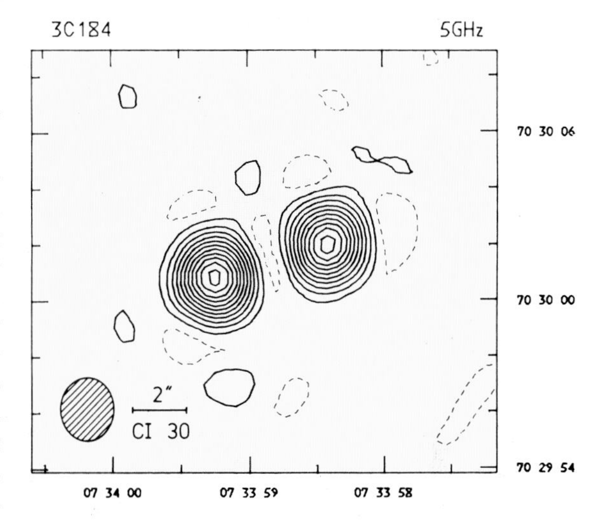

Using information gained from the radio observations shown opposite, the appropriate region of the sky was photographed by the 5-m (200-inch) telescope of the Hale Observatories, and then examined by Edward Kibblewhite's Automatic Plate Measuring machine (see page 238) back at Cambridge University. Ignoring all bright objects on the plate and concentrating on the dimmest images, a trace was produced. This is reproduced above in negative. The radio galaxy 3C 184 lies between the two crosses, which represent the positions of the galaxy's radio lobes shown in the small picture opposite. 3C 184 has a magnitude of 22.5 and is one of the faintest radio galaxies so far identified.

A star such as the Sun, while in the hydrogen-burning stage, would have had a greater luminosity in the past, and this could explain the long period of time between the first appearance of living cells on Earth and the arrival of more complex forms. Other consequences are that our Galaxy would have been more luminous in the past, and smaller too, while there would be some wider cosmological consequences. The relation between redshifts and apparent magnitude would be altered, and so would the predicted movement of Mercury's perihelion and the deflection of starlight by the Sun.

To explain why the gravitational constant should change, Dicke and his colleagues have evoked **Mach's principle** (named after the German physicist and philosopher Ernst Mach) that the inertial properties in any localized area depend on the rest of the matter in the universe; if this matter changes it will affect our local measurements. Dirac, Hoyle and his colleague Jayant Narlikar suggest the variation of gravitation could be due to the existence of two time scales in the universe, one an atomic scale, the other a cosmological one, which although once in step now no longer coincide.

There is also the question of the **isotropy of the universe**. The microwave background is isotropic – it appears the same in all directions – but it could be that the universe did not present this aspect in its early stages and was very anisotropic. The initial big-bang singularity might have been cigar shaped, or even like a disc. Charles Misner has suggested that the universe did start out in this uneven state, but that interaction between neutrinos caused a viscous treacly effect that allowed protogalaxies and collections of protogalaxies to form. On the other hand, Russian cosmologists like Ya.B. Zel'Dovich suggest that the production of nuclear particles caused by tidal pulls in the early stages of the universe caused the evenness we observe now. These are attractive hypotheses which overcome the difficulty in an isotropic big bang of explaining the presence of galaxies and clusters of galaxies which we now observe.

Even the very isotropy of the universe causes problems. This arises from the fact that particle physics, in particular the search for theories which unify the various forces controlling the behaviour of

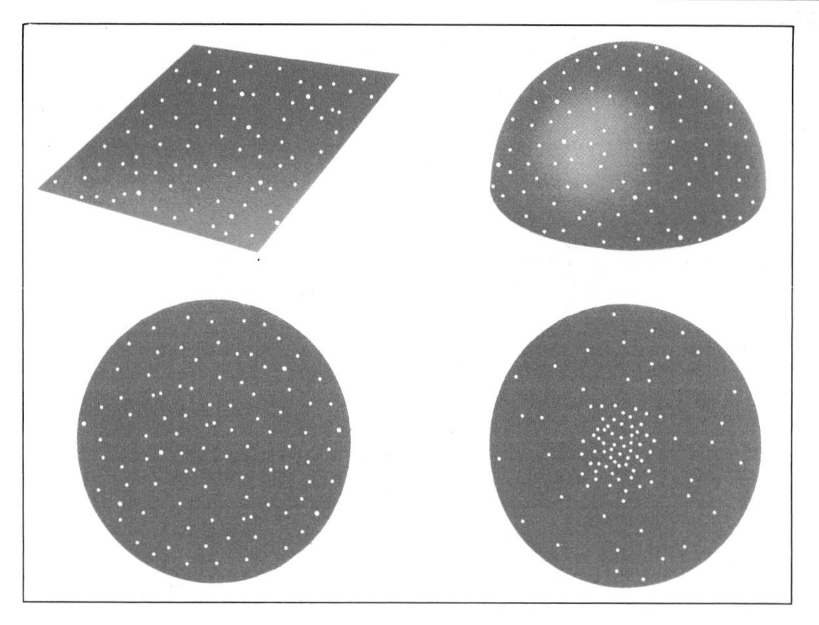

Fig 8·13
According to the shape of space and the model universe we choose, the number of sources we observe at great distances will be different.

the universe, can make predictions about the conditions in the very earliest instants following the Big Bang. Even at such an early stage – and we are talking of only 10^{-39} second after the Big Bang – it can be shown that the universe would have consisted of 10^{80} regions which could not have been causally connected (that is to say that none could have had any influence upon the others). The problem remains how to explain the fact that the relict radiation is so homogenous – to at least one part in 10 000. Some progress is being made with tentative theories – for example that known as the Inflationary Universe - which go some way towards meeting these obstacles. They even indicate that our observable universe may be no more than a very tiny part of a much larger Big Bang universe, some 10^{60} times greater in volume!

Unanswered questions

There are still many unanswered problems. We may not know whether we live in an open (ever-expanding) universe or a closed one, but certainly cosmology seems to be an open-ended subject. There is, for instance, the problem of the nature of what are sometimes called Eddington's 'magic numbers'. Eddington, in the 1930s and early 1940s, pointed out that the ratio of the electromagnetic force between a proton and an electron divided by the gravitational force between these two particles was $0{\cdot}23 \times 10^{40}$, and also that the ratio of the radius of the (observable) universe divided by the radius of the electron was 8×10^{40}, two huge numbers that are of the same order of magnitude, while 10^{40} is approximately the square root of the number of particles (5×10^{79}) which the universe is thought to contain. Are these relationships nothing more than coincidences; or are we perhaps at that age in the evolution of the universe when they happen to coincide; or do these numbers remain the same all the time, and some other fundamental 'constants' (gravity, for instance) change? These magic numbers may have the most profound significance – or they may have none.

Fig. 8·14
A plot of observations using different model universes.

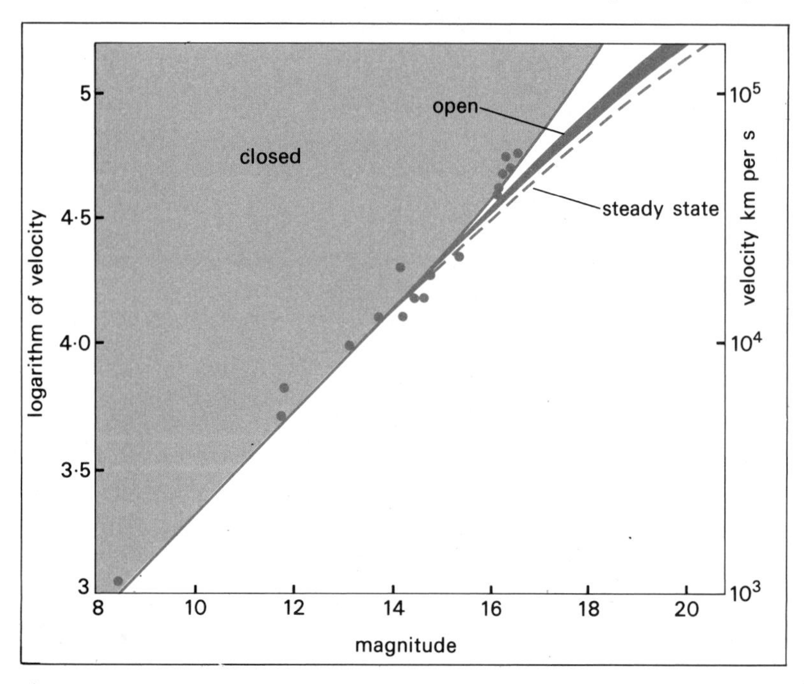

Another point which must be realized is that the hot big-bang theory is only concerned with the earliest stages of the formation of the universe as we now observe it. It does not tell us how the singularity from which the material poured came into being. In other words, the presence of a super atom is taken as our starting point. Such a question seems, on the face of it, to lie outside the realms of scientific enquiry; or perhaps the question has no meaning, for it implies an existence in time before the matter of the universe appeared and, in a space-time universe, space and time are inextricably linked – time would be formed when matter was formed and would have no previous existence.

Again, in curved space-time there is no 'outside' to the universe; space-time expands outwards as the universe evolves, but there is no outside space waiting there to receive it, just as there was no time ticking by waiting for the creation of the universe to begin. It makes no sense then to ask what lies outside the universe because the very question implies that one is thinking of something there already waiting to receive the galaxies as expansion proceeds. But there is the question of what is meant by the word 'universe'. By definition it signifies everything, 'the whole body of things and phenomena observed or postulated' (*Webster's Dictionary*) or 'the whole of created or existing things regarded collectively; . . .' (*Shorter Oxford Dictionary*). However, according to relativity there is a limit to the observable universe – we can never observe at distances where the speed of recession is equal to the speed of light – so there is the obvious question of whether our section of the universe is just part of something larger, more comprehensive. Is our expanding universe no more than one of many bubbles expanding outwards into a vaster universe? Such a suggestion has indeed been made by Narlikar, but although this is an imaginative idea, there seems at present no way of testing whether or not it could be so. The very suggestion pinpoints a danger – the danger of postulating a universe inaccessible to observation, for this is the path that leads to unbridled and unscientific speculation.

Observing the universe

Observing from Earth

The modern astronomer's traditional instrument is the telescope, and this can take two forms – the refractor and the reflector. In both the basic principle is the same: to collect radiation from a distance and bring it to a focus where it can be examined by eye, camera or other equipment.

The refractor

Here light collection and focus are carried out by a lens or **object-glass** at the front end of the telescope tube. Such an object-glass now always has two components, one of 'crown' glass and the other of 'flint' (a denser type than crown glass), to form an image with an acceptable minimum of spherical and chromatic aberration (Fig. 9·1). Yet such a two component object-glass has four lens surfaces which have to be figured, while the lenses themselves must be thick enough to support their own weight and be made of clear homogenous glass, free from bubbles or strains. Object-glasses are, therefore, very expensive, difficult to construct, and at large apertures absorption of light by the necessarily thick lenses becomes a problem. In consequence, an aperture of 1 m seems the maximum practicable.

The light-grasping power of a telescope – its ability to detect faint objects and thus probe far out into space – depends on the square of its aperture. Thus, the light-grasp of a 2-m aperture instrument is not twice but four times that of a 1-m, and with the present desire for large apertures, the refractor has given place to the reflector for all professional work.

The reflector

The reflector collects and focuses incoming radiation by using a concave mirror, on the surface of which is a highly reflective coating, usually of aluminium (Fig. 9·2). The film has a thickness of only some 300 nm and gives a reflecting surface that mirrors accurately the carefully ground and polished surface beneath it. The glass now most used for mirrors (whose optical quality has only to be retained on one surface) is one of the new zero-expansion, glass-ceramic materials which, over the temperature range at which the telescope is used, keeps its shape within very close tolerances – typically one part in 10^7 for each degree Celsius (C) change in temperature.

The aluminium film does not remain permanently highly reflective. When freshly applied it reflects about 90 per cent of the light falling on it, but after one or two years, this has dropped to 85 per cent, and realuminizing is called for. Like the original aluminizing, this has to be done in a vacuum, but observatories carry the necessary equipment for this since it is impracticable to send the mirror away for this purpose.

The basic optical layout of a modern telescope makes use of a number of different focal lengths (Fig. 9·3). For photographing or detecting very distant objects the **prime focus** is used; here radiation suffers only one reflection and so there is the least loss of the incoming signals. In large optical telescopes an observer's cage is built at the prime focus but in smaller instruments, where such a large obstruction would be impracticable, the light receives a second reflection bringing the focus to the side of the tube. This is the method adopted originally by Newton, and such a **Newtonian focus** is commonly used by amateur astronomers (Fig. 9·4).

Modern observing techniques often require the use of ancillary equipment that is unsuitable for use at the prime focus, and alternative foci are provided. One, the **Cassegrain** (named after the seventeenth century Frenchman Cassegrain), lies below the primary mirror, light being reflected by a convex mirror shaped to a HYPERBOLA, below the prime focus back down the telescope tube and through a hole in the centre of the primary. The other is the elbow or **coudé focus**: here additional reflections bend the light beam from a small secondary mirror below the prime focus to a room below the telescope where very massive ancillary equipment is situated. The additional mirrors rotate to compensate for movement of the telescope, with the result that the coudé focus always remains fixed.

The normal optical arrangements of a reflector use a primary mirror whose concave surface is curved in the shape of a PARABOLA, but some variations are used, and in particular a number of modern professional telescopes are of the Ritchey-Chrétien design (originally devised in the 1920s by George Ritchey and Henri Chrétien). Here the primary has a hyperbolic curve and the telescope not only gives a good optical performance but also a short tube length, which is also an advantage since it leads to a greater rigidity of the whole assembly, which is of particular importance in research fields such as astrometry – the determination of stellar positions to a high degree of precision.

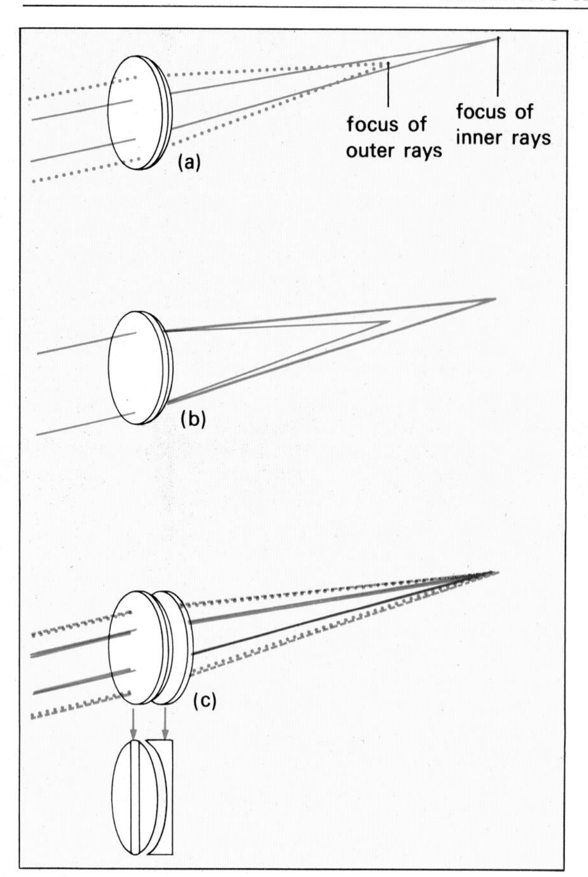

Fig. 9·1 right: *A single lens brings rays passing close to its edge to focus at a different point from those passing near the centre (a). This is because the surfaces of the lens are spherically curved, and this fault is known as* **spherical aberration.** *A single lens also brings rays of different colours to focus at different points (b). This is because the lens disperses white light into its separate colours. This is known as* **chromatic aberration.** *A two component lens, with one component having convex surfaces and the other a plane surface and a concave one (c) may be designed to minimize spherical aberration and also to overcome chromatic aberration for two specific colours.*

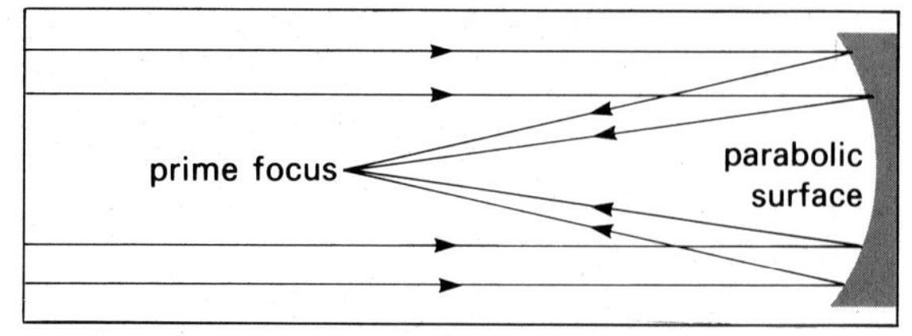

Fig. 9·2 top right: *Use of a concave mirror to bring incoming light to a focus.*

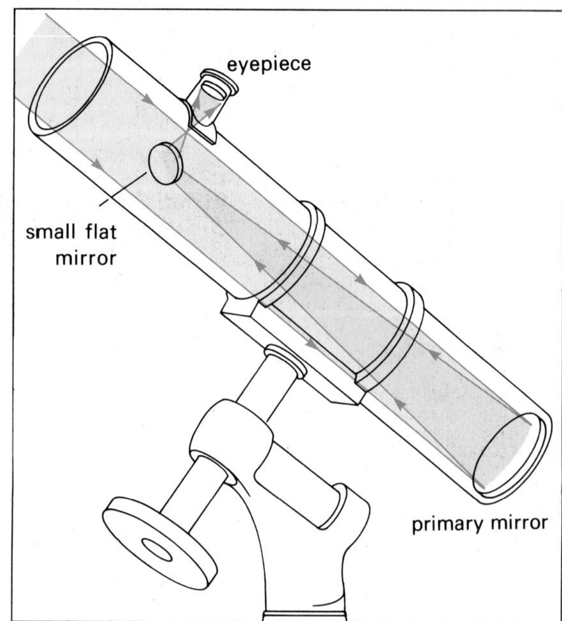

Fig. 9·4 far right, centre: *The Newtonian focus as used in many amateur reflecting telescopes.*

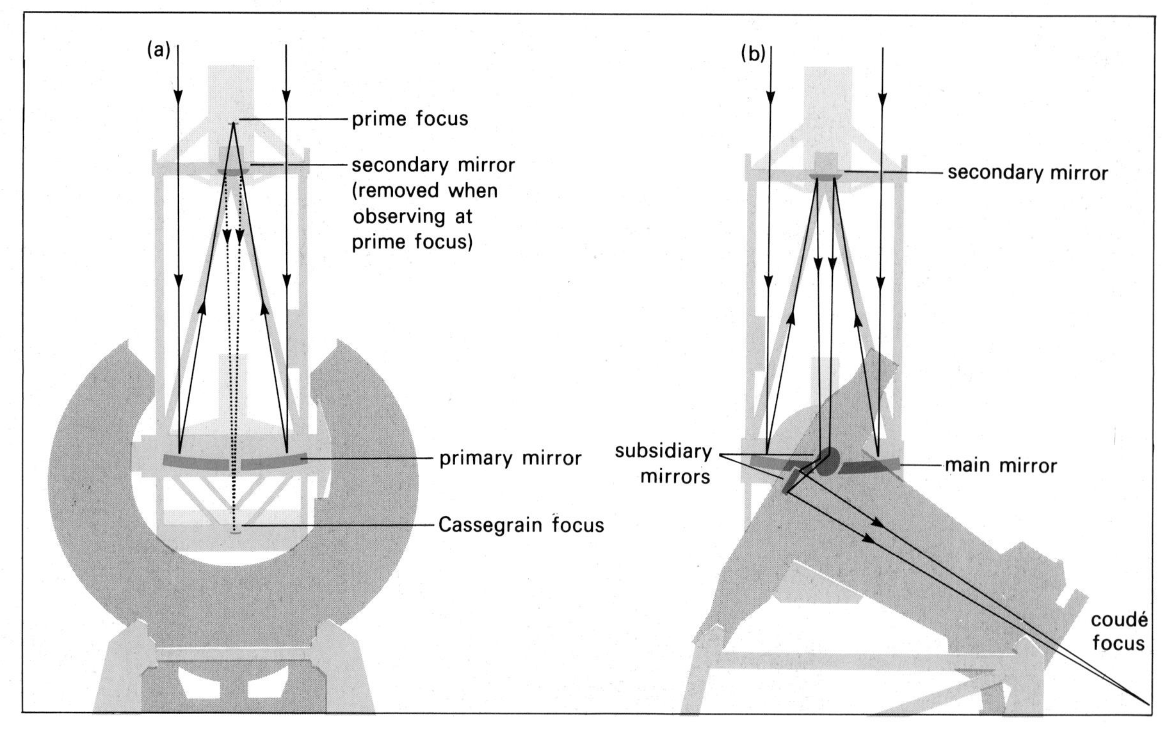

Fig. 9·3
Two drawings of a modern telescope showing the prime focus and Cassegrain focus (which lies below the primary mirror) in (a), and the fixed coudé focus (b).

The Schmidt and other telescopes

Even with all modern refinements, the field of view of a large reflecting telescope is small. The area of really sharp definition on a photograph is about 1° square and although this is perfectly adequate for the detailed examination of specific objects, it is too small an area for any large scale surveys or statistical work. Something with a wider field of view is required for these. In the 1930s the Estonian optician Bernhard Schmidt developed a new wide-angle telescope especially for photographic work, and since then the Schmidt telescope has become one of the optical astronomer's most powerful tools. The instrument uses a spherical primary mirror and at the front of the tube there is a correcting plate. This has a complex shape (Fig. 9·5, page 232) and can be said to parabolize the light as it enters the tube. The spherical primary gives a wide field coverage, typically some 40° square and although the telescope cannot

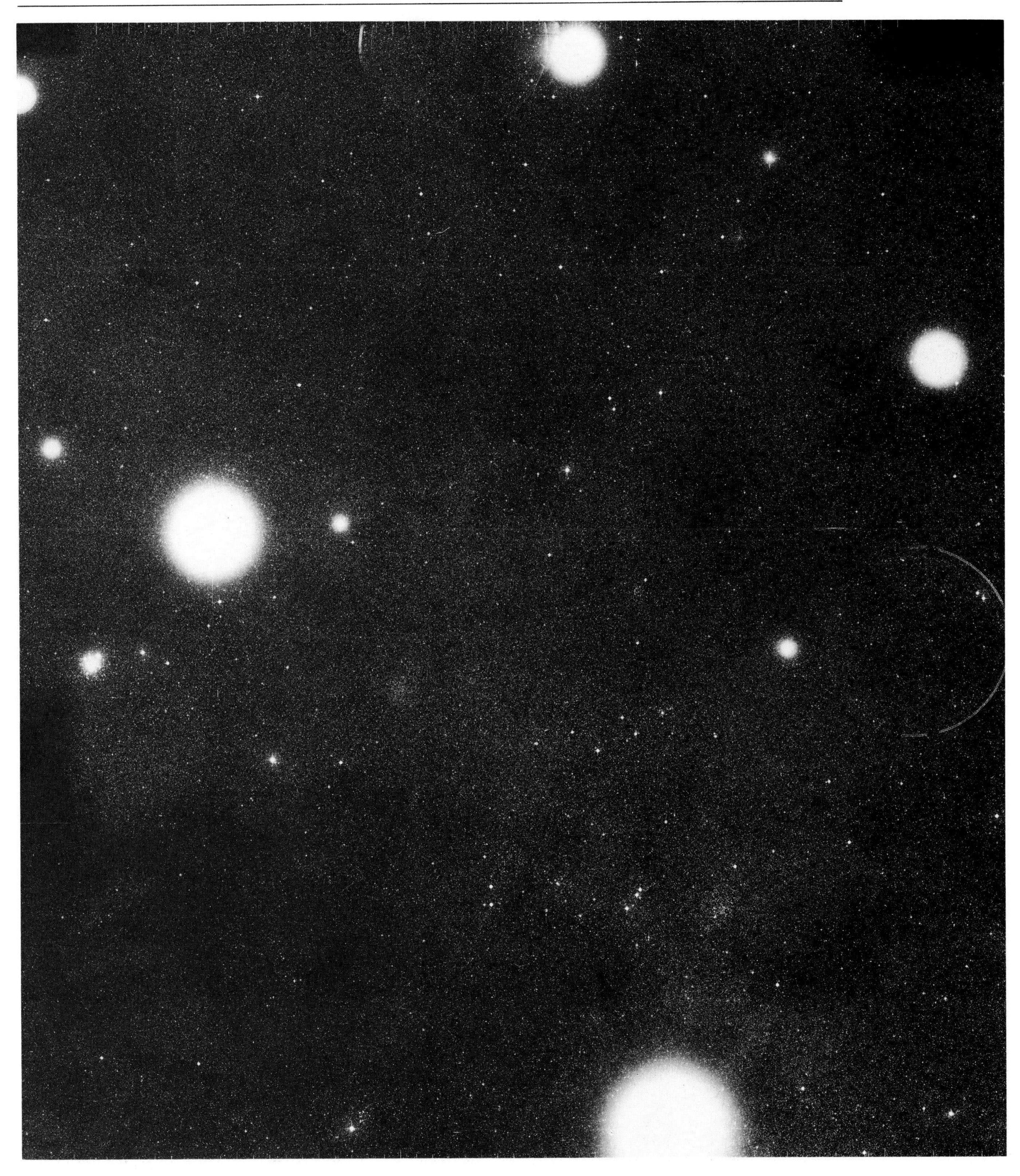

The Southern Cross. A star shines because its internal energy source is due to the annihilation of nuclear particles. The energy is immense because it follows the Einstein formula $E = mc^2$.

be used visually, such an instrument turns out to be a most powerful astronomical camera. This is not only because of its wide field but also because of its low focal ratio. As all photographers know, a lens opened up to aperture f/2·5 is faster than one at f/3·5; in fact, the shorter the exposure required the lower the focal ratio must be. Yet even the latest modern reflectors can never operate at lower than f/3·5, but Schmidts of f/2.5 are not uncommon. Another catadioptric type of reflector with a spherical primary and using a correcting lens at the front of a very short tube, is that designed by Dmitry Maksutov in the 1940s. First developed as a photographic telescope, it can also be used visually at f/7 and f/8 and is now favoured by some amateurs because of its extreme portability. Short tube lengths and fast focal ratios are only one aspect in the design of optical telescopes. Light-grasp necessitates a large aperture, and large aperture is also bound up with the vital question of resolution, since the power of a telescope to

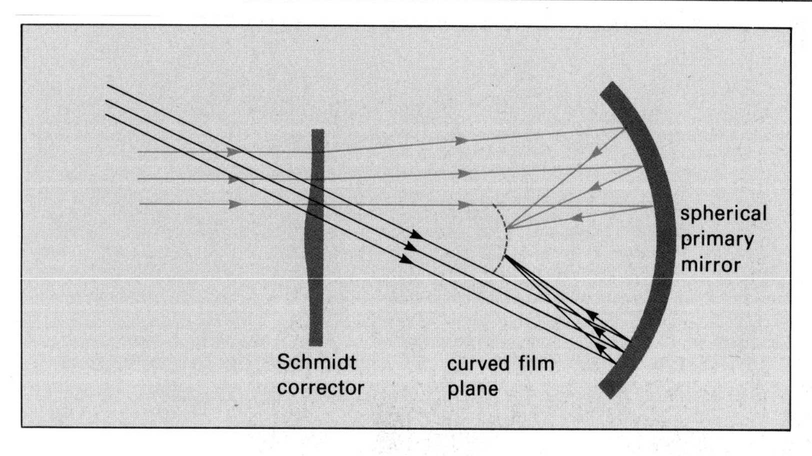

Fig. 9·5
The optical layout of a Schmidt telescope. The 'wavy' surface of the Schmidt corrector is exaggerated here for the sake of clarity.

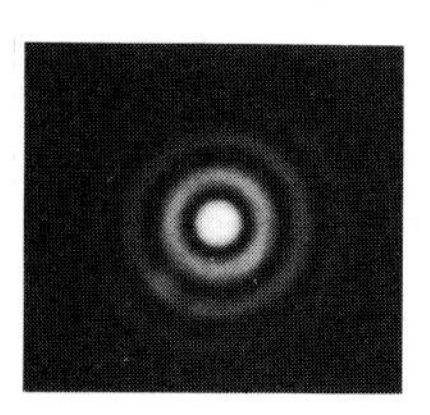

The spurious disc formed by light in a telescope. This out-of-focus image shows the rings caused by interference of the light waves as they approach the focus.

resolve or pick out fine detail is directly proportional to its aperture, a 2-m reflector having twice the **resolving power** of a 1-m instrument. However, a new approach has been taken in a multiple mirror telescope (MMT) completed at Mount Hopkins, Arizona in 1978. Here a battery of six 1·8 m reflectors are mounted together round a 0·76-m reflector which acts as a guide telescope. The light from the 1·8-m reflectors is fed to the same focus, and the resolution of the six apertures is equivalent to that of a 4·47-m telescope, but it is far easier optically to construct than a 4·47-m instrument would be and only about a third of the cost.

A number of other similar telescopes to the MMT are under construction or being designed, none of which would be feasible without modern computer techniques to maintain the correct degree of alignment of the complicated optical paths. Computers are also essential for the next generation of telescopes now being planned, where to achieve large apertures without massive structural engineering problems, designers are turning to the idea of tesselated reflecting surfaces – that is, surfaces composed of many separate elements, like the tiles on a floor. Such a reflecting surface can be made very thin and light, but all the individual mirror elements must be capable of precise adjustment to provide the necessary optical alignment. Moreover, as the telescope turns to follow objects across the sky, compensation has to be made for the changes in shape due to the whole mirror's alteration in position and gravitational loading. Only with computer control has such sophistication become possible.

Interferometers

The resolution that one can obtain theoretically – 0·16 arc sec. with the MMT – is never realized in practice from Earth-based telescopes. This is because the Earth's atmosphere is never still; turbulence breaks up an image, with the result that the most probable resolution for the MMT is 0·7 arc sec., equivalent to a 0·17-m telescope outside the atmosphere, not a 4·47-m one. If one wishes to measure stellar diameters or resolve very close binary stars, one requires a resolution of 0·03 arc sec. or better and the problem would appear insoluble, for even the 5-m (200-inch) reflector at Palomar only has a practical resolving power of rather more than 1 arc sec. Yet there is a solution, and this is to make use of interferometry, a technique well-known to radioastronomers as we shall see later.

Interferometry is based on the fact that light is a wave disturbance. When waves from a bright point source are brought to a focus in a telescope, although the waves from different parts of the lens are brought together at a point, the routes by which they travel to the focus are not the same. Thus a ray from near the rim of a lens will have further to go than one passing straight through the centre of the lens. In consequence the crests and troughs of the light waves will not be in step or **in phase** with one another. When they meet at the focus they will interfere: where crest meets crest they will reinforce each other to give a bright image; where crest meets trough, they will cancel out and there will be darkness. This is why a highly magnified star image never appears as a point of light but as a false or **spurious disc**, bright in the centre and dimmer towards the edge, and surrounded by alternate dark and bright rings.

The phenomenon of the spurious disc can be made use of if we turn the telescope into an interferometer. To do this the incoming light is separated into two components, and then the two beams are made to interfere. The spurious disc is then seen to be crossed by alternate light and dark bands. This technique, first used on the 2·5-m (100-inch) reflector at Mount Wilson in the 1920s has recently been developed by Hanbury Brown who, with Richard Twiss, worked out the theory of the **intensity interferometer**. Essentially this consists of two large mirrors, each 6·5 m diameter, but composed of 252 small hexagonal mirrors fitted together, and mounted on a special railroad track. These mirrors, whose separation may be altered and can be as great as 188 m – almost twenty-eight times greater than was possible with the 1·5 m – feed light directly into photoelectric detectors and the results are computer processed to give a measurement of the fringes. Using this instrument, diameters of stars as small as 0·00041 arc sec. have been obtained.

Although no ordinary telescope could do as well as this, it is still possible to use interferometry to improve its performance. Theoretically, a telescope like the 5-m (200-inch) should be able to resolve detail down to 0·02 arc sec. However, because of the turbulence or continual movement of the air above a telescope – a movement which we see as the twinkling of starlight – a telescope never achieves this theoretical limit; its results are usually 100 times less precise. In 1970 the French astronomer Antoine Labeyrie suggested a method of overcoming atmospheric turbulences. An analysis of the formation of images showed that the telescope image is composed of a host of tiny images caused by small pockets of air, each about 10 cm across. In some of these 'fly's eye' images or **speckles**, the light waves are in phase, in others out of phase. The speckles are continually changing, but by using short exposure of 0·02 s or less, they may be 'frozen'. Those speckles in which the light is in phase show interference, and by taking a large number of exposures and restricting the light to a very narrow band of wavelengths, it is possible to produce a composite picture. To obtain useful

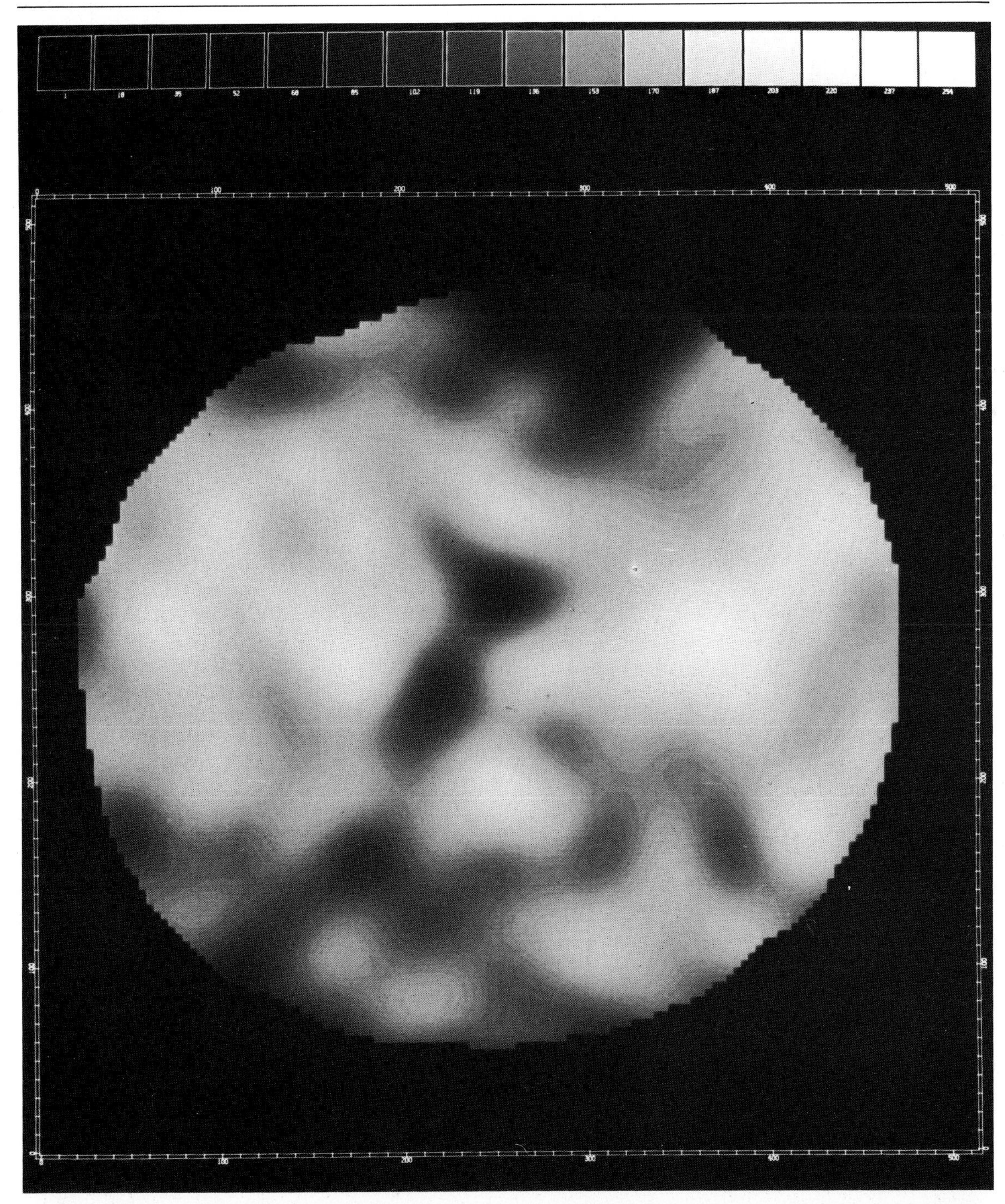

A composite picture of Betelgeuse built up by speckle interferometry.

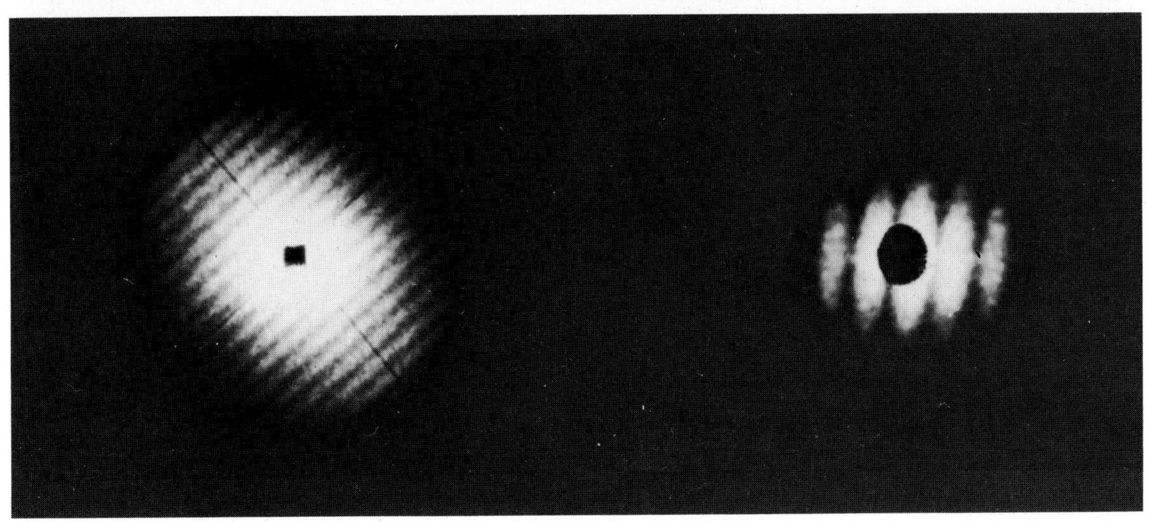

Speckle interferometry patterns of two binary stars, β Cephei (left), whose components are separated by about 0·25 arc seconds, and ι Serpentis (right), whose separation is 0·1 arc second.

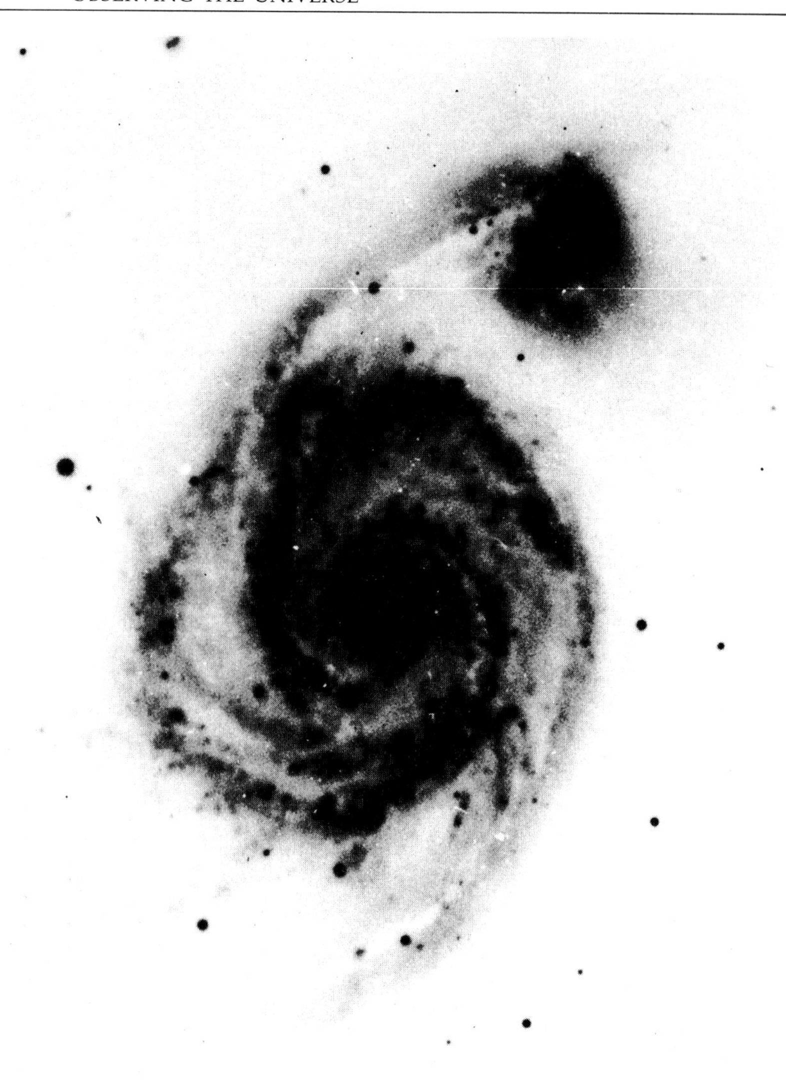

A comparison of an optical photograph (left) and an electronograph of the spiral galaxy M 51. Note how the second shows the large gaseous envelope which surrounds the satellite galaxy.

results this composite picture must be analysed, either by computer or by mapping it with a laser beam. In either case this technique of **speckle interferometry** gives results compatible with a telescope's theoretical resolving power. A picture of Betelgeuse (α Orionis) built up by computer from 2 000 speckle elements is shown on page 233 (*top*), and below it the interference pattern caused by having a double image – that of a binary star. Here the composite picture was mapped by laser, and measurement of the spacing and tilt of the fringes allowed the binary's orbit to be computed.

Speckle interferometry can improve on the results obtained at Mount Wilson in 1920, although the huge separations possible with Hanbury Brown's intensity interferometer are the only means of measuring diameters of any but larger nearby stars. Yet neither would be practicable without the use of electronic techniques: the short 0·02 s or less exposures used in speckle interferometry, for instance, would not be possible unless the telescope images were first electronically enhanced in brightness. The methods used to do this vary in detail but basically all depend on allowing the light to fall on a photo-sensitive material, that is a material which gives off electrons when light strikes it, and then multiplying the number of electrons by electronic means. In this way, a bright picture may be built up of a dim optical source.

Just as radio telescopes may be used together for interferometry, so it has been proposed that such techniques could be applied to optical telescopes. With the increasing use of electronic detectors and computer processing of images (see page 235) it is almost certain that this method could be applied to give very high resolutions of many astronomical objects.

Electronic detectors

Besides the interferometers mentioned above, there are two other devices which are achieving important results using electronic techniques – the **electronographic camera** and image photon counting system. The electronographic camera (Fig. 9·6) has a photosensitive cathode or photocathode on which the telescope image falls. This emits electrons which pass down an **electron multiplier** in which each electron triggers off the emission of others. These electrons finally hit a special photographic plate. The electrons need a vacuum in which to move but, although in the original camera of this type designed in 1952 by André Lallemand the photographic plate had to be used in a vacuum too, in the new design by James McGee and David McMullan this is not necessary. Such cameras are some five times more efficient than those of ordinary type.

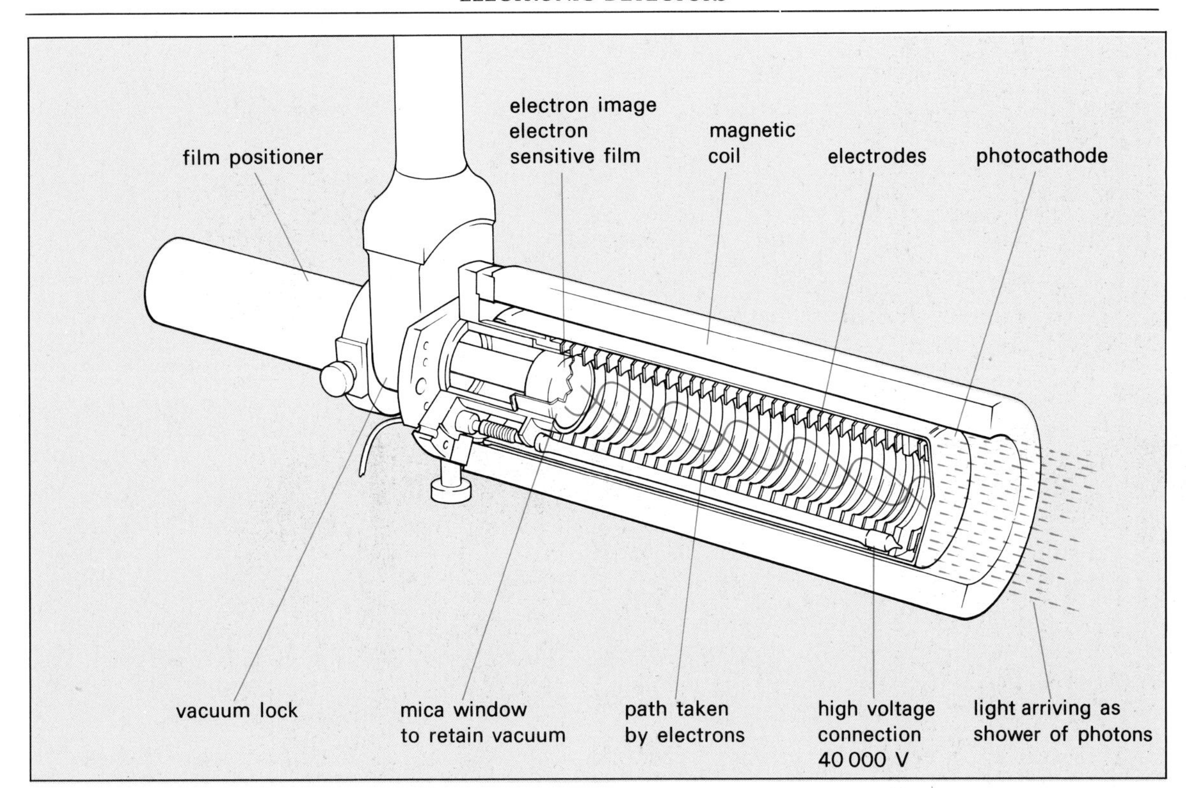

Fig. 9·6. Cut-away drawing of the electronographic camera.

The **image photon counting system**, developed by Alexander Boksenberg, uses a highly sensitive photocathode and electron multiplier which feed a television camera. The electrical signals derived from the incoming light are now dealt with 'digitally', that is, they are converted into a binary code and then analysed by specially developed computer techniques, which eliminate both the unwanted electrical signals generated in the equipment, and also light from the night sky. The equipment is supersensitive, multiplying the incoming light about 7×10^5 times, has good resolving power and measures brightness accurately, while it also terminates observations automatically, so that it can save telescope observing time. With astronomers queueing up to use large telescopes, such saving of observing time has very real advantages.

A further very significant development has been the introduction of the electronic devices known as charge-coupled devices (CCDs). These are microcircuits constructed on a slice of semiconducting silicon in such a way that the active area is divided into an array of picture elements (or pixels), 250 000 or more in a single device. Incoming light photons cause electrons (which are much easier to process) to become trapped in an appropriate 'well' within each pixel. These may then be read out electronically, and, if necessary, subsequently amplified and otherwise manipulated to give a final image which may be displayed on a monitor screen or turned into a standard photographic print. Such tiny devices – no more than a few millimetres square – may be cooled to very low temperatures to reduce 'noise' which degrades the image, and are exceptionally efficient – so much so that they are already beginning to displace the photographic plate in many applications.

Devices such as CCDs, as well as more conventional video equipment (such as the cameras carried by Voyager and other space-probes) and other systems of which the output can be processed electronically, permit the use of computer image-enhancement techniques. Images produced by such methods are very common nowadays, but should not be thought of as just 'pretty pictures'. The various image-enhancement techniques and the use of false colours enable many details to be discerned, in all sorts of astronomical objects.

The raw video output from less than 1 per cent of the image field of the image photon counting system developed at University College, London by Alex Boksenberg. Single photon 'events' are shown.

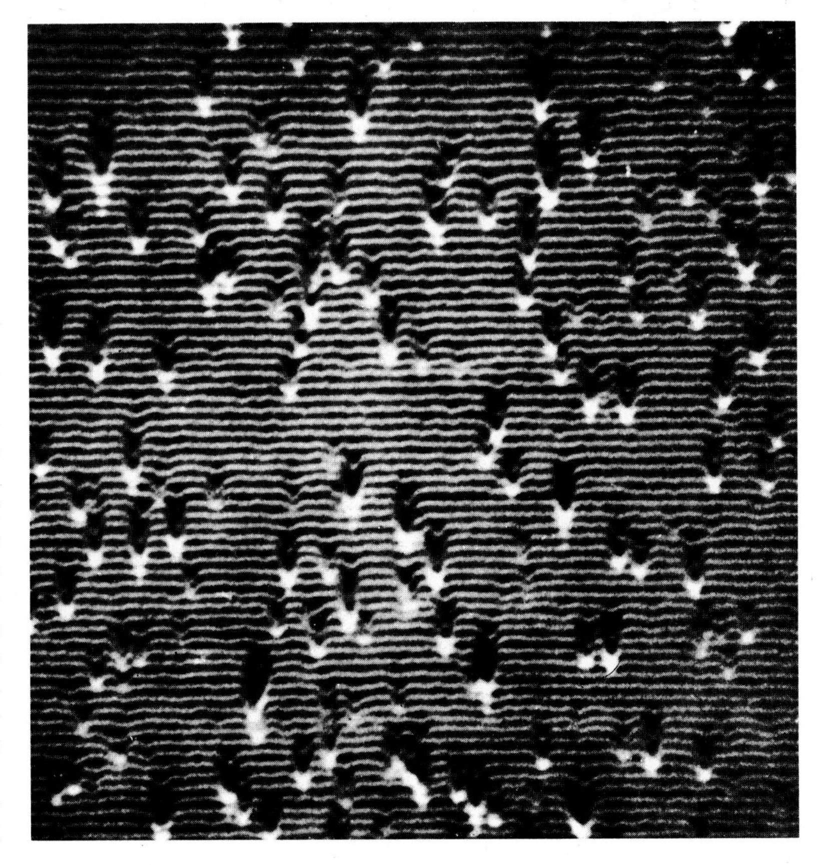

The Serrurier tube in its original form on the 5 m Palomar reflector.

Mountings

A telescope of any kind must be mounted so that its optical parts remain in alignment and it can be directed to any part of the sky, requirements which may seem obvious enough but which are not always easy to achieve. In a refracting telescope, a solid tube is used with the object-glass fitted to a 'cell' at one end, the component lenses being adjustable by set-screws on the cell. In a reflector, the mirror must be supported at its back and its sides so that it does not bend or distort in any other way when the tube points in different directions, and either a compensated lever system or an arrangement of hydraulic pads is used. Forced air circulation is also employed around and below the mirror, but is unnecessary above it because in a reflector the tube is always open. Nowadays the open tube is of **Serrurier** type, which is a framework with a ring at the upper end and a triangular arrangement of struts, named after Mark Serrurier who first designed such a tube for the Palomar 5-m (200-inch) reflector.

The optical system, complete with its tube, has to be supported not only so that it may be directed to any point in the sky but also so that it may readily track the curved apparent path of a celestial object as it moves across the sky due to the Earth's rotation. There are two main types of mounting which achieve this: the altazimuth and the equatorial. The **altazimuth** mounting, as its name implies, allows the telescope to move round in azimuth and up and down in altitude. It is a simple mounting (Fig. 9·7), inexpensive to construct, and has long been used for amateur telescopes. Unfortunately, it suffers from one great disadvantage – two movements, one in azimuth and one in altitude, have to be made to follow a celestial object as it drifts across the field-of-view of the telescope – and for this reason it has dropped out of use for large instruments since the early nineteenth century. Only now, in the late twentieth, with the widespread use of electronics, has automatic control of two movements rather than one become a practical possibility, and beginning with two large optical telescopes – the Russian 6 m reflector and the American MMT (page 232) – an increasing number of telescopes are being mounted in this way. An altazimuth mounting cannot readily follow a star across the zenith – in theory the azimuth and plate rotation would have to be infinitely fast – but this is a very minor disadvantage. The convenience of being able to carry heavy auxiliary instrumentation and, even more important, the much smaller cost for both the mounting itself and for the protective enclosure, mean that it is likely to be used on all new large telescopes. In professional telescopes this means the end of the alternative, which has held supremacy up to now, the equatorial mounting.

The **equatorial** can take a variety of forms, but basically it is an altazimuth tilted over so that the vertical or azimuth axis is parallel to the Earth's axis at that point on the Earth's surface where the telescope is situated. At the north or south geographical poles the equatorial would, in fact, be indistinguishable from an altazimuth, but elsewhere the difference is clear. Because of the tilt, rotation about one axis – the polar axis – allows a celestial body to be tracked with one movement only. The way the polar axis is supported and the telescope fixed to it can vary (overleaf), but the 'horseshoe' and 'English' types are the most common among modern optical instruments. Now such mountings always make provision for having observing equipment fixed at a number of different positions, as Fig. 9·3 shows. For an equatorial large aperture, observations may be made directly at the prime focus when the utmost sensitivity is required, but in a smaller instrument an observing chamber at the front would block out too much light from the main mirror, and the Newtonian focus is used. When heavy ancillary equipment is being used the professional often makes use of the Cassegrain focus (page 229) since quite heavy instruments can be supported below the primary mirror. The third position, the coudé focus, brings the light out to a stationary point. (In an altazimuth mounting, this is known as the **Nasmyth focus**, after its nineteenth century inventor James Nasmyth.) Here, very heavy equipment – a large high-resolution spectroscope, for instance – can be set up, possibly in a chamber of its own. Changing the foci is carried out by changing a mirror at the prime focus, and bringing into operation any other mirrors as necessary. The mirrors at the prime focus – and the observing cage, if there is one – are suspended by narrow section

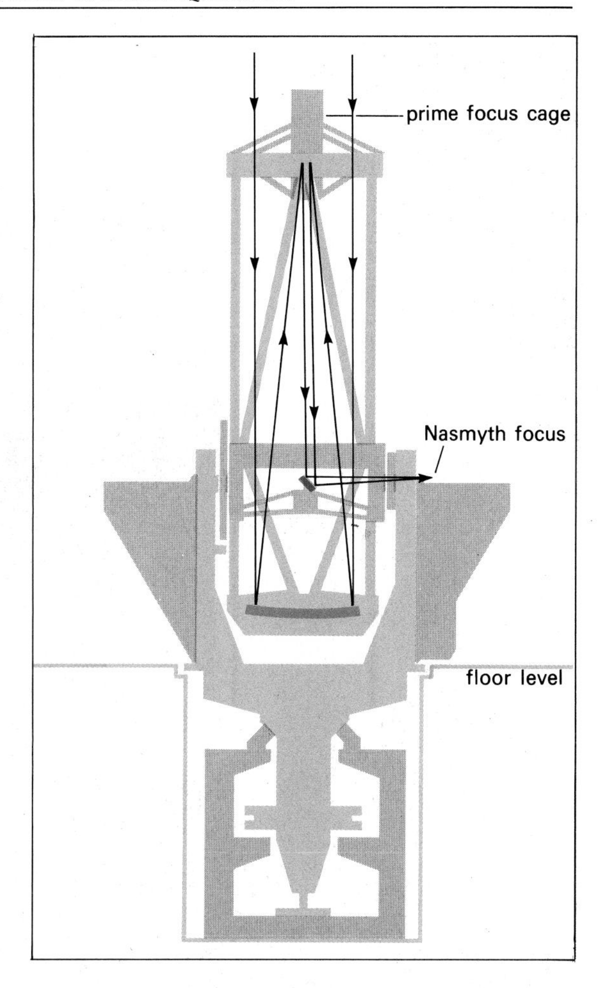

Fig. 9·7 Altazimuth mounting of a modern telescope, showing the prime focus and the Nasmyth focus.

supports, and diffraction patterns from these give rise to the spikes seen crossing bright stars on photographs.

Other optical equipment

Besides cameras, telescopes may feed other ancillary equipment, of which spectroscopes are the chief. These are now made in two forms, either to take photographs or to use photosensitive devices to permit the results to be given in digital form for immediate processing by a computer. This additional use of electronics is an example of a much wider revolution that has come in the handling of observational results. No more than thirty years ago, one could say that for every night spent at the telescope, one spent five nights measuring the plates or otherwise 'reducing' one's data, but now the situation has entirely changed. Plates taken at the telescope are now automatically processed in a machine like the Edinburgh Observatory's COSMOS – so called because it measures the Co-Ordinates, Sizes, Magnitudes, Orientation and Shape of every object on a plate – which now makes it possible to make full use of the large photographic plates (356 mm × 356 mm) taken on a modern Schmidt like the United Kingdom 1·2 m instrument which is surveying the southern skies down to magnitude 23·5 (in blue light). Each plate may contain half a million images and to measure only positions by hand would take many years, so for purely practical reasons the astronomer once had to select which images he would measure.

Opposite page, top:
The Wyoming 2.34-m infrared telescope is typical of modern designs, allthough its English Yoke equatorial mounting is not frequently encountered.

Now, all five kinds of information can be obtained in no more than 18 hours. At the University of Cambridge there is a still newer Automatic Plate Measuring machine, designed by Edward Kibble-white. It utilises laser scanning and has elaborate electronic circuitry. A result of its work is shown on page 226.

In one field of astronomical research – the Sun – some specialist equipment has been devised, since here there is plenty of light available and special forms of analysis may be carried out. Usually the

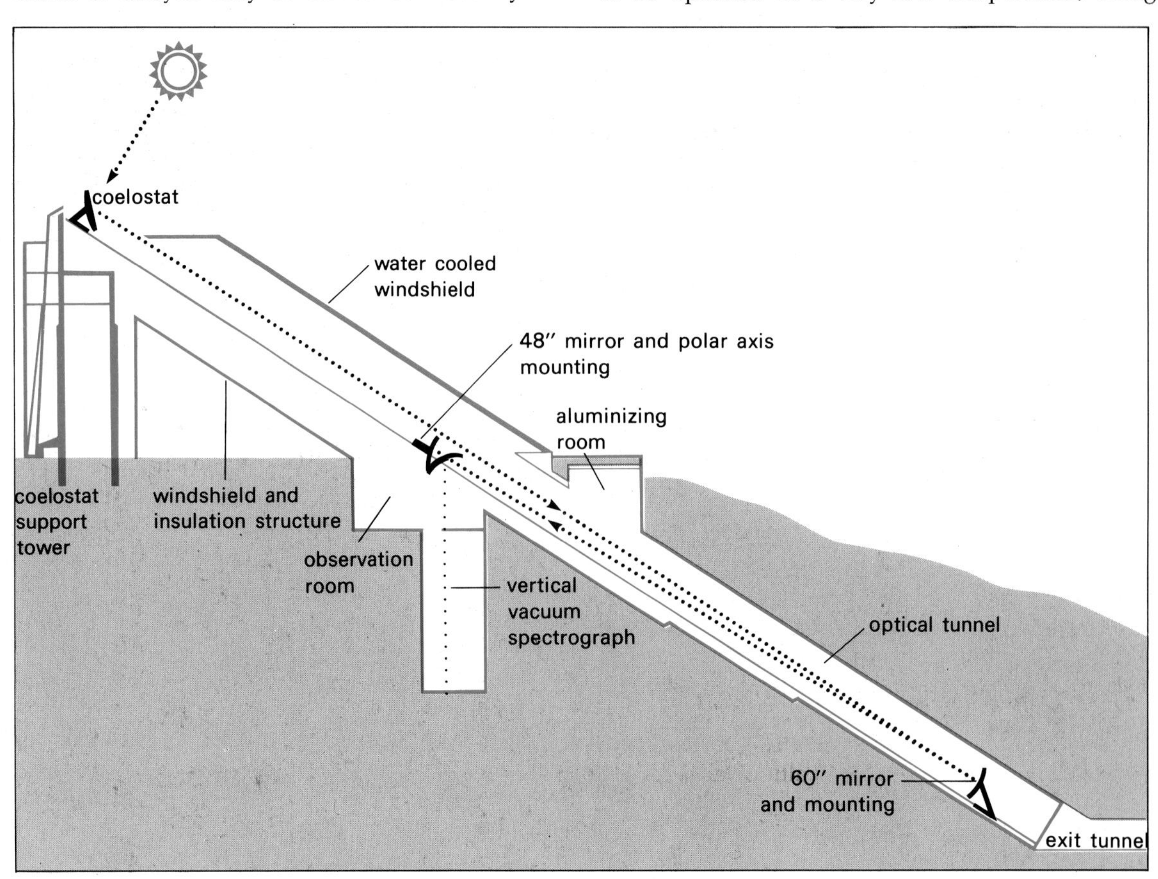

Fig. 9·8 A sectional drawing of the Kitt Peak solar telescope which has a focal length of 91·4 m, and is fed by a coelostat.

Sun's light is fed by a **coelostat** – an arrangement of two mirrors, one of which is equatorially mounted – to a large high-resolution spectroscope. The mirrors may be mounted – as at Mount Hamilton in California – on a tower and made to feed sunlight down to a cool underground chamber where the spectroscope is situated, or to feed the spectroscope by way of a sloping tunnel as at Kitt Peak Observatory in Arizona (Fig. 9·8). Another development is the **spectroheliscope** – a spectroscope which scans the Sun's image in the light of a particular spectral line, and effectively acts like a very selective filter; it allows the observer to examine some way down into the photosphere (page 76). Highly selective **narrow band filters** are also available for solar work. Yet even the spectrohelioscope and the narrow band filter can not show up the corona (page 77) and to do this without waiting for a total solar eclipse the coronagraph must be used. This is essentially a refractor in which an occulting disc acts in place of the Moon to cause a total eclipse, blotting out the light from the photosphere. Great attention has to be paid to avoiding stray light, the object glass has to be extraordinarily highly polished, and the instrument used at high-altitude observatories, where the atmosphere is clean. Even here special precautions have to be taken against dust which otherwise scatters light and degrades the image of the corona.

Another field of recent advancement has been infrared astronomy. Its growth has really been due to the development of highly efficient infrared detectors, typically small semiconductors of germanium or indium antimonide, which, for full efficiency, have to be operated at a very low temperature, being cooled by liquid nitrogen to 77 K, or even to 4 K by liquid helium. There are two serious problems in observing infrared from an Earth-based observatory; the absorption of infrared by water vapour in the atmosphere and the infrared radiation of the atmosphere itself which produces a 'fogging' effect, all but obliterating the infrared signals from celestial objects. The first, the absorption by water vapour, can be reduced by siting an infrared telescope on a high mountain, and it is for this reason that the world's largest infrared telescope or flux collector, the British 3·8 m instrument, is sited on Mauna Kea in Hawaii at a height of 4 200 m. Although the entire range of infrared wavelengths (0·75 μm to 1 000 μm) cannot be observed, even from Mauna Kea, the windows in the atmosphere (near 2 μm, 3·5 μm, 5 μm and 10 μm) allow very valuable work to be done. To minimize atmospheric fogging a 'chopping' technique is used; vibrating mirrors compare the section of sky with an infrared source with another section without a source. This method is also used at optical wavelengths for studies of the magnitude of dim sources.

Opposite page, bottom:
The 76 m steerable dish at Jodrell Bank. It is on an altazimuth mounting.

Radio telescopes

Radio telescopes work on the long wavelength end of the spectrum below the infrared, using a window which accepts wavelengths between a few millimetres and 30 m. In essence, radio telescopes are special directional radio aerials or **antennae** designed to receive radio waves over a specific range of wavelengths and pass these to a special radio receiver, whose output is fed either to a chart recorder or to a computer. The antenna may take a number of forms of which the **dipole** is probably familiar from television aerials, although the **Yagi** (invented by Hidetsugu Yagi) will also be familiar for the same reason. The **helix** antenna, invented by John Kraus, is another type, but in radioastronomy none are used alone; all require some additional reflecting system to enhance the signal-gathering ability of the telescope, just as a large mirror enhances the light-gathering power of an optical telescope. The varieties of reflector appear extremely numerous, but there are really only three basic designs. The best known is probably the **dish-type reflector**, of which the 76-m diameter parabolic dish at Jodrell Bank, England is a prime example; this is fully steerable, as also are the 64-m telescope at Parkes, Australia and the 100-m at Effelsburg, Western Germany, to mention only some of the larger instruments. Yet since radio telescopes can build up their plot of signal strength over a long period of time, many instruments rotate in altitude only, using the Earth's rotation to give a movement in azimuth. The most notable of the dishes to do this is the 300-m spherical reflector at Arecibo in Puerto Rico, the bowl of which is constructed out of a natural hollow in the ground; small changes in the position of the aerial at the prime focus can give a result equivalent to tilting the bowl in altitude by amounts of up to 20° from the zenith. On the other hand, the steerable dish is sometimes mounted equatorially – the 43-m dish at Greenbank, Virginia is a particularly

The 110 m × 24 m Kraus radio telescope at Ohio. The longish wavelengths used allow the reflectors to be made of wire mesh stretched across a girder framework.

The two grating interferometer radio telescopes at Potts Hill, Sydney, Australia. These can be used together to make a rotational synthesis instrument.

well-known example – and here the receiving antenna is mounted below the dish, so that the telescope works like a Cassegrain.

Another much less expensive but more unfamiliar form is the **transit radio telescope**. Like the Arecibo dish, this kind of telescope is movable in altitude only, but unlike the Arecibo instrument, it has a long narrow reflector which has a parabolic surface. This reflecting surface is fixed but it is fed by a flat reflector. It is known sometimes as the **Kraus type** radio telescope, since the first instrument of this kind was designed by Kraus and built at Ohio with a parabolic reflector 110 m × 24 m. Other notable examples are at Nançay, France and at the Pulkovo Observatory, Leningrad.

The surface of a dish or other reflecting surface does not have to be made as accurately as one for an optical telescope because radio waves are between 10^4 and some 10^8 times longer than light waves. Whereas one must have optical quality mirrors figured correct to less than 10^{-4} mm, the surface of the parabola for a radio telescope would only need to be correct to a little less than a millimetre for the very shortest radio wavelengths, and with an unevenness amounting to centimetres for many of the wavelengths in common use. Indeed, for any wavelength in the metre range a solid reflecting surface is no longer needed and a wire mesh reflecting surface is perfectly adequate. However, such a large difference in wavelength between optical and

An aerial view of the eight 13m reflectors of the 5 km aperture synthesis radio telescope at the Mullard Radio Astronomy Observatory at Cambridge University.

radio waves means that aperture for aperture, a radio telescope has much less satisfactory resolving power. To quantify it, whereas the theoretical resolving power of the 2·5-m (100-inch) telescope is 0·05 arc sec., that of the 76-m Jodrell Bank radio telescope is a little less than 12 arc minutes at a wavelength of 21 cm. To obtain a resolving power that is useful for detecting single small radio sources it is obviously necessary to use some sort of interferometer and interferometric techniques have become highly developed in radioastronomy.

One type of interferometer is that devised by Bernard Mills in Australia in 1957. This, the **Mills Cross**, consists of two arrays of antennae with long narrow parabolic reflectors, the arrays lying perpendicular to each other. By electronically combining the signals from the arrays, the instrument gives a **pencil beam** resolution centred on the point of intersection of the arrays. With arrays each of 1 km length, such an instrument can give a resolution of 1·4° at a long wavelength like 15 m, whereas the Jodrell Bank instrument would only provide a resolution of 4° at this wavelength. Another development is the **grating interferometer** designed originally by Willem Christiansen for solar work, where the arrays of antennae are composed of a series of small dishes. Such a telescope is good for resolving detail across a strip of sky.

Probably the most notable design of radio interferometer is that due primarily to Sir Martin Ryle and his

Opposite page, top:
Three of the 27 identical radio telescope dishes which form part of the Very Large Array in New Mexico.

Opposite page, bottom:
Skylab silhouetted against the Earth. Astronomical observations, clear of the Earth's atmosphere, can be made from such a craft, as well as from small unmanned orbiting observatories carrying telescopes and ancillary equipment.

colleagues at Cambridge, England, where the **aperture synthesis** technique has been developed. Basically, this is a method of using an interferometer whose antennae can be spaced out at varying distances from each other so that, by making a set of observations at different spacings, it is possible to build up a radio 'picture' of a source equivalent to that obtained by a radio telescope with a huge aperture, equal in size to the area over which the antennae are spread. Thus, by using a series of small but movable antennae one can get the equivalent resolution of a vast instrument but not, of course, the equivalent sensitivity (the radio equivalent of light-grasp). Such radio telescopes take two forms: one in which there is one fixed antenna array and one moveable one, the other in which the antennae – one fixed and the rest moveable – are laid out in a straight line. In the latter, synthesis is achieved by the rotation of the Earth which sweeps the antenna system round. The 5-km instrument at Cambridge is an example of the second form, the results of this rotational synthesis being fed direct to a computer for analysis.

A recent development has been **long base-line interferometry**, where large dish radio telescopes have been used in pairs. Sometimes the distances have been measured in kilometres, but by recording the observations at each telescope and then analysing both together at a later date, it has been possible even to pair telescopes separated by intercontinental distances. With such extremely long base-lines it has been possible to obtain resolutions down to at least 0·001 arc sec. Using one radio telescope mounted in an orbiting spacecraft, even finer resolution should be possible.

The largest single array of telescopes is the Very Large Array (VLA) located on the Plains of St Augustin, near Socorro, New Mexico. Here 27 dishes, each 25 m in diameter, can be arranged in a vast Y-shaped configuration. With 24 possible positions on each arm and maximum separations of 19 km on the north, and 21 km on the south-west and south-east arms, resolution (which depends upon the observing frequency) can reach down to 0.13 arc sec. Dishes identical to those in the VLA form some of the elements in the British MERLIN (Multi-Element Radio-Linked Interferometer Network) array, which also uses the famous Mark 1A 76 m telescope at Jodrell Bank for some of the time. The maximum separation between MERLIN's telescopes is 135 km, and it can not only achieve a resolution greater than the VLA – down to about 0·1 arc sec. – but also can cover a wider range of frequencies.

Even greater baselines, spanning intercontinental distances, are possible with the technique of **very long base-line interferometry** (VLBI). Thanks to the use of atomic clocks for precise timing, signals may be recorded magnetically at each site and later combined, in just the same way as if the telescopes had been physically connected at the time. This technique has permitted radio astronomers to surpass their optical counterparts in resolution. Only with

Below: *The 300 m spherical reflector radio telescope at Arecibo, Puerto Rico.*

the use of equipment such as the Space Telescope will optical astronomers have a chance to catch up slightly, but by then plans will doubtless be laid for interferometry with radio telescopes in space!

One quite different astronomical use of dish-type radio telescopes is to emit very short radio pulses from them to the planets and to meteors. This **radar** technique of emitting pulses and timing the interval from their emission to their return has many applications. In planetary studies it is possible to obtain accurate distances for the Sun, Moon and planets in this way and, by a more elaborate analysis of a multiplicity of signals, build up a relief picture of a planetary surface. When applied to meteors, the method allows not only meteor tracks but also their velocities to be determined, even in daylight.

Other instruments

Recently astronomers have developed special instruments to answer particular research questions, and of these the **neutrino detector** (sometimes called a neutrino telescope) and the gravitational wave detector are the most significant. The first have been established down disused gold mines in South Dakota and Salt Lake City, U.S.A., Witwatersrand in South Africa and Kolar in southern India, and consist of giant tanks holding hundreds of tonnes of dry cleaning fluid (tetrachloroethylene, C_2Cl_4). Neutrinos will pass through any material with little chance of capture, but encounters with the tetrachloroethylene molecules cause change of the chlorine into radioactive argon. Measurements of the number of changed molecules give a measure of the neutrinos present. The tanks are situated at depths of a kilometre or more to shield them from any effects due to cosmic rays (page 184), and gold mines have been chosen since these are the deepest mines known. Results from these detectors show far less neutrinos than expected, and this has brought about the search for modified theories of the processes occurring in the solar interior (page 91). Other forms of detector including gallium-germanium detectors are being built, in the hope of investigating both high- and low-energy neutrinos.

Several theories of gravitation predict the existence of gravity waves. Various forms of **gravity wave detector** have been built, with the original form being a large aluminium cylinder nearly 1 m in diameter, 1·5 m long and weighing just over 3·5 tonnes, the whole contained within a vacuum chamber. Despite the existence of several potential sources of gravity waves, such as the binary pulsars (page 72), no waves have yet been detected, although there have been one or two false alarms. Calculations show that current sensors are probably insufficiently sensitive to detect the waves which may be reasonably expected to exist, and the development of better devices is under way. In addition to the possible use of the Pioneer spacecraft for the detection of perturbing bodies close to the Solar System (page 152–153), it has also been suggested that if the Doppler shift in the signals from these and other spacecraft could be detected, they could function as gravitational wave detectors. Experiments have already been made in this manner using one of the Voyager craft, but without success. If all these schemes fail to provide any evidence for the existence of gravity waves, fundamental revision of gravitational theories will be required.

Observing from space

The importance of being able to overcome the observing limitations imposed by the Earth's atmosphere cannot be overemphasized. Mention of its adverse effects on optical telescope resolution is only one aspect, and far more serious is the fact that some wavelengths never penetrate down to Earth-based observers. The absorption in the infrared region can, to some extent, be overcome by siting telescopes at very high altitude observatories – such as on Hawaii – but even so that only solves part of the problem. Radiation in the extreme ultraviolet (XUV), X-ray and γ-ray end of the spectrum – all radiation, that is, shorter than about 300 nm in wavelength – is filtered out, yet such high-energy radiation is vital for a full investigation of the universe, as too is observation of cosmic rays, those high speed atomic nuclei whose terrestrial effects only (the emission of 'secondary' particles in the atmosphere) can be observed on Earth.

One method of getting above at least the densest parts of the atmosphere is to use high-altitude balloons which can carry telescopes and other equipment to heights of some 25 km; another is the use of sounding rockets which can be launched to heights of around 160 km before they fall back to Earth. But although much more expensive, full investigations over long periods necessitate the launching of spacecraft, both for direct observation as well as for making close approaches to or soft landings on members of the Solar System.

Because of the cost and complexity of large spacecraft, except in the cases of manned exploration and the use of Skylab, all extra-terrestrial observing has one thing in common – the observing equipment has to be as compact and light in weight as possible, and able to produce its results in a form suitable for radio transmission back to Earth. Every spacecraft, manned or unmanned, is launched by rocket since a rocket receives its impulses from the reaction of its hot gases as they escape from it; it does not have to push on anything to be propelled, as an aircraft, ship, or land vehicle does on Earth. Indeed, a rocket works most efficiently in the airless regions of inter-planetary or interstellar space. To get away from the Earth – to reach **escape velocity** – a spacecraft must reach a speed of 11·18 km per s, and multiple stage launching rockets are used so that once some of the heavy fuel tanks have been exhausted, they may be jettisoned, so that the mass of the vehicle continually decreases and the escape velocity may be more readily reached. The majority of spacecraft devoted to purely astronomical research are placed into Earth orbits (where a minimum velocity of 7·9 km per s must be maintained), but a number are placed into a solar orbit (as, of course, are the planetary probes). In some cases, such as with the Pioneer and Voyager spacecraft, interaction with the gravitational fields of the planets may considerably modify the orbits, and even cause the probes to be ejected from the Solar System.

Although there has been a lull since the phenomenal success of the many planetary missions such as the Mariner, Viking, Pioneer and Voyager probes, activity has by no means ceased. Various nations have many different spacecraft either under construction or being designed, and we can expect missions to examine the Sun and comets, as well as Venus radar and atmospheric probes and a return to Jupiter with the Galileo orbiter and atmospheric entry capsule. In addition, plans are being made for the adaption of relatively 'ordinary' satellites, such as are used for communications and meteorological and remote-sensing purposes in Earth orbit, to missions further out into the Solar System.

Many of these space probes are expected to be launched by the Space Shuttle, which will also have the task of recovering satellites that have reached the end of their useful (or design) lifetimes. The Shuttle will also carry many short-term experiments into space, each usually being devoted to one aspect of the study of astronomical problems, as well as the Spacelab laboratory. The forseeable manned space stations in Earth orbit will also undoubtedly carry many astronomical experiments, such as the solar instrumentation carried aboard Skylab.

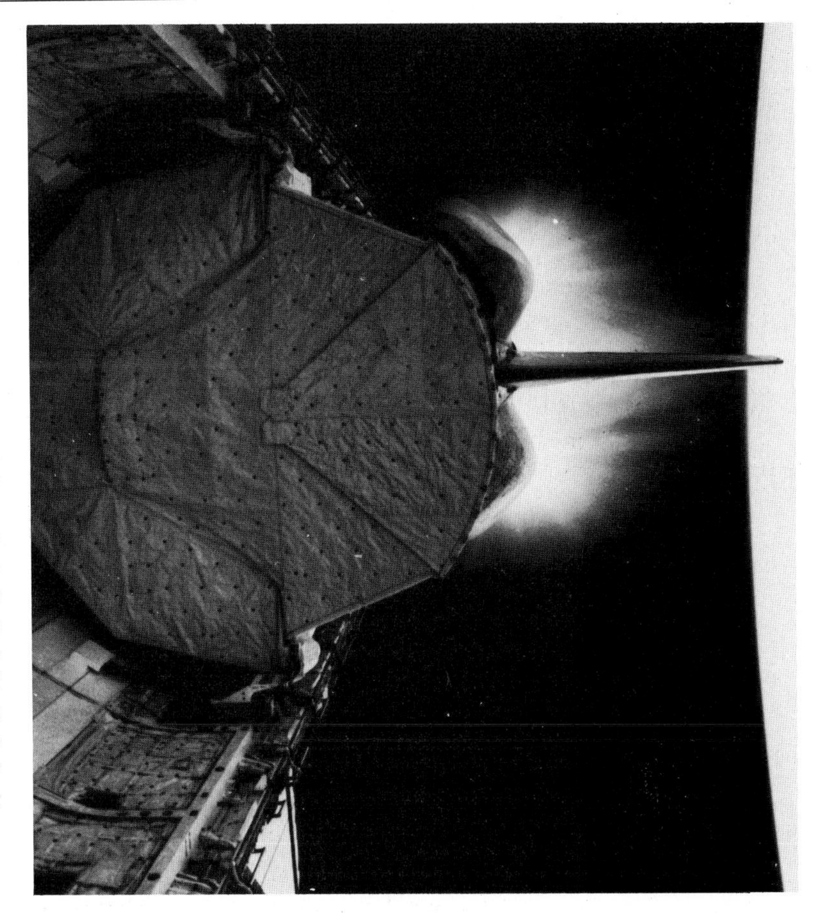

The Space Shuttle, here shown manoeuvring in orbit, is capable of transporting many astronomical satellites and experiments into space.

High-energy astronomy

This is astronomy concerned with the high-energy radiation bands – XUV, X-rays and γ-rays – and with primary cosmic ray particles, and special observing techniques have to be used. For cosmic ray particles, both those received in space and those which reach the Earth, a similar technique is used because both can cause ionization in a liquid, a gas, or even in a suitable solid semi-conductor type of material. In space, detectors are either gas-filled or solid. Ionization is caused, too, by γ-rays and X-rays, and similar detectors are used, the gas-filled being particularly suitable for the longer wavelength, less energetic X-rays. However, detection of the presence and strength of any high energy particles or radiation is only part of the problem: the astrophysicist also wants to know the direction of the source. A conventional telescope is fine for ultraviolet studies, but X-rays and γ-rays will penetrate the mirrors of a reflector, so some other directional system is required for an X-ray or γ-ray 'telescope'. The most effective method is to use the principle of grazing incidence, whereby a beam of X-rays will be reflected from a metal surface (or metallic film on glass) if the beam grazes it at a steep angle of the order of 87° or more. Using special hyperboloidal and paraboloidal surfaces (one after the other) it is possible to make a telescope that will form an image. The highly successful HEAO-2 satellite (renamed Einstein) used a set of such surfaces, arranged concentrically, in its X-ray telescope. This has given many new insights into the high-energy processes which are occurring in all sorts of astronomical objects, from nearby stars to the most distant quasars. However, this method is only of use down to certain wavelengths, typically around 1 nm. For shorter wavelengths a series of circular rings, alternately opaque and clear, is used to give an image by diffraction. Such a **zone plate** device is small and needs a strong influx of X-rays to work, so that at present it is used only in solar observation. Other devices using crystals can determine the degree of polarization of X-rays and also act as X-ray spectroscopes because certain crystals reflect X-rays at specific angles depending on their wavelength.

More conventional optical techniques – albeit using special material – can be employed for most ultraviolet work. Here again phenomenal success has been achieved by one spacecraft, the International Ultraviolet Explorer (IUE) satellite, which over its long lifetime studied many energetic objects. This has led to a new understanding of the processes at work in both single stars and binary systems.

The extension of work into the γ-ray region has been slow, because of the difficulty of devising suitable detectors. However, the COS-B satellite has proved a pioneer in this field, and it has already been mentioned (page 70) how the detection of γ-ray bursts by various spacecraft in different parts of the Solar System, even though they were relatively insensitive to direction, has led to the discovery of the enigmatic gamma-ray bursters.

Longer wavelengths

At lower energies and longer wavelengths, spacecraft have begun to show their potential. Particularly successful has been the Infrared Astronomical Satellite (IRAS) which carried out the first survey of the whole sky at infrared wavelengths, as well as examining specific objects.

An even more sophisticated spacecraft is scheduled to be launched at a later date to continue this work in an even more detailed manner. Infrared satellites have limited lifetimes due to the fact that their detectors (and parts of the structure) have to be cooled to low temperatures by the use of liquid helium, and this gradually boils away so that the supply becomes exhausted. The advent of the Space Shuttle should mean that these and other satellites can be either replenished or recovered, with a considerable saving in costs and a greatly extended working lifetime.

The Space Shuttle will be used in the launch of the Space Telescope. The size of its primary (2·4 m aperture), and the highly sophisticated auxiliary equipment which it is to carry, could well lead to an explosion of knowledge about all manner of astronomical objects. Free from the disturbing effects of the Earth's atmosphere it should achieve phenomenal resolution and possibly even be capable of directly detecting planets of other stellar systems, as well as detecting the faintest objects and reaching to the depths of space. But it will not be alone. Other satellites will be probing the universe throughout the electromagnetic spectrum from radio waves to gamma-rays, achieving a precision impossible from the surface of the Earth. Undoubtedly these will all add greatly to astronomical knowledge and, it is hoped, will help in obtaining solutions to some of the more perplexing problems facing astronomers today.

The large aluminium bar of the Stamford University gravitational wave detector is suspended in a chamber which is cooled to very low temperatures to increase its sensitivity.

Appendices

Abbreviations

In addition to the abbreviations for chemical elements, SI units and multiples, which are given in Appendices 9 and 2–4, many abbreviations are commonly used in astronomy and in this book. These are given below.

Abbreviation	Meaning
A	atomic mass number
Å	Angstrom unit
au	astronomical unit
c	velocity of light
d	day
d, D	distance
e	eccentricity
e^-	electron
e^+	positron
eV	electron volt
E	energy
G	gravitational constant
H_0	Hubble's constant
i	inclination
IR	infrared
L	luminosity
$L_\odot$	solar luminosity
m	apparent magnitude
m_{pe}	photoelectric magnitude
m_{pg}	photographic magnitude
m_{vis}	visual magnitude
M	absolute magnitude
m_v, M_v, etc	magnitude through filters
m, M, M	mass
$M_\odot$	solar mass
n	neutron
N	number
p	proton
P	pressure
P_{rad}	radiation pressure
pc	parsec
PSR	pulsar
q	perihelion distance
QSO	quasi-stellar object (quasar)
QSS	quasi-stellar source
R	gas constant
R	radius
R	Zurich sunspot number
$R_\odot$	solar radius
snu	solar neutrino unit
S	flux density
t, T	time
T	temperature
T_{eff}	effective temperature
UV	ultraviolet
v	velocity
V	volume
V	potential energy
XUV	extreme ultraviolet
z	redshift
γ	photon
$\delta\lambda$	wavelength change
θ	angle
λ	wavelength
λ_0	rest wavelength
ν	neutrino
ω	angle of perihelion
Ω	angle of ascending node
[ ]	'forbidden' transitions

Some Derived SI Units with Special Names

Physical quantity	SI unit	Symbol
frequency	hertz	Hz
energy	joule	J
force	newton	N
power	watt	W
electric charge	coulomb	C
potential difference	volt	V
resistance	ohm	Ω
capacitance	farad	F
magnetic flux	weber	Wb
inductance	henry	H
magnetic flux density	tesla	T
luminous flux	lumen	lm
illumination	lux	lx

Greek alphabet

upper case	lower case	name	upper case	lower case	name
A	α	alpha	N	ν	nu
B	β	beta	Ξ	ξ	xi
Γ	γ	gamma	O	o	omicron
Δ	δ	delta	Π	π	pi
E	ε	epsilon	P	ϱ	rho
Z	ζ	zeta	Σ	σ	sigma
H	η	eta	T	τ	tau
Θ	θ	theta	Y	υ	upsilon
I	ι	iota	Φ	ϕ	phi
K	$\varkappa$	kappa	X	χ	chi
Λ	λ	lambda	Ψ	ψ	psi
M	μ	mu	Ω	ω	omega

Some other units used in astronomy

Length

1 micron (μm) = 10^{-6}m
1 Angstrom (Å or A) = 10^{-10}m
= 10^{-8}cm
1 au = $1{\cdot}4960 \times 10^8$ km
= $9{\cdot}2956 \times 10^7$ miles
= $4{\cdot}8481 \times 10^{-6}$pc
= $1{\cdot}5813 \times 10^{-5}$ light years
1 pc = $3{\cdot}0856 \times 10^{13}$ km
= $1{\cdot}92 \times 10^{13}$ miles
= 3·2616 light years
= $2{\cdot}0626 \times 10^5$ au
1 light year = $9{\cdot}4605 \times 10^{12}$km
= $5{\cdot}88 \times 10^{12}$ miles
= $6{\cdot}324 \times 10^4$au
= 0·3066 pc

Mass and radii

1 solar radius ($R_\odot$) = $6{\cdot}96 \times 10^5$ km
= $4{\cdot}325 \times 10^5$ miles
1 solar mass ($M_\odot$) = $1{\cdot}99 \times 10^{30}$kg
1 Earth radius = $6{\cdot}378 \times 10^3$ km
= $3{\cdot}963 \times 10^3$ miles
1 Earth mass = $5{\cdot}977 \times 10^{24}$ kg

Time

1 tropical year = 365·2422 mean solar days
1 sidereal year = 365·2564 mean solar days
1 sidereal day = $23^h56^m4{\cdot}1^s$

Energy

1 joule (J) = 10^7 ergs
1 electron volt (eV) = $1{\cdot}60207 \times 10^{-19}$J

Conversion factors

1 cm = 0·3937 inch
1 m = 1·0936 yard
1 km = 0·6214 mile
1 kg = 2·2046 pound
1 gm = 0·0353 ounce

1 inch = 25·4 mm
= 2·54 cm
1 foot = 0·3048 m
1 yard = 0·9144 m
1 mile = 1·6093 km
1 pound = 0·4536 kg

Basic SI Units

Basic physical quantity	SI unit	Symbol
length	metre	m
mass	kilogram	kg
time	second	s
electric current	ampere	A
thermodynamic temperature	kelvin	K
amount of substance	mole	mol
luminous intensity	candela	cd

The chemical elements

Each of the chemical elements is represented by a letter or combination of letters, consisting of the initial or an abbreviation of the English or Latin name of the element.

symbol	name	atomic number
Ac	actinium	89
Ag	silver (L *argentum*)	47
Al	aluminium	13
Am	americium	95
Ar	argon	18
As	arsenic	33
At	astatine	85
Au	gold (L *aurum*)	79
B	boron	5
Ba	barium	56
Be	beryllium	4
Bi	bismuth	83
Bk	berkelium	97
Br	bromine	35
C	carbon	6
Ca	calcium	20
Cd	cadmium	48
Ce	cerium	58
Cf	californium	98
Cl	chlorine	17
Cm	curium	96
Co	cobalt	27
Cr	chromium	24
Cs	caesium	55
Cu	copper (L *cuprum*)	29
Dy	dysprosium	66
Er	erbium	68
Es	einsteinium	99
Eu	europium	63
F	fluorine	9
Fe	iron (L *ferrum*)	26
Fm	fermium	100
Fr	francium	87
Ga	gallium	31
Gd	gadolinium	64
Ge	germanium	32
H	hydrogen	1
He	helium	2
Hf	hafnium	72
Hg	mercury (L *hydrargyrum*)	80
Ho	holmium	67
I	iodine	53
In	indium	49
Ir	iridium	77
K	potassium (L *kalium*)	19
Kr	krypton	36
La	lanthanum	57
Li	lithium	3
Lu	lutetium	71
Lw	lawrencium	103
Md	mendelevium	101
Mg	magnesium	12
Mn	manganese	25
Mo	molybdenum	42
N	nitrogen	7
Na	sodium (L *natrium*)	11
Nb	niobium	41
Nd	neodymium	60
Ne	neon	10
Ni	nickel	28
No	nobelium	102
Np	neptunium	93
O	oxygen	8
Os	osmium	76
P	phosphorus	15
Pa	protactinium	91
Pb	lead (L *plumbum*)	82
Pd	palladium	46
Pm	promethium	61
Po	polonium	84
Pr	praseodymium	59
Pt	platinum	78
Pu	plutonium	94
Ra	radium	88
Rb	rubidium	37
Re	rhenium	75
Rh	rhodium	45
Rn	radon	86
Ru	ruthenium	44
S	sulphur	16
Sb	antimony (L *stibium*)	51
Sc	scandium	21
Se	selenium	34
Si	silicon	14
Sm	samarium	62
Sn	tin (L *stannum*)	50
Sr	strontium	38
Ta	tantalum	73
Tb	terbium	65
Tc	technetium	43
Te	tellurium	52
Th	thorium	90
Ti	titanium	22
Tl	thallium	81
Tm	thulium	69
U	uranium	92
V	vanadium	23
W	tungsten (G *Wolfram*)	74
Xe	xenon	54
Y	yttrium	39
Yb	ytterbium	70
Zn	zinc	30
Zr	zirconium	40

Glossary

absolute temperature scale
A temperature scale which has the same divisions as the Celsius (centigrade) scale, but where 0° lies at absolute zero (*q.v.*). Degrees on this scale are written K, short for degrees Kelvin, after Lord Kelvin, the physicist who devised the scale.

absolute zero
The point at 0K or –273·16°C at which all motion of atoms and molecules ceases, and no heat is radiated.

age gradient
A graduation in age from young to old.

Ångstrom
A unit of wavelength often used in astronomy: 1Å = 10^{-10} metres = 0·1 nanometres.

angular momentum
The property of a system due to its revolution or spin. The angular momentum of a body may be found by multiplying its mass by its velocity times its distance from the point of revolution.

angular resolution
The angular distance between the closest details which a telescope can pick out or resolve.

antapex
The point in the sky opposite the solar apex. It lies at right ascension 6 hours and declination –30°, in the constellation of Columba.

apex, solar
The point in the sky to which the Sun appears to be moving. It lies at right ascension 18 hours and declination +30°, in the constellation of Hercules.

atmosphere
A unit of pressure. One atmosphere is the normal pressure of air at sea-level on Earth.

bolometric magnitude
A measure of the total energy emitted by a celestial body, taking into account all wavelengths radiated. Since a celestial body always emits other radiation besides light, the bolometric magnitude is always a smaller number than that expressing the visual magnitude. Thus the absolute visual magnitude of the Sun is +4·83, and its absolute bolometric magnitude is +4·72. The bolometric and visual magnitudes of very blue and very red stars may differ by 4 or 5 magnitudes, indicating that in these stars most of the radiation is non-visual.

catalyst
A substance which assists a reaction to proceed but does not appear in the final product and is found to be unchanged at the end of the reaction.

continuum
A continuous spectrum, or a set of points which form a line, a plane, or some other surface.

dynamo action
The action of a moving electric field, which generates a magnetic field.

electromagnetic spectrum
The full range of electromagnetic radiation, from wavelengths of one hundred million millionths of a metre (10^{-14} metre) to ten thousand (10^4) metres.

electron (e^-)
A stable atomic particle with a negative electric charge (1·60219 × 10^{-19} coulombs). In a simple model of the atom, the electron is conceived of as orbiting the atomic nucleus in certain specific orbits. The rest mass of the electron is 9·108 × 10^{-32} kg.

electrostatic unit
A unit of electric charge of such magnitude that it exerts a force of 1 × 10^{-5} newtons (or 1 gm per cm per sec every sec) on a charge of equal magnitude at a distance of 1 cm (or 10^{-2} m).

electron volt (eV)
A unit in nuclear, atomic and high-energy physics. It is the kinetic energy acquired by one electron passing through a potential difference of 1 volt. (1 eV = 1·60 × 10^{-19} joules).

eccentricity
A quantity indicating the shape of an ellipse or other similar figure. Mathematically, for an elliptical orbit, $e = d/a$ where d is the distance from the centre of the ellipse to one of the foci, and a is half the length of the major axis,

ellipsoid
A solid shape generated by rotating an ellipse around either its major axis (prolate ellipsoid) or its minor axis (oblate ellipsoid).

flux density
A term much used in radio astronomy, it is defined as the amount of power (watts) per unit of frequency (Hertz) falling on a unit area (square metre) of the receiving equipment. Where the source is an extended one spread over an area of sky, the flux density is then taken to describe the power per unit solid angle (steradian).

frequency
The frequency with which a wave fluctuates between crest and crest. Frequency and wavelength are connected since, if a wave is travelling with a fixed speed, then the longer it is, the less frequently its crests will impinge on the observer. In electromagnetic radiation which travels with the speed of light (c), the frequency (f) is equal to c divided by the wavelength λ (i.e. $f = c/\lambda$).

gamma-rays (γ-rays)
Electromagnetic waves of the shortest wavelengths below about 0·1 nm.

gauss
A unit of measurement of the strength of a magnetic field, and often used in astronomy. One gauss is defined as the magnetic field strength which will induce 1 × 10^{-8} volts in a line of wire 1 centimetre long moving sideways at right-angles through the field at a rate of 1 cm per second. The Earth's magnetic field is about 1 gauss.

gravitational potential energy
The energy possessed by a body due to its shape and size. It may be regarded as the energy required to separate all the particles of the body to infinity. Because the reverse process must have taken place for the body to form, it can then be said to possess a certain amount of energy. It is called potential because the energy can only be released when for instance the size of the body alters, i.e. if the body shrinks then energy is liberated which may go into heating it. Energy must be supplied to expand the body.

great circle
Any circle on a sphere which passes through the ends of a diameter.

ground state
That condition of the atom in which all electrons are in their lowest possible energy state.

half-life
For a disintegrating atom or nuclear particle, the average length of time required for half the members of a sample to disintegrate.

hydrostatic equilibrium
Hydrostatic equilibrium in a star is the internal balance between the inward directed gravitational force and the outward gas and radiative forces.

hyperbolic orbit
An open orbit where the eccentricity is greater than 1. The path followed by any body which escapes from the gravitational field of a larger body is an hyperbolic orbit.

igneous rock
Any rock which has formed by cooling from a molten state either on, or beneath, the surface. Distinct from rocks which are metamorphic (altered by heat or pressure) or sedimentary.

isotope
An atomic nucleus having the same atomic number as another nucleus, but a different mass number. Because the mass number, A, is the number of protons plus the number of neutrons, isotopes of the same element have the same number of protons (Z), but different numbers of neutrons (A–Z); e.g. deuterium is an isotope of hydrogen as both have a single proton (Z = 1) but deuterium has a single neutron whereas hydrogen has none.

joule
An internationally adopted unit of energy and corresponds to the work done when a force of one newton moves through a distance of one metre. It is also equivalent to 0·24 calories and 6·25 × 10^{18} electron volts.

Kelvin temperature scale
See absolute temperature scale.

kinetic motions
The small-scale local motions of, for example, stars within a galaxy, as distinct from large-scale bodily motion of the galaxy itself through space or in its overall rotation.

laser
A maser (*q.v.*) which emits its energy at visible wavelengths.

lines of force *See* magnetic lines of force.

logarithmic
Increasing by a 'power' series instead of by adding equal increments, which gives a 'linear' scale. Thus 0, 10, 20, 30, 40, etc. is a linear scale formed by adding tens, while 0, 10, 10^2, 10^3, 10^4, etc. is a logarithmic scale, rising in this case in ascending powers of ten.

magma
The subterranean molten mass from which igneous rocks may later form, or lavas be erupted.

magnetic lines of force
Lines indicating the direction of a magnetic field in space, such as would be mapped by a magnetic needle.

maser
An acronym for *m*icrowave *a*mplification by *s*timulated *e*mission of *r*adiation, whereby the natural oscillations of an atom or a molecule are used for amplifying electromagnetic radiation in the microwave band (1 mm to 30 cm wavelengths).

moment of inertia
A measure of the resistance of a body to a change in its rotation.

muon
Originally known as the μ-meson, this is an atomic particle which seems to be like an electron but with 207 times its mass.

neutrino (v)
A nuclear particle with no electric charge and no mass, but with angular momentum or spin. It carries away energy in nuclear reactions.

neutron (n)
A nuclear particle with no electric charge, and a mass ($1{\cdot}6749 \times 10^{-27}$ kg) rather larger than that of the proton. A free neutron decays after a half-life of 10·6 minutes into a proton, an electron, and an antineutrino (a neutrino with the opposite spin to a neutrino). In the nucleus of an atom neutrons and protons form a stable combination.

newton
The force required to give a mass of one kilogram an acceleration of one metre per second every second.

neucleus, atomic
The positively charged core of an atom, consisting of protons and neutrons and containing the major portion of the atom's mass.

optical continuum
The continuous spectrum in optical (visual) wavelengths (400 nanometres or 4×10^{-7} m, to just over 700 nm or 7×10^{-7} m).

parabolic orbit
An open orbit where the eccentricity is exactly equal to 1. Very unlikely to occur in nature, it is frequently used for computational purposes.

perturbation
The disturbance of the orbit of a body by some outside force or forces.

photo-ionization
The ionization of an atom or molecule caused by absorbing an high-speed (i.e high energy) photon.

photon (γ)
A quantum or discrete quantity of electromagnetic energy.

pion
An unstable nuclear particle, sometimes called a π-meson, with a mass which lies between that of an electron and a proton. (Positive or negative pions have a mass 273 times that of an electron, zero charged pions a mass 264 times that of an electron). A charged pion decays into a muon and a neutrino: an uncharged pion into two γ-rays. Pions are probably exchanged between protons and neutrons and have to do with the strong force binding together the particles of an atomic nucleus.

plasma
A completely ionized gas in which the temperature is too high for atoms to exist as such; it consists of free electrons and free atomic nuclei.

positron (e^+)
The antiparticle to the electron, having the same mass but opposite electric charge to the electron.

polarization
In a theoretical description, the propagation of light involves a wave oscillation at right-angles or sideways to the direction of motion. Ordinary *unpolarized* light has an equal mixture of oscillations in all possible sideways directions. In *plane polarized* light all the oscillations are in the same sideways direction. Light is said to be, for example, 30 per cent polarized if it is a mixture of 30 per cent plane polarized and 70 per cent unpolarized light. Other electromagnetic radiation besides light may also be polarized.

potential difference
The difference in electrical states between two points in an electric circuit; it causes an electrical current to flow between the points. Potential difference is measured in volts.

potential energy
Energy which is stored in the relative positions of bodies, as distinct from kinetic energy which is their energy of motion. For an isolated system the total energy, which is the sum of kinetic and potential energies, is constant. When two bodies move closer, their potential energy decreases and as they move faster the kinetic energy increases. By definition, the potential energy in a stable binary system is a negative number; for a stable system of smaller separation, the potential energy is a larger negative number.

proton (p)
A positively charged nuclear particle with a mass 1836 times that of an electron and is found in all atomic nuclei. The number of protons in the nucleus defines the atomic number, Z, of the element. The mass of the proton may also be expressed in energy units and equals 938 million electron volts (938 MeV).

quantum theory
The theory that radiation is emitted only in discrete units or quanta.

radian
A unit of angular measure; it is the angle subtended at the centre of a circle by a section of the circumference equal in length to the radius. There are 2π radians in a circle (360°), so 1 radian = 57°·2957795131.

radioisotope dating
From a knowledge of the rate at which radioactive decay occurs, and measurements of the relative amounts of the initial and final elements present, the age of material samples may be estimated.

radioactive decay
The spontaneous break-up or fission of the atomic nuclei of certain chemical elements into lighter, more stable, nuclei. It is usually accompanied by the emission of charged particles or γ-rays.

resonance
The response of an oscillating system to an external effect at the same frequency as the oscillating system's natural frequency. In orbital theory the term is applied to orbits where gravitational interaction has caused the periods to assume a simple ratio, such as 2:1 or 3:2.

rotational transition
A small change in the energy of a molecule caused by the rotation of its constituent atoms around their centre of mass.

sedimentary rocks
Rocks formed from deposits of sediment laid down under water, or (more rarely) by wind action.

seismometer
An instrument for measuring the intensity of waves generated by earthquakes or seismic disturbances.

shock front
The boundary of a shock wave, i.e. between the normal pressure, density and temperature of a gas, and the increased values present in a shock wave.

spectroscopy
The technique of producing spectra and analyzing their constituent wavelengths to determine such quantities as chemical composition, temperature and density of the emitting region.

sphere
A solid figure obtained by rotating a circle around a diameter.

spheroid
A solid figure obtained by rotating an ellipse around a diameter. *See* ellipsoid.

steradian
The unit of solid angular measure: it is one radian by one radian.

UTC
Co-ordinated universal time. The internationally agreed Universal Time (UT), it is equivalent to Greenwich Mean Time (GMT), the regular civil time determined by the motion of a mean fictitious Sun around the Earth, and independent of the seasons.

Universal Time *See* UTC.

Watt
An internationally adopted unit of power, the rate at which work is performed. It corresponds to a rate of dissipation of energy of one joule per second.

X-rays
Electromagnetic waves between the extreme ultraviolet and γ-ray regions, the wavelength range being generally taken as about 50 nm to 0·5 pm.

Zeeman effect
The broadening of spectral lines due to the presence of magnetic fields.

Bibliography

Although books are listed under general chapter headings, it should be noted that many contain material of interest which relates to other chapters of this book.

CHAPTER ONE

Introduction

Brandt, J. C. and Maran, C. P., *New Horizons in Astronomy*, 2nd edn., Freeman, San Francisco, 1979.
Illingworth, V. (ed.). *The Macmillan Dictionary of Astronomy*, Macmillan, London, 1979.
Mitton, S. (ed.), *Cambridge Encyclopedia of Astronomy*, Jonathan Cape, London, 1977.
Nicolson, I., *Astronomy, a dictionary of space and the Universe*, Arrow Books, London, 1977.
Unsold, A. *The New Cosmos*, 2nd ed., Springer, New York, 1977.

There is a large number of specialized journals on astronomy, but the following periodicals are of a more general nature.

Astronomy, Astro Media Corp., Milwaukee (monthly).
Journal of the British Astronomical Association, London (bimonthly).
Scientific American, W. H. Freeman, New York (monthly; frequently contains astronomical articles of a very high standard).
Sky and Telescope, Sky Publishing Corp., Cambridge, Mass. (monthly).

CHAPTER THREE

The Stars

Gingerich, O. (ed.), *New Frontiers in Astronomy, readings from Scientific American*, Freeman, New York, 1975.
Kaufman, W. J. III, *Black Holes and Warped Spacetime*, Freeman, San Francisco, 1979.
Meadows, A. J., *Stellar Evolution*, 2nd ed., Pergamon, Oxford, 1978.
Shklovskii, I., *Stars: Their Birth, Life and Death*, Freeman, New York, 1978.
Strohmeier, W., *Variable Stars*, Pergamon, Oxford, 1972.
Tayler, R. J., *The Stars, their structure and evolution*, Wykeham, London, 1972.

CHAPTER FOUR

The Sun

Mitton, S., *Daytime Star*, Faber, London, 1981.
Nicolson, I., *The Sun*, Mitchell Beazley, London, 1982.

CHAPTER FIVE

The Solar System

Beatty, J. K., O'Leary, B. and Chaikin, A., *The New Solar System*, 2nd edn., Sky Publishers, Cambridge, Mass. & Cambridge University Press, 1983.
Brandt, J. C. and Chapman, R. D., *Introduction to Comets*, Cambridge University Press, 1981.
Clark, S. P. Jr., *Structure of the Earth*, Prentice-Hall, Englewood Cliffs, N.J., 1971.
Guest, J. E., and Greeley, R., *Geology on the Moon*, Wykeham, London, 1977.
Morrison, D. and Samz, J., *Voyage to Jupiter*, NASA, Washington, 1980.
Morrison, D., *Voyages to Saturn*, NASA, Washington, 1982.
Mutch, T. A. (ed.), *The Geology of Mars*, Princeton University Press, Princeton, 1976 (pre-Viking, but very well illustrated and informative).
Scientific American, *The Solar System*, W. H. Freeman, San Francisco, 1975.
Wood, J. A., *Meteorites and the Origin of Planets*, McGraw-Hill, New York, 1968.

CHAPTER SIX

The Galaxy

Bok, B. J., & Bok, P. F., *The Milky Way*, Harvard University Press, 5th edn., London 1981.
Gribbin, J., *Galaxy Formation*, Macmillan, London, 1972.
Kaufmann, W. J. III, *Stars and Nebulas*, Freeman, San Francisco, 1978.
Whitney, C. A., *The Discovery of our Galaxy*, Angus & Robertson, London, 1972.

CHAPTER SEVEN

Extragalactic Astronomy

Kaufmann, W. J. III, *Galaxies and Quasars*, Freeman, San Francisco, 1979.
Mitton, S., *Exploring the Galaxies*, Faber, London, 1976.
Shapley, H., *Galaxies*, 3rd edn., Harvard University Press, 1972.
Tayler, R. J., *Galaxies: Structure and Evolution*, Wykeham, London, 1978.

CHAPTER EIGHT

Theories of the Universe

Davies, P. C. W., *Space and time in the modern universe*, Cambridge University Press, Cambridge, 1977.
Gingerich I. (ed.) *Cosmology + 1, readings from Scientific American*, Freeman, New York, 1977.
Harrison, E. R., *Cosmology: The Science of the Universe*, Cambridge University Press, 1981.
Narlikar, J., *The Structure of the Universe*, Oxford, 1977.
Silk, J., *The Big Bang*, Freeman, San Francisco, 1980.
Weinberg, S., *The First Three Minutes*, Deutsch, London, 1977.

CHAPTER NINE

Observing the Universe

Barlow, B. V., *The Astronomical Telescope*, Wykeham, London, 1975.
King, H. C., *The History of the Telescope*, Griffin, London, 1955.
Learner, R., *Astronomy through the Telescope*, Evans, London, 1981.
Smith, F. G., *Radio Astronomy*, 4th edn., Penguin, London, 1974.

General practical books

Baxter, W. M., *The Sun and the Amateur Astronomer*, (revised edn.), David & Charles, Newton Abbot, 1972.
British Astronomical Association: *Nature, Aims and Methods*, (revised edn.), British Astronomical Association, London, 1973 (describes the scope for observers who would like to participate in programmes of co-operative work under the auspices of the Association's sections).
British Astronomical Association: *Star Charts*, B.A.A., London, 1981 (larger, sheet version of star charts in this book, designed for observers' use).
Burnham, R. Jr., *Celestial Handbook*, 3 vols., Dover, New York, 1978.
Chartrand, M. and Wimmer, H., *Skyguide*, Golden Press, New York, 1982.
Duffett-Smith, *Practical Astronomy with your Calculator*, 2nd edn., Cambridge University Press, 1981.
Glasby, J. S., *The Variable Star Observer's Handbook*, Sidgwick and Jackson, London, 1971.
Howard, N. E., *Handbook for Telescope Making*, (revised edn.), Faber and Faber, London, 1969.
Klepesta, J. and Rukl, A., *Constellations*, Hamlyn, London, 1969.
Mallas, J. H. and Kreimer, E., *The Messier Album*, *Sky & Telescope*, Cambridge, Mass., 1978.
Miles, H. (ed.), *Satellite Observers Manual*, British Astronomical Association, London, 1974.
Moore, P. (ed.), *Astronomical Telescopes and Observatories for Amateurs*, David & Charles, Newton Abbott, 1973.
Moore, P., *Naked-Eye Astronomy*, 4th edn., Lutterworth Press, Guildford, 1976.
Moore, P. *et al.*, *Guide for Observers of the Moon*, (2nd edn.), British Astronomical Association, London, 1974.
Moore, P. (ed.), *Practical Amateur Astronomy*, 4th edn., Lutterworth Press, Guildford, 1975.
Norton, A. P., *Norton's Star Atlas and Reference Handbook*, (17th edn., edited by P. Satterthwaite), Gall & Inglis, Edinburgh, 1978.
Roth, G. D. (ed.,) *Astronomy: A Handbook*, (2nd edn., translated and revised by A. Beer), Springer-Verlag, New York, 1975.
Roth, G. D., *Handbook for Planet Observers*, (translated by A. Helm), Faber & Faber, London, 1970.
Sidgwick, J. B., *Observational Astronomy for Amateurs*, (4th edn., revised by J. Muirden), Pelham, London, 1982.
Tirion, W., *Sky Atlas 2000*, *Sky & Telescope*, Cambridge, Mass., 1982 (27 charts drawn for epoch 2000).
Webb, T. W., *Celestial Objects for Common Telescopes*, Vol. 2, (6th edn., revised by M. W. Mayall), Dover Publications, New York, 1962.

Index

Page numbers in **bold type** indicate definitions and principal entries. Page numbers in *italic* indicate illustrations and material mentioned in captions. Page numbers followed by the letter T indicate that the entry will be found in a table.

Acknowledgements

Photographs
Frans Alkemade, Stanford, California 246; Anglo-Australian Observatory - UK Schmidt Unit, Royal Observatory, Edinburgh 47, 187; Ron Arbour, Bishopstoke 31; Stephen Benson, London 242; Big Bear Solar Observatory, Pasadena, California 81 top; British Antarctic Survey, Cambridge 103 centre; British Museum, London 10; D. Buczynski, Lancaster 111; W. Cobley, Cleethorpes 102, 103 bottom; Horace Dall, Luton 232; David Dunlap Observatory - University of Toronto, Ontario 65; European Space Agency (ESA), Darmstadt 98; Peter Gill, London 22; Hale Observatories, Pasadena, California 48, 193 top, 193 bottom left, 193 bottom right; Hale Observatories - Halton Arp 217; Hale Observatories - California Institute of Technology and Carnegie Institute of Washington 63 bottom, 71 top, 171 bottom; Hale Observatories - Charles T. Kowal 154, 192; Harvard University, Cambridge, Massachusetts 233 bottom; Commander H. R. Hatfield, Sevenoaks 108; Alan W. Heath, Nottingham 110 top left, 110 top right; Harold Hill, Wigan Institute for Astronomy, Haleakala Observatory, Hawaii 80 centre; The Jet Propulsion Laboratory, Pasadena, California - NASA 118, 120, 131 top, 132, 134, 138 bottom; Jodrell Bank, Macclesfield - Walton Sound and Film Services, London 239 bottom; Kari Kaila, Finland 129; Kapteyn Astronomical Institute, Rijksuniversiteit Groningen - Dr R. Sancisi 199; Kitt Peak National Observatory, Tucson, Arizona 233 top; Peter Kwentus, East Detroit, Michigan 79 bottom; Leiden Observatory Lick Observatory - Regents, University of California Lockheed Solar Laboratory, Palo Alto, California - Dr Sara F. Martin 58 top; Lunar and Planetary Laboratory, University of Arizona, Tucson - Stephen Larson 142; Lund Observatory 168-9; C. R. Martys 176, 212; M. Maunder 87, 89 top left; Robert McNaught, Prestwick Mullard Radio Astronomy Observatory - BICC Ltd. 241; Mullard Radio Astronomy Observatory - Dr S. Jocelyn Burnell 71 bottom; NASA, Washington DC 29 top, 91, 94 top, 94 bottom, 95, 105 top, 126 top, 126 bottom, 127, 130-1, 135, 138, 141, 143 top, 143 bottom left, 143 bottom right, 144 right, 147, 219, 243 bottom, 245; National Space Science Data Center, Greenbelt, Maryland - NASA 99, 104, 105 bottom, 106-7, 116 top, 116 bottom, 117, 125, 131 top left; Novosti Press Agency, London 122 bottom; Observatoires du Pic-du-Midi et de Toulouse, Bagnères de Bigorre 80-1; Ohio State University Radio Observatory, Columbus - John Kraus 240 top; P. Parviainen, Finland 85 bottom, 115 bottom, 159 bottom; R. J. Poole, Richmond, British Columbia 83; Radio Astronomy Laboratory, University of California - Professor Carl Heiles 186 centre; H. B. Ridley, West Chinnock, Somerset 149 top left, 149 top right, 150; Ann Ronan Picture Library, Loughton 11; Royal Astronomical Society, London - Hale Observatories 64, 171 top, 181, 185 top, 234 left, 236; Royal Astronomical Society - Lick Observatory 71 sequence, 222; Royal Astronomical Society - Alan McClure 156; Royal Astronomical Society - Mullard Radio Astronomy Observatory 185 bottom; Royal Astronomical Society - Royal Greenwich Observatory 77; Royal Greenwich Observatory - Herstmonceux Castle 80 bottom left, 81 bottom, 234 right; Royal Greenwich Observatory - Malcolm Currie 202; Royal Observatory, Edinburgh 167, 179, 184, 203, 231; Sacramento Peak Solar Observatory, Sunspot, New Mexico - Air Force Cambridge Geophysical Laboratories 78 bottom; SPL (Science Photo Library), London - Professor R. J. Allen 215 bottom; SPL, London - Dr Kris Davidson 67; SPL, London - Dr Fred Espenak 90, 114, 115 top; SPL, London - Professor R. Gehrz 239 top; SPL, London - Dr Steve Gull 210, 211 top; SPL, London - Harvard-Smithsonian Centre for Astrophysics 74-5; SPL, London - Dr Jean Lorre 206; SPL, London - Los Alamos National Laboratory 207; SPL, London - NASA 175 SPL, London - David Parker 14 bottom; SPL, London - Dr D. H. Roberts 223, 243 top; SPL, London - Ronald Royer 155; SPL, London - Dr M. Schnepp 55; SPL, London - Steele/Hutcheon 159 top; SPL, London - Dr S. C. Unwin 211 bottom; SPL, London - US Naval Observatory 96; SPL, London - X-ray Astronomy Group, Leicester University 190, 215 top; Space Frontiers Ltd, Havant - AG Astrofotografie 166, 194-5; Space Frontiers Ltd - NASA 130 top; Space Frontiers Ltd - US Naval Research Laboratory 79 top; US Naval Observatory, Washington DC 152; US Naval Observatory - Dr Gart Westerhout 188; University College, London - Department of Physics and Astronomy 235; University of Bochum 183; University of California, Santa Cruz - Lick Observatory Photographs 63 top, 63 centre, 71 sequence; University of London - Dr Alexander Boksenberg 191; University of Michigan, Ann Arbor - Department of Astronomy 47; University of Sydney, New South Wales 240 centre; US Dept of Interior, Geological Survey, Flagstaff, Arizona 122 top; Yerkes Observatory, Williams Bay, Wisconsin 174, 208.

Drawings, diagrams and artist's references
British Astronomical Association - Bert Chapman 112; W. H. Freeman & Co, San Francisco, *Scientific American* - Iben, I. Jr, Figs. 3.16, 3.17 & 3.18; ibid - Thorne, K. S., Fig. 3.28; D. Fuller, Fullerscopes, London page 24 bottom; Peter Gill, London 84 bottom right; Harvard College Observatory - Shapley, H., Fig. 7.1; Harold Hill, Wigan, Lancashire, pages 82, 82-3, 109; R. Miles, Mouldsworth, Cheshire 148 right; Pergamon Press Ltd., Oxford - Meeus, J., Grosjean, C. C. & Vanderleen, W., page 14 top; George Philip & Son, Ltd., London page 25 bottom; D. Reidel Publishing Co, Dordrecht, Holland - Houck, N. & Fesen, R., Fig. 3.8; Royal Astronomical Society, London - Culhane, J. L., Fig. 7.7; ibid - Jenkins, C. J., Pooley, C. G. & Riley, J. M., page 227; Royal Greenwich Observatory - Nautical Almanac Office page 86; Springer Verlag, New York - de Vaucouleurs, G., Fig. 7.2a; ibid - *Astronomy & Astrophysics*, Simonson, S. C., Fig 7.10; Professor J. Toomre, University of Colorado, page 203; University of Chicago Press - *Astrophysical Journal* - Kristian, J., Sandage, A. & Westphal J. A., Fig. 7.3; ibid - Cleary, M. N. & Murray, J. D., Fig. 7.6; ibid - Abell, G., Fig. 7.8; Professor G. de Vaucouleurs, McDonald Observatory, Texas, Fig. 7.2b. The Preliminary Geological Terrain Map of Mercury on page 119 is published by permission of Trask, N. J. & Guest, J. E., from *Journal of Geophysical Research*, vol 80, no. 17, p. 2461-2477, 1975.